Third Edition

Essential Mathematics

for **Games** and **Interactive**

Applications

Third Edition

Essential Mathematics for Games and Interactive Applications

James M. Van Verth
Lars M. Bishop

CRC Press
Taylor & Francis Group
Boca Raton London New York

CRC Press is an imprint of the
Taylor & Francis Group, an **Informa** business

AN A K PETERS BOOK

CRC Press
Taylor & Francis Group
6000 Broken Sound Parkway NW, Suite 300
Boca Raton, FL 33487-2742

First issued in hardback 2019

© 2016 by Taylor & Francis Group, LLC
CRC Press is an imprint of Taylor & Francis Group, an Informa business

No claim to original U.S. Government works

ISBN-13: 978-1-4822-5092-3 (hbk)

Library of Congress Cataloging-in-Publication Data

Van Verth, James M., author.
  Essential mathematics for games and interactive applications / James M. Van Verth, Lars M. Bishop. -- Third edition.
    pages cm
  Includes bibliographical references and index.
  ISBN 978-1-4822-5092-3 (hardcover : alk. paper) 1. Computer games--Programming--Mathematics. 2. Three-dimensional display systems--Mathematics. 3. Computer graphics--Mathematics. I. Bishop, Lars M., author. II. Title.

  QA76.76.C672V47 2015
  794.8'1526--dc23   2015033146

Visit the Taylor & Francis Web site at
http://www.taylorandfrancis.com

and the CRC Press Web site at
http://www.crcpress.com

To Dad and Mom,
for all your love and support. —*Jim*

To Jen,
who constantly helps me get everything done, and
to Nadia and Tasha,
who constantly help me avoid getting any of it done on time. —*Lars*

Contents

Preface

Writing a book is an adventure. To begin with, it is a toy and an amusement; then it becomes a mistress, and then it becomes a master, and then a tyrant. The last phase is that just as you are about to be reconciled to your servitude, you kill the monster, and fling him out to the public.

Sir Winston Churchill

The Adventure Begins

As humorous as Churchill's statement is, there is a certain amount of truth to it; writing this book was indeed an adventure. There is something about the process of writing, particularly a nonfiction work like this, that forces you to test and expand the limits of your knowledge. We hope that you, the reader, benefit from our hard work.

How does a book like this come about? Many of Churchill's books began with his experience—particularly his experience as a world leader in wartime. This book had a more mundane beginning: two engineers at Red Storm Entertainment, separately, asked Jim to teach them about vectors. These engineers were 2D game programmers, and 3D was not new, but was starting to replace 2D at that point. Jim's project was in a crunch period, so he didn't have time to do much about it until proposals were requested for the annual Game Developers Conference. Remembering the engineers' request, he thought back to the classic "Math for SIGGRAPH" course from SIGGRAPH 1989, which he had attended and enjoyed. Jim figured that a similar course, at that time titled "Math for Game Programmers," could help 2D programmers become 3D programmers.

The course was accepted, and together with a cospeaker, Marcus Nordenstam, Jim presented it at GDC 2000. The following years (2001–2002) Jim taught the course alone, as Marcus had moved from the game industry to the film industry. The subject matter changed slightly as well, adding more advanced material such as curves, collision detection, and basic physical simulation.

It was in 2002 that the seeds of what you hold in your hand were truly planted. At GDC 2002, another GDC speaker, whose name, alas, is lost to time, recommended that Jim turn his course into a book. This was an interesting idea, but how to get it published? As it happened, Jim ran into Dave Eberly at SIGGRAPH 2002, and he was looking for someone to write just that book for Morgan Kaufmann. At the same time, Lars, who was working at Numeric Design Limited at the time, was presenting some of the basics of rendering on handheld devices as part of a SIGGRAPH course. Jim and Lars discussed the fact that handheld 3D rendering had brought back some of the "lost arts" of 3D programming, and that this might be included in a book on mathematics for game programming.

Thus, a coauthorship was formed. Lars joined Jim in teaching the GDC 2003 version of what was now called "Essential Math for Game Programmers," and simultaneously joined Jim to help with the book, helping to expand the topics covered to include numerical representations. As we began to flesh out the latter chapters of the outline, Lars was finding that the advent of programmable shaders on consumer 3D hardware was bringing more and more low-level lighting, shading, and texturing questions into his office at NDL. Accordingly, the planned single chapter on texturing and antialiasing became three, covering a wider selection of these rendering topics.

By early 2003, we were furiously typing the first full draft of the first edition of this book, and by GDC 2004 the book was out. Having defeated the dragon, we retired to our homes to live out the rest of our quiet little lives.

Or so we thought.

The Adventure Continues

Response to the first edition was quite positive and the book continued to sell well beyond the initial release. Naturally, thoughts turned to what we could do to improve the book beyond what we already created.

In reviewing the topic list, it was obvious what the most necessary change was. Within a year or so of the publication of the first edition, programmable shading had revolutionized the creation of 3D applications on game consoles and on PC. While the first edition had provided readers with many of the fundamentals behind the mathematics used in shaders, it stopped short of actually discussing them in detail. It was clear that the second edition needed to embrace shaders completely, applying the mathematics of the earlier chapters to an entirely new set of rendering content. So the single biggest change in the second edition was a move to a purely shader-based rendering pipeline.

We also sent out the book to reviewers to ask them what they would like to see added. The most common request was information about random numbers. So a brand new chapter on probability and random numbers was added.

And so, with those changes, we again wearily returned home.

The Adventurers Resume the Quest

And here we are yet once more. For the third edition, further refinements were necessary. With the advent of DirectX 10 and the OpenGL Core Profile, the trend is to remove even more of the old nonshader graphics pipelines, and the text needed to be updated to reflect that. In real-time graphics, there is a push for more realistic materials and lighting, and that chapter has been updated as well. In addition, the entire book was revised to add corrections and make the content flow better. We hope you'll find our efforts worthwhile.

All three times, the experience has been fascinating, sometimes frustrating, but ultimately deeply rewarding. Hopefully, this fascination and respect for the material will be conveyed to you, the reader. The topics in this book can each take a lifetime to study to a truly great depth; we hope you will be convinced to try just that, nonetheless!

Enjoy as you do so, as one of the few things more rewarding than programming and seeing a correctly animated, simulated, and rendered scene on a screen is the confidence of understanding *how* and *why* everything worked. When something in a 3D system goes wrong (and it *always* does), the best programmers are never satisfied with "I fixed it, but I'm not sure how"; without understanding, there can be no confidence in the solution, and nothing new is learned. Such programmers are driven by the desire to understand what went wrong, how to fix it, and learning from the experience. No other tool in 3D programming is quite as important to this process as the mathematical bases[1] behind it.

Those Who Helped Us along the Road

In a traditional adventure the protagonists are assisted by various characters that pass in and out of the pages. Similarly, while this book bears the names of two people on the cover, the material between its covers bears the mark of many, many more. We thank a few of them here.

The folks at our publisher, AK Peters, were extremely patient with both of us as we made up for being more experienced this time around by being busier and less responsive! Special thanks go to our executive editor, Rick Adams, our production editor, Laurie Oknowsky, project editor, Jennifer Stair, and project manager, Viswanath Prasanna, and all the great staff at CRC Press and Datapage.

In addition, we acknowledge the editors at Morgan Kaufman who helped us publish the first two editions: Tim Cox, Troy Lilly, Stacie Pierce, and Richard Camp were patient and helpful in the daunting task of leading two first-time authors through the process. Laura Lewin, Chris Simpson, Georgia Kennedy, and Paul Gottehrer were all patient, professional, and flexible when we most needed it.

Our reviewers were top-notch. Fabian Giesen and Robin Green provided thorough feedback for the third edition. Erin Catto and Chad Robertson reviewed the entire second edition of the book. Robert Brown, Matthew McCallus, Greg Stelmack, and Melinda Theilbar were invaluable for their comments on the random numbers chapter. Dave Eberly provided an in-depth review of the first edition. Ian Ashdown, Steven Woodcock, John O'Brien, J.R. Parker, Neil Kirby, John Funge, Michael van Lent, Peter Norvig, Tomas Akenine-Möller, Wes Hunt, Peter Lipson, Jon McAllister, Travis Young, Clark Gibson, Joe Sauder, and Chris Stoy each reviewed parts of the first edition or the proposals for them. Despite having tight deadlines, they all provided page after page of useful feedback, keeping us honest and helping us generate a better arc to the material. Several of them went *well* above and beyond the call of duty, providing detailed comments and even rereading sections of the book that required significant changes. Finally, thanks also to Victor Brueggemann and Garner Halloran, who asked Jim the questions that started this whole thing off 11 years ago.

Jim and Lars acknowledge the folks at their jobs at Google and NVIDIA Corporation, who were *very* understanding with respect to the time-consuming process of creating a book.

---

[1] Vector or otherwise.

Also, thanks to the talented engineers at these and previous companies who provided the probing discussions and great questions that led to and continually fed this book.

In addition, Jim thanks Mur and Fiona, his wife and daughter, who were willing to put up with this a third time after his long absences the first and second times through; his sister, Liz, who provided illustrations for an early draft of this text; and his parents, Jim and Pat, who gave him the resources to make it in the world and introduced him to the world of computers so long ago.

Lars thanks Jen, his wife, who somehow had the courage to survive a second and a third edition of the book even after being promised that the first edition "was it"; and his parents, Steve and Helene, who supported, nurtured, and taught him so much about the value of constant learning and steadfast love.

And lastly, we once again thank you, the reader, for joining us on this adventure. May the teeth of this monster find fertile ground in your minds, and yield a new army of 3D programmers.

Additional material is available from the authors' own website: www.essentialmath.com.

Authors

James M. Van Verth is a software engineer at Google, Chapel Hill, NC, where he works on GPU support for the Skia 2D Graphics Library. Prior to that, he worked for Insomniac Games, Durham, NC, NVIDIA; Durham, NC; and Red Storm Entertainment, Cary, NC, in various roles. For the past 17 years he also has been a regular speaker at the Game Developers Conference, teaching the all-day tutorials "Math for Game Programmers" and "Physics for Game Programmers," on which this book is based. His background includes a BA in math/computer science from Dartmouth College, Hanover, NH, an MS in computer science from the State University of New York at Buffalo, and an MS in computer science from the University of North Carolina at Chapel Hill.

Lars M. Bishop is an engineer in the Handheld Developer Technologies group at NVIDIA, Durham, NC. Prior to joining NVIDIA, Lars was the chief technology officer at Numerical Design Limited, Chapel Hill, NC, leading the development of the Gamebryo3D cross-platform game engine. He received a BS in math/computer science from Brown University, Providence, RI, and an MS in computer science from the University of North Carolina at Chapel Hill. His outside interests include photography, drumming, and playing bass guitar.

Introduction

The (Continued) Rise of 3D Games

Over the past two decades or so (driven by increasingly powerful computer and video game console hardware), three-dimensional (3D) games have expanded from custom-hardware arcade machines to the realm of hardcore PC games, to consumer set-top video game consoles, and even to handheld devices such as personal digital assistants (PDAs) and cellular telephones. This explosion in popularity has led to a corresponding need for programmers with the ability to program these games. As a result, programmers are entering the field of 3D games and graphics by teaching themselves the basics, rather than a classic college-level graphics and mathematics education. At the same time, many college students are looking to move directly from school into the industry. These different groups of programmers each have their own set of skills and needs in order to make the transition. While every programmer's situation is different, we describe here some of the more common situations.

Many existing, self-taught 3D game programmers have strong game experience and an excellent practical approach to programming, stressing visual results and strong optimization skills that can be lacking in college-level computer science programs. However, these programmers are sometimes less comfortable with the conceptual mathematics that form the underlying basis of 3D graphics and games. This can make developing, debugging, and optimizing these systems more of a trial-and-error exercise than would be desired.

Programmers who are already established in other specializations in the game industry, such as networking or user interfaces, are now finding that they want to expand their abilities into core 3D programming. While having experience with a wide range of game concepts, these programmers often need to learn or refresh the basic mathematics behind 3D games before continuing on to learn the applications of the principles of rendering and animation.

On the other hand, college students entering (or hoping to enter) the 3D games industry often ask what material they need to know in order to be prepared to work on these games. Younger students often ask what courses they should attend in order to gain the most useful background for a programmer in the industry. Recent graduates, on the other hand, often ask how their computer graphics knowledge best relates to the way games are developed for today's computers and game consoles.

We have designed this book to provide something for each of these groups of readers. We attempt to provide readers with a conceptual understanding of the mathematics needed to create 3D games, as well as an understanding of how these mathematical bases actually *apply* to games and graphics. The book provides not only theoretical mathematical background, but also many examples of how these concepts are used to affect how a game looks (how it is *rendered*) and plays (how objects move and react to users). Each type of reader is likely to find sections of the book that, for him or her, provide mainly refresher courses, a new understanding of the applications of basic mathematical concepts, or even completely new information. The specific sections that fall into each category for a particular reader will, of course, depend on the reader.

How to Read This Book

Perhaps the best way to discuss any reader's approach to reading this book is to think in terms of how a 3D game or other interactive application works at the highest level. Most readers of this book likely intend to apply what they learn from it to create, extend, or fix a 3D game or other 3D application. Each chapter in this book deals with a different topic that has applicability to some or all of the major parts of a 3D game.

Game Engines

An interactive 3D application such as a game requires quite a large amount of code to do all of the many things asked of it. This includes representing the virtual world, animating parts of it, drawing that virtual world, and dealing with user interaction in a game-relevant manner. The bulk of the code required to implement these features is generally known as a *game engine*. Game engines have evolved from small, simple, low-level rendering systems in the early 1990s to massive and complex software systems in modern games, capable of rendering detailed and expansive worlds, animating realistic characters, and simulating complex physics. At their core, these game engines are really implementations of the concepts discussed throughout this book.

Initially, game engines were custom affairs, written for a single use as a part of the game itself, and thrown away after being used for that single game project. Today, game developers have several options when considering an engine. They may purchase a commercial engine from another company and use it unmodified for their project. They may purchase an engine and modify it very heavily to customize their application. Finally, they may write their own, although most programmers choose to use such an internally developed engine for multiple games to offset the large cost of creating the engine.

In any of these cases, the developer must still understand the basic concepts of the game engine. Whether as a user, a modifier, or an author of a game engine, the developer must understand at least a majority of the concepts presented in this book. To better understand how the various chapters in this book surface in game engines, we first present some common modules one might find in a game engine (though not necessarily in this order):

1. Move and place objects in the scene.

2. Animate the characters in the scene based on animator-created sequences (e.g., soccer players running downfield).

3. Draw the current configuration of the game's scene to the screen.

4. Detect collisions between the characters and objects (e.g., the soccer ball entering the goal or two players sliding into one another).

5. React to these collisions and basic forces, such as gravity in the scene, in a physically correct manner (e.g., the soccer ball in flight).

All of these steps will need to be done for each frame to present the player with a convincing game experience. Thus, the code to implement the steps above must be correct and optimal.

Chapters 1–6: The Basics

Perhaps the most core parts of any game engine are the low-level mathematical and geometric representations and algorithms. The pieces of code will be used by each and every step listed above. Chapter 1 provides the lowest-level basis for this. It discusses the practicalities of representing real numbers on a computer, with a focus on the issues most likely to affect the development of a 3D game engine for a PC, console, or handheld device.

Chapter 2 provides a focused review of vectors and points, objects that are used in all game engines to represent locations, directions, velocities, and other geometric quantities in all aspects of a 3D application. Chapters 3 and 4 review the basics of linear and affine algebra as they relate to orienting, moving, and distorting the objects and spaces that make up a virtual world. And Chapter 5 introduces the quaternion, a very powerful nonmatrix representation of object orientation that will be pivotal to the later chapters on animation and simulation.

The game engine's module 2, animating characters and other objects based on data created by computer animators or motion-captured data, is introduced in Chapter 6. This chapter discusses methods for smoothly animating the position, orientation, and appearance of objects in the virtual game world. The importance of good, complex character and object animation in modern engines continues to grow as new games attempt to create smoother, more convincing representations of athletes, rock stars, soldiers, and other human characters.

Three-dimensional engine code that implements all of these fundamental objects must be built carefully and with a good understanding of both the underlying mathematics and programming issues. Otherwise, the game engine built on top of these basic objects or functions will be based upon a poor foundation. Many game programmers' multiday debugging sessions have ended with the realization that the complex bug was rooted in an error in the engine's basic mathematics code.

Some readers will have a passing familiarity with the topics in these chapters. However, most readers will want to start with these chapters, as many of the topics are covered in more conceptual detail than is often discussed in basic graphics texts. Readers new to the material will want to read in detail, while those who already know some linear algebra can use the chapters to fill in any missing background. All of these chapters form a basis for the rest of the book, and an understanding of these topics, whether existing or new, will be key to successful 3D programming.

Chapters 7–10: Rendering

Chapters 7–10 apply the foundational objects detailed in Chapters 1–6 to explain module 3 of the game engine: the rendering or drawing pipeline, perhaps the best-known part of any game engine. In some game engines, more time and effort are spent designing, programming, and tuning the rendering pipeline than the rest of the engine in its entirety. Chapter 7 describes the mathematics and geometry behind the virtual cameras used to view the scene or game world. Chapter 8 describes the representation of color and the concept of *shaders*, which are short programs that allow modern graphics hardware to draw the scene objects to the display device. Chapter 9 explains how to use these programmable shaders to implement simple approximations of real-world lighting. The rendering section concludes with Chapter 10, which details the methods used by low-level rendering systems to draw to the screen. An understanding of these details can allow programmers to create much more efficient and artifact-free rendering code for their engines.

Chapters 11–13: Randomness, Collision, and Physics

Chapter 11 covers one element for adding realism to games: random numbers. Everything up to this point has been carefully determined and planned by the programmer or artist. Adding randomness adds the unexpected behavior that we see in real life. Gunshots are not always exact, clouds are not perfectly spherical, and walls are not pristine. This chapter discusses how to handle randomness in a game, and how we can get effects such as those discussed above.

Module 4, detecting collisions, is discussed in Chapter 12. This chapter describes the mathematics and programming behind detecting when two game objects touch, intersect, or penetrate. Many genres of game have exacting requirements when it comes to collision, be it a racing game, a sports title, or a military simulation.

Finally, module 5, reacting in a realistic manner to physical forces and collisions, is covered in Chapter 13. This chapter describes how to make game objects behave and react in physically convincing ways.

Put together, the chapters that form this book are designed to give a good basis for the foundations of a game engine, details on the ways that engines represent and draw their virtual worlds, and an introduction to making those worlds seem real and active.

Interactive Demo Applications and Support Libraries

Source Code
Demo
Name

Three-dimensional games and graphics are, by their nature, not only visual but also dynamic. While figures are indeed a welcome necessity in a book about 3D applications, interactive demos can be even more important. It is difficult to truly understand such topics as lighting, quaternion interpolation, or physical simulation without being able to see them work firsthand and to interact with these complex systems. The website for this book (www.essentialmath.com) includes a collection of source code and demonstrations that are designed to illustrate the concepts in a way that is analogous to the static figures in the book itself. Whenever a topic is illustrated with an interactive demo, a special icon like the one seen next to this paragraph will appear in the margin.

Source Code
Library
Name

In addition to the source code for each of the demos, the website includes the supporting libraries used to create the demos, with full source code. Often, code from these supporting libraries is excerpted in the book itself in order to explain how the particular concept is implemented. In such situations, again an icon will appear in the margin to note this. This source code is designed to allow readers to modify and experiment themselves, as a way of better understanding the way the code works.

The source code is written entirely in C++, a language that is likely to be familiar to most game developers. C++ was chosen because it is one of the most commonly used languages in 3D game development and because vectors, matrices, quaternions, and graphics algorithms decompose very well into C++ classes. In addition, C++'s support of operator overloading means that the math library can be implemented in a way that makes the code look very similar to the mathematical derivations in the text. However, in some sections of the text, the class declarations as printed in the book are not complete with respect to the code on the website. Often, class members that are not relevant to the particular discussion (especially member variable accessor and "housekeeping" functions) have been omitted for clarity. These other functions may be found in the full class declarations/definitions in the source code on the website.

Note that we have modified our mathematical notation slightly to allow our equations to be as compatible as possible with the code. Mathematicians normally start indexing with 1, for example, $P_1, P_2, \ldots, P_n$. This does not match how indexing is done in C++: $P[0]$ is the first element in the array P. To avoid this disconnect, in our equations we will be using the convention that the starting element in a list is indexed as 0; thus, $P_0, P_1, \ldots, P_{n-1}$. This should allow for a direct translation from equation to code.

Math Libraries

All of the demos use a shared core math library called IvMath, which includes C++ classes that implement vectors and matrices of different dimensions, along with a few other basic mathematical objects discussed in the book. This library is designed to be useful to readers beyond the examples supplied with the book, as the library includes a wide range of functions and operators for each of these objects, some of which are beyond the scope of the book's demos.

The animation demos use a shared library called IvCurves, which includes classes that implement spline curves, the basic objects used to animate position. IvCurves is built upon IvMath, extending this basic functionality to include animation. As with IvMath, the IvCurves library is likely to be useful beyond the scope of the book, as these classes are flexible enough to be used (along with IvMath) in other applications.

Finally, the simulation demos use a shared library called IvCollision, which implements basic object intersection (collision) data structures and algorithms. Building on the IvMath library, this set of classes and functions not only forms the basis for the later demos in the book, but also is an excellent starting point for experimentation with other forms of object collision and physics modeling.

Engine and Rendering Libraries

In addition to the math libraries, the website includes a set of classes that implement a simple gamelike application framework, basic rendering, input handling, and timer functionality. All of these functions are grouped under the heading of game engine functionality, and are located in the IvEngine library. The engine's rendering code takes the form of a set of renderer-abstraction classes that simplify the interfaces between the C++ classes in `IvMath` and the C-based, low-level rendering application programmer interfaces (APIs). This code is included as a part of the rendering library `IvGraphics`. It includes renderer setup, basic render-state management, and rendering of simple geometric primitives, such as spheres, cubes, and boxes.

Since this book focuses on the mathematics and concepts behind 3D games, we chose not to center the discussion around a large-scale, general 3D game engine. Doing so would introduce an extra layer of indirection that would not serve the conceptual requirements of the book. Valuable real estate in chapters would be spent on background in the use of a particular engine — the one written for the book. For an example and discussion of a full game engine, the reader is encouraged to read Jason Gregory's *Game Engine Architecture* [63].

We have opted to implement our rendering system and examples using two standard SDKs: the multiplatform OpenGL [135] and the popular Direct3D DX11. We also use the GLFW utility toolkit to implement windowing setup and input handling for OpenGL, neither of which is a core topic of this book.

Supplementary Material

In addition to the sample code, we have included some useful reading material on the website for those who haven't absorbed enough of our luminous prose. This includes supplemental material that unfortunately didn't make its way into the book proper due to space considerations, plus slides and notes from years of GDC tutorials that cover topics well beyond those presented in this book. Again, these can be found at www.essentialmath.com.

References and Further Reading

Hopefully, this book will leave readers with a desire to learn even more details and the breadth of the mathematics involved in creating high-performance, high-quality 3D games. Wherever possible, we have included references to other books, articles, papers, and web sites that detail particular subtopics that fall outside the scope of this book. The full set of references may be found at the back of the book.

We have attempted to include references that the vast majority of readers should be able to locate. When possible, we have referenced recent or standard industry texts and well-known conference proceedings. However, in some cases we have included references to older magazine articles and technical reports when we found those references to be particularly complete, seminal, or well written. In some cases, older references can be easier for the less experienced reader to understand, as they often tend to assume less common knowledge when it comes to computer graphics and game topics.

In the past, older magazine articles and technical reports were notoriously difficult for the average reader to locate. However, the Internet and digital publishing have made great strides toward reversing this trend. For example, the following sources have made several classes of resources far more accessible:

- The magazine most commonly referenced in this book, *Game Developer*, is sadly now defunct. However, many of the articles from past issues are available on

Gamasutra, at www.gamasutra.com/topic/game-developer. Several other technical magazines also offer such websites.

- Most papers and technical reports are often available on the Internet. The two most common methods of finding these resources are via publication portals such as Citeseer (www.citeseer.com) and via the authors' personal homepages (if they have them). Most of the technical reports referenced in this book are available online from such sources. Owing to the dynamic nature of the Internet, we suggest using a search engine if the publication portals do not succeed in finding the desired article.

- Technical societies are now placing major historical publications into their "digital libraries," which are often made accessible to members. The Association for Computing Machinery (ACM) has done this via its ACM Digital Library, which is available to ACM members. As an example, the full text of the entire collection of papers from all SIGGRAPH conferences (the conference proceedings most frequently referenced in this book) is available electronically to ACM SIGGRAPH members. Many of these papers are available at the authors' websites as well.

For further reading, we suggest several books that cover topics related to this book in much greater detail. In most cases they assume that the reader is familiar with the concepts discussed in this book. Jason Gregory's *Game Engine Architecture* [63] discusses the design and implementation of a full game engine. For graphics and physically based lighting, Akenine-Möller et al. [1] is an excellent source for real-time systems, and Pharr and Humphreys [121] for offline systems. Books by Gino van den Bergen [149] and Christer Ericson [41] cover topics in interactive collision detection. Finally, Millington [109] provides a more advanced discussion of a wide range of physical simulation topics.

1 Representing Real Numbers

1.1 Introduction

Most basic undergraduate computer architecture books [139] present the basics of integral data types (e.g., `int` and `unsigned int`, `short`, etc., in C/C++) but give only brief introductions to floating-point and other nonintegral number representations. Since the mathematics of 3D graphics are generally real-valued (as we shall see from the ubiquity of $\mathbb{R}$, $\mathbb{R}^2$, and $\mathbb{R}^3$ in the coming chapters), it is important for anyone in the field to understand the features, limitations, and idiosyncrasies of the standard computer representations of these nonintegral types. While it's possible to use real numbers successfully without any understanding of how they are implemented, this can lead to subtle bugs and performance problems at inopportune stages during the development of an application.

This chapter will focus on the IEEE floating-point format and its variants, which is the standard computer representation for real numbers. It will cover the associated bitwise formats, basic operations, features, and limitations. By design, we'll begin with general mathematical discussions of number representation and transition toward implementation-related topics of specific relevance to 3D programmers. Most of the chapter will be spent on the ubiquitous IEEE floating-point numbers, especially discussions of floating-point limitations that often cause issues in 3D pipelines. It will also present a brief case study of floating-point–related performance issues in a real application.

We assume that the reader is familiar with the basic concepts of integer and whole-number representations on modern computers, including signed representation via two's complement, range, overflow, common storage lengths (8, 16, and 32 bits), standard C and C++ basic types (`int`, `unsigned int`, `short`, etc.), and type conversion. For an introduction to these concepts of binary number representation, we refer the reader to a basic computer architecture text, such as Stallings [139].

1.2 Preliminary Concepts

The issues described above relating to storage of integers (such as overflow) are still important when considering real numbers. However, real-number representations add additional complexities that will result in implementation trade-offs, subtle errors, and difficult-to-trace performance issues that can easily confuse the programmer.

1.2.1 Fixed-Point Numbers

Fixed-point numbers are based on a very simple observation with respect to computer representation of integers. In the standard binary representation, each bit represents twice the value of the bit to its right, with the least significant bit representing 1. The following diagram shows these powers of 2 for a standard 8-bit unsigned value:

2^7	2^6	2^5	2^4	2^3	2^2	2^1	2^0
128	64	32	16	8	4	2	1

Just as a decimal number can have a decimal point, which represents the break between integral and fractional values, a binary value can have a binary point, or more generally a radix point (a decimal number is referred to as radix 10, a binary number as radix 2). In the common integer number layout, we can imagine the radix point being to the right of the last digit. However, it does not have to be placed there. For example, let us place the radix point in the middle of the number (between the fourth and fifth bits). The diagram would then look like this:

2^3	2^2	2^1	2^0 . 2^{-1}	2^{-2}	2^{-3}	2^{-4}
8	4	2	1 . $\frac{1}{2}$	$\frac{1}{4}$	$\frac{1}{8}$	$\frac{1}{16}$

Now, the least significant bit represents 1/16. The basic idea behind fixed point is one of scaling. A fixed-point value and an integer with the same bit pattern are related by an implicit scaling factor. This scaling factor is fixed for a given fixed-point format and is the value of the least significant bit in the representation. In the case of the preceding format, the scaling factor is 1/16. The spacing between each representable number can be seen in Figure 1.1, where a representable value lies on each vertical line. Each value is spaced exactly 1/16 apart, which is consistent with our scaling factor.

The standard nomenclature for a fixed-point format is A-dot-B, where A is the number of integral bits (to the left of the radix point) and B is the number of fractional bits (to the right of the radix point). For example, the 8-bit format described above would be referred

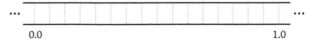

Figure 1.1. Representable values for 4-dot-4 fixed point between 0 and 1.

to as 4-dot-4. Regular 32-bit integers could be referred to as 32-dot-0 because they have no fractional bits. More generally, the scaling factor for an A-dot-B format is simply 2^{-B}. Note that, as expected, the scaling factor for a 32-dot-0 format (integers) is $2^0 = 1$. No matter what the format, the radix point is locked, or fixed, at B bits from the least significant bit—thus the name "fixed point."

1.2.2 Approximations

While computer representations of whole numbers (unsigned int) and integers (int) are limited to a finite subset of their pure counterparts, in each case the finite set is contiguous; that is, if i and $i+2$ are both representable, then $i+1$ is also representable. So inside the range defined by the minimum and maximum representable integer values, all integers can be represented exactly. This is possible because any finitely bounded range of integers contains a finite number of elements.

When dealing with real numbers, however, this is no longer true. A subset of real numbers can have infinitely many elements even when bounded by finite minimal and maximal values. As a result, no matter how tightly we bound the range of real numbers (other than the trivial case of $R_{min} = R_{max}$) that we choose to represent, we will be unable to represent all the values within that range.

As an example, suppose that we choose to represent real numbers as a fixed-point number, where the first m bits represent the integer value, and the last n bits represent the fractional value. The smallest possible fraction we can represent is $1/2^n$, and the next largest is $2/2^n$. We can't represent $3/2^{n+1}$—it falls between those two values, so we'll either need to round up or round down. Hence, issues of both range and precision will be constant companions over the course of our discussion of real-number representations.

1.2.3 Precision and Error

In order to adequately understand the representations of real numbers, we need to understand the concept of precision and error. Imagine a generic function $Rep(A)$, which returns the value in a numerical representation that is closest to the value A. In a perfect system, $Rep(A) = A$ for all values of A. As we've seen, however, even limiting our range to finite extremes will not allow us to represent all real numbers in the bounded range exactly. $Rep(A)$ will be a many-to-one mapping, with infinitely many real numbers A mapping to each distinct value returned by $Rep(A)$, and only one represented exactly. In other words, for almost all real values A, $Rep(A) \neq A$, or $|Rep(A) - A| \neq 0$. The representation in such a case is an approximation of the actual value.

By making use of $|Rep(A) - A|$, we can measure the error in the approximation. The simplest way to represent this is *absolute error*, which is defined as

$$AbsError = |Rep(A) - A|$$

This is simply the number line distance between the actual value and its representation. While this value does correctly signify the difference between the two values, it doesn't quantify another important factor in representation error—the scale at which the error affects computation.

To better understand this, imagine a system of measurement that is accurate to within a kilometer. Such a system might be considered suitably accurate for measuring the distance

(149,597,871 km) between the earth and the sun. However, it would likely be woefully inaccurate at measuring the size of an apple (0.00011 km), which would be rounded to 0 km! Intuitively, this is obvious, but in both cases, the absolute error of representation is less than 1 km. Clearly, absolute error is not sufficient in all cases.

Relative error takes the scale of the value being approximated into account. It does so by dividing the absolute error by the actual value being represented, making it a measure of the ratio of the error to the magnitude of the value being approximated. More specifically, it is defined as

$$RelError = \left| \frac{Rep(A) - A}{A} \right|$$

As such, relative error is dimensionless; even if the values being approximated have units (such as kilometers), the relative error has no units. However, due to the division, relative error cannot be computed for a value that approximates 0.

Revisiting our previous example, the relative errors in each case would be (approximately)

$$RelError_{Sun} = \left| \frac{1 \text{ km}}{149,597,871 \text{ km}} \right| \approx 7 \times 10^{-9}$$

$$RelError_{Apple} = \left| \frac{0.00011 \text{ km}}{0.00011 \text{ km}} \right| = 1.0$$

Clearly, relative error is a much more useful error metric in this case. The earth–sun distance error is tiny (compared to the distance being measured), while the size of the apple was estimated so poorly that the error had the same magnitude as the actual value. In the former case, a relatively close representation was found, while in the latter case, the representation is all but useless.

Taking another example, let's look at the absolute and relative errors for our 4-dot-4 fixed-point representation. We know that our representable values are separated by a step size of $1/16$. That means that any real value within the range of this fixed-point format is no more than $1/32$ from a representable value. And so our absolute error is bounded by a constant, or

$$0 \leq |Rep(A) - A| \leq \frac{1}{32}$$

Dividing this by A, we can see that our relative error is not well bounded:

$$0 \leq \left| \frac{Rep(A) - A}{A} \right| \leq \frac{1}{32|A|}$$

Our relative error grows in inverse proportion to the magnitude of A. This is one characteristic of fixed-point numbers: absolute error is bounded but relative error can vary greatly.

1.3 Floating-Point Numbers

1.3.1 Review: Scientific Notation

In order to better introduce floating-point numbers, it is instructive to review the well-known standard representation for real numbers in science and engineering: scientific notation. Computer floating point is very much analogous to scientific notation.

Scientific notation (in its strictest, so-called normalized form) consists of two parts:

1. A decimal number, called the *mantissa*, such that

$$1.0 \leq |mantissa| < 10.0$$

2. An integer, called the *exponent*

Together, the exponent and mantissa are combined to create the number

$$mantissa \times 10^{exponent}$$

Any decimal number can be represented in this notation (other than 0, which is simply represented as 0.0), and the representation is unique for each number. In other words, for two numbers written in this form of scientific notation, the numbers are equal if and only if their mantissas and exponents are equal. This uniqueness is a result of the requirements that the exponent be an integer and that the mantissa be "normalized" (i.e., have magnitude in the range [1.0, 10.0]). Examples of numbers written in scientific notation include

$$102 = 1.02 \times 10^2$$

$$243{,}000 = 2.43 \times 10^5$$

$$-0.0034 = -3.4 \times 10^{-3}$$

Examples of numbers that constitute incorrect scientific notation and their correct form include

Incorrect	Correct
11.02×10^3	1.102×10^4
0.92×10^{-2}	9.2×10^{-3}

1.3.2 Restricted Scientific Notation

For the purpose of introducing the concept of finiteness of representation, we will briefly discuss a contrived, restricted scientific notation. We extend the rules for scientific notation:

1. The mantissa must be written with a single, nonzero integral digit.

2. The mantissa must be written with a fixed number of fractional digits (we define this as M digits).

3. The exponent must be written with a fixed number of digits (we define this as *E* digits).

4. The mantissa and the exponent each have individual signs.

For example, the following number is in a format with $M = 3, E = 2$:

$$\pm \boxed{1.1\,|2|3} \times 10^{\pm \boxed{1|2}}$$

Limiting the number of digits allocated to the mantissa and exponent means that any value that can be represented by this system can be represented uniquely by six decimal digits and two signs. However, this also implies that there are a limited number of values that could ever be represented exactly by this system, namely:

$$(\text{exponents}) \times (\text{mantissas}) \times (\text{exponent signs}) \times (\text{mantissa signs})$$
$$= (10^2) \times (9 \times 10^3) \times (2) \times (2)$$
$$= 3{,}600{,}000$$

Note that the leading digit of the mantissa must be nonzero (since the mantissa is normalized), so that there are only nine choices for its value [1, 9], leading to $9 \times 10 \times 10 \times 10 = 9{,}000$ possible mantissas.

This makes both the range and precision finite. The minimum and maximum exponents are

$$\pm(10^E - 1) = \pm(10^2 - 1) = \pm 99$$

The largest mantissa value is

$$10.0 - (10^{-M}) = 10.0 - (10^{-3}) = 10.0 - 0.001 = 9.999$$

Note that the smallest allowed nonzero mantissa value is still 1.000 due to the requirement for normalization. This format has the following numerical limitations:

Maximum representable value: 9.999×10^{99}

Minimum representable value: -9.999×10^{99}

Smallest positive value: 1.000×10^{-99}

While one would likely never use such a restricted form of scientific notation in practice, it demonstrates the basic building blocks of binary floating point, the most commonly used computer representation of real numbers in modern computers.

1.3.3 Binary Scientific Notation

There is no reason that scientific notation must be written in base-10. In fact, in its most basic form, the real-number representation known as floating point is similar to a base-2 version of the restricted scientific notation given previously. In base-2, our restricted scientific notation would become

$$SignM \times mantissa \times 2^{SignE \times exponent}$$

where *exponent* is an E-bit integer, and *SignM* and *SignE* are independent bits representing the signs of the mantissa and exponent, respectively.

Mantissa is a bit more complicated. It is a 1-dot-M fixed-point number whose most significant bit is always 1. Hence, the resulting *Mantissa* is in the range

$$1.0 \leq Mantissa \leq \left(2.0 - \frac{1}{2^M} \right)$$

Put together, the format involves $M + E + 3$ bits ($M + 1$ for the mantissa, E for the exponent, and 2 for the signs). Creating an example that is analogous to the preceding decimal case, we analyze the case of $M = 3, E = 2$:

$$\pm \boxed{1.\,0\,1\,0} \times 2^{\pm \boxed{0\,1}}$$

Any value that can be represented by this system can be represented uniquely by 8 bits. The number of values that ever could be represented exactly by this system is

$$(\text{exponents}) \times (\text{mantissas}) \times (\text{exponent signs}) \times (\text{mantissa signs})$$
$$= (2^2) \times (1 \times 2^3) \times (2) \times (2)$$
$$= 2^7 = 128$$

This seems odd, as an 8-bit number should have 256 different values. However, note that the leading bit of the mantissa must be 1, since the mantissa is normalized (and the only choices for a bit's value are 0 and 1). This effectively fixes one of the bits and cuts the number of possible values in half. We shall see that the most common binary floating-point format takes advantage of the fact that the integral bit of the mantissa is fixed at 1.

In this case, the minimum and maximum exponents are

$$\pm(2^E - 1) = \pm(2^2 - 1) = \pm 3$$

The largest mantissa value is

$$2.0 - 2^{-M} = 2.0 - 2^{-3} = 1.875$$

This format has the following numerical limitations:

Maximum representable value: $1.875 \times 2^3 = 15$

Minimum representable value: $-1.875 \times 2^3 = -15$

Smallest positive value: $1.000 \times 2^{-3} = 0.125$

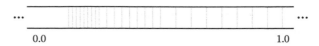

Figure 1.2. Representable values for our 8-bit floating point between 0 and 1.

If we look at the representable values of this format between 0 and 1 (Figure 1.2), we see that the relative spacing is much different than our fixed-point format. The values are close together near 0, and spread farther and farther apart as we move away from 0. Also notice that we cannot represent 0 with our current format. There is also a large gap between the smallest representable positive number and 0—this is known as the *hole at zero*. We will discuss how to handle both of these cases later.

From the listed limits, it is quite clear that a floating-point format based on this simple 8-bit binary notation would not be useful to most real-world applications. However, it does introduce the basic concepts that are shared by real floating-point representations. While there are countless possible floating-point formats, the universal popularity of a single set of formats (those described in the IEEE 754 specification [4]) makes it the obvious choice for any discussion of the details of floating-point representation. In the remainder of this chapter, we will explain the major concepts of floating-point representation as evidenced by the IEEE standard format.

1.4 IEEE 754 Floating-Point Standard

By the early to mid-1970s, scientists and engineers were very frequently using floating point formats to represent real numbers; at the time, higher powered computers even included special hardware to accelerate floating-point calculations. However, these same scientists and engineers were finding the lack of a floating-point standard to be problematic. Their complex (and often very important) numerical simulations were producing different results, depending only on the make and model of computer upon which the simulation was run. Numerical code that had to run on multiple platforms became riddled with platform-specific code to deal with the differences between different floating-point processors and libraries.

In order for cross-platform numerical computing to become a reality, a standard was needed. Over the course of the next decade, a draft standard for floating-point formats and behaviors became the de facto standard on most floating-point hardware. Once adopted, it became known as the IEEE 754 floating-point standard [83]. It was further revised in 2008 [4], and it forms the basis of almost every hardware and software floating-point system on the market.

While the history of the standard is fascinating [87], this section will focus on explaining part of the standard itself, as well as using the standard and one of its specified formats to explain the concepts of modern floating-point arithmetic.

1.4.1 Basic Representation

The IEEE 754-2008 standard specifies three binary basic formats, two decimal basic formats, one binary interchange format, and one decimal interchange format. For our purposes, we will focus on three of these: a 32-bit *single-precision* format for floating-point numbers (also known as *binary32*), a 64-bit *double-precision* format (*binary64*), and a 16-bit half-precision interchange format (*binary16*). The first two define both the format and mathematical operations; the last is only intended for use as a storage format.

It is the single-precision format that is of greatest interest for most games and interactive applications and is thus the format that will form the basis of most of the floating-point discussion in this text. The three formats are fundamentally similar, so all of the concepts regarding single precision are applicable to double precision and half precision as well.

The following diagram shows the basic memory layout of the IEEE single-precision format, including the location and size of the three components of any floating-point system: sign, exponent, and mantissa:

Sign	Exponent	Mantissa
1 bit	8 bits	23 bits

The sign in the IEEE floating-point format is represented as an explicit bit (the high-order bit). Note that this is the sign of the number itself (the mantissa), not the sign of the exponent. Differentiating between positive and negative exponents is handled in the exponent itself (and is discussed next). The only difference between X and $-X$ in IEEE floating point is the high-order bit. A sign bit of 0 indicates a positive number, and a sign bit of 1 indicates a negative number.

This sign bit format allows for some efficiencies in creating a floating-point math system in either hardware or software. To negate a floating-point number, simply flip the sign bit, leaving the rest of the bits unchanged. To compute the absolute value of a floating-point number, simply set the sign bit to 0 and leave the other bits unchanged. In addition, the sign bits of the result of a multiplication or division are simply the exclusive-OR of the sign bits of the operands.

The exponent in this case is stored as a biased number. Biased numbers represent both positive and negative integers (inside of a fixed range) as whole numbers by adding a fixed, positive bias. To represent an integer I, we add a positive bias B (that is constant for the biased format), storing the result as the whole number (nonnegative integer) W. To decode the represented value I from its biased representation W, the formula is simply

$$I = W - B$$

To encode an integer value, the formula is

$$W = I + B$$

Clearly, the minimum integer value that can be represented is

$$I = 0 - B = -B$$

The maximal value that can be represented is related to the maximum whole number that can be represented, W_{max}. For example, with an 8-bit biased number, that value is

$$I = W_{max} - B = (2^8 - 1) - B$$

Most frequently, the bias chosen is as close as possible to $W_{max}/2$, giving a range that is equally distributed to about zero. Over the course of this chapter, when we are referring to a biased number, the term *value* will refer to I, while the term *bits* will refer to W.

Such is the case with the IEEE single-precision floating-point exponent, which uses 8 bits of representation and a bias of 127. This would seem to lead to minimum and maximum exponents of $-127 (= 0 - 127)$ and $128 (= 255 - 127)$, respectively. However, for reasons that will be explained, the minimum and maximum values (-127 and 128) are reserved for special cases, leading to an exponent range of $[-126, 127]$. As a reference, these base-2 exponents correspond to base-10 exponents of approximately $[-37, 38]$.

The mantissa is normalized (in almost all cases), as in our discussion of decimal scientific notation (where the units digit was required to have magnitude in the range $[1, 9]$). However, the meaning of "normalized" in the context of a binary system means that the leading bit of the mantissa is always 1. Unlike a decimal digit, a binary digit has only one nonzero value. To optimize storage in the floating-point format, this leading bit is omitted, or hidden, freeing all 23 explicit mantissa bits to represent fractional values (and thus these explicit bits are often called the "fractional" mantissa bits). To decode the entire mantissa into a rational number (ignoring for the moment the exponent), assuming the fractional bits (as a 23-bit unsigned integer) are in F, the conversion is

$$1.0 + \frac{F}{2.0^{23}}$$

So, for example, the fractional mantissa bits

$$11100000000000000000000_2 = 7340032_{10}$$

become the rational number

$$1.0 + \frac{7340032.0}{2.0^{23}} = 1.875$$

1.4.2   Range and Precision

The range of single-precision floating point is by definition symmetric, as the system uses an explicit sign bit—every positive value has a corresponding negative value. This leaves the questions of maximal exponent and mantissa, which when combined will represent the explicit values of greatest magnitude. In the previous section, we found that the maximum base-2 exponent in single-precision floating point is 127. The largest mantissa would be equal to setting all 23 explicit fractional mantissa bits, resulting (along with the implicit 1.0 from the hidden bit) in a mantissa of

$$1.0 + \sum_{i=1}^{23} \frac{1}{2^i} = 1.0 + 1.0 - \frac{1}{2^{23}} = 2.0 - \frac{1}{2^{23}} \approx 2.0$$

The minimum and maximum single-precision floating-point values are then

$$\pm \left(2.0 - \frac{1}{2^{23}} \right) \times 2^{127} \approx \pm 3.402823466 \times 10^{38}$$

The precision of single-precision floating point can be loosely approximated as follows: for a given normalized mantissa, the difference between it and its nearest neighbor is 2^{-23}.

To determine the actual spacing between a floating-point number and its neighbor, the exponent must be known. Given an exponent E, the difference between two neighboring single-precision values is

$$\delta_{fp} = 2^E \times 2^{-23} = 2^{E-23}$$

However, we note that in order to represent a value A in single precision, we must find the exponent E_A such that the mantissa is normalized (i.e., the mantissa M_A is in the range $1.0 \leq M_A < 2.0$), or

$$1.0 \leq \frac{|A|}{2^{E_A}} < 2.0$$

Multiplying through, we can bound $|A|$ in terms of 2^{E_A}:

$$1.0 \leq \frac{|A|}{2^{E_A}} < 2.0$$
$$2^{E_A} \leq |A| < 2^{E_A} \times 2.0$$
$$2^{E_A} \leq |A| < 2^{E_A+1}$$

As a result of this bound, we can roughly approximate the entire exponent term 2^{E_A} with $|A|$ and substitute to find an approximation of (δ_{fp}), the distance between neighboring floating-point values around $|A|$, as

$$\delta_{fp} = 2^{E_A-23} = \frac{2^{E_A}}{2^{23}} \approx \frac{|A|}{2^{23}}$$

From our initial discussion on absolute error, we use a general bound on the absolute error equal to half the distance between neighboring representation values:

$$AbsError_A \approx \delta_{fp} \times \frac{1}{2} = \frac{|A|}{2^{23}} \times \frac{1}{2} = \frac{|A|}{2^{24}}$$

This approximation shows that the absolute error of representation in a floating-point number is directly proportional to the magnitude of the value being represented. Having approximated the absolute error, we can approximate the relative error as

$$RelError_A = \frac{AbsError_A}{|A|} \approx \frac{|A|}{2^{24} \times |A|} = \frac{1}{2^{24}} \approx 6 \times 10^{-8}$$

The relative error of representation is thus generally constant, regardless of the magnitude of A. This is the opposite of what we saw with fixed-point numbers.

1.4.3   Arithmetic Operations

In the subsequent sections, we discuss the basic methods used to perform common arithmetic operations upon floating-point numbers. While few users of floating point will ever need to implement these operations at a bitwise level themselves, a basic understanding of the methods is a pivotal step toward being able to understand the limitations of floating point. The methods shown are designed for ease of understanding and do not represent the actual, optimized algorithms that are implemented in hardware.

The IEEE standard specifies that the basic floating-point operations of a compliant floating-point system must return values that are exactly equivalent to the result computed and *then* rounded to the available precision. The following sections are designed as an introduction to the basics of floating-point operations and do not discuss the exact methods used for rounding the results. At the end of the section, there is a discussion of the programmer-selectable rounding modes specified by the IEEE standard.

The intervening sections include information regarding common issues that arise from these operations, because each operation can produce problematic results in specific situations.

1.4.3.1 Addition and Subtraction

In order to add a pair of floating-point numbers, the mantissas of the two addends first must be shifted such that their radix points are lined up. In a floating-point number, the radix points are aligned if and only if their exponents are equal. If we raise the exponent of a number by 1, we must shift its mantissa to the right by 1 bit. For simplicity, we will first discuss addition of a pair of positive numbers. The standard floating-point addition method works (basically) as follows to add two positive numbers $A = S_A \times M_A \times 2^{E_A}$ and $B = S_B \times M_B \times 2^{E_B}$, where $S_A = S_B = 1.0$ due to the current assumption that A and B are nonnegative.

1. Swap A and B if needed so that $E_A \geq E_B$.

2. Shift M_B to the right by $E_A - E_B$ bits. If $E_A \neq E_B$, then this shifted M_B will not be normalized, and M_B will be less than 1.0. This is needed to align the radix points.

3. Compute M_{A+B} by directly adding the shifted mantissas M_A and M_B.

4. Set $E_{A+B} = E_A$.

5. The resulting mantissa M_{A+B} may not be normalized (it may have an integral value of 2 or 3). If this is the case, shift M_{A+B} to the right 1 bit and add 1 to E_{A+B}.

Note that there are some interesting special cases implicit in this method. For example, we are shifting the smaller number's mantissa to the right to align the radix points. If the two numbers differ in exponents by more than the number of mantissa bits, then the smaller number will have all of its mantissa shifted away, and the method will add 0 to the larger value. This is important to note, as it can lead to some very strange behavior in applications. Specifically, if an application repeatedly adds a small value to an accumulator, as the accumulator grows there will come a point at which adding the small value to the accumulator will result in no change to the accumulator's value (the delta value being added will be shifted to zero each iteration)!

Floating-point addition must take negative numbers into account as well. There are three distinct cases here:

- *Both operands positive.* Add the two mantissas as is and set the result sign to positive.

- *Both operands negative.* Add the two mantissas as is and set the result sign to negative.

- *One positive operand and one negative operand.* Negate (two's complement) the mantissa of the negative number and add.

In the case of subtraction (or addition of numbers of opposite sign), the result may have a magnitude that is significantly smaller than either of the operands, including a result of 0. If this is the case, there may be considerable shifting required to reestablish the normalization of the result, shifting the mantissa to the left (and shifting zeros into the lowest-precision bits) until the integral bit is 1. This shifting can lead to precision issues (see Section 1.4.6) and can even lead to nonzero numbers that cannot be represented by the normalized format discussed so far (see Section 1.4.5).

We have purposefully omitted discussion of rounding, as rounding the result of an addition is rather complex to compute quickly. This complexity is due to the fact that one of the operands (the one with the smaller exponent) may have bits that are shifted out of the operation, but must still be considered to meet the IEEE standard of "exact result, then rounded." If the method were simply to ignore the shifted bits of the smaller operand, the result could be incorrect. You may want to refer to Hennessy and Patterson [78] for details on the floating-point addition algorithm.

1.4.3.2 Multiplication

Multiplication is actually rather straightforward with IEEE floating-point numbers. Once again, the three components that must be computed are the sign, the exponent, and the mantissa. As in the previous section, we will give the example of multiplying two floating-point numbers, A and B.

Owing to the fact that an explicit sign bit is used, the sign of the result may be computed simply by computing the exclusive-OR of the sign bits, producing a positive result if the signs are equal and a negative result otherwise. The result of the multiplication algorithm is sign-invariant.

To compute the initial exponent (this initial estimate may need to be adjusted at the end of the method if the initial mantissa of the result is not normalized), we simply sum the exponents. However, since both E_A and E_B contain a bias value of 127, the sum will contain a bias of 254. We must subtract 127 from the result to reestablish the correct bias:

$$E_{A \times B} = E_A + E_B - 127$$

To compute the result's mantissa, we multiply the normalized source mantissas M_A and M_B as 1-dot-23 format fixed-point numbers. The method for multiplying two X-dot-Y bit format fixed-point numbers is to multiply them using the standard integer multiplication method and then divide the result by 2^Y (which can be done by shifting the result to the right by Y bits). For 1-dot-23 format source operands, this produces a (possibly unnormalized) 3-dot-46 result. Note from the format that the number of integral bits may be 3, as the resulting mantissa could be rounded up to 4.0. Since the source mantissas are normalized, the resulting mantissa (if it is not 0) must be ≥ 1.0, leading to three possibilities for the mantissa $M_{A \times B}$: it may be normalized, it may be too large by 1 bit, or it may be too large by 2 bits. In the latter two cases, we add either 1 or 2 to $E_{A \times B}$ and shift $M_{A \times B}$ to the right by 1 or 2 bits until it is normalized.

1.4.3.3 Rounding Modes

The IEEE 754-2008 specification defines five rounding modes that an implementation must support. These rounding modes are

- Round toward nearest, ties to even.

- Round toward nearest, ties away from 0.

- Round toward 0 (truncation).

- Round toward $-\infty$ (floor).

- Round toward ∞ (ceiling).

The specification defines these modes with specific references to bitwise rounding methods that we will not discuss here, but the basic ideas are quite simple.

Round toward nearest, ties to even (also known as *bankers' rounding*) is the default rounding mode. This rounds to the nearest representable value—if there is a tie, it chooses the one with a zero least significant bit 0. Round toward nearest, ties away from 0 is similar, but chooses the value with the largest magnitude.

For the last three, we break the mantissa into the part that can be represented (the leading 1 along with the next 23 most significant bits), which we call M, and the remaining lower order bits, which we call R. Round toward 0 is also known as *truncation* and is the simplest to understand; in this mode, M is used and R is simply ignored or truncated. Round toward $\pm\infty$ are modes that round toward positive (∞) or negative ($-\infty$) based on the sign of the result and whether $R = 0$ or not, as shown in the following tables.

Round toward ∞		
	$R = 0$	$R \neq 0$
$M \geq 0$	M	$M + 1$
$M < 0$	M	M

Round toward $-\infty$		
	$R = 0$	$R \neq 0$
$M \geq 0$	M	M
$M < 0$	M	$M + 1$

One possible use for setting the rounding mode is for interval arithmetic, where you track an upper and lower bound for each calculation, and use ceiling and floor modes, respectively. It also is used in converting floating-point values to a string. However, in most cases, there's no reason to change the rounding mode from the default.

1.4.4 Special Values

One of the most important parts of the IEEE floating-point specification is its definition of numerous special values. While these special values co-opt bit patterns that would otherwise represent specific floating-point numbers, this trade-off is accepted as worthwhile, owing to the nature and importance of these special values.

1.4.4.1 Zero

The representation of 0.0 in floating point is more complex than one might think. Since the high-order bit of the mantissa is assumed to be 1 (and has no explicit bit in the representation), it is not enough to simply set the 23 explicit mantissa bits to 0, as that would simply represent the number $1.0 \times 2E^{-127}$. It is necessary to define zero explicitly, in this case as a number whose exponent bits are all 0 *and* whose explicit mantissa bits are 0. This is sensible, as this value would otherwise represent the smallest possible normalized value. Note that the exponent bits of 0 map to an exponent value of -127, which is reserved for special values such as 0. All other numbers with exponent value -127 (i.e., those with nonzero mantissa bits) are reserved for a class of very small numbers called *denormals*, which will be described later.

Another issue with respect to floating-point zero arises from the fact that IEEE floating-point numbers have an explicit sign bit. The IEEE specification defines both positive and negative 0, differentiated by only the sign bit. This allows certain computations to continue to propagate the sign bit even for results of 0. To avoid very messy code, the specification does require that $+0$ and -0 are treated as equal in comparisons. However, the bitwise representations are distinct, which means that applications should never use bitwise equality tests with floating-point numbers! The bitwise representations of both zeros are

$$+0.0 =$$

0	00000000	00000000000000000000000
S	Exponent	Mantissa

$$-0.0 =$$

1	00000000	00000000000000000000000
S	Exponent	Mantissa

The standard does list the behavior of positive and negative zero explicitly, including the definitions:

$$(+0) - (+0) = (+0)$$
$$-(+0) = (-0)$$

Also, the standard defines the sign of the result of a multiplication or division operation as negative if and only if exactly one of the signs of the operands is negative. This includes zeros. Thus,

$$(+0)(+0) = +0$$
$$(-0)(-0) = +0$$
$$(-0)(+0) = -0$$
$$(-0)P = -0$$
$$(+0)P = +0$$
$$(-0)N = +0$$
$$(+0)N = -0$$

where $P > 0$ and $N < 0$.

1.4.4.2   Infinity

At the other end of the spectrum from zero, the standard also defines positive infinity (∞_{fp}) and negative infinity ($-\infty_{fp}$), along with rules for the behavior of these values. In a sense the infinities are not pure mathematical values. Rather, they are used to represent values that fall outside of the range of valid exponents. For example, 1.0×10^{38} is just within the range of single-precision floating point, but in single precision,

$$(1.0 \times 10^{38})^2 = 1.0 \times 10^{76} \approx \infty_{fp}$$

The behavior of infinity is defined by the standard as follows (the standard covers many more cases, but these are representative):

$$\infty_{fp} - P = \infty_{fp}$$
$$\frac{P}{\infty_{fp}} = +0$$
$$\frac{-P}{\infty_{fp}} = -0$$
$$\frac{P}{+0} = \infty_{fp}$$
$$\frac{P}{-0} = -\infty_{fp}$$

where

$$0 < P < \infty_{fp}$$

The bitwise representations of $\pm\infty_{fp}$ use the reserved exponent value 128 and all explicit mantissa bit zeros. The only difference between the representations of the two infinities is, of course, the sign bit. The representations are diagrammed as follows:

$$\infty_{fp} =$$

0	11111111	00000000000000000000000
S	Exponent	Mantissa

$$-\infty_{fp} =$$

1	11111111	00000000000000000000000
S	Exponent	Mantissa

To test for infinite values, in C99 you can use the `isinf()` macro (also provided by many C++ compilers), and in C++11 the `std::isinf()` function.

Floating-point numbers with exponent values of 128 and nonzero mantissa bits do not represent infinities. They represent the next class of special values—*nonnumerics*.

1.4.4.3 Nonnumeric Values

All the following function call examples represent exceptional cases:

Function Call	Issue
arcsine(2.0)	Function not defined for argument.
sqrt(−1.0)	Result is imaginary.
0.0/0.0	Result is indeterminate.
$\infty - \infty$	Result is indeterminate.

In each of these cases, none of the floating-point values we have discussed will accurately represent the situation. Here we need a value that indicates the fact that the desired computation cannot be represented as a real number. The IEEE specification includes a special set of values for these cases, known collectively as not a numbers (NaNs). The general format is

$$NaN =$$

0	11111111	[23 low-order bits not all 0]
S	Exponent	Mantissa

The leading bit of the mantissa controls which type of NaN it is; the remainder of the bits can be used to indicate which error triggered the NaN.

If the leading bit of the mantissa is 1, then they are quiet NaNs, or QNaNs (Kahan [87] simply calls them NaNs). These represent indeterminate values and are quietly passed through later computations (generally as QNaNs). They are not supposed to signal an exception, but rather allow floating-point code to return the fact that the result of the desired operation was indeterminate. Floating-point implementations (hardware or software) will generate QNaNs in cases such as those in our comparison.

If the leading bit is a 0, and the remaining bits are not all 0, then this is a signalling NaN, or SNaN. These represent unrecoverable mathematical errors and signal an exception (if exceptions are disabled, they are converted to QNaNs). These are less interesting to us, as most floating-point units (FPUs) are designed not to generate SNaNs, and there have been issues in the support for SNaNs in current compilers [87]. As a result, SNaNs are encountered very rarely.

Any calculation with a NaN will generate a NaN as a result, and any comparison will return false. In the latter case, this means you can run into some odd situations, as this table shows:

Comparison	Result
NaN >= P	False
NaN <= P	False
NaN == NaN	False

In this case, NaN is neither less than, greater than, nor equal to P, and NaN is not even equal to itself! If you see a situation like this, it's very likely a NaN has ended up in your calculations somewhere. To test for NaN values, in C99 you can use the `isnan()` macro (also provided by many C++ compilers), and in C++11 the `std::isnan()` function.

1.4.5 Very Small Values

1.4.5.1 Normalized Mantissas and the Hole at Zero

As we saw in Figure 1.2, one side effect of the normalized mantissa is very interesting behavior near 0. To better understand this, let us look at the smallest normalized value (we will look at the positive case; the negative case is analogous) in single-precision floating point, which we will call F_{min}. This would have an exponent of -126 and zeros in all explicit mantissa bits. The resulting mantissa would have only the implicit leading bit set, producing a value of

$$F_{min} = 2^0 \times 2^{-126} = 2^{-126}$$

The largest value smaller than this in a normalized floating-point system would be 0.0. The smallest value larger than F_{min} is found by setting its least significant mantissa bit to 1. This value, which we will call F_{next}, would be simply

$$F_{next} = (2^0 + 2^{-23}) \times 2^{-126} = 2^{-126} + 2^{-149} = F_{min} + 2^{-149}$$

So the distance between F_{min} and its nearest smaller neighbor (0.0) is 2^{-126}, and the distance between F_{min} and F_{next} is only 2^{-149}.

In fact, F_{min} has a sequence of approximately 2^{23} larger neighbors that are each a distance of 2^{-149} from the previous. This leaves a large set of numbers between 0.0 and F_{min} that cannot be represented with nearly the accuracy as the numbers slightly larger than F_{min}. This gap in the representation is often referred to as the *hole at zero*. The operation of representing numbers in the range $(-F_{min}, F_{min})$ with 0 is often called *flushing to zero*.

One problem with flush-to-zero is that the subtraction of two numbers that are not equal can result in 0. In other words, with flush-to-zero,

$$A - B = 0 \nRightarrow A = B$$

How can this be? See the following example:

$$A = 2^{-126} \times (2^0 + 2^{-2} + 2^{-3})$$
$$B = 2^{-126} \times (2^0)$$

Both of these are valid single-precision floating-point numbers. In fact, they have equal exponents: -126. Clearly, they are also not equal floating-point numbers: A's mantissa has two additional 1 bits. However, their subtraction produces:

$$
\begin{aligned}
A - B &= (2^{-126} \times (2^0 + 2^{-2} + 2^{-3}) - (2^{-126} \times (2^0))) \\
&= 2^{-126} \times ((2^0 + 2^{-2} + 2^{-3}) - (2^0)) \\
&= 2^{-126} \times (2^{-2} + 2^{-3}) \\
&= 2^{-128} \times (2^0 + 2^{-1})
\end{aligned}
$$

which would be returned as zero on a flush-to-zero floating-point system. While this is a contrived example, it can be seen that any pair of nonequal numbers whose difference has a magnitude less than 2^{-126} would demonstrate this problem. There is a solution to this and other flush-to-zero issues, however.

1.4.5.2   Denormals and Gradual Underflow

To solve the problem of the hole at zero, the IEEE specification specifies behavior for very small numbers known as *gradual underflow*. The idea is quite simple. Rather than require every floating-point number to be normalized, the specification reserves numbers called *denormals* (or *denormalized numbers* or *subnormals*) with nonzero explicit mantissa bits and an exponent of -127. In a denormal, the implicit high-order bit of the mantissa is 0. This allows numbers with magnitude smaller than 1.0×2^{-126} to be represented. In a denormal, the exponent is assumed to be -126 (even though the actual bits would represent -127), and the mantissa is in the range $[\frac{1}{2^{23}}, 1 - \frac{1}{2^{23}}]$. The smallest nonzero value that can be represented with a denormal is $2^{-23} \times 2^{-126} = 2^{-149}$, filling in the hole at zero. Note that all nonzero floating-point values are still unique, as the specification only allows denormalized mantissas with an exponent of -126, the minimum valid exponent. Figure 1.3 shows the resulting spacing in our 8-bit floating-point format. Note that support for denormals is, technically, a trade-off. By co-opting what would be the smallest exponent (-127) to represent denormals, we have traded away some precision in the range $[2^{-127}, 2^{-126})$ in order to avoid a huge precision loss in the range $(0, 2^{-127})$.

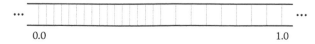

Figure 1.3. Representable values for our 8-bit floating point including 0 and denormals.

As a historical note, gradual underflow and denormalized value handling were perhaps the most hotly contested of all sections in the IEEE floating-point specification. Flush-to-zero is much simpler to implement in hardware, which also tends to mean that it performs faster and makes the hardware cheaper to produce. When the IEEE floating-point standard was being formulated in the late 1970s, several major computer manufacturers were using the flush-to-zero method for dealing with underflow. Changing to the use of gradual underflow required these manufacturers to design FPU hardware or software that could handle the unnormalized mantissas that are generated by denormalization. This would lead to either more complex FPU hardware or a system that emulated some or all of the denormalized computations in software or microcode. The former could make the FPUs more expensive to produce, while the latter could lead to greatly decreased performance of the floating-point system when denormals are generated. However, several manufacturers showed that it could be implemented in floating-point hardware, paving the way for this more accurate method to become part of the de facto (and later, official) standard. However, performance of denormalized values is still an issue, even today. We will discuss a real-world example of denormal performance in Section 1.5.2.

1.4.6 Catastrophic Cancelation

We have used relative error as a metric of the validity of the floating-point representation of a given number. As we have already seen, converting real numbers A and B into the closest floating-point approximations A_{fp} and B_{fp} generally results in some amount of relative representation error, which we compute as

$$RelErr_A = \left| \frac{A - A_{fp}}{A} \right|$$

$$RelErr_B = \left| \frac{B - B_{fp}}{B} \right|$$

These relative representation errors accurately represent how well A_{fp} and B_{fp} represent A and B, but the result of adding or subtracting A_{fp} and B_{fp} may contain a much greater level of relative error. The addition or subtraction of a pair of floating-point numbers can lead to a result with magnitude much smaller than either of the operands. Subtracting two nearly (but not exactly) equal values will result in a value whose magnitude is much smaller than either of the operands.

Recall that the last step in adding or subtracting two floating-point numbers is to renormalize the result so that the leading mantissa bit is 1. If the result of an addition or subtraction has much lower magnitude (smaller exponent) than the operands, then there will be some number N of leading mantissa bits that are all 0. The mantissa must be shifted left N bits so that the leading bit is 1 (and the exponent decremented by N, of course), renormalizing the number. Zeros will be shifted into all of the N lowest-order (explicit) mantissa bits. It is these zeros that are the cause of the error; that is, the zeros that are shifted into the lower order bits are not actual data. Thus, the N least significant mantissa bits may all be wrong. This can greatly compound relative error.

As an example, imagine that we are measuring the distances between pairs of points on the real-number line. Each of these pairs might represent the observed positions of two characters A and B in a game at two different times, t and $t + 1$. We will move each character by the same amount δ between t and $t + 1$. Thus, $A' = A + \delta$ and $B' = B + \delta$. If we use the values

$$A = 1.5$$
$$B = 10^7$$
$$\delta = 1.5$$

we can clearly see that in terms of real numbers,

$$A' = 3.0$$
$$B' = 10,000,001.5$$

However, if we look at the single-precision floating-point representations, we get

$$A'_{fp} = 3.0$$
$$B'_{fp} = 10,000,002.0$$

A' is represented exactly, but B' is not, giving a relative error of representation for B'_{fp} of

$$RelErr_B = \left| \frac{0.5}{10^7} \right|$$
$$= 5 \times 10^{-8}$$

This is quite a small relative error. However, if we compute the distances $A' - A$ and $B' - B$ in floating point, the story is very different:

$$A'_{fp} - A_{fp} = 3.0 - 1.5 \qquad\qquad = 1.5 = \delta$$
$$B'_{fp} - B_{fp} = 10,000,002.0 - 10^7 \quad = 2.0$$

In the case of $A' - A$, we get the expected value, δ. But in the case of $B' - B$, we get a relative error of

$$RelErr = \left| \frac{2.0 - 1.5}{1.5} \right|$$
$$= 0.\overline{3}$$

The resulting error is much larger in the B case, even though $A' - A = B' - B$. What is happening here can be seen by looking at the bitwise representations:

		Exponent	Mantissa Bits
B	$=$	23	10011000100101101010000000
B'	$=$	23	10011000100101101010000010
$B' - B$	$=$	23	00000000000000000000000010
Normalized	$=$	1	10000000000000000000000000

In the case of $B' - B$, almost all of the original mantissa bits in the operands were canceled out in the subtraction, leaving the least significant bits of the operands as the *most* significant bits of the result. Basically none of the fractional bits of the resulting mantissa were actual data—the system simply shifted in zeros. The precision of such a result is very low, indeed. This is catastrophic cancelation; the significant bits are all canceled, causing a catastrophically large growth in the representation error of the result.

The best way to handle catastrophic cancelation in a floating-point system is to avoid it. Numerical methods that involve computing a small value as the subtraction or addition of two potentially large values should be reformulated to remove the operation. An example of a common numerical method that uses such a subtraction is the well-known quadratic formula:

$$\frac{-B \pm \sqrt{B^2 - 4AC}}{2A}$$

Both of the subtractions in the numerator can involve large numbers whose addition/subtraction can lead to small results. However, revising the formula can lead to better-conditioned results. One possible refactoring tries to avoid cancelation by replacing the subtraction with an addition:

$$\frac{2C}{-B \mp \sqrt{B^2 - 4AC}}$$

However, if B^2 is large compared to $4AC$, for the second root we'll again end up dividing by a very small number. The solution is to rewrite the equation as

$$Q = \frac{1}{2}\left(B + \text{sgn}(B)\sqrt{B^2 - 4AC}\right)$$
$$x_0 = Q/A$$
$$x_1 = C/Q$$

where x_0 and x_1 are the roots, and sgn() is the sign of B.

1.4.7 Comparing Floating-Point Numbers

Another consequence of relative representation error is that even without catastrophic cancelation, there will still be some error that creeps into our calculations over time. This is known as *floating-point drift*. In general, we have enough precision (in particular for games) that even if our answer is not precise to the least significant bit, it won't affect the placement of an object or the animation of a character. However, often we want to compare a result v with a fixed value k to see if we've hit a certain value (a common example is 0). We may want to catch a situation where we're dividing by that value and so avoid catastrophic cancelation, or we may want to see if an object has hit a certain position or speed. Because of the error, checking equality directly in this case will not work—instead we want to add a certain uncertainty to the check.

One possibility is to check against a constant small value ϵ:

$$|v - k| \leq \epsilon$$

However, as we've seen, this is absolute error. With floating point, absolute error can vary greatly depending on the magnitude of the value, so choosing a single ϵ for all possible values will not work. In that case, doing a check using relative error seems appropriate:

$$\left| \frac{v - k}{k} \right| \leq \epsilon$$

We can remove the division by multiplying by $|k|$:

$$|v - k| \leq |k|\epsilon$$

It's possible that v is much larger than k, so we should choose the maximum of both, or

$$|v - k| \leq \max(|v|, |k|)\epsilon$$

For many cases this will work. However, Ericson [41] points out that for values less than 1 we'll be scaling our value ϵ down, thereby reducing the uncertainty we were trying to introduce. His solution is to do an absolute check for values less than 1, which can be achieved by

$$|v - k| \leq \max(|v|, |k|, 1)\epsilon$$

For systems without a fast max function, the following approximation is suggested:

$$|v - k| \leq (|v| + |k| + 1)\epsilon$$

Our final function becomes

```
inline bool IvAreEqual( float a, float b, float epsilon )
{
    return (IvAbs(a - b) <= epsilon*(IvAbs(a) + IvAbs(b) + 1.0f));
}
```

Dawson [29] further refines this by noting that strictly speaking, we need a different ϵ for the absolute check, based on the relative expected magnitudes of v and k. His solution performs the absolute check first, and then if it fails, tries the relative check. In our library, we use the Ericson approximation because it is fast and does not require knowing the expected magnitude of the inputs, making it more general. However, those who have better knowledge of their inputs and need tighter control over their floating-point error should consider Dawson's approach. See [29] for more details.

Finally, it's possible that in performance-critical code, the relative check may still be too slow. Performing the absolute check can be fine as long as you know the relative magnitude of your inputs and can choose an appropriate epsilon. The key point to remember is that comparing values directly is never appropriate—always perform some sort of bounded check.

1.4.8   Double Precision

As mentioned, the IEEE 754 specification supports a 64-bit double-precision floating-point value, known in C/C++ as the intrinsic `double` type. The format is completely analogous to the single-precision format, with the following bitwise layout:

Sign	Exponent	Mantissa
1 bit	11 bits	52 bits

Double-precision values have a range of approximately 10^{308} and can represent values smaller than 10^{-308}. A programmer's common response to the onset of precision or range issues is to switch the code to use double-precision floating-point values in the offending section of code (or sometimes even throughout the entire system). While double precision can solve almost all range issues and many precision issues (though catastrophic cancelation can still persist) in interactive 3D applications, there are several drawbacks that should be considered prior to its use:

- *Memory.* Since double-precision values require twice the storage of single-precision values, memory requirements for an application can grow quickly, especially if arrays of values must be stored as double precision.

- *Performance.* Operations on some hardware FPUs can be significantly slower when computing double-precision results. Additional expense can be incurred for conversion between single- and double-precision values.

- *Platform issues.* Not all platforms support double precision.

The latter two are less of an issue than they were in the past for desktop central processing units (CPUs), but can still cause problems on certain mobile platforms and on graphics processing units (GPUs). The key takeway is when developing a real-time application, it is good to be aware whether your particular platform supports double precision, and what the relative cost is compared to single precision.

1.4.9   Half Precision

Half-precision floating-point numbers refer to floating-point values that can fit in 16 bits of storage. They were originally introduced by Nvidia in the Cg shading language as `fp16`, and became a de facto standard on GPUs and for many graphics SDKs because they offer a dynamic range of values, but without the storage cost of single precision. Since GPUs often handle large amounts of parallel computations on large datasets, minimizing memory usage and bus traffic by using a data type that is half the size is a significant optimization.

With the release of IEEE 754-2008, IEEE floating point now supports this 16-bit floating-point format, calling it `binary16`. In the standard, it is defined as an interchange format—it is meant for storage, not for computation. As one might expect, it has a similar layout as the other IEEE 754 formats:

Sign	Exponent	Mantissa
1 bit	5 bits	10 bits

It has a biased exponent, normalized mantissa, and the standard extrema values of $\pm\infty$, NaN, and denormals. Note that while support for IEEE-style specials and denormals is common on current GPU half-precision implementations, it is not universal, especially if your application must run on older GPUs as well. Applications need to be mindful that very small and very large values may be handled differently on different GPU platforms, not unlike the pre-IEEE 754 floating-point situation. A discussion of how a real application had to deal with these differences in exceptional behaviors may be found in "GPU Image Processing in Apple's Motion" in Pharr [120].

The reduced size of half precision comes with significantly reduced precision and range when compared to even a single-precision 32-bit floating-point format. Assuming IEEE-style specials and denormals, the extrema of `binary16` are

Maximum representable value: 65,504

Smallest positive value: $2^{-25} \approx 3.0 \times 10^{-8}$

Largest consecutive integer: 2,048

These limits can be reached with surprising ease if one is not careful, especially when considering the squaring of numbers. The square of any value greater than around 255 will map to infinity when stored in half precision. And if one is using half precision as a form of integer storage, repeated incrementation will fail at a value that is well within common usage: 2,048. Above 2,048, odd integers simply cannot be represented, with these holes in the set of integers getting larger and larger as the value rises. Thus, `binary16` values are not recommended for counting items. Some of the issues associated with these reduced limits may be found in the article "A Toolkit for Computation on GPUs" in Fernando [45].

How then are half-precision values usable? The answer is one of real-world use cases. `binary16` values are most frequently used on GPUs in shader code that generates the final color drawn to the screen. In these cases, color-related values were historically limited to values between 0.0 and 1.0. The ability to use the much larger range afforded by floating values makes possible high-end rendering effects as bright light "blooming," glare, and other so-called high dynamic range (HDR) effects.

That said, many GPUs these days support single-precision IEEE 754 32-bit floating-point values, so it may seem that `binary16` is no longer necessary. However, on mobile hardware there is a high level of customer expectation for 3D graphics, combined with a huge rise in screen pixel densities and an unfortunate lack of increased memory. With a certain technique called deferred rendering (see Chapter 8), the buffer representing the screen data can get quite large, and having a format that is half the size makes this approach possible. It also helps with the number one performance issue on mobile today: memory bandwidth. Because of this, the `binary16` format is likely to continue to be popular for some time on mobile GPUs.

1.5 Real-World Floating Point

While the IEEE floating-point specification does set the exact behavior for a wide range of the possible cases that occur in real-world situations, in real-world applications on real-world platforms, the specification cannot tell the entire story. The following sections will discuss some issues that are of particular interest to 3D game developers.

1.5.1 Internal FPU Precision

Some readers will likely try some of the exceptional cases themselves in small test applications. In doing so, they may find some surprising behavior. For example, examine the following code:

```
main()
{
    float fHuge = 1.0e30f; // valid single precision
    fHuge *= 1.0e38f; // result = infinity
    fHuge /= 1.0e38f; // ????
}
```

Stepping in a debugger, the following can happen on many major compilers and systems:

1. After the initial assignment, `fHuge` = `1.0e30`, as expected.

2. After the multiplication, `fHuge` = ∞_{fp}, as expected.

3. After the division, `fHuge` = `1.0e30`!

This seems magical. How can the system divide the single value ∞_{fp} and get back the original number? A look at the assembly code gives a hint. The basic steps the compiler generates are as follows:

1. Load `1.0e30` and `1.0e38` into the FPU.

2. Multiply the two loaded values and return ∞_{fp}, keeping the result in the FPU as well.

3. Divide the previous result (still in the FPU) by `1.0e38` (still in the FPU), returning the correct result.

The important item to note is that the result of each computation was both returned and kept in the FPU for later computation. This step is where the apparent magic occurs. The FPU in this case uses high-precision (sometimes as long as `long double`) registers in the FPU. The conversion to single precision happens during the transfer of values from the FPU into memory. While the returned value in `fBig` was indeed ∞_{fp}, the value retained in the FPU was higher precision and was the correct value, `1.0e68`. When the division occurs, the result is correct, not ∞_{fp}.

However, an application cannot count on this result. If the FPU had to flush the intermediate values out of its registers, then the result of the three lines above would have been quite different. For example, if significant floating-point work had to be computed between the above multiplication and the final division, the FPU might have run out of registers and had to evict the high-precision version of `fHuge`. This can lead to odd behavior differences, sometimes even between optimized and debugging builds of the same source code.

In addition, there is a trend away from using higher precision registers in doing floating-point computations. For example, the Intel SSE (Streaming SIMD Extensions) and AVX (Advanced Vector eXtensions) coprocessors store temporary results with the same precision as the source data: 32 bits for single precision and 64 bits for double precision. While the x87 FPU does have 80-bit registers, the default for the Microsoft 64-bit compiler is to use SSE/AVX for general floating-point computations. Similarly, some GPUs and mobile CPUs have no extra internal precision for calculations. For these reasons, it is more likely that what you see in the debugger is what is stored internally.

1.5.2 Performance

The IEEE floating-point standard specifies behavior for floating-point systems; it does not specify information regarding performance. Just because a floating-point implementation is correct does not mean that it is fast. Furthermore, the speed of one floating-point operation (e.g., addition) does not imply much about the speed of another (e.g., square root). Finally, not all input data are to be considered equal in terms of performance. The following sections describe examples of some real-world performance pitfalls found in floating-point implementations.

1.5.2.1 Performance of Denormalized Numbers

Around the time that the first edition of this book was being created, one of the authors was in the midst of creating a demo for a major commercial 3D game engine. He found that in some conditions the performance of the demo dropped almost instantaneously by as much as 20 percent. The code was profiled and it was found that one section of animation code was suddenly running 10 to 100 times slower than in the previous frames. An examination of the offending code determined that it consisted of nothing more than basic floating-point operations, specifically, multiplications and divisions. Moreover, there were no loops in the code, and the number of calls to the code was not increasing. The code itself was simply taking 10 to 100 times longer to execute.

Further experiments outside of the demo found that a fixed set of input data (captured from tests of the demo) could always reproduce the problem. The developers examined the code more closely and found that very small nonzero values were creeping into the system. In fact, these numbers were denormalized. Adjusting the numbers by hand even slightly outside of the range of denormals and into normalized floating-point values instantly returned the performance to the original levels. The immediate thought was that exceptions were causing the problem (floating-point exceptions can cause multiple-order-of-magnitude performance drops in floating-point code). However, all floating-point exceptions were disabled (masked) in the test application.

To verify the situation, the developers wrote an extremely simple test application. Summarized, it was as follows:

```
float TestFunction(float fValue)
{
    return fValue;
}

main()
{
    int i;
    float fTest;
    // Start "normal" timer here
    for (i = 0; i < 10000; i++)
    {
        // 1.0e-36f is normalized in single precision
        fTest = TestFunction(1.0e-36f);
    }
    // End "normal" timer here
    // Start "denormal" timer here
    for (i = 0; i < 10000; i++)
    {
        // 1.0e-40f is denormalized in single precision
        fTest = TestFunction(1.0e-40f);
    }
    // End "denormal" timer here
}
```

Having verified that the assembly code generated by the optimizer did indeed call the desired function the correct number of times with the desired arguments, they found that the denormal loop took 30 times as long as the normal loop (even with exceptions masked). A careful reading of Intel's performance recommendations [84] for the Pentium series of CPUs found that *any* operation (including simply loading to a floating-point register) that produced or accepted as an operand a denormal value was run using so-called assist microcode, which is known to be much slower than standard FPU instructions. Intel had followed the IEEE 754 specification, but had made the design decision to allow exceptional cases such as denormals to cause very significant performance degradation.

One possible solution is to disable denormals in your application. However, doing this across the board may cause unexpected precision errors. The correct solution is to determine where the denormals are coming from and change the calculation if necessary. In this case, the values in question were known to be normalized to be between 0.0 and 1.0. As a result, it was more than safe to simply clamp small values to 0.

This historical example was run on a older generation Intel CPU, but even on modern processors denormal performance can still be an issue. However, starting with the Sandy Bridge series, Intel has taken steps to improve this by adding some hardware support for denormals in the AVX floating-point path. At the time of printing, it is still not as fast as disabling them entirely [30], but it may mean that, in the future, denormals will not be as much of a performance issue.

1.5.2.2 *Software Floating-Point Emulation*

Applications should take extreme care on new platforms to determine whether or not the platform supports hardware-assisted floating point. In order to ensure that code from other platforms ports and executes without major rewriting, some compilers supply software floating-point emulation libraries for platforms that do not support floating point in hardware. This lack of support is less true than it used to be, but was especially common on popular embedded and handheld chipsets such as the early ARM processors [7]. These processors have no FPUs, but C/C++ floating-point code compiled for these devices will generate valid, working emulation code. The compilers will often do this silently, leaving the uninformed developer with a working program that exhibits horrible floating-point performance, in some cases hundreds of times slower than could be expected from a hardware FPU. And even though the current ARM processors do have floating-point hardware, for backwards compatibility certain compiler settings still use the old software library.

It's also worth reiterating that not all FPUs support both single and double precision at the same performance. For example, on the ARMv7, the NEON SIMD extension (similar to SSE) is a fast path that supports only single-precision floating point. The only way to get double-precision floating support in hardware is to use the slower VFP floating-point processor. Even worse, other processors have only provided single-precision hardware support, with software support for double precision. As a result, use of double precision can lead to much slower code. Because of this, it is important to remember that double precision can be introduced into an expression in subtle ways. For example, remember that in C/C++, floating-point constants are double precision by default, so whenever possible, explicitly specify constants as single precision, using the f suffix. The difference between double- and single-precision performance can be as simple as 1.0 instead of 1.0f.

1.5.3 IEEE Specification Compliance

While major floating-point errors in modern processors are relatively rare (even Intel was caught off guard by the magnitude of public reaction to what it considered minor and rare errors in the floating-point divider on the original Pentium chips), this does not mean that it is safe to assume that all FPU in modern CPUs are always fully compliant to IEEE specifications and support both single and double precision. The greatest lurking risk to modern developers assuming full IEEE compliance is conscious design decisions, not errors on the part of hardware engineers. However, in most cases, for the careful and attentive programmer, these processors offer the possibilities of great performance increases to 3D games.

As more and more FPUs were designed and built for multimedia and 3D applications (rather than the historically important scientific computation applications for which earlier FPUs were designed), manufacturers began to deviate from the IEEE specification, optimizing for high performance over accuracy. This is especially true with respect to the "exceptional" cases in the spec, such as denormals, infinity, and NaNs.

Hardware vendors make the argument that while these special values are critically important to scientific applications, for 3D games and multimedia they generally occur only in error cases that are best handled by avoiding them in the first place.

An important example of such design decisions involved Intel's Streaming SIMD Extensions (SSE) [84], a coprocessor that was added to the Pentium series with the advent of the Pentium III. The original coprocessor was a special vector processor that executes parallel math operations on four floating-point values, packed into a 128-bit register. The SSE instructions were specifically targeted at 3D games and multimedia, and this is evident from even a cursory view of the design. Several design decisions related to the special-purpose FPU merit mentioning here:

- The original SSE (Pentium III) instructions can only support 32-bit floating-point values, not doubles.

- Denormal values can be (optionally) rounded to 0 (flushed to zero), disabling gradual underflow.

- Full IEEE 754 behavior can be supported as an option, but at less than peak performance.

With the move from SSE to AVX, these differences from the standard have for the most part been removed, but as we mentioned above, performance of denormals can still be an issue [30].

A similar example is the ARM NEON coprocessor, which is also a SIMD design. While VFP, the scalar floating-point coprocessor, is IEEE 754 compliant, NEON's vector design does not support double-precision floating point, denormals, or NaNs.

1.5.4   Precision in Graphics Processing Units

As mentioned in Section 1.4.9, GPUs were early adopters of half-precision formats for reducing bus traffic and processing time. For the most part (other than storage formats), these have been replaced in the desktop world with single-precision floating point and in some cases double-precision, particularly with the increase in GPU computing. In the mobile world, however, formats smaller than single precision are still common. The lower precision can be a sizable performance upgrade not just because of the cost of operations, but because the shader compiler may be able to fit more values in limited register space. This can cause big differences in the compiled result.

Because of this, in OpenGL and DirectX11, programmers can specify the minimum precision they need for program variables on the GPU. The GPU will then use this to schedule tasks appropriately. In OpenGL ES [129], these are specified as

Identifier	Range	Precision
lowp	$(-2, 2)$	2^{-8} (absolute)
mediump	$(-2^{14}, 2^{14})$	2^{-10} (relative)
highp	$(-2^{62}, 2^{62})$	2^{-6} (relative)

Note that lowp has an absolute precision bound—this is because it's most often implemented as a fixed-point format. Both mediump and highp are usually implemented using floating point, with mediump being roughly half precision and highp most likely single precision. However, note that the range and precision required are less than the half- and single-precision formats, respectively, so the GPU manufacturer may use a nonstandard format, such as 20- or 24-bit. Alternatively, they could be implemented in the same way, using either single precision for both or a slightly smaller format. The best one can know is that mediump *may* be faster than highp—the only way to be sure is to consult your hardware guides or test in a GPU program.

1.6 Code

Source Code
Library
IvMath

While this text's companion CD-ROM and web site do not include specific code that demonstrates the concepts in this chapter, source code that deals with issues of floating-point representation may be found throughout the math library IvMath. For example, the source code for IvMatrix33, IvMatrix44, IvVector3, IvVector4, and IvQuat includes sections of code that avoid denormalized numbers and comparisons to exact floating-point zero.

CPU chipset manufacturers Intel, AMD, and ARM have been focused on 3D graphics and game performance and have made public many code examples, presentations, and software libraries that detail how to write high-performance floating-point code for their processors. Many of these resources may be found on their developer web sites [3, 7, 84].

1.7 Chapter Summary

In this chapter, we have discussed the details of how computers represent real numbers. These representations have inherent limitations that any serious programmer must understand in order to use them efficiently and correctly. Floating point presents subtle limitations, especially issues of limited precision. We have also discussed the basics of error metrics for number representations.

Hopefully, this chapter has instilled two important pieces of information in the reader. The first and most basic piece of information is an understanding of the inner workings of the number systems that pervade 3D games. This should allow the programmer to truly comprehend the reasons *why* the math-related code behaves (or, more importantly, why it *misbehaves*) as it does. The second piece of information is an appreciation of why one should pay attention to the topic of floating-point representation in the first place—namely, to better prepare the 3D game developer to do what is needed at some point in the development of a game: optimize or fix a section of slow or incorrect math code. Better yet, it can assist the developer to avoid writing this potentially problematic code in the first place.

For further reading, Kahan's papers on the history and status of the IEEE floating-point standard ([87] and related papers and lectures by Kahan, available from the same source) offer fascinating insights into the background of modern floating-point computation. In addition, back issues of *Game Developer* magazine (such as [76]) provide frequent discussion of number representations as they relate to computer games. Dawson's series of blog posts on floating point ([29, 30], and others) cover many of the topics we have presented here, with a focus on optimization. Finally, for a general discussion of optimization including floating point, see [48].

2 Vectors and Points

2.1 Introduction

The two building blocks of most objects in our interactive digital world are points and vectors. Points represent locations in space, which can be used either as measurements on the surface of an object to approximate the object's shape (this approximation is called a *model*), or as simply the position of a particular object. We can manipulate an object indirectly through its position or by modifying its points directly. Vectors, on the other hand, represent the difference or displacement between two points. Both have some very simple properties that make them extremely useful throughout computer graphics and simulation.

In this chapter, we'll discuss the properties and representation of vectors and points, as well as the relationship between them. We'll present how they can be used to build up other familiar entities from geometry classes: in particular, lines, planes, and polygons. Because many problems in computer games boil down to examples in applied algebra, having computer representations of standard geometric objects built on basic primitives is extremely useful.

It is likely that the reader has a basic understanding of these entities from basic math classes, but the symbolic representations used by the mathematician may be unfamiliar or forgotten. We will review them in detail here. We will also cover linear algebra concepts—properties of vectors in particular—that are essential for manipulating three-dimensional (3D) objects. Without a thorough understanding of this fundamental material, any work in programming 3D games and applications will be quite confusing.

2.2 Vectors

One might expect that we would cover points first since they are the building blocks of our standard model, but in actuality the basic unit of most of the mathematics we'll discuss in this book is the vector. We'll begin by discussing the vector as a geometric entity since that's primarily how we'll be using it and it's more intuitive to think of it that way. From there we'll present a set of vectors known as a vector space and show how using the properties of vector spaces allows us to manipulate vectors in the computer. We'll conclude by discussing operations that we can perform on vectors and how we can use them to solve certain problems in 3D programming.

2.2.1   Geometric Vectors

A *geometric vector* **v** is an entity with magnitude (also called length) and direction and is represented graphically as a line segment with an arrowhead on one end (Figure 2.1). The length of the segment represents the magnitude of the vector, and the arrowhead indicates its direction. A vector whose magnitude is 1 is a *unit* or *normalized* vector and is shown as $\hat{\mathbf{v}}$. The *zero* vector **0** has a magnitude of 0 but no direction.

A vector does not have a location or position in space. To make some geometric calculations easier to understand, we may draw two vectors as if they were attached or place a vector relative to a location in space. Despite this, it is important to remember that two vectors with the same magnitude and direction are equal, no matter where they are drawn on a page. For example, in Figure 2.1, the leftmost and rightmost vectors are equal.

In games, we use vectors in one of two ways. The first is as a representation of direction. For example, a vector may indicate direction toward an enemy, toward a light, or perpendicular to a plane. The second meaning represents change. If we have an object moving through space, we can assign a velocity vector to the object, which represents a change in position. We can displace the object by adding the velocity vector to the object's location to get a new location. Vectors also can be used to represent change in other vectors. For example, we can modify our velocity vector by adding another to it; the second vector is called *acceleration*.

We can perform arithmetic operations on vectors just as we can with real numbers. One basic operation is addition. Geometrically, addition combines two vectors together into a new vector. If we think of a vector as an agent that changes position, then the new vector **u** = **v** + **w** combines the position-changing effect of **v** and **w** into one entity.

As an example, in Figure 2.2, we have three locations, *P*, *Q*, and *R*. There is a vector **v** that represents the change in position or displacement from *P* to *Q* and a vector **w** that

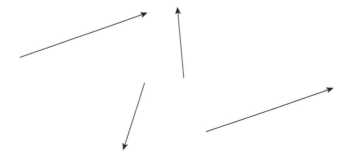

Figure 2.1. Vectors.

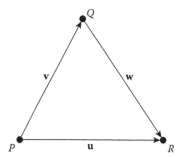

Figure 2.2. Vector addition.

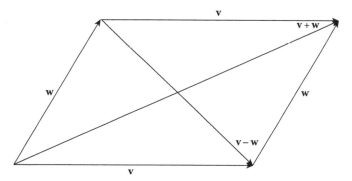

Figure 2.3. Vector addition and subtraction.

represents the displacement from Q to R. If we want to know the vector that represents the displacement from P to R, then we add **v** and **w** to get the resulting vector **u**.

Figure 2.3 shows another approach, which is to treat the two vectors as the sides of a parallelogram. In this case, the sum of the two vectors is the long diagonal of the parallelogram. Subtraction, or $\mathbf{v} - \mathbf{w}$, is shown by the other vector crossing the parallelogram. Remember that the difference vector is drawn from the second vector head to the first vector head—the opposite of what one might expect.

The algebraic rules for vector addition are very similar to those for real numbers:

1. $\mathbf{v} + \mathbf{w} = \mathbf{w} + \mathbf{v}$ (commutative property).

2. $\mathbf{u} + (\mathbf{v} + \mathbf{w}) = (\mathbf{u} + \mathbf{v}) + \mathbf{w}$ (associative property).

3. $\mathbf{v} + \mathbf{0} = \mathbf{v}$ (additive identity).

4. For every **v**, there is a vector $-\mathbf{v}$ such that $\mathbf{v} + (-\mathbf{v}) = \mathbf{0}$ (additive inverse).

We can verify this informally by drawing a few test cases. For example, if we examine Figure 2.3 again, we can see that one path along the outside of the parallelogram represents $\mathbf{v} + \mathbf{w}$ and the other represents $\mathbf{w} + \mathbf{v}$. The resulting vector is the same in both cases. Figure 2.4 presents the associative property in a similar fashion.

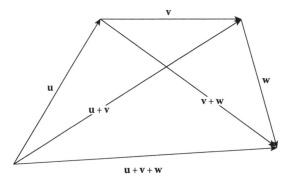

Figure 2.4. Associative property of vector addition.

Figure 2.5. Scalar multiplication.

The other basic operation is scalar multiplication, which changes the length of a vector by multiplying it by a single real value, also known as a *scalar* (Figure 2.5). Multiplying a vector by 2, for example, makes it twice as long. Multiplying by a negative value changes the length and points the vector in the opposite direction (the length remains nonnegative). Multiplying by 0 always produces the zero-length vector **0**.

The algebraic rules for scalar multiplication should also look familiar:

5. $(ab)\mathbf{v} = a(b\mathbf{v})$ (associative property).

6. $(a+b)\mathbf{v} = a\mathbf{v} + b\mathbf{v}$ (distributive property).

7. $a(\mathbf{v} + \mathbf{w}) = a\mathbf{v} + a\mathbf{w}$ (distributive property).

8. $1 \cdot \mathbf{v} = \mathbf{v}$ (multiplicative identity).

As with the additive rules, diagrams can be created that provide a certain amount of intuitive understanding.

2.2.2   Real Vector Spaces

2.2.2.1   Definition

Mathematicians like to find patterns and create abstractions. So it was with geometric vectors—they noticed that other areas of mathematics had similar properties, so they created the notion of a *linear space* or *vector space* to generalize these properties. Informally, a vector space is a collection of entities (called vectors, naturally) that meet the rules for addition and scalar multiplication that we just defined for geometric vectors.

Why is this important to us? First of all, since it is an abstraction, we can use it for manipulating higher dimensional vectors than we might be able to conceive of geometrically. It also can be used for representing entities that we wouldn't normally consider as vectors but that follow the same algebraic rules, which can be quite powerful. There are certain properties of vector spaces that will prove to be quite useful when we cover matrices and linear transformations. Finally, it provides a means for representing our geometric entities symbolically, which allows us to do algebra and provides us a way to store vectors in the computer.

To simplify our approach, we are going to concentrate on a subset of vector spaces known as *real vector spaces*, so called because their fundamental components are drawn from $\mathbb{R}$, the set of all real numbers. We usually say that such a vector space V is *over* $\mathbb{R}$. We also formally define an element of $\mathbb{R}$ in this case as a *scalar*.

In order for a set of elements V to be a vector space, we need to define two specific operations on the elements that follow certain algebraic axioms. As we indicated, the two operations are addition and scalar multiplication, and the axioms are properties 1–8, presented above. We'll define these operations so that the vector space V has *closure* with respect to them; that is, applying this operation to elements in our vector space ends up with a result in our vector space. More formally,

1. For any $\mathbf{u}$ and $\mathbf{v}$ in V, $\mathbf{u} + \mathbf{v}$ is in V (additive closure).

2. For any a in $\mathbb{R}$ and $\mathbf{v}$ in V, $a\mathbf{v}$ is in V (multiplicative closure).

So we formally define a real vector space as a set V over $\mathbb{R}$ with closure with respect to addition and scalar multiplication on its elements, where axioms 1–8 hold for all $\mathbf{u}, \mathbf{v}, \mathbf{w}, \mathbf{0}$ in V and all a, b in $\mathbb{R}$.

2.2.2.2 Examples

As we might expect, by this definition, our geometric vectors do make up a vector space. Another example of a real vector space is simply $\mathbb{R}$. Real numbers are closed under addition and multiplication, and those operations have exactly the properties described above.

A vector space we'll be using throughout the book is the set of all ordered pairs of real numbers, called $\mathbb{R}^2$. For now we can think of this as informally representing two-dimensional (2D) space—for example, diagrams on an infinitely extending, flat page. Symbolically, this is represented by

$$\mathbb{R}^2 = \{(x, y) \mid x, y \in \mathbb{R}\}$$

In this context, the symbol $\mid$ means "such that" and the symbol $\in$ means "is a member of." So we read this as "the set of all possible pairs (x, y), such that x and y are members of the set of real numbers." As mentioned, this is a set of ordered pairs; $(1.0, -0.5)$ is a different member of the set from $(-0.5, 1.0)$.

For this to be a vector space, we must also define addition:

$$(x_0, y_0) + (x_1, y_1) = (x_0 + x_1, y_0 + y_1)$$

and scalar multiplication:

$$a(x_0, y_0) = (ax_0, ay_0)$$

Using these definitions and the preceding algebraic axioms, it can be shown that $\mathbb{R}^2$ is a vector space.

Two more vector spaces are $\mathbb{R}^3$ and $\mathbb{R}^4$, defined as follows:

$$\mathbb{R}^3 = \{(x, y, z) \mid x, y, z \in \mathbb{R}\}$$
$$\mathbb{R}^4 = \{(w, x, y, z) \mid w, x, y, z \in \mathbb{R}\}$$

with addition and scalar multiplication operations similar to those of $\mathbb{R}^2$.

Like $\mathbb{R}^2$ these are ordered lists, where two members with the same values but differing orders are not the same. Again informally, we can think of elements in $\mathbb{R}^3$ as representing positions in 3D space, which is where we will be spending most of our time. Correspondingly, $\mathbb{R}^4$ can be thought of as representing a fourth-dimensional space, which is difficult to visualize spatially[1] (hence our need for an abstract representation) but is extremely useful for certain computer graphics concepts.

We can extend our definitions to $\mathbb{R}^n$, a generalized n-dimensional space over $\mathbb{R}$:

$$\mathbb{R}^n = \{(x_0, \ldots, x_{n-1}) \mid x_0, \ldots, x_{n-1} \in \mathbb{R}\}$$

The members of $\mathbb{R}^n$ are referred to as an n-tuple.

Generalized over $\mathbb{R}^n$, we have

$$\mathbf{u} + \mathbf{v} = (u_0, \ldots, u_{n-1}) + (v_0, \ldots, v_{n-1})$$
$$= (u_0 + v_0, \ldots, u_{n-1} + v_{n-1})$$

and

$$a\mathbf{v} = a(v_0, \ldots, v_{n-1})$$
$$= (av_0, \ldots, av_{n-1})$$

It is important to understand that—despite the name—a vector space does not necessarily have to be made up of geometric vectors. What we have described is a series of sets of ordered lists, possibly with no relation to a geometric construct. While they *can* be related to geometry, the term *vector*, when used in describing members of vector spaces, is an abstract concept. As long as a set of elements can be shown to have the preceding arithmetic properties, we define it as a vector space and any element of a vector space as a vector. It is perhaps more correct to say that the geometric representations of 2D and 3D vectors that we use are visualizations that help us better understand the abstract nature of $\mathbb{R}^2$ and $\mathbb{R}^3$, rather than the other way around.

---

[1] Unless you are one of a particularly gifted pair of children [116].

2.2.2.3 Subspaces

Suppose we have a subset W of a vector space V. We call W a *subspace* if it is itself a vector space when using the same definition for addition and multiplication operations. In order to show that a given subset W is a vector space, we only need to show that closure under addition and scalar multiplication holds; the rest of the properties are satisfied because W is a subset of V. For example, the subset of all vectors in $\mathbb{R}^3$ with $z = 0$ is a subspace, since

$$(x_0, y_0, 0) + (x_1, y_1, 0) = (x_0 + x_1, y_0 + y_1, 0)$$
$$a(x_0, y_0, 0) = (ax_0, ay_0, 0)$$

The resulting vectors still lie in the subspace $\mathbb{R}^3$ with $z = 0$.

Note that any subspace must contain $\mathbf{0}$ in order to meet the conditions for a vector space. So the subset of all vectors in $\mathbb{R}^3$ with $z = 1$ is not a subspace since $\mathbf{0}$ cannot be represented. And while $\mathbb{R}^2$ is not a subspace of $\mathbb{R}^3$ (since the former is a set of pairs and the latter a set of triples), it can be embedded in a subspace of $\mathbb{R}^3$ by a mapping, for example, $(x, y) \rightarrow (x, y, 0)$.

2.2.3 Linear Combinations and Basis Vectors

Our definitions of vector addition and scalar multiplication can be used to describe some special properties of vectors. Suppose we have a set S of n vectors, where $S = \{\mathbf{v}_0, \ldots, \mathbf{v}_{n-1}\}$. We can combine these to create a new vector $\mathbf{v}$ using the expression

$$\mathbf{v} = a_0 \mathbf{v}_0 + a_1 \mathbf{v}_1 + \cdots + a_{n-1} \mathbf{v}_{n-1}$$

for some arbitrary real scalars $a_0, \ldots, a_{n-1}$. This is known as a *linear combination* of all vectors $\mathbf{v}_i$ in S.

If we take all the possible linear combinations of all vectors in S, then the set T of vectors thus created is the *span* of S. We can also say that the set S spans the set T. For example, vectors $\mathbf{v}_0$ and $\mathbf{v}_1$ in Figure 2.6 span the set of vectors that lie on the surface of the page (assuming your book is held flat).

We can use linear combinations to define some properties of our initial set S. Suppose we can find a nonzero vector $\mathbf{v}_i$ in S that can be represented by a linear combination of other members of S. In other words,

$$\mathbf{v}_i = a_0 \mathbf{v}_0 + \cdots + a_{i-1} \mathbf{v}_{i-1} + a_{i+1} \mathbf{v}_{i+1} + \cdots + a_{n-1} \mathbf{v}_{n-1}$$

Another way to think of this is that if we were to remove $\mathbf{v}_i$, then S would still span the same space. If such a $\mathbf{v}_i$ exists, then we say that S is *linearly dependent*. If we can't find any such $\mathbf{v}_i$, then the vectors $\mathbf{v}_0, \ldots, \mathbf{v}_{n-1}$ are *linearly independent*.

An example of a linearly dependent set of vectors can be seen in Figure 2.7. Vector $\mathbf{v}_0$ is equal to the linear combination $-1 \cdot \mathbf{v}_1 + 0 \cdot \mathbf{v}_2$, or just $-\mathbf{v}_1$. Two linearly dependent vectors $\mathbf{v}$ and $\mathbf{w}$ are said to be *parallel*, that is, $\mathbf{w} = a\mathbf{v}$.

A set of vectors β in a vector space V is defined as a *basis* if β both spans V and is linearly independent. Each element of β is called a *basis vector*. A basis is special for two reasons. First, because it spans its vector space, we can produce any vector in V by using

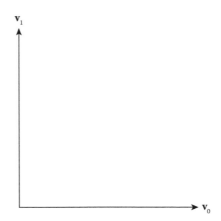

Figure 2.6. Two vectors spanning a plane.

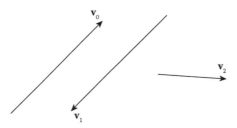

Figure 2.7. Linearly dependent set of vectors.

linear combinations of the vectors in the basis. Second, because the vectors in a basis are linearly independent, there is only one set of coefficients $a_0, a_1, \ldots, a_n$ that can produce a given vector in V.

There is an infinite number of bases for a given vector space, but they will always have the same number of elements. We formally define a vector space's *dimension* as equal to the number of basis vectors required to span it. So, for example, any basis for $\mathbb{R}^3$ will contain three basis vectors, and so it is (as we'd expect) a 3D space.

Among the many bases for a vector space, we define one as the *standard basis*. In general for $\mathbb{R}^n$, this is represented as $\{\mathbf{e}_0, \ldots, \mathbf{e}_{n-1}\}$, where

$$\mathbf{e}_0 = (1, 0, \ldots, 0)$$
$$\mathbf{e}_1 = (0, 1, \ldots, 0)$$
$$\vdots$$
$$\mathbf{e}_{n-1} = (0, 0, \ldots, 1)$$

As mentioned above, one property of a basis β is that for every vector $\mathbf{v}$ in V, there is a unique linear combination of the vectors in β that equal $\mathbf{v}$. So, using a general basis $\beta = \{\mathbf{b}_0, \mathbf{b}_1, \ldots, \mathbf{b}_{n-1}\}$, there is only one list of coefficients $a_0, \ldots, a_{n-1}$ such that

$$\mathbf{v} = a_0\mathbf{b}_0 + a_1\mathbf{b}_1 + \cdots + a_{n-1}\mathbf{b}_{n-1}$$

Because of this, instead of using the full equation to represent $\mathbf{v}$, we can abbreviate it by using only the coefficients $a_0, \ldots, a_{n-1}$ and store them in an ordered n-tuple as $(a_0, \ldots, a_{n-1})$. Note that the coefficient values will be dependent on which basis we're using and will almost certainly be different from basis to basis. The ordering of the basis vectors is important: a different ordering will not necessarily generate the same coefficients for a given vector. For most cases, though, we'll be assuming the standard basis.

Let's take as an example $\mathbb{R}^3$, the vector space we'll be using most often. In this case, the standard basis is three vectors $\{\mathbf{e}_0, \mathbf{e}_1, \mathbf{e}_2\}$, or as this basis is often represented, $\mathbf{i}$, $\mathbf{j}$, and $\mathbf{k}$. Their corresponding geometric representations can be seen in Figure 2.8. Note that these vectors are of unit length and perpendicular to each other (we will define *perpendicular* more formally when we discuss dot products).

Using this basis, we can uniquely represent any vector $\mathbf{v}$ in $\mathbb{R}^3$ by using the formula $\mathbf{v} = a_0\mathbf{i} + a_1\mathbf{j} + a_2\mathbf{k}$. As with the basis vectors, in $\mathbb{R}^3$ we usually replace the general coefficients a_0, a_1, and a_2 with x, y, and z, so

$$\mathbf{v} = x\mathbf{i} + y\mathbf{j} + z\mathbf{k}$$

We can think of x, y, and z as the amounts we move in the $\mathbf{i}$, $\mathbf{j}$, and $\mathbf{k}$ directions, from the tail of $\mathbf{v}$ to its tip (see Figure 2.8). Since the $\mathbf{i}$, $\mathbf{j}$, and $\mathbf{k}$ vectors are known and fixed, we just store the x, y, and z values and use them to represent our vector numerically. In this way, a 3D vector $\mathbf{v}$ is represented by an ordered triple (x, y, z). These are known as the vector *components*.

We can do the same for $\mathbb{R}^2$ by using as our basis $\{\mathbf{i}, \mathbf{j}\}$, where $\mathbf{i} = (1, 0)$ and $\mathbf{j} = (0, 1)$, and representing a 2D vector as the ordered pair (x, y).

By doing this, we have also neatly solved the problem of representing our geometric vectors algebraically. By using a standard basis, we can use an ordered triple to represent

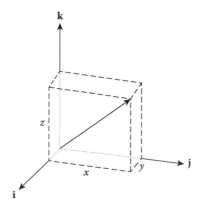

Figure 2.8. Standard 3D basis vectors.

the same concept as a line segment with an arrowhead. And by setting a correspondence between our algebraic basis and our geometric representation, we can guarantee that the ordered triple we use in one circumstance will be the same as the one we use in the other. Because of this, when working with vectors in $\mathbb{R}^2$ and $\mathbb{R}^3$, we will use the two representations interchangeably.

Using our new knowledge of bases, it's now possible to show that our previous definitions of addition and scalar multiplication for $\mathbb{R}^3$ are valid. For example, if we add two 3D vectors $\mathbf{v}_0$ and $\mathbf{v}_1$ together and expand and rearrange terms, we get

$$\begin{aligned}
\mathbf{v}_0 + \mathbf{v}_1 &= (x_0\mathbf{i} + y_0\mathbf{j} + z_0\mathbf{k}) + (x_1\mathbf{i} + y_1\mathbf{j} + z_1\mathbf{k}) \\
&= x_0\mathbf{i} + x_1\mathbf{i} + y_0\mathbf{j} + y_1\mathbf{j} + z_0\mathbf{k} + z_1\mathbf{k} \\
&= (x_0 + x_1)\mathbf{i} + (y_0 + y_1)\mathbf{j} + (z_0 + z_1)\mathbf{k}
\end{aligned}$$

If we remove $\mathbf{i}$, $\mathbf{j}$, and $\mathbf{k}$ to create ordered triples, we find the expected result that to add two vectors, we take each component in xyz order and add them:

$$(x_0, y_0, z_0) + (x_1, y_1, z_1) = (x_0 + x_1, y_0 + y_1, z_0 + z_1) \tag{2.1}$$

Scalar multiplication works similarly:

$$\begin{aligned}
a\mathbf{v} &= a(x\mathbf{i} + y\mathbf{j} + z\mathbf{k}) \\
&= a(x\mathbf{i}) + a(y\mathbf{j}) + a(z\mathbf{k}) \\
&= (ax)\mathbf{i} + (ay)\mathbf{j} + (az)\mathbf{k}
\end{aligned}$$

And again, pulling out $\mathbf{i}$, $\mathbf{j}$, and $\mathbf{k}$ follows what we defined previously:

$$a(x, y, z) = (ax, ay, az) \tag{2.2}$$

2.2.4 Basic Vector Class Implementation

Source Code
Library
IvMath
Filename
IvVector3

Now that we've presented an algebraic representation for vectors, we can talk about how we will store them in the computer. As we've mentioned many times, if we know the basis we're using to span our vector space, all we need to represent a vector are the coefficients of the linear combination. In our case, we'll assume the standard basis and thus store the components x, y, and z.

The following are some excerpts from the included C++ math library. For a vector in $\mathbb{R}^3$, our bare-bones class definition is

```
class IvVector3
{
public:
    inline IvVector3() {}
    inline IvVector3( float xVal, float yVal, float zVal ) :
        x( xVal ),
        y( yVal ),
        z( zVal )
    {
    }
```

```
inline ~IvVector3() {}
IvVector3( const IvVector3& vector );
IvVector3& operator=( const IvVector3& vector );
float x,y,z;
...
};
```

We can observe a few things about this declaration. First, we declared our member variables as a type `float`. This is the single-precision IEEE floating-point representation for real numbers, as discussed in Chapter 1. While not as precise as double-precision floating point, it has the advantage of being compact and compatible with standard representations on most graphics hardware. We are also explicitly specifying the dimension in the class name. Each vector class will have slightly different operations available, so we have made the choice to clearly distinguish between them. See [128] for a different approach that uses templates to abstract both dimension and underlying type.

Another thing to notice is that, like many vector libraries, we're making our member variables public. This is not usually recommended practice in C++; usually the data are hidden and only made available through an accessor. One motivation for such data hiding is to avoid unexpected side effects when changing a member variable, but this not an issue here since the data are so simple. However, this breaks another motivation for data hiding, which is that you can change your underlying representation without modifying nonlibrary code. For example, you may wish to use one of the platform-specific instruction sets for performing floating operations in parallel, such as SSE/AVX on Intel-based processors and NEON on ARM-based processors. In general, however, it's been found that using a standard serial implementation for vectors works well for most cases, and it's better to save this kind of optimization for specific instances. For this reason, and for clarity, we have chosen not to implement an SSE or NEON implementation, but see [47] for one possible solution.

The class has a default constructor and destructor, which do nothing. The constructor could initialize the components to 0, but doing so takes time, which adds up when we have large arrays of vectors (a common occurrence), and in most cases we'll be setting the values to something else anyway. For this purpose, there is an additional constructor that takes three floating-point values and uses them to set the components. We can use the copy constructor and assignment operator as well.

Now that we have the data set up for our class, we can add some operations to them. The corresponding operator for vector addition is

```
IvVector3 operator+(const IvVector3& v0, const IvVector3& v1)
{
    return IvVector3( v0.x + v1.x, v0.y + v1.y, v0.z  + v1.z );
}
```

Scalar multiplication is also straightforward:

```
IvVector3
operator*( float a, const IvVector3& vector)
{
    return IvVector3( a*vector.x, a*vector.y, a*vector.z );
}
```

Similar operators for postmultiplication and -division by a scalar are also provided within the library; their declarations are

```
IvVector3 operator*( const IvVector3& vector, float scalar );
IvVector3 operator/( const IvVector3& vector, float scalar );
IvVector3& operator*=( IvVector3& vector, float scalar );
IvVector3& operator/=( IvVector3& vector, float scalar );
```

Now that we have a numeric representation for vectors and have covered the algebraic form of addition and scaling, we can add some new vector operations as well. As before, we'll focus primarily on the case of $\mathbb{R}^3$. Vectors in $\mathbb{R}^2$ and $\mathbb{R}^4$ have similar properties; any exceptions will be discussed in the particular parts.

2.2.5   Vector Length

We have mentioned that a vector is an entity with length and direction but so far haven't provided any means of measuring or comparing these quantities in two vectors. We'll see shortly how the dot product provides a way to compare vector directions. First, however, we'll consider how to measure a vector's magnitude.

There is a general class of size-measuring functions known as *norms*. A norm $\|\mathbf{v}\|$ is defined as a real-valued function on a vector $\mathbf{v}$ with the following properties:

1. $\|\mathbf{v}\| \geq 0$, and $\|\mathbf{v}\| = 0$ if and only if $\mathbf{v} = \mathbf{0}$.

2. $\|a\mathbf{v}\| = |a|\|\mathbf{v}\|$.

3. $\|\mathbf{v} + \mathbf{w}\| \leq \|\mathbf{v}\| + \|\mathbf{w}\|$.

We use the $\|\mathbf{v}\|$ notation to distinguish a norm from the absolute value function $|a|$.

An example of a norm is the Manhattan distance, also called the ℓ_1 norm, which is just the sum of the absolute values of the given vector's components:

$$\|\mathbf{v}\|_{\ell_1} = \sum_i |v_i|$$

This measures the distance from one end of the vector to the other as if we were traveling along a grid of city streets.

One that we'll use more often is the Euclidean norm, also known as the ℓ_2 norm or just *length*. If we give no indication of which type of norm we're using, this is usually what we mean.

We derive the Euclidean norm as follows. Suppose we have a 2D vector $\mathbf{u} = x\mathbf{i} + y\mathbf{j}$ (Figure 2.9). Recall the Pythagorean theorem $x^2 + y^2 = d^2$. Since x is the distance along $\mathbf{i}$ and y is the distance along $\mathbf{j}$, then the length d of $\mathbf{u}$ is

$$\|\mathbf{u}\| = d = \sqrt{x^2 + y^2}$$

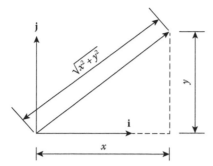

Figure 2.9. Length of 2D vector.

A similar formula is used for a vector $\mathbf{v} = (x, y, z)$, using the standard basis in $\mathbb{R}^3$:

$$\|\mathbf{v}\| = \sqrt{x^2 + y^2 + z^2} \tag{2.3}$$

And the general form in $\mathbb{R}^n$ with respect to the standard basis is

$$\|\mathbf{v}\| = \sqrt{v_0^2 + v_1^2 + \cdots + v_{n-1}^2}$$

We've mentioned the use of unit-length vectors as pure indicators of direction, for example, in determining the viewing direction or relative location of a light source. Often, though, the process we'll use to generate our direction vector will not automatically create one of unit length. To create a unit vector $\hat{\mathbf{v}}$ from a general vector $\mathbf{v}$, we *normalize* $\mathbf{v}$ by multiplying it by 1 over its length, or

$$\hat{\mathbf{v}} = \frac{\mathbf{v}}{\|\mathbf{v}\|}$$

This sets the length of the vector to $\|\mathbf{v}\| \cdot 1/\|\mathbf{v}\|$, which is our desired value 1.

Our implementations of length methods (for $\mathbb{R}^3$) are as follows:

```
float
IvVector3::Length() const
{
    return IvSqrt( x*x + y*y + z*z );
}


float
IvVector3::LengthSquared() const
{
    return x*x + y*y + z*z;
}


IvVector3&
IvVector3::Normalize()
{
    float lengthsq = x*x + y*y + z*z;
```

```
ASSERT( !IsZero( lengthsq ) );
if ( IsZero( lengthsq ) )
{
    x = y = z = 0.0f;
    return *this;
}

float recip = IvRecipSqrt( lengthsq );
x *= recip;
y *= recip;
z *= recip;

return *this;
}
```

Note that in addition to the mathematical operations we've just described, we have defined a LengthSquared() method. Performing the square root can be a costly operation, even on systems that have a special machine instruction to compute it. Often we're only doing a comparison between lengths, so it is better and certainly faster in those cases to compute and compare length squared instead. Both length and length squared are increasing functions starting at 0, so when only performing a comparison, the results will be the same.

The length methods also introduce some new functions that will be useful to us throughout the math library. We use our own square root functions IvSqrt() and IvRecipSqrt() instead of sqrtf() and 1.0f/sqrtf(). There are a number of reasons for this choice. As mentioned, the standard library implementation of square root is often slow. Rather than use it, we can use an approximation on some platforms, which is faster and accurate enough for our purpose. On other platforms, there are internal assembly instructions that are not used by the standard library. In particular, there may be an instruction that performs 1.0f/sqrtf() (also known as the reciprocal square root) directly, which is faster than calculating the square root and performing the floating-point divide. Defining our own layer of indirection gives us flexibility and ensures that we can guarantee ourselves the best performance.

2.2.6 Dot Product

Now that we've considered vector length, we can look at vector direction. The function we will use for this is called the *dot product*, or less commonly, the *Euclidean inner product* (see below for the formal definition of inner products). It is probably the most useful vector operation for 3D games and applications.

Given two vectors $\mathbf{v}$ and $\mathbf{w}$ with an angle θ between them, the dot product $\mathbf{v} \cdot \mathbf{w}$ is defined as

$$\mathbf{v} \cdot \mathbf{w} = \|\mathbf{v}\| \|\mathbf{w}\| \cos \theta \tag{2.4}$$

Using Equation 2.4, we can find a coordinate-dependent definition in $\mathbb{R}^3$ by examining a triangle formed by $\mathbf{v}$, $\mathbf{w}$, and $\mathbf{v} - \mathbf{w}$ (Figure 2.10). The law of cosines gives us

$$\|\mathbf{v} - \mathbf{w}\|^2 = \|\mathbf{v}\|^2 + \|\mathbf{w}\|^2 - 2\|\mathbf{v}\| \|\mathbf{w}\| \cos \theta$$

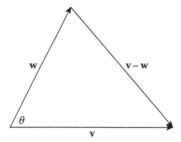

Figure 2.10. Law of cosines.

We can rewrite this as

$$-2\|\mathbf{v}\|\|\mathbf{w}\|\cos\theta = \|\mathbf{v}-\mathbf{w}\|^2 - \|\mathbf{v}\|^2 - \|\mathbf{w}\|^2$$

Substituting in the definition of vector length in $\mathbb{R}^3$ and expanding, we get

$$-2\|\mathbf{v}\|\|\mathbf{w}\|\cos\theta = (v_x - w_x)^2 + (v_y - w_y)^2 + (v_z - w_z)^2$$
$$- (v_x^2 + v_y^2 + v_z^2) - (w_x^2 + w_y^2 + w_z^2)$$
$$-2\|\mathbf{v}\|\|\mathbf{w}\|\cos\theta = -2v_x w_x - 2v_y w_y - 2v_z w_z$$
$$\|\mathbf{v}\|\|\mathbf{w}\|\cos\theta = v_x w_x + v_y w_y + v_z w_z$$

So, to compute the dot product in $\mathbb{R}^3$, multiply the vectors componentwise, and then add

$$\mathbf{v} \cdot \mathbf{w} = v_x w_x + v_y w_y + v_z w_z$$

Note that for this definition to hold, vectors $\mathbf{v}$ and $\mathbf{w}$ need to be represented with respect to the standard basis $\{\mathbf{i}, \mathbf{j}, \mathbf{k}\}$. The general form for vectors $\mathbf{v}$ and $\mathbf{w}$ in $\mathbb{R}^n$, again with respect to the standard basis, is

$$\mathbf{v} \cdot \mathbf{w} = v_0 w_0 + v_1 w_1 + \cdots + v_{n-1} w_{n-1}$$

For vectors $\mathbf{u}$, $\mathbf{v}$, $\mathbf{w}$, and scalar a, the following algebraic rules apply:

1. $\mathbf{v} \cdot \mathbf{w} = \mathbf{w} \cdot \mathbf{v}$ (symmetry).

2. $(\mathbf{u} + \mathbf{v}) \cdot \mathbf{w} = \mathbf{u} \cdot \mathbf{w} + \mathbf{v} \cdot \mathbf{w}$ (additivity).

3. $a(\mathbf{v} \cdot \mathbf{w}) = (a\mathbf{v}) \cdot \mathbf{w} = \mathbf{v} \cdot (a\mathbf{w})$ (homogeneity).[2]

4. $\mathbf{v} \cdot \mathbf{v} \geq 0$ (positivity).

5. $\mathbf{v} \cdot \mathbf{v} = 0$ if and only if $\mathbf{v} = \mathbf{0}$ (definiteness).

---

[2] Note that the leading scalar does not apply to both terms on the right-hand side; assuming so is a common mistake.

Also note that we can relate the dot product to the length function by noting that

$$\mathbf{v} \bullet \mathbf{v} = \|\mathbf{v}\|^2 \tag{2.5}$$

As mentioned, the dot product has many uses. By Equation 2.4, if the angle between two vectors $\mathbf{v}$ and $\mathbf{w}$ in standard Euclidean space is 90 degrees, then $\mathbf{v} \bullet \mathbf{w} = 0$. So, we define that two vectors $\mathbf{v}$ and $\mathbf{w}$ are perpendicular, or *orthogonal*, when $\mathbf{v} \bullet \mathbf{w} = 0$. Recall that we stated that our standard basis vectors for $\mathbb{R}^3$ are orthogonal. We can now demonstrate this. For example, taking $\mathbf{i} \bullet \mathbf{j}$ we get

$$\begin{aligned}
\mathbf{i} \bullet \mathbf{j} &= (1,0,0) \bullet (0,1,0) \\
&= 0+0+0 \\
&= 0
\end{aligned}$$

It is possible, although not always recommended, to use Equation 2.4 to test whether two unit vectors $\hat{\mathbf{v}}$ and $\hat{\mathbf{w}}$ are pointing generally in the same direction. If they are, $\cos \theta$ is close to 1, so $1 - \hat{\mathbf{v}} \bullet \hat{\mathbf{w}}$ is close to 0 (we use this formula to avoid problems with floating-point precision). Similarly, if $1 + \hat{\mathbf{v}} \bullet \hat{\mathbf{w}}$ is close to 0, they are pointing in opposite directions. Performing this test only takes six floating-point addition and multiplication operations. However, if $\mathbf{v}$ and $\mathbf{w}$ are not known to be normalized, then we need a different test: $\|\mathbf{v}\|^2 \|\mathbf{w}\|^2 - (\mathbf{v} \bullet \mathbf{w})^2$. This takes 18 operations.

Note that for unit vectors,

$$\begin{aligned}
1 - (\hat{\mathbf{v}} \bullet \hat{\mathbf{w}})^2 &= 1 - \cos^2 \theta \\
&= \sin^2 \theta
\end{aligned}$$

and for nonunit vectors,

$$\begin{aligned}
\|\mathbf{v}\|^2 \|\mathbf{w}\|^2 - (\mathbf{v} \bullet \mathbf{w})^2 &= \|\mathbf{v}\|^2 \|\mathbf{w}\|^2 (1 - \cos^2 \theta) \\
&= \|\mathbf{v}\|^2 \|\mathbf{w}\|^2 \sin^2 \theta
\end{aligned}$$

So assuming we use this, the method we use to test closeness to zero will have to be different for both cases.

In any case, using dot product for this test is not really recommended unless your vectors are prenormalized *and* speed is of the essence. As θ gets close to 0, $\cos \theta$ changes very little. Due to the lack of floating-point precision, the set of angles that might be considered 0 is actually broader than one might expect. As we will see, there is another method to test for parallel vectors that is faster with nonunit vectors and has fewer problems with near-zero angles.

A more common use of the dot product is to classify values of the angle between two vectors. We know that if $\mathbf{v} \bullet \mathbf{w} > 0$, then the angle is less than 90 degrees; if $\mathbf{v} \bullet \mathbf{w} < 0$, then the angle is greater than 90 degrees; and if $\mathbf{v} \bullet \mathbf{w} = 0$, then the angle is exactly 90 degrees (Figure 2.11). As opposed to testing for parallel vectors, this will work with vectors of *any* length.

For example, suppose that we have an artificial intelligence (AI) agent that is looking for enemy agents in the game. We can compute a vector $\mathbf{v}$ pointing in the direction the AI is

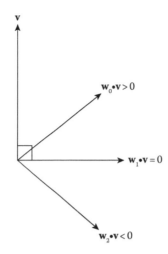

Figure 2.11. Dot product as measurement of angle.

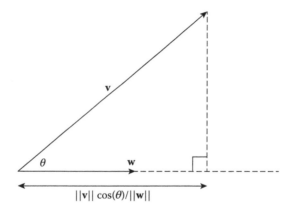

Figure 2.12. Dot product as projection.

looking and a vector $\mathbf{t}$ that points toward an object in our scene. If $\mathbf{v} \cdot \mathbf{t} < 0$, then the object is behind us and therefore not visible to our AI.

Equation 2.4 allows us to use the dot product in another manner. Suppose we have two vectors $\mathbf{v}$ and $\mathbf{w}$, where $\mathbf{w} \neq \mathbf{0}$. We define the *projection* of $\mathbf{v}$ onto $\mathbf{w}$ as

$$\text{proj}_{\mathbf{w}}\mathbf{v} = \frac{\mathbf{v} \cdot \mathbf{w}}{\|\mathbf{w}\|^2}\mathbf{w}$$

This gives the part of $\mathbf{v}$ that is parallel to $\mathbf{w}$, which is the same as dropping a perpendicular from the end of $\mathbf{v}$ onto $\mathbf{w}$ (Figure 2.12).

We can get the part of $\mathbf{v}$ that is perpendicular to $\mathbf{w}$ by subtracting the projection:

$$\text{perp}_{\mathbf{w}}\mathbf{v} = \mathbf{v} - \frac{\mathbf{v} \cdot \mathbf{w}}{\|\mathbf{w}\|^2}\mathbf{w}$$

Both of these equations will be very useful to us. Note that if $\mathbf{w}$ is normalized, then the projection simplifies to

$$\mathrm{proj}_{\hat{\mathbf{w}}}\mathbf{v} = (\mathbf{v} \cdot \hat{\mathbf{w}})\hat{\mathbf{w}}$$

The corresponding library implementation of dot product in $\mathbb{R}^3$ is as follows:

```
float
IvVector3::Dot( const IvVector3& other )
{
    return x*other.x + y*other.y + z*other.z;
}
```

2.2.7 Gram–Schmidt Orthogonalization

The combination of dot product and normalization allows us to define a particularly useful class of vectors. If a set of vectors β are all unit vectors and pairwise orthogonal, we say that they are *orthonormal*. Our standard basis $\{\mathbf{i}, \mathbf{j}, \mathbf{k}\}$ is an example of an orthonormal set of vectors.

In many cases, we start with a general set of vectors and want to generate the closest possible orthonormal one. One example of this is when we perform operations on currently orthonormal vectors. Even if the pure mathematical result should not change their length or relative orientation, due to floating-point precision problems the resulting vectors may be no longer orthonormal. The process that allows us to create orthonormal vectors from possibly nonorthonormal vectors is called *Gram–Schmidt orthogonalization*.

This works as follows. Suppose we have a set of nonorthogonal vectors $\mathbf{v}_0, \ldots, \mathbf{v}_{n-1}$ in $\mathbb{R}^n$, and from them we want to create an orthonormal set $\mathbf{w}_0, \ldots, \mathbf{w}_{n-1}$. We'll use the first vector from our original set as the starting vector for our new set so

$$\mathbf{w}_0 = \mathbf{v}_0$$

Now we want to create a vector orthogonal to $\mathbf{w}_0$, which points generally in the direction of $\mathbf{v}_1$. We can do this by computing the projection of $\mathbf{v}_1$ on $\mathbf{w}_0$, which produces the component vector of $\mathbf{v}_1$ parallel to $\mathbf{w}_0$. The remainder of $\mathbf{v}_1$ will be orthogonal to $\mathbf{w}_0$, so

$$\begin{aligned}
\mathbf{w}_1 &= \mathbf{v}_1 - \mathrm{proj}_{\mathbf{w}_0}\mathbf{v}_1 \\
&= \mathbf{v}_1 - \frac{\mathbf{v}_1 \cdot \mathbf{w}_0}{\|\mathbf{w}_0\|^2}\mathbf{w}_0
\end{aligned}$$

We perform the same process for $\mathbf{w}_2$: we project $\mathbf{v}_2$ on $\mathbf{w}_0$ and $\mathbf{w}_1$ to compute the parallel components and then subtract those from $\mathbf{v}_2$ to generate a vector orthogonal to both $\mathbf{w}_0$ and $\mathbf{w}_1$:

$$\begin{aligned}
\mathbf{w}_2 &= \mathbf{v}_2 - \mathrm{proj}_{\mathbf{w}_0}\mathbf{v}_2 - \mathrm{proj}_{\mathbf{w}_1}\mathbf{v}_2 \\
&= \mathbf{v}_2 - \frac{\mathbf{v}_2 \cdot \mathbf{w}_0}{\|\mathbf{w}_0\|^2}\mathbf{w}_0 - \frac{\mathbf{v}_2 \cdot \mathbf{w}_1}{\|\mathbf{w}_1\|^2}\mathbf{w}_1
\end{aligned}$$

In general, we have

$$\mathbf{w}_i = \mathbf{v}_i - \sum_{j=0}^{i-1} \text{proj}_{\mathbf{w}_j} \mathbf{v}_i$$

$$= \mathbf{v}_i - \sum_{j=0}^{i-1} \frac{\mathbf{v}_i \cdot \mathbf{w}_j}{\|\mathbf{w}_j\|^2} \mathbf{w}_j$$

Unfortunately, when working with floating-point arithmetic, the standard Gram–Schmidt method amplifies rounding error and is not numerically stable, producing nonorthogonal vectors. A better approach (known as *modified Gram–Schmidt*, or MGS) projects our target vector against the first vector, subtracts that from the target, and then takes the result and projects that against the next vector, and so on. For example, for $\mathbf{w}_2$ we would use

$$\mathbf{t}_1 = \mathbf{v}_2 - \text{proj}_{\mathbf{w}_0} \mathbf{v}_2$$
$$\mathbf{w}_2 = \mathbf{t}_1 - \text{proj}_{\mathbf{w}_1} \mathbf{t}_1$$

Or in general, to produce $\mathbf{w}_i$:

$$\mathbf{t}_1 = \mathbf{v}_i - \text{proj}_{\mathbf{w}_0} \mathbf{v}_i$$
$$\mathbf{t}_2 = \mathbf{t}_1 - \text{proj}_{\mathbf{w}_1} \mathbf{t}_1$$
$$\vdots$$
$$\mathbf{w}_i = \mathbf{t}_{i-1} - \text{proj}_{\mathbf{w}_{i-1}} \mathbf{t}_{i-1}$$

The end result will be an orthogonal set of possibly non–unit-length vectors. To create an orthonormal set, we can either normalize the resulting $\mathbf{w}_j$ vectors at the end or normalize as we go, the latter of which simplifies the projection calculation to $(\mathbf{v}_i \cdot \hat{\mathbf{w}}_j) \hat{\mathbf{w}}_j$.

One final note: while Gram–Schmidt orthogonalization is relatively efficient, a faster alternative for vectors in $\mathbb{R}^3$ is to use a vector triple product. See Section 2.2.9 for more details.

2.2.8 Cross Product

Suppose we have two vectors $\mathbf{v}$ and $\mathbf{w}$ and want to find a new vector $\mathbf{u}$ orthogonal to both. The operation that computes this is the *cross product*, also known as the *vector product*. There are two possible choices for the direction of the vector, each the negation of the other (Figure 2.13); the one chosen is determined by the right-hand rule. Hold your right hand so that your forefinger points forward, your middle finger points out to the left, and your thumb points up. If you roughly align your forefinger with $\mathbf{v}$, and your middle finger with $\mathbf{w}$, then the cross product will point in the direction of your thumb (Figure 2.14). The length of the cross product is equal to the area of a parallelogram bordered by the two vectors (Figure 2.15). This can be computed using the formula

$$\|\mathbf{v} \times \mathbf{w}\| = \|\mathbf{v}\| \|\mathbf{w}\| \sin \theta \qquad (2.6)$$

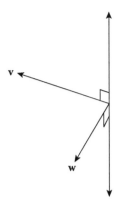

Figure 2.13. Two directions of orthogonal 3D vectors.

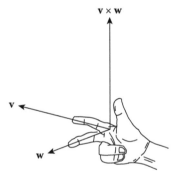

Figure 2.14. Cross product direction.

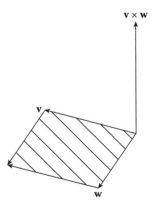

Figure 2.15. Cross product length equals area of parallelogram.

where θ is the angle between $\mathbf{v}$ and $\mathbf{w}$. Note that the cross product is not commutative, so order is important:

$$\mathbf{v} \times \mathbf{w} = -(\mathbf{w} \times \mathbf{v})$$

Also, if the two vectors are parallel, $\sin \theta = 0$, so we end up with the zero vector.

It is a common mistake to believe that if $\mathbf{v}$ and $\mathbf{w}$ are unit vectors, the cross product will also be a unit vector. A quick look at Equation 2.6 shows this is true only if $\sin \theta$ is 1, in which case θ is 90 degrees.

The formula for the cross product is

$$\mathbf{v} \times \mathbf{w} = (v_y w_z - w_y v_z, v_z w_x - w_z v_x, v_x w_y - w_x v_y)$$

Certain processors can implement this as a two-step operation, by creating two vectors and performing the subtraction in parallel:

$$\mathbf{v} \times \mathbf{w} = (v_y w_z, v_z w_x, v_x w_y) - (w_y v_z, w_z v_x, w_x v_y)$$

For vectors $\mathbf{u}$, $\mathbf{v}$, $\mathbf{w}$, and scalar a, the following algebraic rules apply:

1. $\mathbf{v} \times \mathbf{w} = -\mathbf{w} \times \mathbf{v}$.

2. $\mathbf{u} \times (\mathbf{v} + \mathbf{w}) = (\mathbf{u} \times \mathbf{v}) + (\mathbf{u} \times \mathbf{w})$.

3. $(\mathbf{u} + \mathbf{v}) \times \mathbf{w} = (\mathbf{u} \times \mathbf{w}) + (\mathbf{v} \times \mathbf{w})$.

4. $a(\mathbf{v} \times \mathbf{w}) = (a\mathbf{v}) \times \mathbf{w} = \mathbf{v} \times (a\mathbf{w})$.

5. $\mathbf{v} \times \mathbf{0} = \mathbf{0} \times \mathbf{v} = \mathbf{0}$.

6. $\mathbf{v} \times \mathbf{v} = \mathbf{0}$.

The cross product is not associative. For example, in general

$$\mathbf{v} \times (\mathbf{v} \times \mathbf{w}) \neq (\mathbf{v} \times \mathbf{v}) \times \mathbf{w} = \mathbf{0}$$

There are two common uses for the cross product. The first, and most used, is to generate a vector orthogonal to two others. Suppose we have three points P, Q, and R, and we want to generate a unit vector $\mathbf{n}$ that is orthogonal to the plane formed by the three points (this is known as a normal vector). Begin by computing $\mathbf{v} = (Q - P)$ and $\mathbf{w} = (R - P)$. Now we have a decision to make. Computing $\mathbf{v} \times \mathbf{w}$ and normalizing will generate a normal in one direction, whereas $\mathbf{w} \times \mathbf{v}$ and normalizing will generate one in the opposite direction (Figure 2.16). Usually we'll set things up so that the normal points from the inside toward the outside of our object.

Like the dot product, the cross product can also be used to determine if two vectors are parallel by checking whether the resulting vector is close to the zero vector. Deciding whether to use this test as opposed to the dot product depends on what your data are. The cross product takes nine operations. We can test for zero by examining the dot product of

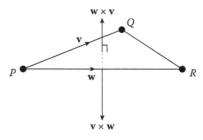

Figure 2.16. Computing normal for triangle.

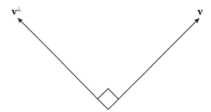

Figure 2.17. Perpendicular vector.

the result with itself $((\mathbf{v} \times \mathbf{w}) \cdot (\mathbf{v} \times \mathbf{w}))$. If it is close to 0, then we know the vectors are nearly parallel. The dot product takes an additional 5 operations, or total of 14, for our test. Recall that testing for parallel vectors using the dot product of nonnormalized vectors takes 18 operations; in this case, the cross product test is faster.

The cross product of two vectors is defined only for vectors in $\mathbb{R}^3$. However, in $\mathbb{R}^2$, we can define a similar operation on a single vector $\mathbf{v}$, called the *perpendicular*. This is represented as $\mathbf{v}^\perp$. The result of the perpendicular is the vector rotated by 90 degrees. As with the cross product, we have two choices: in this case, counterclockwise or clockwise rotation. The standard definition is to rotate counterclockwise (Figure 2.17), so if $\mathbf{v} = (x, y)$, $\mathbf{v}^\perp = (-y, x)$.

The perpendicular has similar properties to the cross product. First, it produces a vector orthogonal to the original. Also, when used in combination with the dot product in $\mathbb{R}^2$ (also known as the *perpendicular dot product*),

$$\mathbf{v}^\perp \cdot \mathbf{w} = \|\mathbf{v}\| \|\mathbf{w}\| \sin \theta$$

where θ is the *signed* angle between $\mathbf{v}$ and $\mathbf{w}$. That is, if the shortest rotation to get from $\mathbf{v}$ to $\mathbf{w}$ is in a clockwise direction, then θ is negative. And similar to the cross product, the absolute value of the perpendicular dot product is equal to the area of a parallelogram bordered by the two vectors.

It is possible to take pseudo–cross products in dimensions greater than three by using $n - 1$ vectors in an n-dimensional space,[3] but in general they won't be useful to us.

---

[3] "You can take the cross product of three vectors in 4-space, but you'll need a bigger hand" [81].

Our `IvVector3` cross product method is

```
IvVector3
IvVector3::Cross( const IvVector3& other )
{
    return IvVector3( y*other.z - other.y*z,
                      z*other.x - other.z*x,
                      x*other.y - other.x*y );
}
```

2.2.9 Triple Products

In $\mathbb{R}^3$, there are two extensions of the two single operation products called *triple products*. The first is the *vector triple product*, which returns a vector and is computed as $\mathbf{u} \times (\mathbf{v} \times \mathbf{w})$.

A special case is $\mathbf{w} \times (\mathbf{v} \times \mathbf{w})$ (Figure 2.18). Examining this, $\mathbf{v} \times \mathbf{w}$ is perpendicular to both $\mathbf{v}$ and $\mathbf{w}$. The result of $\mathbf{w} \times (\mathbf{v} \times \mathbf{w})$ is a vector perpendicular to both $\mathbf{w}$ and $(\mathbf{v} \times \mathbf{w})$. Therefore, if we combine normalized versions of $\mathbf{w}$, $(\mathbf{v} \times \mathbf{w})$, and $\mathbf{w} \times (\mathbf{v} \times \mathbf{w})$, we have an orthonormal basis (all are perpendicular and of unit length). As mentioned, this can be more efficient than Gram–Schmidt for producing orthogonal vectors, but of course it only works in $\mathbb{R}^3$.

The second triple product is called the *scalar triple product*. It (naturally) returns a scalar value, and its formula is $\mathbf{u} \bullet (\mathbf{v} \times \mathbf{w})$. To understand this geometrically, suppose we treat these three vectors as the edges of a slanted box, or parallelopiped (Figure 2.19). Then the area of the base equals $\|\mathbf{v} \times \mathbf{w}\|$, and $\|\mathbf{u}\| \cos\theta$ gives the (possibly negative) height of the box. So,

$$\mathbf{u} \bullet (\mathbf{v} \times \mathbf{w}) = \|\mathbf{u}\| \|\mathbf{v} \times \mathbf{w}\| \cos\theta$$

or area times height equals the signed volume of the box.

In addition to computing volume, the scalar triple product can be used to test the direction of the angle between two vectors $\mathbf{v}$ and $\mathbf{w}$, relative to a third vector $\mathbf{u}$ that is linearly independent of both. If $\mathbf{u} \bullet (\mathbf{v} \times \mathbf{w}) > 0$, then the shortest rotation from $\mathbf{v}$ to $\mathbf{w}$ is

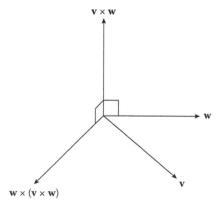

Figure 2.18. Vector triple product.

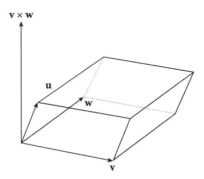

Figure 2.19. Scalar triple product equals signed volume of parallelopiped.

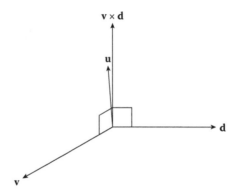

Figure 2.20. Scalar triple product indicates left turn.

in a counterclockwise direction (assuming our vectors are right-handed, as we will discuss shortly) around **u**. Similarly, if $\mathbf{u} \cdot (\mathbf{v} \times \mathbf{w}) < 0$, the shortest rotation is in a relative clockwise direction.

For example, suppose we have a tank with current velocity **v** and desired direction **d** of travel. Our tank is oriented so that its current up direction points along a vector **u**. We take the cross product $\mathbf{v} \times \mathbf{d}$ and dot it with **u**. If the result is positive, then we know that **d** lies to the left of **v** (counterclockwise rotation), and we turn left. Similarly, if the value is less than 0, then we know we must turn right to match **d** (Figures 2.20 and 2.21).

If we know that the tank is always oriented so that it lies on the xy plane, we can simplify this considerably. Vectors **v** and **d** will always have z values of 0, and **u** will always point in the same direction as the standard basis vector **k**. In this case, the result of $\mathbf{u} \cdot (\mathbf{v} \times \mathbf{d})$ is equal to the z value of $\mathbf{v} \times \mathbf{d}$. So the problem simplifies to taking the cross product of **v** and **d** and checking the sign of the resulting z value to determine our turn direction.

Finally, we can use the scalar triple product to test whether ordered vectors in $\mathbb{R}^3$ are left-handed or right-handed. We can test this informally for our standard basis by using the right-hand rule. Take your right hand and point the thumb along **k** and your fingers along **i**. Now, rotating around your thumb, sweep your fingers counterclockwise into **j** (Figure 2.22). This 90-degree rotation of **i** into **j** shows that the basis is right-handed.

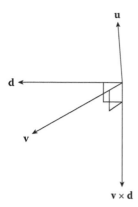

Figure 2.21. Scalar triple product indicates right turn.

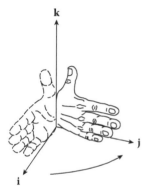

Figure 2.22. Right-handed rotation.

We can do the same trick with the left hand rotating clockwise to show that a set of vectors is left-handed.

Formally, if we have three basis vectors $\{\mathbf{v}_0, \mathbf{v}_1, \mathbf{v}_2\}$, then they are right-handed if $\mathbf{v}_0 \cdot (\mathbf{v}_1 \times \mathbf{v}_2) > 0$, and left-handed if $\mathbf{v}_0 \cdot (\mathbf{v}_1 \times \mathbf{v}_2) < 0$. If $\mathbf{v}_0 \cdot (\mathbf{v}_1 \times \mathbf{v}_2) = 0$, we've got a problem—our vectors are linearly dependent and hence not a basis.

While the scalar triple product only applies to vectors in $\mathbb{R}^3$, we can use the perpendicular dot product to test vectors in $\mathbb{R}^2$ for both turning direction and right- or left-handedness. For example, if we have two basis vectors $\{\mathbf{v}_0, \mathbf{v}_1\}$ in $\mathbb{R}^2$, then they are right-handed if $\mathbf{v}_0^{\perp} \cdot \mathbf{v}_1 > 0$ and left-handed if $\mathbf{v}_0^{\perp} \cdot \mathbf{v}_1 < 0$.

For vectors $\mathbf{u}$, $\mathbf{v}$, and $\mathbf{w}$ in $\mathbb{R}^3$, the following algebraic rules regarding the triple products apply:

1. $\mathbf{u} \times (\mathbf{v} \times \mathbf{w}) = (\mathbf{u} \cdot \mathbf{w})\mathbf{v} - (\mathbf{u} \cdot \mathbf{v})\mathbf{w}$.

2. $(\mathbf{u} \times \mathbf{v}) \times \mathbf{w} = (\mathbf{u} \cdot \mathbf{w})\mathbf{v} - (\mathbf{v} \cdot \mathbf{w})\mathbf{u}$.

3. $\mathbf{u} \cdot (\mathbf{v} \times \mathbf{w}) = \mathbf{w} \cdot (\mathbf{u} \times \mathbf{v}) = \mathbf{v} \cdot (\mathbf{w} \times \mathbf{u})$.

2.2.10 Grassman Algebra

An alternative to the cross product was developed by Grassman in 1844 and further refined by Clifford in 1878. This is called the exterior product or *wedge product*. The wedge product of two vectors in $\mathbb{R}^3$ produces—rather than a vector—a new entity called a bivector, which can be thought to represent the parallelogram in Figure 2.15 or, more specifically, a family of parallelograms with equal orientation and area. The wedge product of a vector and bivector produces a trivector, which represents a family of signed volumes. As one might expect, these operations are closely related to the cross product and scalar triple product, respectively. The combination of scalars, vectors, bivectors, and trivectors plus the wedge operator is called an exterior or *Grassman algebra* for $\mathbb{R}^3$.

Advantages of Grassman algebra are that it can provide a different geometric understanding of a problem, and it allows us to have an associative algebra that includes vectors. It's also possible to extend the wedge product to dimensions beyond $\mathbb{R}^3$, which we cannot do with the cross product. That said, Grassman algebra has not yet been widely adopted in the game industry or graphics circles, and while it can provide a new perspective on a problem, it is generally not necessary for the simple cases we'll be discussing in this book. For those with further interest, we recommend [95] as a good introduction.

2.3 Points

Now that we have covered vectors and vector operations in some detail, we turn our attention to a related entity: the point. While the reader probably has some intuitive notion of what a point is, in this section we'll provide a mathematical representation and discuss the relationship between vectors and points. We'll also discuss some special operations that can be performed on points and alternatives to the standard Cartesian coordinate system.

Within this section, it is also assumed that the reader has some general sense of what lines and planes are. More information on these topics follows in subsequent sections.

2.3.1 Points as Geometry

Everyone who has been through a first-year geometry course should be familiar with the notion of a *point*. Euclid describes the point in his work *Elements* [43] as "that which has no part." Points have also been presented as the cross section of a line, or the intersection of two lines. A less vague but still not satisfactory definition is to describe them as an infinitely small entity that has only the property of location. In games, we use points for two primary purposes: to represent the position of game objects and as the basic building block of their geometric representation. Points are represented graphically by a dot.

Euclid did not present a means for representing position numerically, although later Greek mathematicians used latitude, longitude, and altitude. The primary system we use now—Cartesian coordinates—was originally published by Rene Descartes in his 1637 work *La geometrie* [33] and further revised by Newton and Leibniz.

In this system, we measure a point's location relative to a special, anchored point, called the *origin*, which is represented by the letter O. In $\mathbb{R}^2$, we informally define two perpendicular real-number lines or axes—known as the x- and y-axes—that pass through the origin. We indicate the location of a point P by a pair (x, y) in $\mathbb{R}^2$, where x is the distance from the point to the y-axis, and y is the distance from the point to the x-axis.

Another way to think of it is that we count x units along the x-axis and then y units up parallel to the y-axis to reach the point's location. This combination of origin and axes is called the *Cartesian coordinate system* (Figure 2.23).

For $\mathbb{R}^3$, three perpendicular coordinate axes—x, y, and z—intersect at the origin. There are corresponding coordinate planes xy, yz, and xz that also intersect at the origin. Take the room you're sitting in as our space, with one corner of the room as the origin, and think

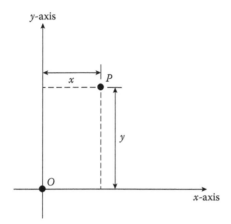

Figure 2.23. Two-dimensional Cartesian coordinate system.

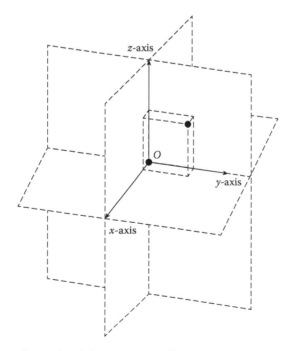

Figure 2.24. Three-dimensional Cartesian coordinate system.

of the walls and floor as the three coordinate planes (assume they extend infinitely). The edges where the walls and floor join together correspond to the axes. We can think of a 3D position as being a real-number triple (x, y, z) corresponding to the distance of the point to the three planes, or counting along each axis as before.

In Figure 2.24, you can see an example of a 3D coordinate system. Here the axis pointing up is called the z-axis, the one to the side is the y-axis, and the one aimed slightly out of the page is the x-axis. Another system that is commonly used in graphic books has the y-axis pointing up, the x-axis to the right, and the z-axis out of the page. Some graphics developers favor this because the x- and y-axes match the relative axes of the 2D screen, but most of the time we'll be using the former convention for this book.

Both of the 3D coordinate systems we have described are right-handed. As before, we can test this via the right-hand rule. This time point your thumb along the z-axis, your fingers along the x-axis, and rotate counterclockwise into the y-axis. As with left-handed bases, we can have left-handed coordinate systems (and will be using them later in this book), but the majority of our work will be done in a right-handed coordinate system because of convention.

2.3.2 Affine Spaces

We can provide a more formal definition of coordinate systems based on what we already know of vectors and vector spaces. Before we can do so, though, we need to define the relationship between vectors and points. Points can be related to vectors by means of an *affine space*. An affine space consists of a set of points W and a vector space V. The relation between the points and vectors is defined using the following two operations: For every pair of points P and Q in W, there is a unique vector $\mathbf{v}$ in V such that

$$\mathbf{v} = Q - P$$

Correspondingly, for every point P in W and every vector $\mathbf{v}$ in V, there is a unique point Q such that

$$Q = P + \mathbf{v} \tag{2.7}$$

This relationship can be seen in Figure 2.25. We can think of the vector $\mathbf{v}$ as acting as a displacement between the two points P and Q. To determine the displacement between two points, we subtract one from another. To displace a point, we add a vector to it and that gives us a new point.

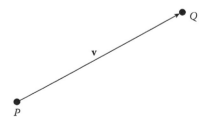

Figure 2.25. Affine relationship between points and vectors.

We can define a fixed-point O in W, known as the *origin*. Then using Equation 2.7, we can represent any point P in W as

$$P = O + \mathbf{v}$$

or, expanding our vector using n basis vectors that span V:

$$P = O + a_0\mathbf{v}_0 + a_1\mathbf{v}_1 + \cdots + a_{n-1}\mathbf{v}_{n-1} \tag{2.8}$$

Using this, we can represent our point using an n-tuple $(a_0, \ldots, a_{n-1})$ just as we do for vectors. The combination of the origin O and our basis vectors $(\mathbf{v}_0, \ldots, \mathbf{v}_{n-1})$ is known as a *coordinate frame*.

Note that we can use any point in W as our origin and—for an n-dimensional affine space—any n linearly independent vectors as our basis. Unlike the Cartesian axes, this basis does not have to be orthonormal, but using an orthonormal basis (as with vectors) does make matching our physical geometry with our abstract representation more straightforward. Because of this, we will work with the standard origin $(0, 0, \ldots, 0)$, and the standard basis $\{(1, 0, \ldots, 0), (0, 1, \ldots, 0), \ldots, (0, 0, \ldots, 1)\}$. This is known as the *Cartesian frame*.

In $\mathbb{R}^3$, our Cartesian frame will be the origin $O = (0, 0, 0)$ and the standard ordered basis $\{\mathbf{i}, \mathbf{j}, \mathbf{k}\}$ as before. Our basis vectors will lie along the x-, y-, and z-axes, respectively. By using this system, we can use the same triple (x, y, z) to represent a point and the corresponding vector from the origin to the point (Figure 2.26).

To compute the distance between two points, we use the length of the vector that is their difference. So, if we have two points $P_0 = (x_0, y_0, z_0)$ and $P_1 = (x_1, y_1, z_1)$ in $\mathbb{R}^3$, the difference is

$$\mathbf{v} = P_1 - P_0 = (x_1 - x_0, y_1 - y_0, z_1 - z_0)$$

and the distance between them is

$$\text{dist}(P_1, P_0) = \|v\| = \sqrt{(x_1 - x_0)^2 + (y_1 - y_0)^2 + (z_1 - z_0)^2}$$

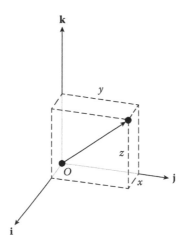

Figure 2.26. Relationship between points and vectors in Cartesian affine frame.

This is also known as the *Euclidean distance*. In the $\mathbb{R}^3$ Cartesian frame, the distance between a point $P = (x, y, z)$ and the origin is

$$\text{dist}(P, O) = \sqrt{x^2 + y^2 + z^2}$$

2.3.3 Affine Combinations

So far the only operation that we've defined on points alone is subtraction, which results in a vector. However, there is a limited addition operation that we can perform on points that gives us a point as a result. It is known as an *affine combination*, and has the form

$$P = a_0 P_0 + a_1 P_1 + \cdots + a_k P_k \tag{2.9}$$

where

$$a_0 + a_1 + \cdots + a_k = 1 \tag{2.10}$$

So, an affine combination of points is like a linear combination of vectors, with the added restriction that all the coefficients need to add up to 1. We can show why this restriction allows us to perform this operation by rewriting Equation 2.10 as

$$a_0 = 1 - a_1 - \cdots - a_k$$

and substituting into Equation 2.9 to get

$$\begin{aligned} P &= (1 - a_1 - \cdots - a_k)P_0 + a_1 P_1 + \cdots + a_k P_k \\ &= P_0 + a_1(P_1 - P_0) + \cdots + a_k(P_k - P_0) \end{aligned} \tag{2.11}$$

If we set $\mathbf{u}_1 = (P_1 - P_0)$, $\mathbf{u}_2 = (P_2 - P_0)$, and so on, we can rewrite this as

$$P = P_0 + a_1\mathbf{u}_1 + a_2\mathbf{u}_2 + \cdots + a_k\mathbf{u}_k$$

So, by restricting our coefficients in this manner, it allows us to rewrite the affine combination as a point plus a linear combination of vectors, a perfectly legal operation.

Looking back at our coordinate frame Equation 2.8, we can see that it too is an affine combination. Just as we use the coefficients in a linear combination of basis vectors to represent a general vector, we can use the coefficients of an affine combination of origin and basis vectors to represent a general point.

An affine combination spans an affine space, just as a linear combination spans a vector space. If the vectors in Equation 2.11 are linearly independent, we can represent any point in the spanned affine space using the coefficients of the affine combination, just as we did before with vectors. In this case, we say that the points $P_0, P_1, \ldots, P_k$ are *affinely independent*, and the ordered points are called a *simplex*. The coefficients are called *barycentric coordinates*. For example, we can create an affine combination of a simplex made of three affinely independent points P_0, P_1, and P_2. The affine space spanned by the affine combination $a_0 P_0 + a_1 P_1 + a_2 P_2$ is a plane, and any point in the plane can be specified by the coordinates (a_0, a_1, a_2).

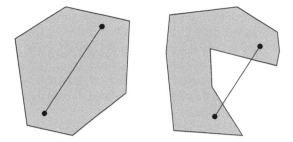

Figure 2.27. Convex versus nonconvex set of points.

We can further restrict the set of points spanned by the affine combination by considering properties of convex sets. A *convex set* of points is defined such that a line drawn between any pair of points in the set remains within the set (Figure 2.27). The *convex hull* of a set of points is the smallest convex set that includes all the points. If we restrict our coefficients $(a_0, \ldots, a_{n-1})$ such that $0 \le a_0, \ldots, a_{n-1} \le 1$, then we have a *convex combination*, and the span of the convex combination is the convex hull of the points. For example, the convex combination of three affinely independent points spans a triangle. We will discuss the usefulness of this in more detail when we cover triangles in Section 2.6.

If the barycentric coordinates in a convex combination of n points are all $1/n$, then the point produced is called the *centroid*, which is the mean of a set of points.

2.3.4 Point Implementation

Source Code
Library
IvMath
Filename
IvVector3

Using the Cartesian frame and standard basis in $\mathbb{R}^3$, the x, y, and z values of a point P in $\mathbb{R}^3$ match the x, y, and z values of the corresponding vector $P - O$, where O is the origin of the frame. This also means that we can use one class to represent both, since one can be easily converted to the other. Because of this, many math libraries don't even bother implementing a point class and just treat points as vectors.

Other libraries indicate the difference by treating them both as 4-tuples and indicate a point as $(x, y, z, 1)$ and a vector as $(x, y, z, 0)$. In this system if we subtract a point from a point, we automatically get a vector:

$$(x_0, y_0, z_0, 1) - (x_1, y_1, z_1, 1) = (x_0 - x_1, y_0 - y_1, z_0 - z_1, 0)$$

Similarly, a point plus a vector produces a point:

$$(x_0, y_0, z_0, 1) + (x_1, y_1, z_1, 0) = (x_0 + x_1, y_0 + y_1, z_0 + z_1, 1)$$

Even affine combinations give the expected results:

$$\sum_{i=0}^{n-1} a_i(x_i, y_i, z_i, 1) = \left(\sum_{i=0}^{n-1} a_i x_i, \sum_{i=0}^{n-1} a_i y_i, \sum_{i=0}^{n-1} a_i z_i, \sum_{i=0}^{n-1} a_i \right)$$

$$= \left(\sum_{i=0}^{n-1} a_i x_i, \sum_{i=0}^{n-1} a_i y_i, \sum_{i=0}^{n-1} a_i z_i, 1 \right)$$

In our case, we will not be using a separate class for points. There would be a certain amount of code duplication, since the `IvPoint3` class would end up being very similar to the `IvVector3` class. Also to be considered is the performance cost of converting points to vectors and back again. Further, to maintain type correctness, we may end up distorting equations unnecessarily; this obfuscates the code and can lead to a loss in performance as well. Finally, most production game engines don't make the distinction, and we wish to remain compatible with the overall state of the industry.

Despite not making the distinction in the class structure, it is important to remember that points and vectors are not the same. One has direction and length and the other position, so not all operations apply to both. For example, we can add two vectors together to get a new vector. As we've seen, adding two points together is only allowed in certain circumstances. So, while we will be using a single class, we will be maintaining mathematical correctness in the text and writing the code to reflect this.

As mentioned, most of what we need for points is already in the `IvVector3` class. The only additional code we'll have to implement is for distance and distance squared operations:

```
float
Distance( const IvVector3& point1,
          const IvVector3& point2 )
{
    float x = point1.x - point2.x;
    float y = point1.y - point2.y;
    float z = point1.z - point2.z;

    return IvSqrt( x*x + y*y + z*z );
}


float
DistanceSquared( const IvVector3& point1,
                 const IvVector3& point2 )
{
    float x = point1.x - point2.x;
    float y = point1.y - point2.y;
    float z = point1.z - point2.z;

    return ( x*x + y*y + z*z );
}
```

2.3.5 Polar and Spherical Coordinates

Cartesian coordinates are not the only way of measuring location. We've already mentioned latitude, longitude, and altitude, and there are other, related systems. Take a 2D point P and compute the vector $\mathbf{v} = P - O$. We can specify the location of P using the distance r from P to the origin, which is the length of $\mathbf{v}$, and the angle θ between $\mathbf{v}$ and the positive x-axis, where $\theta > 0$ corresponds to a counterclockwise rotation from the axis. The components (r, θ) are known as *polar coordinates*. They're particularly useful when thinking about rotations of points, as we'll discuss in Chapter 4.

It is easy to convert from polar to Cartesian coordinates. We begin by forming a right triangle using the x-axis, a line from P to O, and the perpendicular from P to the

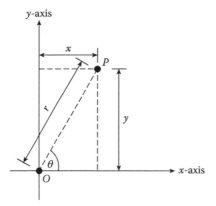

Figure 2.28. Relationship between polar and Cartesian coordinates.

x-axis (Figure 2.28). The hypotenuse has the length r and is θ degrees from the x-axis. Using simple trigonometry, the lengths of the other two sides of the triangle, x and y, can be computed as

$$x = r\cos\theta \tag{2.12}$$
$$y = r\sin\theta$$

From Cartesian to polar coordinates, we reverse the process. It's easy enough to generate r by computing the distance between P and O. Finding θ is not as straightforward. The naive approach is to solve Equation 2.12 for θ, which gives us $\theta = \arccos(x/r)$. However, the `acos()` function under C++ only returns an angle in the range of $[0, \pi)$, so we've lost the sign of the angle. Since

$$\frac{y}{x} = \frac{r\sin\theta}{r\cos\theta}$$
$$= \frac{\sin\theta}{\cos\theta}$$
$$= \tan\theta$$

an alternate choice would be $\arctan(y/x)$, but this doesn't handle the case when $x = 0$. To manage this, C++ provides a library function called `atan2()`, which takes y and x as separate arguments and computes $\arctan(y/x)$. It has no problems with division by 0 and maintains the signed angle with a range of $[-\pi, \pi]$. We'll represent the use of this function in our equations as $\arctan 2(y, x)$. The final result is

$$r = \sqrt{x^2 + y^2}$$
$$\theta = \arctan 2(y, x)$$

If r is 0, θ may be set arbitrarily.

The system that extends this to three dimensions is called *spherical coordinates*. In this system, we call the distance from the point to the origin ρ instead of r. We create a sphere of radius ρ centered on the origin and define where the point lies on the sphere by two angles, ϕ and θ. If we take a vector **v** from the origin to the point and project it down to the xy plane, θ is the angle between the x-axis and rotating counterclockwise around z. The other quantity, ϕ, measures the angle between **v** and the z-axis. The three values, ρ, ϕ, and θ, represent the location of our point (Figure 2.29).

Spherical coordinates can be converted to Cartesian coordinates as follows. Begin by building a right triangle as before, except with its hypotenuse along ρ and base along the z-axis (Figure 2.30). The length z is then $\rho \cos \phi$. To compute x and y, we project the vector **v** down onto the xy plane, and then use polar coordinates. The length r of the projected vector **v**′ is $\rho \sin \phi$, so we have

$$x = \rho \sin \phi \cos \theta$$
$$y = \rho \sin \phi \sin \theta$$
$$z = \rho \cos \phi$$

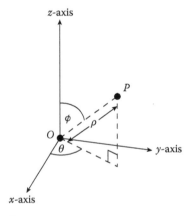

Figure 2.29. Spherical coordinates.

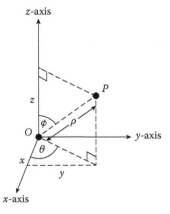

Figure 2.30. Relationship between spherical and Cartesian coordinates.

To convert from Cartesian to spherical coordinates, we begin by computing ρ, which again is the distance from the point to the origin. To find ϕ, we need to find the value of $\rho \sin \phi$. This is equal to the projected xy length r since

$$
\begin{aligned}
r &= \sqrt{x^2 + y^2} \\
&= \sqrt{(\rho \sin \phi \cos \theta)^2 + (\rho \sin \phi \sin \theta)^2} \\
&= \sqrt{(\rho \sin \phi)^2 (\cos^2 \theta + \sin^2 \theta)} \\
&= \rho \sin \phi
\end{aligned}
$$

And since, as with polar coordinates,

$$
\begin{aligned}
\frac{r}{z} &= \frac{\rho \sin \phi}{\rho \cos \phi} \\
&= \tan \phi
\end{aligned}
$$

we can compute $\phi = \arctan 2(r, z)$. Similarly, $\theta = \arctan 2(y, x)$. Summarizing:

$$
\begin{aligned}
\rho &= \sqrt{x^2 + y^2 + z^2} \\
\phi &= \arctan 2 \left(\sqrt{x^2 + y^2}, z \right) \\
\theta &= \arctan 2(y, x)
\end{aligned}
$$

2.4 Lines

2.4.1 Definition

As with the point, a *line* as a geometric concept should be familiar. Euclid [43] defines a line as "breadthless length" and a *straight line* as that "which lies evenly with the points on itself." A straight line also has been referred to as the shortest distance between two points, although in non-Euclidean geometry this is not necessarily true.

From first-year algebra, we know that a line in $\mathbb{R}^2$ can be represented by the slope-intercept form

$$ y = mx + b \tag{2.13} $$

where m is the slope of the line (it describes how y changes with each step of x), and b is the coordinate location where the line crosses the y-axis (called the y-intercept). In this case, x varies over all values and y is represented in terms of x. This general form works for all lines in $\mathbb{R}^2$ except for those that are vertical, since in that case the slope is infinite and the y-intercept is either nonexistent or all values along the y-axis.

Equation 2.13 has a few problems. First of all, as mentioned, we can't easily represent a vertical line—it has infinite slope. And, it isn't obvious how to transform this equation into one useful for three dimensions. We will need a different representation.

2.4.2 Parameterized Lines

One possible representation is known as a *parametric equation*. Instead of representing the line as a single equation with a number of variables, each coordinate value is calculated by a separate function. This allows us to use one form for a line that is generalizable across all dimensions. As an example, we will take Equation 2.13 and parameterize it.

To compute the parametric equation for a line, we need two points on our line. We can take the *y*-intercept $(0, b)$ as one of our points, and then take one step in the positive x direction, or $(1, m + b)$, to get the other. Subtracting point 1 from point 2, we get a 2D vector $\mathbf{d} = (1, m)$, which is oriented in the same direction as the line (Figure 2.31). If we take this vector and add all the possible scalar multiples of it to the starting point $(0, b)$, then the points generated will lie along the line. We can express this in one of the following forms:

$$L(t) = P_0 + t(P_1 - P_0) \tag{2.14}$$
$$= (1 - t)P_0 + tP_1$$
$$= P_0 + t\mathbf{d} \tag{2.15}$$

The variable t in this case is called a *parameter*.

We started with a 2D example, but the formulas we just derived work beyond two dimensions. As long as we have two points, we can just substitute them into the preceding equations to represent a line. More formally, if we examine Equation 2.14, we see it matches Equation 2.11. The affine combination of two unequal or noncoincident points span a line. Equation 2.15 makes this even clearer. If we think of P_0 as our origin and $\mathbf{d}$ as a basis vector, they span a one-dimensional (1D) affine space, which is the line.

Since our line is spanned by an affine combination of our two points, the logical next question is: What is spanned by the convex combination? The convex combination requires that t and $(1 - t)$ lie between 0 and 1, which holds only if t lies in the interval [0, 1]. Clamping t to this range gives us a *line segment* (Figure 2.32). The edges of polygons are line segments, and we'll also be using line segments when we talk about bounding objects and collision detection.

If we clamp t to only one end of the range, usually specifying that $t \geq 0$, then we end up with a *ray* (Figure 2.33) that starts at P_0 and extends infinitely along the line in the direction

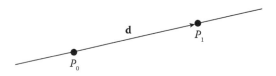

Figure 2.31. Line.

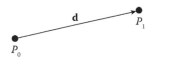

Figure 2.32. Line segment.

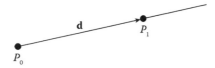

Figure 2.33. Ray.

of **d**. Rays are useful for intersection and visibility tests. For example, P_0 may represent the position of a camera, and **d** is the viewing direction.

In code, we'll be representing our lines, rays, and line segments as a point on the line P and a vector **d**; so, for example, the class definition for a line in $\mathbb{R}^3$ is

Source Code
Library
IvMath
Filename
IvLine3
IvLineSegment3
IvRay3

```
class IvLine3
{
public:
   IvLine3( const IvVector3& direction, const IvVector3& origin );

   IvVector3 mDirection;
   IvVector3  mOrigin;
};
```

2.4.3 Generalized Line Equation

There is another formulation of our 2D line that can be useful. Let's start by writing out the equations for both x and y in terms of t:

$$x = P_x + td_x$$
$$y = P_y + td_y$$

Solving for t in terms of x, we have

$$t = \frac{(x - P_x)}{d_x}$$

Substituting this into the y equation, we get

$$y = d_y \frac{(x - P_x)}{d_x} + P_y$$

We can rewrite this as

$$0 = \frac{(y - P_y)}{d_y} - \frac{(x - P_x)}{d_x}$$
$$= (-d_y)x + (d_x)y + (d_y P_x - d_x P_y)$$
$$= ax + by + c \tag{2.16}$$

where

$$a = -d_y$$
$$b = d_x$$
$$c = d_y P_x - d_x P_y = -a P_x - b P_y$$

We can think of a and b as the components of a 2D vector $\mathbf{n}$, which is perpendicular to the direction vector $\mathbf{d}$, and so is orthogonal to the direction of the line (Figure 2.34). This gives us a way of testing where a 2D point lies relative to a 2D line. If we substitute the coordinates of the point into the x and y values of the equation, then a value of 0 indicates it's on the line, a positive value indicates that it's on the side where the vector is pointing, and a negative value indicates that it's on the opposite side. If we normalize our vector, we can use the value returned by the line equation to indicate the distance from the point to the line.

To see why this is so, suppose we have a test point Q. We begin by constructing the vector between Q and our line point P, or $Q - P$. There are two possibilities. If Q lies on the side of the line where $\mathbf{n}$ is pointing, then the distance between Q and the line is

$$d = \|Q - P\| \cos \theta$$

where θ is the angle between $\mathbf{n}$ and $Q - P$. But since $\mathbf{n} \cdot (Q - P) = \|\mathbf{n}\| \|Q - P\| \cos \theta$, we can rewrite this as

$$d = \frac{\mathbf{n} \cdot (Q - P)}{\|\mathbf{n}\|}$$

If Q is lying on the opposite side of the line, then we take the dot product with the negative of $\mathbf{n}$, so

$$d = \frac{-\mathbf{n} \cdot (Q - P)}{\| -\mathbf{n}\|}$$
$$= -\frac{\mathbf{n} \cdot (Q - P)}{\|\mathbf{n}\|}$$

Since d is always positive, we can just take the absolute value of $\mathbf{n} \cdot (Q - P)$ to get

$$d = \frac{|\mathbf{n} \cdot (Q - P)|}{\|\mathbf{n}\|} \tag{2.17}$$

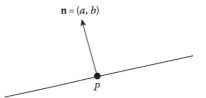

Figure 2.34. Normal form of 2D line.

If we know that **n** is normalized, we can drop the denominator. If $Q = (x, y)$ and (as we've stated) $\mathbf{n} = (a, b)$, we can expand our values to get

$$d = a(x - P_x) + b(y - P_y)$$
$$= ax + by - aP_x - bP_y$$
$$= ax + by + c$$

If our **n** is not normalized, then we need to remember to divide by $\|\mathbf{n}\|$ to get the correct distance.

2.4.4 Collinear Points

Three or more points are said to be *collinear* if they all lie on a line. Another way to think of this is that despite there being more than two points, the affine space that they span is only 1D.

To determine whether three points P_0, P_1, and P_2 are collinear, we take the cross product of $P_1 - P_0$ and $P_2 - P_0$ and test whether the result is close to the zero vector. This is equivalent to testing whether basis vectors for the affine space are parallel.

2.5 Planes

Euclid [43] defines a surface as "that which has length and breadth only," and a plane surface, or just a *plane*, as "a surface which lies evenly with the straight lines on itself." Another way of thinking of this is that a plane is created by taking a straight line and sweeping each point on it along a second straight line. It is a flat, limitless, infinitely thin surface.

2.5.1 Parameterized Planes

As with lines, we can express a plane algebraically in a number of ways. The first follows from our parameterized line. From basic geometry, we know that two noncoincident points form a line and three noncollinear points form a plane. So, if we can parameterize a line as an affine combination of two points, then it makes sense that we can parameterize a plane as an affine combination of three points P_0, P_1, and P_2, or

$$P(s, t) = (1 - s - t)P_0 + sP_1 + tP_2$$

Alternatively, we can represent this as an origin point plus the linear combination of two vectors:

$$P(s, t) = P_0 + s(P_1 - P_0) + t(P_2 - P_0)$$
$$= P_0 + s\mathbf{u} + t\mathbf{v}$$

As with the parameterized line equation, if our points are of higher dimension, we can create planes in higher dimensions from them. However, in most cases, our planes will be firmly entrenched in $\mathbb{R}^3$.

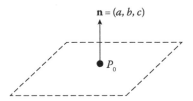

Figure 2.35. Normal form of plane.

2.5.2 Generalized Plane Equation

We can define an alternate representation for a plane in $\mathbb{R}^3$, just as we did for a line in $\mathbb{R}^2$. In this form, a plane is defined as the set of points perpendicular to a normal vector $\mathbf{n} = (a, b, c)$ that also contains the point $P_0 = (x_0, y_0, z_0)$, as shown in Figure 2.35. If a point P lies on the plane, then the vector $\mathbf{v} = P - P_0$ also lies on the plane. For $\mathbf{v}$ and $\mathbf{n}$ to be orthogonal, then $\mathbf{n} \cdot \mathbf{v} = 0$. Expanding this gives us the *normal-point* form of the plane equation, or

$$a(x - x_0) + b(y - y_0) + c(z - z_0) = 0$$

We can pull all the constants into one term to get

$$0 = ax + by + cz - (ax_0 + by_0 + cz_0)$$
$$= ax + by + cz + d$$

So, extending Equation 2.16 to three dimensions gives us the equation for a plane in $\mathbb{R}^3$.

This is the *generalized plane equation*. As with the generalized line equation, this equation can be used to test where a point lies relative to either side of a plane. Again, comparable to the line equation, it can be proven that if $\mathbf{n}$ is normalized, $|ax + by + cz + d|$ returns the distance from the point to the plane.

Testing points versus planes using the general plane equation happens quite often. For example, to detect whether a point lies inside a convex polyhedron, you can do a plane test for every face of the polyhedron. Assuming the plane normals point away from the center of the polyhedron, if the point is on the negative side of all the planes, then it lies inside. We may also use planes as culling devices that cut our world into half-spaces. If an object lies on one side of a plane, we consider it (say, for rendering purposes); otherwise, we ignore it. The distance property can be used to test whether a sphere is intersecting a plane. If the distance between the sphere's center and the plane is less than the sphere's radius, then the sphere is intersecting the plane.

Given three points in $\mathbb{R}^3$, P, Q, and R, we generate the generalized plane equation as follows. First, we compute two vectors $\mathbf{u}$ and $\mathbf{v}$, where

$$\mathbf{u} = Q - P$$
$$\mathbf{v} = R - P$$

Now we take the cross product of these two vectors to get the normal to the plane:

$$\mathbf{n} = \mathbf{u} \times \mathbf{v}$$

We usually normalize **n** at this point so that we can take advantage of the distance-measuring properties of the plane equation. This gives us our values a, b, and c. Taking P as the point on the plane, we compute d by

$$d = -(aP_x + bP_y + cP_z)$$

We can also use this to convert our parameterized form to the generalized form by starting with the cross product step.

Since we'll be working in $\mathbb{R}^3$ most of the time and because of its useful properties, we'll be using the generalized plane equation as the basis for our class:

Source Code
Library
IvMath
Filename
IvPlane

```
class IvPlane
{
public:
  IvPlane( float a, float b, float c, float d );

  IvVector3 mNormal;
  float     mOffset;
};
```

And while we'll be using this as our standard plane, from time to time we'll be making use of the parameterized form, so it's good to keep it in mind.

2.5.3 Coplanar Points

Four or more points are said to be *coplanar* if they all lie on a plane. Another way to think of this is that despite the number of points being greater than three, the affine space that they span is only 2D.

To determine whether four points P_0, P_1, P_2, and P_3 are coplanar, we create vectors $P_1 - P_0$, $P_2 - P_0$, and $P_3 - P_0$, and compute their triple scalar product. If the result is near 0, then they may be coplanar, if they're not collinear. To determine if they are collinear, take the cross products $(P_1 - P_0) \times (P_2 - P_0)$ and $(P_1 - P_0) \times (P_3 - P_0)$. If both results are near 0, then the points are collinear instead.

2.6 Polygons and Triangles

Source Code
Library
IvMath
Filename
IvTriangle

The current class of graphics processors wants their geometric data in primarily one form: points. However, having just a collection of points is not enough. We need to organize these points into smaller groups, for both rendering and computational purposes.

A *polygon* is made up of a set of *vertices* (which are represented by points) and *edges* (which are represented by line segments). The edges define how the vertices are connected together. A *convex polygon* is one where the set of points enclosed by the vertices and edges is a convex set; otherwise, it's a *concave polygon*.

The most commonly used polygons for storing geometric data are *triangles* (three vertices) and *quadrilaterals* (four vertices). While some rendering systems accept quadrilaterals (also referred to as just *quads*) as data, most want geometry grouped in triangles, so we'll follow that convention throughout the remainder of the book. One advantage triangles have over quadrilaterals is that three noncollinear vertices are guaranteed to be coplanar, so they can be used to define a single plane. If the three vertices of a triangle are collinear, then we have a *degenerate triangle*. Degenerate triangles can cause problems on some hardware and with some geometric algorithms, so it's often desirable to remove them by checking for collinearity of the triangle vertices, using the technique described previously.

If the points are not collinear, then as we've stated, the three vertices P_0, P_1, and P_2 can be used to find the triangle's incident plane. If we set $\mathbf{u} = P_1 - P_0$ and $\mathbf{v} = P_2 - P_0$, then we can define this via the parameterized plane equation $P(s, t) = P_0 + s\mathbf{u} + t\mathbf{v}$. Alternately, we can compute the generalized plane equation by computing the cross product of $\mathbf{u}$ and $\mathbf{v}$, normalizing to get the normal $\hat{\mathbf{n}}$, and then computing d as described in Section 2.5.2.

It's often necessary to test whether a 3D point lying on the triangle plane is inside or outside of the triangle itself (Figure 2.36). We begin by computing three vectors $\mathbf{v}_0$, $\mathbf{v}_1$, and $\mathbf{v}_2$, where

$$\mathbf{v}_0 = P_1 - P_0$$
$$\mathbf{v}_1 = P_2 - P_1$$
$$\mathbf{v}_2 = P_0 - P_2$$

We take the cross product of $\mathbf{v}_0$ and $\mathbf{v}_1$ to get a normal vector $\mathbf{n}$ to the triangle. We then compute three vectors from each vertex to the test point:

$$\mathbf{w}_0 = P - P_0$$
$$\mathbf{w}_1 = P - P_1$$
$$\mathbf{w}_2 = P - P_2$$

If the point lies inside the triangle, then the cross product of each $\mathbf{v}_i$ with each $\mathbf{w}_i$ will point in the same direction as $\mathbf{n}$, which we can test by using a dot product. If the result is negative, then we know they're pointing in opposite directions, and the point lies outside. For example, in Figure 2.36, the normal vector to the triangle, computed as $\mathbf{v}_0 \times \mathbf{v}_1$, points out of the page. But the cross product $\mathbf{v}_0 \times \mathbf{w}_0$ points into the page, so the point lies outside.

We can speed up this operation by projecting the point and triangle to one of the xy, xz, or yz planes and treating it as a 2D problem. To improve our accuracy, we'll choose the one that provides the maximum area for the projection of the triangle. If we look at the normal $\mathbf{n}$

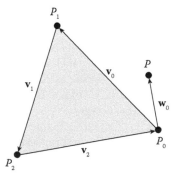

Figure 2.36. Point in triangle test.

for the triangle, one of the coordinate values (x, y, z) will have the maximum absolute value; that is, the normal is pointing generally along that axis. If we drop that coordinate and keep the other two, that will give us the maximum projected area. We can then throw out a number of zero terms and end up with a considerably faster test. This is equivalent to using the perpendicular dot product instead of the cross product. More detail on this technique can be found in Section 12.3.5.

Another advantage that triangles have over quads is that (again, assuming the vertices aren't collinear) they are convex polygons. In particular, the convex combination of the three triangle vertices spans all the points that make up the triangle. Given a point P inside the triangle and on the triangle plane, it is possible to compute its particular barycentric coordinates (s, t), as used in the parameterized plane equation $P(s, t) = P_0 + s\mathbf{u} + t\mathbf{v}$. If we compute a vector $\mathbf{w} = P - P_0$, then we can rewrite the plane equation as

$$P = P_0 + s\mathbf{u} + t\mathbf{v}$$
$$\mathbf{w} = s\mathbf{u} + t\mathbf{v}$$

If we take the cross product of $\mathbf{v}$ with $\mathbf{w}$, we get

$$\mathbf{v} \times \mathbf{w} = \mathbf{v} \times (s\mathbf{u} + t\mathbf{v})$$
$$= s(\mathbf{v} \times \mathbf{u}) + t(\mathbf{v} \times \mathbf{v})$$
$$= s(\mathbf{v} \times \mathbf{u})$$

Taking the length of both sides gives

$$\|\mathbf{v} \times \mathbf{w}\| = |s| \|\mathbf{v} \times \mathbf{u}\|$$

The quantity $\|\mathbf{v} \times \mathbf{u}\| = \|\mathbf{u} \times \mathbf{v}\|$. And since P is inside the triangle, we know that to meet the requirements of a convex combination, $s \geq 0$; thus,

$$s = \frac{\|\mathbf{v} \times \mathbf{w}\|}{\|\mathbf{u} \times \mathbf{v}\|}$$

A similar construction finds that

$$t = \frac{\|\mathbf{u} \times \mathbf{w}\|}{\|\mathbf{u} \times \mathbf{v}\|}$$

Note that this is equivalent to computing the areas a and b of the two subtriangles shown in Figure 2.37 and dividing by the total area of the triangle c, so

$$s = \frac{b}{c}$$
$$t = \frac{a}{c}$$

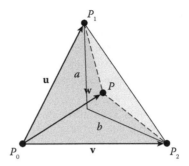

Figure 2.37. Computing barycentric coordinates for points in a triangle.

where

$$a = \frac{1}{2}\|\mathbf{u} \times \mathbf{w}\|$$

$$b = \frac{1}{2}\|\mathbf{v} \times \mathbf{w}\|$$

$$c = \frac{1}{2}\|\mathbf{u} \times \mathbf{v}\|$$

These simple examples are only a taste of how we can use triangles in mathematical calculations. More details on the use and implementation of triangles can be found throughout the text, particularly in Chapters 8 and 12.

2.7   Chapter Summary

In this chapter, we have covered some basic geometric entities: vectors and points. We have discussed linear and affine spaces, the relationships between them, and how we can use affine combinations of vectors and points to define other entities, like lines and planes. We've also shown how we can use our knowledge of affine spaces and vector properties to compute some simple tests on triangles. These skills will prove useful to us throughout the remainder of the text.

For those interested in reading further, Anton and Rorres [6] is a standard reference suggested for many first courses in linear algebra. Other texts with slightly different approaches are Axler [8] and Friedberg et al. [51]. Information on points and affine spaces can be found in deRose [32], as well as in Schneider and Eberly [133].

3 Linear Transformations and Matrices

3.1 Introduction

In Chapter 2 we discussed vectors and points and some simple operations we can apply to them. Now we'll begin to expand our discussion to cover specific functions that we can apply to vectors and points, functions known as transformations. In this chapter, we'll discuss a class of transformations that we can apply to vectors called *linear transformations*. These encompass nearly all of the common operations we might want to perform on vectors and points, so understanding what they are and how to apply them is important. We'll define these functions and how distinguished they are from the other more general transformations.

Properties of linear transformations allow us to use a structure called a *matrix* as a compact representation for transforming vectors. A matrix is a simple two-dimensional (2D) array of values, but within it lies all the power of a linear transformation. Through simple operations we can use the matrix to apply linear transformations to vectors. We can also combine the two transformation matrices to create a new one that has the same effect of the first two. Using matrices effectively lies at the heart of the pipeline for manipulating virtual objects and rendering them on the screen.

Matrices have other applications as well. Examining the structure of a matrix can tell us something about the transformation it represents, for example, whether it can be reversed, what that reverse transformation might be, or whether it distorts the data that it is given. Matrices can also be used to solve systems of linear equations, which is useful to know for certain algorithms in graphics and physical simulation. For all of these reasons, matrices are primary data structures in graphics application programmer interfaces (APIs).

3.2 Linear Transformations

Linear transformations are a very useful and important concept in linear algebra. As one of a class of functions known as transformations, they map vector spaces to vector spaces. This allows us to apply complex functions to, or *transform*, vectors. Linear transformations perform this mapping while also having the additional property of preserving linear combinations. We will see how this permits us to describe a linear transformation in terms of how it affects the basis vectors of a vector space. Later sections will show how this in turn allows us to use matrices to represent linear transformations.

3.2.1 Definitions

Before we can begin to discuss transformations and linear transformations in particular, we need to define a few terms. A *relation* maps a set X of values (known as the *domain*) to another set Y of values (known as the *range*). A *function* is a relation where every value in the first set maps to one and only one value in the second set, for example, $f(x) = \sin x$. An example of a relation that is not a function is $\pm\sqrt{x}$, because there are two possible results for a positive value of x, either positive or negative.

A function whose domain is an n-dimensional space and whose range is an m-dimensional space is known as a *transformation*. A transformation that maps from $\mathbb{R}^n$ to $\mathbb{R}^m$ is expressed as $T : \mathbb{R}^n \to \mathbb{R}^m$. If the domain and the range of a transformation are equal (i.e., $T : \mathbb{R}^n \to \mathbb{R}^n$), then the transformation is sometimes called an *operator*.

An example of a transformation is the function

$$f(x, y) = x^2 + 2y$$

which maps from $\mathbb{R}^2$ to $\mathbb{R}$. Another example is

$$f(x, y, z) = x^2 + 2y + \sqrt{z}$$

which maps from $\mathbb{R}^3$ to $\mathbb{R}$.

For an example with a multidimensional range, we can define a transformation from $\mathbb{R}^2$ to $\mathbb{R}^2$ by using two functions $f(a, b)$ and $g(a, b)$ as follows:

$$T(a, b) = (f(a, b), g(a, b)) \tag{3.1}$$

A *linear transformation* T is a mapping between two vector spaces V and W, where for all $\mathbf{v}$ in V and for all scalars a:

1. $T(\mathbf{v}_0 + \mathbf{v}_1) = T(\mathbf{v}_0) + T(\mathbf{v}_1)$ for all $\mathbf{v}_0, \mathbf{v}_1$ in V.

2. $T(a\mathbf{v}) = aT(\mathbf{v})$ for all $\mathbf{v}$ in V.

To determine whether a transformation is linear, it is sufficient to show that

$$T(a\mathbf{x} + \mathbf{y}) = aT(\mathbf{x}) + T(\mathbf{y}) \tag{3.2}$$

An example of a linear transformation is $T(\mathbf{x}) = k\mathbf{x}$, where k is any fixed scalar. We can show this by

$$\begin{aligned} T(a\mathbf{x} + \mathbf{y}) &= k(a\mathbf{x} + \mathbf{y}) \\ &= ak\mathbf{x} + k\mathbf{y} \\ &= aT(\mathbf{x}) + T(\mathbf{y}) \end{aligned}$$

On the other hand, the function $g(x) = x^2$ is not linear because, for $a = 2$, $x = 1$, and $y = 1$:

$$\begin{aligned} g(2(1) + 1) &= (2(1) + 1)^2 \\ &= 3^2 = 9 \\ &\neq 2(g(1)) + g(1) \\ &= 2(1^2) + 1^2 = 3 \end{aligned}$$

As we might expect, the only operations possible in a linear transformation are multiplication by a constant and addition.

3.2.2 Linear Transformations and Basis Vectors

Using standard function notation to represent linear transformations (as in Equation 3.1) is not the most convenient or compact format, particularly for transformations between higher-dimensional vector spaces. Let's examine the properties of vectors as they undergo a linear transformation and see how that can lead us to a better representation.

Recall that we can represent any vector $\mathbf{x}$ in an n-dimensional vector space V as

$$\mathbf{x} = x_0 \mathbf{v}_0 + x_1 \mathbf{v}_1 + \cdots + x_{n-1} \mathbf{v}_{n-1}$$

where $\{\mathbf{v}_0, \mathbf{v}_1, \ldots, \mathbf{v}_{n-1}\}$ is a basis for V.

Now suppose we have a linear transformation $T : V \rightarrow W$ that maps from V to an m-dimensional vector space W. If we apply our transformation to our arbitrary vector $\mathbf{x}$, then we can use Equation 3.2 to rewrite it as follows:

$$\begin{aligned} T(\mathbf{x}) &= T(x_0 \mathbf{v}_0 + x_1 \mathbf{v}_1 + \cdots + x_{n-1} \mathbf{v}_{n-1}) \\ &= x_0 T(\mathbf{v}_0) + x_1 T(\mathbf{v}_1) + \cdots + x_{n-1} T(\mathbf{v}_{n-1}) \end{aligned} \qquad (3.3)$$

We end up with a linear combination using the original components and the transformed basis vectors. So, if we know how our linear transformation affects our basis for V, then we can calculate the result of the linear transformation for any arbitrary vector in V given only the components for that vector. This is extraordinarily powerful, as we'll see when we discuss matrices.

Let's break this down further. For a member $\mathbf{v}_j$ of V's basis, we can represent $T(\mathbf{v}_j)$ in terms of the basis $\{\mathbf{w}_0, \mathbf{w}_1, \ldots, \mathbf{w}_{m-1}\}$ for W, again as a linear combination:

$$T(\mathbf{v}_j) = a_{0,j} \mathbf{w}_0 + a_{1,j} \mathbf{w}_1 + \cdots + a_{m-1,j} \mathbf{w}_{m-1}$$

We will be assuming that $\{\mathbf{w}_0, \ldots, \mathbf{w}_{m-1}\}$ is the standard basis for W. In that case, this simplifies to

$$T(\mathbf{v}_j) = (a_{0,j}, a_{1,j}, \ldots, a_{m-1,j}) \qquad (3.4)$$

Combining Equations 3.3 and 3.4 gives us

$$\begin{aligned}
T(\mathbf{x}) = {} & x_0(a_{0,0}, a_{1,0}, \ldots, a_{m-1,0}) \\
& + x_1(a_{0,1}, a_{1,1}, \ldots, a_{m-1,1}) \\
& \vdots \\
& + x_{n-1}(a_{0,n-1}, a_{1,n-1}, \ldots, a_{m-1,n-1})
\end{aligned} \qquad (3.5)$$

If we set $\mathbf{b} = T(\mathbf{x})$, then for a given component of $\mathbf{b}$

$$b_i = a_{i,0}x_0 + a_{i,1}x_1 + \cdots + a_{i,n-1}x_{n-1} \qquad (3.6)$$

Knowing this, we can precalculate and store the components $(a_{0,j}, a_{1,j}, \ldots, a_{m-1,j})$ for each of the n transformed basis vectors and use Equation 3.5 to transform a general vector $\mathbf{x}$.

Let's look at an example taking a transformation from $\mathbb{R}^2$ to $\mathbb{R}^2$, using the standard basis for both vector spaces:

$$T(a, b) = (a + b, b)$$

If we look at how this affects our standard basis for $\mathbb{R}^2$, we get

$$\begin{aligned}
T(1, 0) &= (1 + 0, 0) = (1, 0) \\
T(0, 1) &= (0 + 1, 1) = (1, 1)
\end{aligned}$$

Transforming an arbitrary vector in $\mathbb{R}^2$, say $(2, 3)$, we get

$$\begin{aligned}
T(2, 3) &= 2T(1, 0) + 3T(0, 1) \\
&= 2(1, 0) + 3(1, 1) \\
&= (5, 3)
\end{aligned}$$

which is what we expect.

3.2.3   Range and Null Space

It should be made clear that applying a linear transformation to a basis does not necessarily produce a basis for the new vector space. It only shows where the basis vectors end up in the new vector space—in our case in terms of the standard basis. In fact, a transformed basis may be no longer linearly independent. Take as another example

$$\mathcal{T}(a, b) = (a + b, 0) \tag{3.7}$$

Applying this to our standard basis for $\mathbb{R}^2$, we get

$$\mathcal{T}(1, 0) = (1 + 0, 0) = (1, 0)$$
$$\mathcal{T}(0, 1) = (0 + 1, 0) = (1, 0)$$

The two resulting vectors are clearly linearly dependent, and as we expect, they only span the space of 2D vectors with a zero y component.

So while a transformation maps from one vector space to another, it's possible that it may not map to all the vectors in the destination. We formally define the *range* $R(\mathcal{T})$ of a linear transformation $\mathcal{T}: V \rightarrow W$ as the set of all vectors in W that are mapped to by at least one vector in V, or

$$R(\mathcal{T}) = \{\mathcal{T}(\mathbf{x}) | \mathbf{x} \in V\}$$

The dimension of $R(\mathcal{T})$ is called the *rank* of the transformation.

Correspondingly, we define the *null space* (or *kernel*) $N(\mathcal{T})$ of a linear transformation $\mathcal{T}: V \rightarrow W$ as the set of all vectors in V that map to $\mathbf{0}$, or

$$N(\mathcal{T}) = \{\mathbf{x} \mid \mathcal{T}(\mathbf{x}) = \mathbf{0}\}$$

The dimension of $N(\mathcal{T})$ is called the *nullity* of the transformation.

The range and null space have two important properties. First of all, they are both vector spaces, and in fact, the null space is a subspace of V and the range is a subspace of W. Second,

$$nullity(\mathcal{T}) + rank(\mathcal{T}) = dim(V)$$

To get a better sense of this, let's look again at the transformation in Equation 3.7. It's range space is of the form $(x, 0)$, so it can be spanned by the vector $(1, 0)$ and has dimension 1. The transformation will produce the vector $(0, 0)$ only when $a = -b$. So the null space has a basis of $(1, -1)$ and is also one-dimensional (1D). As we expect, they add up to 2, the dimension of our original vector space (Figure 3.1).

This transformation and the example in the previous section illustrate one useful property. If the rank of a linear transformation $\mathcal{T}$ equals the number of elements in a transformed basis β, then we can say that β is linearly independent. In fact, the rank is equal to the number of linearly independent elements in β, and those linearly independent elements will span the range of $\mathcal{T}$. Knowing when a transformed basis is linearly independent will be important when we discuss linear systems, matrix inverses, and the determinant.

In summary, knowing that we can represent a linear transformation in terms of the transformed basis vectors is a very powerful tool. As we will now see, it is precisely this property of linear transformations that allows us to represent them concisely by using a matrix.

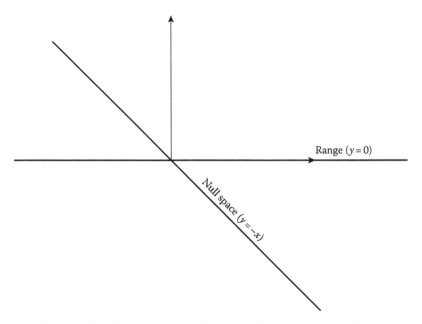

Figure 3.1. Range and null space for transformation $T(a, b) = (a + b, 0)$.

3.3 Matrices

3.3.1 Introduction to Matrices

A matrix is a rectangular, 2D array of values. Throughout this book, most of the values we use will be real numbers, but they could be complex numbers or even vectors. Each individual value in a matrix is called an *element*. Examples of matrices are

$$
\mathbf{A} = \begin{bmatrix} 1 & 0 & 0 \\ 0 & 1 & 0 \\ 0 & 0 & 1 \end{bmatrix}
\quad
\mathbf{B} = \begin{bmatrix} 0 & 35 & -15 \\ 2 & 52 & 1 \end{bmatrix}
\quad
\mathbf{C} = \begin{bmatrix} 2 & -1 \\ 0 & 2 \\ 6 & 3 \end{bmatrix}
$$

A matrix is described as having m rows by n columns, or being an $m \times n$ matrix. A row is a horizontal group of elements from left to right, while a column is a vertical, top-to-bottom group. Matrix $\mathbf{A}$ in our example has three rows and three columns and is a 3×3 matrix, whereas matrix $\mathbf{C}$ is a 3×2 matrix. Rows are numbered 0 to $m - 1$,[1] while columns are numbered 0 to $n - 1$. An individual element of a matrix $\mathbf{A}$ is referenced as either $(\mathbf{A})_{i,j}$ or just $a_{i,j}$, where i is the row number and j is the column. Looking at matrix $\mathbf{B}$, element $b_{1,0}$ contains the value 2 and element $b_{0,1}$ equals 35.

If an individual matrix has an equal number of rows and columns, that is, if $m = n$, then it is called a *square matrix*. In our example, matrix $\mathbf{A}$ is square, whereas matrices $\mathbf{B}$ and $\mathbf{C}$ are not.

---

[1] As a reminder, mathematical convention starts with 1, but we're using 0 to be compatible with C++.

If all elements of a matrix are 0, then it is called a *zero matrix*. We will represent a matrix of this type as **0** and assume a matrix of the appropriate size for the operation we are performing.

If two matrices have an equal number of rows and columns, then they are said to be the same *size*. If they are the same size and their corresponding elements have the same values, then they are *equal*. Below, the two matrices are the same size, but they are not equal.

$$\begin{bmatrix} 0 & 1 \\ 3 & 2 \\ 0 & -3 \end{bmatrix} \neq \begin{bmatrix} 0 & 0 \\ 2 & -3 \\ 1 & 3 \end{bmatrix}$$

The set of elements where the row and column numbers are the same (e.g., row 1, column 1) is called the *main diagonal*. In the next example, the main diagonal is in bold.

$$\mathbf{U} = \begin{bmatrix} \mathbf{3} & -5 & 0 & 1 \\ 0 & \mathbf{2} & 6 & 0 \\ 0 & 0 & \mathbf{1} & -8 \\ 0 & 0 & 0 & \mathbf{1} \end{bmatrix}$$

The *trace* of a matrix is the sum of the main diagonal elements. In this case the trace is $3+2+1+1 = 7$.

In matrix **U**, all elements below the diagonal are equal to 0. This is known as an *upper triangular matrix*. Note that elements above the diagonal don't necessarily have to be nonzero in order for the matrix to be upper triangular, nor does the matrix have to be square.

If elements above the diagonal are 0, then we have a *lower triangular matrix*:

$$\mathbf{L} = \begin{bmatrix} 3 & 0 & 0 & 0 \\ 2 & 2 & 0 & 0 \\ 0 & 3 & 1 & 0 \\ -6 & 1 & 0 & 1 \end{bmatrix}$$

Finally, if a square matrix's nondiagonal elements are all 0, we call the matrix a *diagonal matrix*:

$$\mathbf{D} = \begin{bmatrix} 3 & 0 & 0 & 0 \\ 0 & 2 & 0 & 0 \\ 0 & 0 & 1 & 0 \\ 0 & 0 & 0 & 1 \end{bmatrix}$$

It follows that any diagonal matrix is both an upper triangular and a lower triangular matrix.

3.3.2 Simple Operations

3.3.2.1 Matrix Addition and Scalar Multiplication

We can add and scale matrices just as we can do for vectors. Adding two matrices together,

$$\mathbf{S} = \mathbf{A} + \mathbf{B}$$

is done componentwise like vectors; thus,

$$s_{i,j} = a_{i,j} + b_{i,j}$$

Clearly, in order for this to work, $\mathbf{A}$, $\mathbf{B}$, and $\mathbf{S}$ must all be the same size (also known as *conformable for addition*). Subtraction works similarly but, as with real numbers and vectors, is not commutative.

To scale a matrix,

$$\mathbf{P} = s\mathbf{A}$$

each element is multiplied by the scalar, again like vectors:

$$p_{i,j} = s\, a_{i,j}$$

Matrix addition and scalar multiplication have their algebraic rules, which should seem quite familiar at this point:

1. $\mathbf{A} + \mathbf{B} = \mathbf{B} + \mathbf{A}$.

2. $\mathbf{A} + (\mathbf{B} + \mathbf{C}) = (\mathbf{A} + \mathbf{B}) + \mathbf{C}$.

3. $\mathbf{A} + \mathbf{0} = \mathbf{A}$.

4. $\mathbf{A} + (-\mathbf{A}) = \mathbf{0}$.

5. $a(\mathbf{A} + \mathbf{B}) = a\mathbf{A} + a\mathbf{B}$.

6. $a(b\mathbf{A}) = (ab)\mathbf{A}$.

7. $(a + b)\mathbf{A} = a\mathbf{A} + b\mathbf{A}$.

8. $1\mathbf{A} = \mathbf{A}$.

As we can see, these rules match the requirements for a vector space, and so the set of matrices of a given size is also a vector space.

3.3.2.2 Transpose

The *transpose* of a matrix $\mathbf{A}$ (represented by $\mathbf{A}^T$) interchanges the rows and columns of $\mathbf{A}$. It does this by exchanging elements across the matrix's main diagonal, so $(\mathbf{A}^T)_{i,j} = (\mathbf{A})_{j,i}$.

An example of this is

$$\begin{bmatrix} 2 & -1 \\ 0 & 2 \\ 6 & 3 \end{bmatrix} = \begin{bmatrix} 2 & 0 & 6 \\ -1 & 2 & 3 \end{bmatrix}$$

As we can see, the matrix does not have to be square, so an $m \times n$ matrix becomes an $n \times m$ matrix. Also, the main diagonal doesn't change, or is invariant, since $(\mathbf{A}^T)_{i,i} = (\mathbf{A})_{i,i}$.

A matrix where $(\mathbf{A})_{i,j} = (\mathbf{A})_{j,i}$ (i.e., cross-diagonal entries are equal) is called a *symmetric matrix*. All diagonal matrices are symmetric. Another example of a symmetric matrix is

$$\begin{bmatrix} 3 & 1 & 2 & 3 \\ 1 & 2 & -5 & 0 \\ 2 & -5 & 1 & -9 \\ 3 & 0 & -9 & 1 \end{bmatrix}$$

The transpose of a symmetric matrix is the matrix again, since in this case $(\mathbf{A}^T)_{j,i} = (\mathbf{A})_{i,j} = (\mathbf{A})_{j,i}$.

A matrix where $(\mathbf{A})_{i,j} = -(\mathbf{A})_{j,i}$ (i.e., cross-diagonal entries are negated and the diagonal is 0) is called a *skew symmetric matrix*. An example of a skew symmetric matrix is

$$\begin{bmatrix} 0 & 1 & 2 \\ -1 & 0 & -5 \\ -2 & 5 & 0 \end{bmatrix}$$

The transpose of a skew symmetric matrix is the negation of the original matrix, since in this case $(\mathbf{A}^T)_{j,i} = (\mathbf{A})_{i,j} = -(\mathbf{A})_{j,i}$.

Some useful algebraic rules involving the transpose are

1. $(\mathbf{A}^T)^T = \mathbf{A}$

2. $(a\mathbf{A}^T) = a\mathbf{A}^T$

3. $(\mathbf{A} + \mathbf{B})^T = \mathbf{A}^T + \mathbf{B}^T$

where a is a scalar and $\mathbf{A}$ and $\mathbf{B}$ are conformable for addition.

3.3.3 Vector Representation

If a matrix has only one row or one column, then we have a row or column matrix, respectively:

$$\begin{bmatrix} 0.5 & 0.25 & 1 & -1 \end{bmatrix} \qquad \begin{bmatrix} 5 \\ -3 \\ 6.9 \end{bmatrix}$$

These are often used to represent vectors. There is no particular standard as to which one to use. Historically, the OpenGL specification and its documentation used columns and DirectX used rows. Currently, however, you can use either convention in both APIs. In this text, we will assume that vectors are represented as column matrices (also known as column vectors). First of all, most math texts use column vectors and we wish to remain compatible. In addition, the classical presentation of quaternions (another means for performing some linear transformations) uses a concatenation order consistent with the use of column matrices for vectors.

The choice to represent vectors as column matrices does have some effect on how we construct and multiply our matrices, which we will discuss in more detail in the following parts. In the cases where we do wish to indicate that a vector is represented as a row matrix, we'll display it with a transpose applied, like $\mathbf{b}^T$.

3.3.4   Block Matrices

A matrix also can be represented by submatrices, rather than by individual elements. This is also known as a *block matrix*. For example, the matrix

$$\begin{bmatrix} 2 & 3 & 0 \\ -3 & 2 & 0 \\ 0 & 0 & 1 \end{bmatrix}$$

also can be represented as

$$\begin{bmatrix} \mathbf{A} & \mathbf{0} \\ \mathbf{0}^T & 1 \end{bmatrix}$$

where

$$\mathbf{A} = \begin{bmatrix} 2 & 3 \\ -3 & 2 \end{bmatrix}$$

and

$$\mathbf{0} = \begin{bmatrix} 0 \\ 0 \end{bmatrix}$$

We will sometimes use this to represent a matrix as a set of row or column matrices. For example, if we have a matrix $\mathbf{A}$,

$$\begin{bmatrix} a_{0,0} & a_{0,1} & a_{0,2} \\ a_{1,0} & a_{1,1} & a_{1,2} \\ a_{2,0} & a_{2,1} & a_{2,2} \end{bmatrix}$$

we can represent its rows as three vectors,

$$\mathbf{a}_0^T = \begin{bmatrix} a_{0,0} & a_{0,1} & a_{0,2} \end{bmatrix}$$

$$\mathbf{a}_1^T = \begin{bmatrix} a_{1,0} & a_{1,1} & a_{1,2} \end{bmatrix}$$

$$\mathbf{a}_2^T = \begin{bmatrix} a_{2,0} & a_{2,1} & a_{2,2} \end{bmatrix}$$

and represent $\mathbf{A}$ as

$$\begin{bmatrix} \mathbf{a}_0^T \\ \mathbf{a}_1^T \\ \mathbf{a}_2^T \end{bmatrix}$$

Similarly, we can represent a matrix $\mathbf{B}$ with its columns as three vectors

$$\mathbf{b}_0 = \begin{bmatrix} b_{0,0} \\ b_{1,0} \\ b_{2,0} \end{bmatrix}$$

$$\mathbf{b}_1 = \begin{bmatrix} b_{0,1} \\ b_{1,1} \\ b_{2,1} \end{bmatrix}$$

$$\mathbf{b}_2 = \begin{bmatrix} b_{0,2} \\ b_{1,2} \\ b_{2,2} \end{bmatrix}$$

and subsequently $\mathbf{B}$ as

$$\begin{bmatrix} \mathbf{b}_0 & \mathbf{b}_1 & \mathbf{b}_2 \end{bmatrix}$$

As mentioned earlier, the transpose notation tells us whether we're using row or column vectors.

3.3.5 Matrix Product

The primary operation we will apply to matrices is multiplication, also known as the *matrix product*. The product is important to us because it allows us to do two essential things. First, multiplying a matrix by a compatible vector will *transform* the vector. Second, multiplying matrices together will create a single matrix that performs their combined transformations. We'll discuss exactly what is occurring when we cover vector transformations below, but for now we'll just define how to perform matrix multiplication.

As with real numbers, the product $\mathbf{C}$ of two matrices $\mathbf{A}$ and $\mathbf{B}$ is represented as

$$\mathbf{C} = \mathbf{AB}$$

Computing the matrix product is not as simple as multiplying real numbers, but is not that bad if you understand the process. To calculate a given element $c_{i,j}$ in the product, we take the dot product of row i from $\mathbf{A}$ with column j from $\mathbf{B}$. We can express this symbolically as

$$c_{i,j} = \sum_{k=0}^{n-1} a_{i,k} b_{k,j}$$

As an example, we'll look at computing the first element of a 3×3 matrix:

$$\begin{bmatrix} a_{0,0} & a_{0,1} & a_{0,2} \\ \vdots & \vdots & \vdots \\ \vdots & \vdots & \vdots \end{bmatrix} \begin{bmatrix} b_{0,0} & \cdots & \cdots \\ b_{1,0} & \cdots & \cdots \\ b_{2,0} & \cdots & \cdots \end{bmatrix} = \begin{bmatrix} c_{0,0} & \cdots & \cdots \\ \vdots & \ddots & \vdots \\ \vdots & \cdots & \ddots \end{bmatrix}$$

To compute the value of $c_{0,0}$, we take the dot product of row 0 from $\mathbf{A}$ and column 0 from $\mathbf{B}$:

$$c_{0,0} = a_{0,0} b_{0,0} + a_{0,1} b_{1,0} + a_{0,2} b_{2,0}$$

Expanding this for a 2×2 matrix:

$$\begin{bmatrix} a_{0,0} & a_{0,1} \\ a_{1,0} & a_{1,1} \end{bmatrix} \begin{bmatrix} b_{0,0} & b_{0,1} \\ b_{1,0} & b_{1,1} \end{bmatrix} = \begin{bmatrix} a_{0,0} b_{0,0} + a_{0,1} b_{1,0} & a_{0,0} b_{0,1} + a_{0,1} b_{1,1} \\ a_{1,0} b_{0,0} + a_{1,1} b_{1,0} & a_{1,0} b_{0,1} + a_{1,1} b_{1,1} \end{bmatrix}$$

If we represent $\mathbf{A}$ as a collection of rows and $\mathbf{B}$ as a collection of columns, then

$$\begin{bmatrix} \mathbf{a}_0^T \\ \mathbf{a}_1^T \end{bmatrix} \begin{bmatrix} \mathbf{b}_0 & \mathbf{b}_1 \end{bmatrix} = \begin{bmatrix} \mathbf{a}_0 \cdot \mathbf{b}_0 & \mathbf{a}_0 \cdot \mathbf{b}_1 \\ \mathbf{a}_1 \cdot \mathbf{b}_0 & \mathbf{a}_1 \cdot \mathbf{b}_1 \end{bmatrix}$$

There is a restriction on which matrices can be multiplied together; in order to perform a dot product, the two vectors have to have the same length. So, to multiply together two matrices, the number of columns in the first (i.e., the width of each row) has to be the same as the number of rows in the second (i.e., the height of each column). This is known as being *conformable for multiplication*. Because of this restriction, only square matrices can be multiplied by themselves.

We can also multiply by using block matrices:

$$\begin{bmatrix} \mathbf{A} & \mathbf{B} \\ \mathbf{C} & \mathbf{D} \end{bmatrix} \begin{bmatrix} \mathbf{E} & \mathbf{F} \\ \mathbf{G} & \mathbf{H} \end{bmatrix} = \begin{bmatrix} \mathbf{AE} + \mathbf{BG} & \mathbf{AF} + \mathbf{BH} \\ \mathbf{CE} + \mathbf{DG} & \mathbf{CF} + \mathbf{DH} \end{bmatrix}$$

Note that this is only allowable if the submatrices are conformable for addition and multiplication.

In general, matrix multiplication is not commutative. As an example, if we multiply a row matrix by a column matrix, we perform a dot product:

$$\begin{bmatrix} 1 & 2 \end{bmatrix} \begin{bmatrix} 3 \\ 4 \end{bmatrix} = 1 \cdot 3 + 2 \cdot 4 = 11$$

Because of this, you may often see a dot product represented as

$$\mathbf{a} \cdot \mathbf{b} = \mathbf{a}^T \mathbf{b}$$

If we multiply them in the opposite order, we get a square matrix:

$$\begin{bmatrix} 3 \\ 4 \end{bmatrix} \begin{bmatrix} 1 & 2 \end{bmatrix} = \begin{bmatrix} 3 & 6 \\ 4 & 8 \end{bmatrix}$$

Even multiplication of square matrices is not necessarily commutative:

$$\begin{bmatrix} 3 & 6 \\ 4 & 8 \end{bmatrix} \begin{bmatrix} 1 & 0 \\ 1 & 1 \end{bmatrix} = \begin{bmatrix} 9 & 6 \\ 12 & 8 \end{bmatrix}$$

$$\begin{bmatrix} 1 & 0 \\ 1 & 1 \end{bmatrix} \begin{bmatrix} 3 & 6 \\ 4 & 8 \end{bmatrix} = \begin{bmatrix} 3 & 6 \\ 7 & 14 \end{bmatrix}$$

Aside from the size restriction and not being commutative, the algebraic rules for matrix multiplication are very similar to those for real numbers:

1. $\mathbf{A}(\mathbf{BC}) = (\mathbf{AB})\mathbf{C}$

2. $a(\mathbf{BC}) = (a\mathbf{B})\mathbf{C}$

3. $\mathbf{A}(\mathbf{B} + \mathbf{C}) = \mathbf{AB} + \mathbf{AC}$

4. $(\mathbf{A} + \mathbf{B})\mathbf{C} = \mathbf{AC} + \mathbf{BC}$

5. $(\mathbf{AB})^T = \mathbf{B}^T \mathbf{A}^T$

where $\mathbf{A}$, $\mathbf{B}$, and $\mathbf{C}$ are matrices conformable for multiplication and a is a scalar. Note that matrix multiplication is still associative (rules 1 and 2) and distributive (rules 3 and 4).

3.3.6 Transforming Vectors

As previously indicated, matrices can be used to transform vectors. We do this by multiplying the matrix by a column matrix representing the vector we wish to transform, or simply

$$\mathbf{b} = \mathbf{Ax}$$

Let's expand our terms and examine the components of the matrix and each vector:

$$
\begin{bmatrix} b_0 \\ b_1 \\ \vdots \\ b_{m-1} \end{bmatrix} = \begin{bmatrix} a_{0,0} & a_{0,1} & \cdots & a_{0,n-1} \\ a_{1,0} & a_{1,1} & \cdots & a_{1,n-1} \\ \vdots & \vdots & \ddots & \vdots \\ a_{m-1,0} & a_{m-1,1} & \cdots & a_{m-1,n-1} \end{bmatrix} \begin{bmatrix} x_0 \\ x_1 \\ \vdots \\ x_{n-1} \end{bmatrix}
$$

Note that $\mathbf{x}$ has n components and the resulting vector $\mathbf{b}$ has m. In order for the multiplication to proceed, matrix $\mathbf{A}$ must be $m \times n$. This represents a transformation from an n-dimensional space V to an m-dimensional space W. As with general matrix multiplication, the number of columns in $\mathbf{A}$ must match the number of elements in $\mathbf{x}$, and the number of elements in the result $\mathbf{b}$ will equal the number of rows in $\mathbf{A}$.

Recall that for a linear transformation, if we know where the basis of a vector space is mapped to, we know where the remainder of the vectors are mapped. Let's use this fact to see how this operation performs a linear transformation. Suppose that we know that our standard basis $\{\mathbf{e}_0, \mathbf{e}_1, \ldots, \mathbf{e}_{n-1}\}$ in V is transformed to $\{\mathbf{a}_0, \mathbf{a}_1, \ldots, \mathbf{a}_{n-1}\}$ in W, again using the standard basis. We will store, in order, each of these transformed basis vectors as the columns of $\mathbf{A}$, or

$$
\mathbf{A} = \begin{bmatrix} \mathbf{a}_0 & \mathbf{a}_1 & \cdots & \mathbf{a}_{n-1} \end{bmatrix}
$$

Using our matrix multiplication definition to compute the product of $\mathbf{A}$ and a vector $\mathbf{x}$ in V, we see that the result for element i in $\mathbf{b}$ is

$$
b_i = a_{i,0}x_0 + a_{i,1}x_1 + \cdots + a_{i,n-1}x_{n-1}
$$

This is exactly the same as Equation 3.6. So, by setting up our matrix with the transformed basis vectors in each column, we can use matrix multiplication to perform linear transformations. We will use this important fact throughout the book to build our transformation matrices.

Column vectors aren't the only possibility. We can also premultiply by a vector by treating it as a row matrix:

$$
\mathbf{c}^T = \mathbf{x}^T \mathbf{A}
$$

or, expanded:

$$
\begin{bmatrix} c_0 & c_1 & \cdots & c_{n-1} \end{bmatrix} = \begin{bmatrix} x_0 & x_1 & \cdots & x_{m-1} \end{bmatrix} \begin{bmatrix} a_{0,0} & a_{0,1} & \cdots & a_{0,n-1} \\ a_{1,0} & a_{1,1} & \cdots & a_{1,n-1} \\ \vdots & \vdots & \ddots & \vdots \\ a_{m-1,0} & a_{m-1,1} & \cdots & a_{m-1,n-1} \end{bmatrix}
$$

In this case, the rows of $\mathbf{A}$ are acting as our transformed basis vectors, and the number of components in $\mathbf{x}^T$ must match the number of rows in our matrix.

At this point, we can define some additional properties for matrices. The *column space* of a matrix is the vector space spanned by the matrix's column vectors and is the range of the linear transformation performed by postmultiplying by a column vector. Correspondingly, the *row space* is the vector space spanned by the row vectors of the matrix and, as we'd expect, is the range of the linear transformation performed by premultiplying by a row vector. As it happens, the dimensions of the row space and column space are equal and that value is called the *rank* of the matrix. The matrix rank is equal to the rank of the associated linear transformation.

The column space and row space are not necessarily the same vector space. As an example, take the matrix

$$\begin{bmatrix} 0 & 1 & 0 \\ 0 & 0 & 1 \\ 0 & 0 & 0 \end{bmatrix}$$

When postmultiplied by a column vector, it maps a vector (x, y, z) in $\mathbb{R}^3$ to a vector $(y, z, 0)$ on the xy plane. Premultiplying by a row vector, on the other hand, maps (x, y, z) to $(0, x, y)$ on the yz plane. They have the same dimension, and hence the same rank, but they are not the same vector space.

This makes a certain amount of sense. When we multiply by a row vector, we use the row vectors of the matrix as our transformed basis instead of the column vectors. To achieve the same result as the column vector multiplication, we need to change our matrix's column vectors to row vectors by taking the transpose:

$$\begin{bmatrix} x & y & z \end{bmatrix} \begin{bmatrix} 0 & 0 & 0 \\ 1 & 0 & 0 \\ 0 & 1 & 0 \end{bmatrix} = \begin{bmatrix} y & z & 0 \end{bmatrix}$$

We can now see the purpose of the transpose: it exchanges a matrix's row space with its column space.

Like a linear transformation, a matrix also has a null space, which is all vectors $\mathbf{x}$ in V such that

$$\mathbf{Ax} = \mathbf{0}$$

In the preceding example, the null space N is all vectors with zero y and z components. As with linear transformations, $\dim(N) + \text{rank}(\mathbf{A}) = \dim(V)$.

3.3.7 Combining Linear Transformations

Suppose we have two transformations, $S : U \to V$ and $T : V \to W$, and we want to perform one after the other; namely, for a vector $\mathbf{x}$, we want the result $T(S(\mathbf{x}))$. If we know that we are going to transform a large collection of vectors by S and the resulting vectors by T, it will be more efficient to find a single transformation that generates the same result so that we only have to transform the vectors once. This is known as the *composition* of S and T and is written as

$$(T \circ S)(\mathbf{x}) = T(S(\mathbf{x}))$$

Composition (or alternatively, *concatenation*) of transformations is done via generalized matrix multiplication.

Suppose that matrix $\mathbf{A}$ is the corresponding transformation matrix for S and $\mathbf{B}$ is the corresponding matrix for T. Recall that in order to set up $\mathbf{A}$ for vector transformation, we pretransform the standard basis vectors by S and store them as the columns of $\mathbf{A}$. Now we need to transform those vectors again, this time by T. We could either do this explicitly or use the fact that multiplying by $\mathbf{B}$ will transform vectors in V by T. So we just multiply each column of $\mathbf{A}$ by $\mathbf{B}$ and store the results, in order, as columns in a new matrix $\mathbf{C}$:

$$\mathbf{C} = \mathbf{BA}$$

If U has dimension n, V has dimension m, and W has dimension l, then $\mathbf{A}$ will be an $m \times n$ matrix and $\mathbf{B}$ will be an $l \times m$ matrix. Since the number of columns in $\mathbf{B}$ matches the number of rows in $\mathbf{A}$, the matrix product can proceed, as we'd expect. The result $\mathbf{C}$ will be an $l \times n$ matrix and will apply the transformation of $\mathbf{A}$ followed by the transformation of $\mathbf{B}$ in a single matrix–vector multiplication.

This is the power of using matrices as a representation for linear transformations. By continually concatenating matrices, we can use the result to produce the effect of an entire series of transformations, in order, through a single matrix multiplication. Note that the order does matter. The preceding result $\mathbf{C}$ will perform the result of applying $\mathbf{A}$ followed by $\mathbf{B}$. If we swap the terms (assuming they're still conformable under multiplication),

$$\mathbf{D} = \mathbf{AB}$$

and matrix $\mathbf{D}$ will perform the result of applying $\mathbf{B}$ followed by $\mathbf{A}$. This is almost certainly not the same transformation.

For the discussion thus far, we have assumed that the resulting matrix will be applied to a vector represented as a column matrix. It is good to be aware that the choice of whether to represent a vector as a row matrix or column matrix affects the order of multiplications when combining matrices. Suppose we multiply a column vector $\mathbf{u}$ by three matrices, where the intended transformation order is to apply $\mathbf{M}_0$, then $\mathbf{M}_1$, and finally $\mathbf{M}_2$:

$$\begin{aligned} \mathbf{v} &= \mathbf{M}_0 \mathbf{u} \\ \mathbf{w} &= \mathbf{M}_1 \mathbf{v} \\ \mathbf{x} &= \mathbf{M}_2 \mathbf{w} \end{aligned} \tag{3.8}$$

If we take Equation 3.8 and substitute $\mathbf{M}_1 \mathbf{v}$ for $\mathbf{w}$ and then $\mathbf{M}_0 \mathbf{u}$ for $\mathbf{v}$, we get

$$\begin{aligned} \mathbf{x} &= \mathbf{M}_2 \mathbf{M}_1 \mathbf{v} \\ &= \mathbf{M}_2 \mathbf{M}_1 \mathbf{M}_0 \mathbf{u} \\ &= \mathbf{M}_c \mathbf{u} \end{aligned}$$

Doing something similar for a row vector $\mathbf{a}^T$,

$$\mathbf{b}^T = \mathbf{a}^T \mathbf{N}_0$$
$$\mathbf{c}^T = \mathbf{b}^T \mathbf{N}_1$$
$$\mathbf{d}^T = \mathbf{c}^T \mathbf{N}_2$$

and substituting,

$$\mathbf{d}^T = \mathbf{b}^T \mathbf{N}_1 \mathbf{N}_2$$
$$= \mathbf{a}^T \mathbf{N}_0 \mathbf{N}_1 \mathbf{N}_2$$
$$= \mathbf{a}^T \mathbf{N}_r$$

the order difference is quite clear. When using row vectors and concatenating, matrix order follows the left-to-right progress used in English text. Column vectors work right to left instead, which may not be as intuitive. We will just need to be careful about our matrix order and transpose any matrices that assume we're using row vectors.

There are two other ways to modify transformation matrices that aren't used as often. Instead of concatenating two transformations, we may want to create a new one by adding two together: $\mathcal{Q}(\mathbf{x}) = \mathcal{S}(\mathbf{x}) + \mathcal{T}(\mathbf{x})$. This is easily done by adding the corresponding matrices together, so the matrix that performs $\mathcal{Q}$ is $\mathbf{C} = \mathbf{A} + \mathbf{B}$. Another means we might use for generating a new transformation from an existing one is to scale it: $\mathcal{R}(\mathbf{x}) = s \cdot \mathcal{T}(\mathbf{x})$. The corresponding matrix is created by scaling the original matrix: $\mathbf{D} = s\mathbf{A}$.

3.3.8 Identity Matrix

We know that when we multiply a scalar or vector by 1, the result is the scalar or vector again:

$$1 \cdot x = x$$

Similarly, in matrix multiplication there is a special matrix known as the *identity matrix*, represented by the letter $\mathbf{I}$. Thus,

$$\mathbf{A} \cdot \mathbf{I} = \mathbf{I} \cdot \mathbf{A} = \mathbf{A}$$

The identity matrix maps the basis vectors of the domain to the same vectors in the range; it performs a linear transformation that has no effect on the source vector, also known as the identity transformation.

All identity matrices have a similar form: a diagonal square matrix, where the diagonal is all 1s:

$$\mathbf{I} = \begin{bmatrix} 1 & 0 & \cdots & 0 \\ 0 & 1 & & 0 \\ \vdots & & \ddots & \vdots \\ 0 & 0 & \cdots & 1 \end{bmatrix}$$

If a particular $n \times n$ identity matrix is needed, it is sometimes referred to as $\mathbf{I}_n$. Take as an example $\mathbf{I}_3$:

$$\mathbf{I}_3 = \begin{bmatrix} 1 & 0 & 0 \\ 0 & 1 & 0 \\ 0 & 0 & 1 \end{bmatrix}$$

Rather than referring to it in this way, we'll just use the term $\mathbf{I}$ to represent a general identity matrix and assume it is the correct size in order to allow an operation to proceed.

3.3.9   Performing Vector Operations with Matrices

Recall that if we multiply a row vector by a column vector, it performs a dot product. For example, in $\mathbb{R}^3$:

$$\mathbf{w}^T \mathbf{v} = w_x v_x + w_y v_y + w_z v_z = \mathbf{v} \bullet \mathbf{w}$$

And multiplying them in the opposite order produces a square matrix (again in $\mathbb{R}^3$):

$$\mathbf{T} = \mathbf{v}\mathbf{w}^T = \begin{bmatrix} v_x w_x & v_x w_y & v_x w_z \\ v_y w_x & v_y w_y & v_y w_z \\ v_z w_x & v_z w_y & v_z w_z \end{bmatrix}$$

This square matrix $\mathbf{T}$ is known as the *outer product* or *tensor product* $\mathbf{v} \otimes \mathbf{w}$. We can use it to rewrite vector expressions of the form $(\mathbf{u} \bullet \mathbf{w})\mathbf{v}$ as

$$(\mathbf{u} \bullet \mathbf{w})\mathbf{v} = (\mathbf{v} \otimes \mathbf{w})\mathbf{u}$$

In particular, we can rewrite a projection by a unit vector as

$$(\mathbf{u} \bullet \hat{\mathbf{v}})\hat{\mathbf{v}} = (\hat{\mathbf{v}} \otimes \hat{\mathbf{v}})\mathbf{u}$$

This will prove useful to us in Chapter 4.

We can also perform our other vector product, the cross product, through a matrix multiplication. If we have two vectors $\mathbf{v}$ and $\mathbf{w}$ and we want to compute $\mathbf{v} \times \mathbf{w}$, we can replace $\mathbf{v}$ with a particular skew symmetric matrix, represented as $\tilde{\mathbf{v}}$:

$$\tilde{\mathbf{v}} = \begin{bmatrix} 0 & -v_z & v_y \\ v_z & 0 & -v_x \\ -v_y & v_x & 0 \end{bmatrix}$$

Multiplying by $\mathbf{w}$ gives

$$\begin{bmatrix} 0 & -v_z & v_y \\ v_z & 0 & -v_x \\ -v_y & v_x & 0 \end{bmatrix} \begin{bmatrix} w_x \\ w_y \\ w_z \end{bmatrix} = \begin{bmatrix} v_y w_z - w_y v_z \\ v_z w_x - w_z v_x \\ v_x w_y - w_x v_y \end{bmatrix}$$

which is the formula for the cross product. This will also prove useful to us in subsequent chapters.

3.3.10 Implementation

One might expect that the most natural data format for, say, a 3×3 matrix would be

Source Code
Library
IvMath
Filename
IvMatrix33
IvMatrix44

```
class IvMatrix33
{
    float mData[3][3];
};
```

However, the memory layout of such a matrix may not be ideal for our purposes. In C or C++, 2D arrays are stored in what is called *row major order*, meaning that the matrix is stored in memory in a row-by-row order. If we use a 1D array to represent our matrix data instead,

```
class IvMatrix33
{
    float mV[9];
};
```

the index order for a 3×3 row major matrix is

$$\begin{bmatrix} 0 & 1 & 2 \\ 3 & 4 & 5 \\ 6 & 7 & 8 \end{bmatrix}$$

The indexing operator for a row major matrix (we have to use `operator()` because `operator[]` only works for a single index) is

```
float&
IvMatrix33::operator()(unsigned int row, unsigned int col)
{
    return mV[col + 3*row];
}
```

So why won't this work for us? The problem is that in OpenGL and now by default in DirectX, matrices are stored column by column instead of row by row. This is a format known as *column major order*. Writing out our indices in column major order gives us

$$\begin{bmatrix} 0 & 3 & 6 \\ 1 & 4 & 7 \\ 2 & 5 & 8 \end{bmatrix}$$

Notice that the indices are the transpose of row major order. The indexing operator becomes

```
float&
IvMatrix33::operator()(unsigned int row, unsigned int col)
{
    return mV[row + 3*col];
}
```

Alternatively, if we want to use 2D arrays:

```
float&
IvMatrix33::operator()(unsigned int row, unsigned int col)
{
  return mV[col][row];
}
```

It's common to find references describing a matrix intended to be used with row vectors (i.e., its transformed basis vectors are stored as rows) as being in row major order and, similarly, referring to a matrix intended to be used with column vectors as being in column major order. This is incorrect terminology. Row and column major order refer only to the storage format, namely, where an element $a_{i,j}$ will lie in the 1D representation of the matrix. Whether your matrix library intends for vectors to be pre- or postmultiplied should be independent of the underlying storage.

The reason for the confusion is probably historical. In Silicon Graphics International's first API, IRIS GL, matrices were stored in row major order and the API assumed row vectors. When SGI opened up IRIS GL to create OpenGL, they adopted column vectors instead. However, the underlying hardware didn't change, so in order to remain compatible, they had to transpose every matrix before sending it to the hardware. Rather than do that explicitly, they pretransposed the matrix in the storage format; that is, they used column major order. The end result was that the underlying memory representation was exactly the same. Versions of Direct3D prior to DirectX 10 ended up adopting the IRIS GL convention, so the row vector/row major and column vector/column vector pairings became set in people's minds. However, there is no reason why you couldn't store a matrix intended for use with row vectors in column major order, and vice versa.

Using column major format and column vectors, matrix–vector multiplication becomes

```
IvVector3
IvMatrix33::operator*( const IvVector3& vector ) const
{
    IvVector3 result;

    result.x = mV[0]*vector.x + mV[3]*vector.y + mV[6]*vector.z;
    result.y = mV[1]*vector.x + mV[4]*vector.y + mV[7]*vector.z;
    result.z = mV[2]*vector.x + mV[5]*vector.y + mV[8]*vector.z;

    return result;
}
```

And matrix–matrix multiplication is similar, expanding this across the three columns of the right-hand matrix.

Matrix addition is just

```
IvMatrix33
IvMatrix33::operator+( const IvMatrix33& other ) const
{
    IvMatrix33 result;
```

```
for (int i = 0; i < 9; ++i)
{
    result.mV[i] = mV[i]+other.mV[i];
}
return result;
}
```

And again, scalar multiplication of matrices is similar.

This concludes our main discussion of linear transformations and matrices. The remainder of the chapter will be concerned with other useful properties of matrices: solving systems of linear equations, determinants, and eigenvalues and eigenvectors.

3.4 Systems of Linear Equations

3.4.1 Definition

Other than performing linear transformations, another purpose of matrices is to act as a mechanism for solving systems of linear equations. A general system of m linear equations with n unknowns is represented as

$$\begin{aligned}
b_0 &= a_{0,0}x_0 + a_{0,1}x_1 + \cdots + a_{0,n-1}x_{n-1} \\
b_1 &= a_{1,0}x_0 + a_{1,1}x_1 + \cdots + a_{1,n-1}x_{n-1} \\
&\vdots \quad \vdots \\
b_{m-1} &= a_{m-1,0}x_0 + a_{m-1,1}x_1 + \cdots + a_{m-1,n-1}x_{n-1}
\end{aligned} \tag{3.9}$$

The problem we are trying to solve is: Given $a_{0,0}, \ldots, a_{m-1,n-1}$ and $b_0, \ldots, b_{m-1}$, what are the values of $x_0, \ldots, x_{n-1}$? For a given linear system, the set of all possible solutions is called the *solution set*.

As an example, the system of equations

$$\begin{aligned}
x_0 + 2x_1 &= 1 \\
3x_0 - x_1 &= 2
\end{aligned}$$

has the solution set $\{x_0 = 5/7, x_1 = 1/7\}$.

There may be more than one solution to the linear system. For example, the plane equation

$$ax + by + cz = -d$$

has an infinite number of solutions: the solution set for this example is all the points on the particular plane.

Alternatively, it may not be possible to find any solution to the linear system. Suppose that we have the linear system

$$\begin{aligned}
x_0 + x_1 &= 1 \\
x_0 + x_1 &= 2
\end{aligned}$$

There are clearly no solutions for x and y. The solution set is the empty set.

Let's reexamine Equation 3.9. If we think of $(x_0, \ldots, x_{n-1})$ as elements of an n-dimensional vector $\mathbf{x}$ and $(b_0, \ldots, b_{m-1})$ as elements of an m-dimensional vector $\mathbf{b}$, then this starts to look a lot like matrix multiplication. We can rewrite this as

$$
\begin{bmatrix}
a_{0,0} & a_{0,1} & \cdots & a_{0,n-1} \\
a_{1,0} & a_{1,1} & \cdots & a_{1,n-1} \\
\vdots & \vdots & \ddots & \vdots \\
a_{m-1,0} & a_{m-1,1} & \cdots & a_{m-1,n-1}
\end{bmatrix}
\begin{bmatrix}
x_0 \\
x_1 \\
\vdots \\
x_{n-1}
\end{bmatrix}
=
\begin{bmatrix}
b_0 \\
b_1 \\
\vdots \\
b_{m-1}
\end{bmatrix}
$$

or our old friend

$$\mathbf{Ax} = \mathbf{b}$$

The coefficients of the equation become the elements of matrix $\mathbf{A}$, and matrix multiplication encapsulates our entire linear system. Now the problem becomes one of the form: Given $\mathbf{A}$ and $\mathbf{b}$, what is $\mathbf{x}$?

3.4.2   Solving Linear Systems

One case is very easy to solve. Suppose $\mathbf{A}$ looks like

$$
\begin{bmatrix}
1 & a_{0,1} & \cdots & a_{0,n-1} \\
0 & 1 & \cdots & a_{1,n-1} \\
\vdots & \vdots & \ddots & \vdots \\
0 & 0 & \cdots & 1
\end{bmatrix}
$$

This is equivalent to the linear system

$$b_0 = x_0 + a_{0,1}x_1 + \cdots + a_{0,n-1}x_{n-1}$$
$$b_1 = x_1 + \cdots + a_{1,n-1}x_{n-1}$$
$$\vdots \quad \vdots$$
$$b_{m-1} = x_{n-1}$$

We see that we immediately have the solution to one unknown via $x_{n-1} = b_{m-1}$. We can substitute this value into the previous $m - 1$ equations and possibly solve for another x_i. If so, we can substitute that x_i into the remaining unsolved equations and so on up the chain. If there is a single solution for the system of equations, we will find it; otherwise, we will solve as many terms as possible and derive a solution set for the remainder.

This matrix is said to be in *row echelon form*. The formal definition for row echelon form is

1. If a row is entirely zeros, it will be below any nonzero rows of the matrix; in other words, all zero rows will be at the bottom of the matrix.

2. The first nonzero element of a row (if any) will be 1 (called a *leading 1*).

3. Each leading 1 will be to the right of a leading 1 in any preceding row.

If the following additional condition is met, we say that the matrix is in *reduced row echelon form*.

4. Each column with a leading 1 will be 0 in the other rows.

The process we've described gives us a clue about how to proceed in solving general systems of linear equations. Suppose we can multiply both sides of our equation by a series of matrices so that the left-hand side becomes a matrix in row echelon form. Then we can use this in combination with the right-hand side to give us the solution for our system of equations.

However, we need to use matrices that preserve the properties of the linear system; the solution set for both systems of equations must remain equal. This restricts us to those matrices that perform one of three transformations called *elementary row operations*. These are

1. Multiply a row by a nonzero scalar.

2. Add a nonzero multiple of one row to another.

3. Swap two rows.

These three types of transformations maintain the solution set of the linear system while allowing us to reduce it to a simpler problem. The matrices that perform elementary row operations are called *elementary matrices*.

Some simple examples of elementary matrices include one that multiplies row 2 by a scalar a:

$$\begin{bmatrix} 1 & 0 & 0 \\ 0 & a & 0 \\ 0 & 0 & 1 \end{bmatrix}$$

one that adds k times row 2 to row 1:

$$\begin{bmatrix} 1 & k & 0 \\ 0 & 1 & 0 \\ 0 & 0 & 1 \end{bmatrix}$$

and one that swaps rows 2 and 3:

$$\begin{bmatrix} 1 & 0 & 0 \\ 0 & 0 & 1 \\ 0 & 1 & 0 \end{bmatrix}$$

3.4.3 Gaussian Elimination

Source Code
Library
IvMath
Filename
IvGaussianElim

In practice we don't solve linear systems through matrix multiplication. Instead, it is more efficient to iteratively perform the operations directly on **A** and **b**. The most basic method for solving linear systems is known as *Gaussian elimination*, after Karl Friedrich Gauss, a prolific German mathematician of the eighteenth and nineteenth centuries. It involves concatenating the matrix **A** and vector **b** into a form called an *augmented matrix* and then performing a series of elementary row operations on the augmented matrix, in a particular order. This will either give us a solution to the system of linear equations or tell us that computing a single solution is not possible; that is, there is either no solution or an infinite number of solutions.

To create the augmented matrix, we take the original matrix **A** and combine it with our constant vector **b**, for example,

$$\left[\begin{array}{ccc|c} 1 & 2 & 3 & 3 \\ 4 & 5 & 6 & 2 \\ 7 & 8 & 9 & 1 \end{array} \right]$$

The vertical line within the matrix indicates the separation between **A** and **b**. To this augmented matrix, we will directly apply one or more of our row operations.

The process begins by looking at the first element in the first row. The first step is called a *pivoting* step. At the very least, we need to ensure that we have a nonzero entry in the diagonal position, so if necessary, we will swap this row with one of the lower rows that has a nonzero entry in the same column. The element that we're swapping into place is called the *pivot* element, and swapping two rows to move the pivot element into place is known as *partial pivoting*. For better numerical precision, we usually go one step further and swap with the row that contains the element of largest absolute value. If no pivot element can be found, then there is no single solution and we abort.

Now let's say that the current pivot element value is k. We scale the entry row by $1/k$ to set the diagonal entry to 1. Finally, we set the column elements below the diagonal entry to 0 by adding appropriate multiples of the current row. Then we move on to the next row and look at its diagonal entry. At the end of this process, our matrix will be in row echelon form.

Let's take a look at an example. Suppose we have the following system of linear equations:

$$\begin{array}{rrrrrr} x & -3y & + & z & = 5 \\ 2x & -y & + & 2z & = 5 \\ 3x & +6y & + & 9z & = 3 \end{array}$$

The equivalent augmented matrix is

$$\left[\begin{array}{ccc|c} 1 & -3 & 1 & 5 \\ 2 & -1 & 2 & 5 \\ 3 & 6 & 9 & 3 \end{array} \right]$$

If we look at column 0, the maximal entry is 3, in row 2. So we begin by swapping row 2 with row 0:

$$\begin{bmatrix} 3 & 6 & 9 & 3 \\ 2 & -1 & 2 & 5 \\ 1 & -3 & 1 & 5 \end{bmatrix}$$

We scale the new row 0 by $1/3$ to set the pivot element to 1:

$$\begin{bmatrix} 1 & 2 & 3 & 1 \\ 2 & -1 & 2 & 5 \\ 1 & -3 & 1 & 5 \end{bmatrix}$$

Now we start clearing the lower entries. The first entry in row 1 is 2, so we scale row 0 by -2 and add it to row 1:

$$\begin{bmatrix} 1 & 2 & 3 & 1 \\ 0 & -5 & -4 & 3 \\ 1 & -3 & 1 & 5 \end{bmatrix}$$

We do the same for row 2, scaling by -1 and adding:

$$\begin{bmatrix} 1 & 2 & 3 & 1 \\ 0 & -5 & -4 & 3 \\ 0 & -5 & -2 & 4 \end{bmatrix}$$

We are done with row 0 and move on to row 1. Row 1, column 1, is the maximal entry in the column, so we don't need to swap rows. However, it isn't 1, so we need to scale row 1 by $-1/5$:

$$\begin{bmatrix} 1 & 2 & 3 & 1 \\ 0 & 1 & 4/5 & -3/5 \\ 0 & -5 & -2 & 4 \end{bmatrix}$$

We now need to clear element 1 of row 2 by scaling row 1 by 5 and adding:

$$\begin{bmatrix} 1 & 2 & 3 & 1 \\ 0 & 1 & 4/5 & -3/5 \\ 0 & 0 & 2 & 1 \end{bmatrix}$$

Finally, we scale the bottom row by $1/2$ to set the pivot element in the row to 1:

$$\begin{bmatrix} 1 & 2 & 3 & 1 \\ 0 & 1 & 4/5 & -3/5 \\ 0 & 0 & 1 & 1/2 \end{bmatrix}$$

This matrix is now in row echelon form. We have two possibilities at this point. We could clear the upper triangle of the matrix in a fashion similar to how we cleared the lower triangle, but by working up from the bottom and adding multiples of rows. The solution **x** to the linear system would end up in the right-hand column. This is known as *Gauss–Jordan elimination*.

But let's look at the linear system we have now:

$$x + 2y + 3z = 1$$
$$y + 4/5z = -3/5$$
$$z = 1/2$$

As expected, we already have a known quantity: z. If we plug z into the second equation, we can solve for y:

$$y = -3/5 - 4/5z \qquad (3.10)$$
$$= -3/5 - 4/5(1/2) \qquad (3.11)$$
$$= -1 \qquad (3.12)$$

Once y is known, we can solve for x:

$$x = 1 - 2y - 3z \qquad (3.13)$$
$$= 1 - 2(-1) - 3(1/2) \qquad (3.14)$$
$$= 3/2 \qquad (3.15)$$

So our final solution for **x** is $(3/2, -1, 1/2)$.

This process of substituting known quantities into our equations is called *back substitution*.

A summary of Gaussian elimination with back substitution follows:

```
for p = 1 to n do
    // find the element with largest absolute value in col p

    // if max is zero, stop!

    // if max element not in row p, swap rows

    // set pivot element to 1
    multiply row p by 1/A[p][p]

    // clear lower column entries
    for r = p+1 to n do
        subtract row p times A[r,p] from current row,
            so that element in pivot column becomes 0

// do backwards substitution
for row = n-1 to 1
    for col = row+1 to n
        // subtract out known quantities
        b[row] = b[row] - A[row][col]*b[col]
```

The pseudocode shows what may happen when we encounter a linear system with no single solution. If we can't swap a nonzero entry in the pivot location, then there is a column that is all zeros. This is only possible if the rank of the matrix (i.e., the number of linearly independent column vectors) is less than the number of unknowns. In this case, there is no solution to the linear system and we abort.

In general, we can state that if the rank of the coefficient matrix $\mathbf{A}$ equals the rank of the augmented matrix $\mathbf{A}|\mathbf{b}$, then there will be at least one solution to the linear system. If the two ranks are unequal, then there are no solutions. There is a single solution only if the rank of $\mathbf{A}$ is equal to the minimum of the number of rows or columns of $\mathbf{A}$.

3.5 Matrix Inverse

This may seem like a lot of trouble to go to solve a simple equation like $\mathbf{b} = \mathbf{A}\mathbf{x}$. If this were scalar math, we could simply divide both sides of the equation by $\mathbf{A}$ to get

$$\mathbf{x} = \mathbf{b}/\mathbf{A}$$

Unfortunately, matrices don't have a division operation. However, we can use an equivalent concept: the inverse.

3.5.1 Definition

In scalar multiplication, the inverse is defined as the reciprocal:

$$x \cdot \frac{1}{x} = 1$$

or

$$x \cdot x^{-1} = 1$$

Correspondingly, for a given matrix $\mathbf{A}$, we can define its inverse $\mathbf{A}^{-1}$ as a matrix such that

$$\mathbf{A} \cdot \mathbf{A}^{-1} = \mathbf{I}$$

and

$$\mathbf{A}^{-1} \cdot \mathbf{A} = \mathbf{I}$$

There are a few things that fall out from this definition. First of all, in order for the first multiplication to occur, the number of rows in the inverse must be the same as the number of columns in the original matrix. For the second to occur, the converse is true. So, the matrix and its inverse must be square and the same size. Since not all matrices are square, it's clear that not every matrix has an inverse.

Second, the inverse of the inverse returns the original matrix. Given

$$\mathbf{A}^{-1} \cdot (\mathbf{A}^{-1})^{-1} = \mathbf{I}$$

and

$$\mathbf{A}^{-1} \cdot \mathbf{A} = \mathbf{I}$$

then

$$(\mathbf{A}^{-1})^{-1} = \mathbf{A}$$

Even if a matrix is square, there isn't always an inverse. An extreme example is the zero matrix. Any matrix multiplied by this gives the zero matrix, so there is no matrix multiplication that will produce the identity. Another set of examples is matrices that have a zero row or column vector. Multiplying by such a row or column will return a dot product of 0, so you'll end up with a zero row or column vector in the product as well—again, not the identity matrix. In general, if the null space of the matrix is nonzero, then the matrix is noninvertible; that is, the matrix is only invertible if the rank of the matrix is equal to the number of rows and columns.

Given these identities, we can now solve for our preceding linear system. Recall that the equation was

$$\mathbf{A}\mathbf{x} = \mathbf{b}$$

If we multiply both sides by $\mathbf{A}^{-1}$, then

$$\mathbf{A}^{-1}\mathbf{A}\mathbf{x} = \mathbf{A}^{-1}\mathbf{b}$$
$$\mathbf{I}\mathbf{x} = \mathbf{A}^{-1}\mathbf{b}$$
$$\mathbf{x} = \mathbf{A}^{-1}\mathbf{b}$$

Therefore, if we could find the inverse of $\mathbf{A}$, we could use it to solve for $\mathbf{x}$. This is not usually a good idea, computationally speaking. It's usually cheaper to solve for $\mathbf{x}$ directly, rather than generating the inverse and then performing the matrix multiplication. The latter can also lead to increased numerical error. However, sometimes finding the inverse is a necessary evil.

The left-hand side of the above derivation shows us that we can think of the inverse $\mathbf{A}^{-1}$ as undoing the effect of $\mathbf{A}$. If we start with $\mathbf{A}\mathbf{x}$ and premultiply by $\mathbf{A}^{-1}$, we get back $\mathbf{x}$, our original vector.

We can find the inverse of a matrix using Gaussian elimination to solve for it column by column. Suppose we call the first column of $\mathbf{A}^{-1}$ $\mathbf{x}_0$. We can represent this as

$$\mathbf{x}_0 = \mathbf{A}^{-1}\mathbf{e}_0$$

where, as we recall, $\mathbf{e}_0 = (1, 0, \ldots, 0)$. Multiplying both sides by $\mathbf{A}$ gives

$$\mathbf{A}\mathbf{x}_0 = \mathbf{e}_0$$

Finding the solution to this linear system gives us the first column of $\mathbf{A}^{-1}$. We can do the same for the other columns, but using $\mathbf{e}_1, \mathbf{e}_2$, and so on. Instead of solving these one at a time, though, it is more efficient to create an augmented matrix with $\mathbf{A}$ and $\mathbf{e}_0, \ldots, \mathbf{e}_{n-1}$ as columns on the right, or just $\mathbf{I}$. For example,

$$\left[\begin{array}{ccc|ccc} 2 & 0 & 4 & 1 & 0 & 0 \\ 0 & 3 & -9 & 0 & 1 & 0 \\ 0 & 0 & 1 & 0 & 0 & 1 \end{array}\right]$$

If we use Gauss–Jordan elimination to turn the left-hand side of the augmented matrix into the identity matrix, then we will end up with the inverse (if any) on the right-hand side. From here we perform our elementary row operations as before. The maximal entry is already in the pivot point, so we scale the first row by $1/2$:

$$\begin{bmatrix} 1 & 0 & 2 & 1/2 & 0 & 0 \\ 0 & 3 & -9 & 0 & 1 & 0 \\ 0 & 0 & 1 & 0 & 0 & 1 \end{bmatrix}$$

The nonpivot entries in the first column are 0, so we move to the second column. Scaling the second row by $1/3$ to set the pivot point to 1 gives us

$$\begin{bmatrix} 1 & 0 & 2 & 1/2 & 0 & 0 \\ 0 & 1 & -3 & 0 & 1/3 & 0 \\ 0 & 0 & 1 & 0 & 0 & 1 \end{bmatrix}$$

Again, our nonpivot entries in the second column are 0, so we move to the third column. Our pivot entry is 1, so we don't need to scale. We add -2 times the last row to the first row to clear that entry, then 3 times the last row to the second row to clear that entry, and get

$$\begin{bmatrix} 1 & 0 & 0 & 1/2 & 0 & -2 \\ 0 & 1 & 0 & 0 & 1/3 & 3 \\ 0 & 0 & 1 & 0 & 0 & 1 \end{bmatrix}$$

The inverse of our original matrix is now on the right-hand side of the augmented matrix.

3.5.2 Simple Inverses

Gaussian elimination, while useful, is unnecessary for computing the inverse of many of the matrices we will be using. The majority of matrices that we will encounter in games and three-dimensional (3D) applications have simple inverses, and knowing the form of the matrix can make computing the inverse trivial.

One case is that of an *orthogonal matrix*, where the component row or column vectors are orthonormal. Recall that this means that the vectors are of unit length and perpendicular. If a matrix $\mathbf{A}$ is orthogonal, its inverse is the transpose:

$$\mathbf{A}^{-1} = \mathbf{A}^T$$

One example of an orthogonal matrix is

$$\begin{bmatrix} 0 & 0 & 1 \\ 1 & 0 & 0 \\ 0 & 1 & 0 \end{bmatrix}^{-1} = \begin{bmatrix} 0 & 1 & 0 \\ 0 & 0 & 1 \\ 1 & 0 & 0 \end{bmatrix}$$

Another simple case is a diagonal matrix with nonzero elements in the diagonal. The inverse of such a matrix is also diagonal, where the new diagonal elements are the reciprocal of the original diagonal elements, as shown by the following:

$$
\begin{bmatrix} a & 0 & 0 \\ 0 & b & 0 \\ 0 & 0 & c \end{bmatrix}^{-1} = \begin{bmatrix} 1/a & 0 & 0 \\ 0 & 1/b & 0 \\ 0 & 0 & 1/c \end{bmatrix}
$$

The third case is a modified identity matrix, where the diagonal is all 1s, but one column or row is nonzero. One such 3×3 matrix is

$$
\begin{bmatrix} 1 & 0 & x \\ 0 & 1 & y \\ 0 & 0 & 1 \end{bmatrix}
$$

For a matrix of this form, we simply negate the nonzero nondiagonal elements to invert it. Using the previous example,

$$
\begin{bmatrix} 1 & 0 & x \\ 0 & 1 & y \\ 0 & 0 & 1 \end{bmatrix}^{-1} = \begin{bmatrix} 1 & 0 & -x \\ 0 & 1 & -y \\ 0 & 0 & 1 \end{bmatrix}
$$

Finally, we can combine this knowledge to take advantage of an algebraic property of matrices. If we have two square matrices $\mathbf{A}$ and $\mathbf{B}$, both of which are invertible, then

$$
(\mathbf{AB})^{-1} = \mathbf{B}^{-1}\mathbf{A}^{-1}
$$

So, if we know that our current matrix is the product of any of the cases we've just discussed, we can easily compute its inverse using the preceding formula. This will prove to be useful in subsequent chapters.

3.6 Determinant

3.6.1 Definition

The *determinant* is a scalar quantity created by evaluating the elements of a square matrix. In real vector spaces, it acts as a general measure of how vectors transformed by the matrix change in size. For example, if we take the columns of a 2×2 matrix (i.e., the transformed basis vectors) and use them as the sides of a parallelogram (Figure 3.2), then the absolute value of the determinant is equal to the area of a parallelogram. For a 3×3 matrix, the absolute value of the determinant is equal to the volume of a parallelepiped described by the three transformed basis vectors (Figure 3.3).

The sign of the determinant depends on whether or not we have switched our ordered basis vectors from being relatively right-handed to being left-handed. In Figure 3.2, the shortest angle from $\mathbf{a}_0$ to $\mathbf{a}_1$ is clockwise, so they are left-handed. The determinant, therefore, is negative.

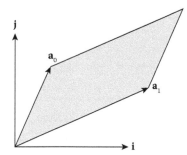

Figure 3.2. Determinant of 2×2 matrix as area of parallelogram bounded by transformed basis vectors $\mathbf{a}_0$ and $\mathbf{a}_1$.

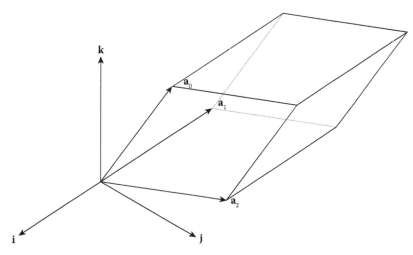

Figure 3.3. Determinant of 3×3 matrix as volume of parallelepiped bounded by transformed basis vectors $\mathbf{a}_0$, $\mathbf{a}_1$, and $\mathbf{a}_2$.

We represent the determinant in one of two ways, either det $(\mathbf{A})$ or $|\mathbf{A}|$. The first is more often used with a symbol, and the second when showing the elements of a matrix:

$$\det (\mathbf{A}) = \begin{vmatrix} 1 & -3 & 1 \\ 2 & -1 & 2 \\ 3 & 6 & 9 \end{vmatrix}$$

The diagrams showing area of a parallelogram and volume of a parallelepiped should look familiar from our discussion of cross product and triple scalar product. In fact, the cross product is sometimes represented as

$$\mathbf{v} \times \mathbf{w} = \begin{vmatrix} \mathbf{i} & \mathbf{j} & \mathbf{k} \\ v_x & v_y & v_z \\ w_x & w_y & w_z \end{vmatrix}$$

while the triple product is represented as

$$\mathbf{u} \cdot (\mathbf{v} \times \mathbf{w}) = \begin{vmatrix} u_x & u_y & u_z \\ v_x & v_y & v_z \\ w_x & w_y & w_z \end{vmatrix}$$

Since $\det(\mathbf{A}^T) = \det(\mathbf{A})$, this representation is equivalent.

3.6.2   Computing the Determinant

There are a few ways of representing the determinant computation for a specific matrix $\mathbf{A}$. A standard recursive definition, choosing any row i, is

$$\det(\mathbf{A}) = \sum_{j=0}^{n-1} a_{i,j}(-1)^{(i+j)} \det(\tilde{\mathbf{A}}_{i,j})$$

Alternatively, we can expand by column j instead:

$$\det(\mathbf{A}) = \sum_{i=0}^{n-1} a_{i,j}(-1)^{(i+j)} \det(\tilde{\mathbf{A}}_{i,j})$$

In both cases, $\tilde{\mathbf{A}}_{i,j}$ is the submatrix formed by removing the ith row and jth column from $\mathbf{A}$. The base case is the determinant of a matrix with a single element, which is the element itself.

The term $\det(\tilde{\mathbf{A}}_{i,j})$ is also referred to as the *minor of entry* $a_{i,j}$, and the term $(-1)^{(i+j)} \det(\tilde{\mathbf{A}}_{i,j})$ is called the *cofactor of entry* $a_{i,j}$.

The first formula tells us that for a given row i, we multiply each row entry $a_{i,j}$ by the determinant of the submatrix formed by removing row i and column j and either add or subtract it to the total depending on its position in the matrix. The second does the same but moves along column j instead of row i.

Let's compute an example determinant, expanding by row 0:

$$\det\left(\begin{bmatrix} 1 & 1 & 2 \\ 2 & 4 & -3 \\ 3 & 6 & -5 \end{bmatrix}\right) = ?$$

The first element of row 0 is 1, and the submatrix with row 0 and column 0 removed is

$$\begin{bmatrix} 4 & -3 \\ 6 & -5 \end{bmatrix}$$

The second element is also 1. However, we negate it since we are considering row 0 and column 1: $0 + 1 = 1$, which is odd. The submatrix is $\mathbf{A}$ with row 0 and column 1 removed:

$$\begin{bmatrix} 2 & -3 \\ 3 & -5 \end{bmatrix}$$

The third element of the row is 2, with the submatrix

$$\begin{bmatrix} 2 & 4 \\ 3 & 6 \end{bmatrix}$$

We don't negate it since we are considering row 0 and column 2: $0 + 2 = 2$, which is even.
So, the determinant is

$$\det(\mathbf{A}) = 1 \cdot \begin{vmatrix} 4 & -3 \\ 6 & -5 \end{vmatrix} - 1 \cdot \begin{vmatrix} 2 & -3 \\ 3 & -5 \end{vmatrix} + 2 \cdot \begin{vmatrix} 2 & 4 \\ 3 & 6 \end{vmatrix}$$
$$= -1$$

In general, the determinant of a 2×2 matrix is

$$\det\left(\begin{bmatrix} a & b \\ c & d \end{bmatrix}\right) = a \cdot \det([d]) - b \cdot \det([c]) = ad - bc$$

And the determinant of a 3×3 matrix is

$$\det\left(\begin{bmatrix} a & b & c \\ d & e & f \\ g & h & i \end{bmatrix}\right) = a \cdot \det\left(\begin{bmatrix} e & f \\ h & i \end{bmatrix}\right) - b \cdot \det\left(\begin{bmatrix} d & f \\ g & i \end{bmatrix}\right)$$
$$+ c \cdot \det\left(\begin{bmatrix} d & e \\ g & h \end{bmatrix}\right)$$

or

$$a(ei - fh) - b(di - fg) + c(dh - eg)$$

There are some additional properties of the determinant that will be useful to us. If we have two $n \times n$ matrices $\mathbf{A}$ and $\mathbf{B}$, the following hold:

1. $\det(\mathbf{AB}) = \det(\mathbf{A})\det(\mathbf{B})$.

2. $\det(\mathbf{A}^{-1}) = \dfrac{1}{\det(\mathbf{A})}$.

We can look at the value of the determinant to tell us some features of our matrix. First of all, as we have mentioned, any matrix that transforms our basis vectors from right-handed to left-handed will have a negative determinant. If the matrix is also orthogonal, we call a matrix of this type a *reflection*. We will learn more about reflection matrices in Chapter 4.

Then there are matrices that have a determinant of 1. The matrices we will encounter most often with this property are orthogonal matrices, where the handedness of the resulting basis stays the same (i.e., a right-handed basis is transformed to a right-handed basis). Figure 3.4

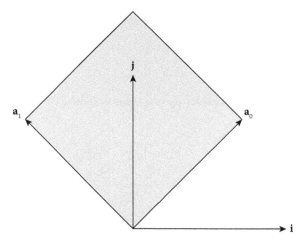

Figure 3.4. Determinant of example 2×2 orthogonal matrix.

provides an example. Our transformed basis vectors are $(-\sqrt{2}/2, \sqrt{2}/2)$ and $(\sqrt{2}/2, \sqrt{2}/2)$. They remain orthonormal, so their area is just the product of the lengths of the two vectors, or 1×1 or 1. This type of matrix is called a *rotation*. As with reflections, we'll see more of rotations in Chapter 4.

Finally, if the determinant is 0, then we know that the matrix has no inverse. The obvious case is if the matrix has a row or column of all 0s. Look again at our formula for the determinant. Suppose row i is all 0s. Multiplying all the submatrices against this row and summing together will clearly give us 0 as a result. The same is true for a zero column. The other and related possibility is that we have a linearly dependent row or column vector. In both cases the rank of the matrix is less than n—the size of the matrix—and therefore the matrix does not have an inverse. So, if the determinant of a matrix is 0, we know the matrix is not invertible.

3.6.3   Determinants and Elementary Row Operations

For 2×2 and 3×3 matrices, computing the determinant in this manner is a simple process. However, for larger and larger matrices, our recursive definition becomes unwieldy, and for large enough n, will take an unreasonable amount of time to compute. In addition, computing the determinant in this manner can lead to floating-point precision problems. Fortunately, there is another way.

Suppose we have an upper triangular matrix $\mathbf{U}$. The first part of the determinant sum is $u_{0,0}\tilde{\mathbf{U}}_{0,0}$. The other terms, however, are 0, because the first column with the first row removed is all 0s. So the determinant is just

$$\det(\mathbf{U}) = u_{0,0}\tilde{\mathbf{U}}_{0,0}$$

If we expand the recursion, we find that the determinant is the product of all the diagonal elements, or

$$\det(\mathbf{U}) = u_{0,0}u_{1,1}\ldots u_{nn}$$

As we did when solving linear systems, we can use Gaussian elimination to change our matrix into row echelon form, which is an upper triangular matrix. However, this assumes that elementary row operations have no effect on the determinant, which is not the case. Let's look at a few examples.

Suppose we have the matrix

$$\begin{bmatrix} 2 & -4 \\ -1 & 1 \end{bmatrix}$$

The determinant of this matrix is -2. If we multiply the first row by $1/2$, we get

$$\begin{bmatrix} 1 & -2 \\ -1 & 1 \end{bmatrix}$$

which has a determinant of -1. Multiplying a row by a scalar k multiplies the determinant by k as well.

Now suppose we add two times the first row to the second one. We get

$$\begin{bmatrix} 1 & -2 \\ 1 & -3 \end{bmatrix}$$

which also has a determinant of -1. Adding a multiple of one row to another has no effect on the determinant.

Finally, we can swap row 1 with row 2:

$$\begin{bmatrix} 1 & -3 \\ 1 & -2 \end{bmatrix}$$

which has a determinant of 1. Swapping two rows or two columns changes the sign of the determinant.

The effect of elementary row operations on the determinant can be summarized as follows:

Operation	Effect
Multiply row by k	Multiplies determinant by k
Add multiple of one row to another	No effect
Swap rows	Changes sign of determinant

Therefore, our approach for calculating the determinant for a general matrix is this: As we perform Gaussian elimination, we keep a running product p of any multiplies we do to create leading 1s and negate p for every row swap. If we find a zero column when we look for a pivot element, we know the determinant is 0 and return such.

Let's suppose our final product is p. This represents what we've multiplied the determinant of our original matrix by to get the determinant of the final matrix $\mathbf{A}'$, or

$$p \cdot \det(\mathbf{A}) = \det(\mathbf{A}')$$

so

$$\det(\mathbf{A}) = \frac{1}{p} \cdot \det(\mathbf{A}')$$

We know that the determinant of $\mathbf{A}'$ is 1, since the diagonal of the row echelon matrix is all 1s. So our final determinant is just $1/p$. However, this is just the product of the multiplies we do to create leading 1s, and -1 for every row swap, or

$$p = \frac{1}{p_{0,0}} \frac{1}{p_{1,1}} \cdots \frac{1}{p_{n,n}} (-1)^k$$

where k is the number of row swaps. Then,

$$\frac{1}{p} = p_{0,0}, p_{1,1}, \ldots, p_{n,n}(-1)^k$$

So all we need to do is multiply our running product by each pivot element and negate for each row swap. At the end of our Gaussian elimination process, our running product will be the determinant we seek.

Source Code
Library
IvMath
Filename
IvMatrix33

3.6.4   Adjoint Matrix and Inverse

Recall that the cofactor of an entry $a_{i,j}$ is

$$C_{i,j} = (-1)^{(i+j)} \det(\tilde{\mathbf{A}}_{i,j})$$

For an $n \times n$ matrix, we can construct a corresponding matrix where we replace each element with its corresponding cofactor, or

$$\begin{bmatrix} C_{0,0} & C_{0,1} & \cdots & C_{0,n-1} \\ C_{1,0} & C_{1,1} & \cdots & C_{1,n-1} \\ \vdots & \vdots & \ddots & \vdots \\ C_{n-1,1} & C_{n-1,2} & \cdots & C_{n-1,n-1} \end{bmatrix}$$

This is called the matrix of cofactors from $\mathbf{A}$, and its transpose is the *adjoint matrix* $\mathbf{A}^{\text{adj}}$.

Gabriel Cramer, a Swiss mathematician, showed that the inverse of a matrix can be computed from the adjoint by

$$\mathbf{A}^{-1} = \frac{1}{\det(\mathbf{A})} \mathbf{A}^{\text{adj}}$$

Many graphics engines use *Cramer's method* to compute the inverse, and for 3×3 and 4×4 matrices it's not a bad choice; for matrices of this size, Cramer's method is actually faster than Gaussian elimination. Because of this, we have chosen to implement `IvMatrix33::Inverse()` using an efficient form of Cramer's method.

However, whether you're using Gaussian elimination or Cramer's method, you're probably doing more work than is necessary for the matrices we will encounter. Most will be in one of the formats described in Section 3.5.2 or a multiple of these matrix types. Using the process described in that section, you can compute the inverse by decomposing the matrix into a set of these types, inverting the simple matrices, and multiplying in reverse order to compute the matrix. This is often faster than either Gaussian elimination or Cramer's method and can be more tolerant of floating-point errors because you can find near-exact solutions for the simple matrices.

3.7 Eigenvalues and Eigenvectors

There are two more properties of a matrix that we can find useful in certain circumstances: the *eigenvalue* and *eigenvector*. If we have an $n \times n$ matrix $\mathbf{A}$, then a nonzero vector $\mathbf{x}$ is called an eigenvector if there is some scalar value λ such that

$$\mathbf{Ax} = \lambda\mathbf{x} \qquad (3.16)$$

In this case, the value λ is the eigenvalue associated with that eigenvector.

We can solve for the eigenvalues of a matrix by rewriting Equation 3.16 as

$$\mathbf{Ax} = \lambda\mathbf{Ix} \qquad (3.17)$$

or

$$(\lambda\mathbf{I} - \mathbf{A})\mathbf{x} = \mathbf{0}$$

It can be shown that there is a nonzero solution of this equation if and only if

$$\det(\lambda\mathbf{I} - \mathbf{A}) = 0$$

This is called the *characteristic equation* of $\mathbf{A}$. Expanding this equation gives us an n-degree polynomial of λ, and solving for the roots of this equation will give us the eigenvalues of the matrix.

Now, for a given eigenvalue there will be an infinite number of associated eigenvectors, all scalar multiples of each other. This is called the *eigenspace* for that eigenvalue. To find the eigenspace for a particular eigenvector, we simply substitute that eigenvalue into Equation 3.17 and solve for $\mathbf{x}$.

In practice, solving the characteristic equation becomes more and more difficult the larger the matrix. However, there is a particular class of matrices called *real symmetric matrices*, so called because they only have real elements and are diagonally symmetric. Such matrices have a few nice properties. First of all, their eigenvectors are orthogonal. Secondly, it is possible to find a matrix $\mathbf{R}$, such that $\mathbf{R}^T\mathbf{AR}$ is a diagonal matrix $\mathbf{D}$. It turns out that the columns of $\mathbf{R}$ are the eigenvectors of $\mathbf{A}$, and the diagonal elements of $\mathbf{D}$ are the corresponding eigenvectors. This process is called *diagonalization*.

There are a number of standard methods for finding **R**. One such is the Jacobi method, which computes a series of matrices to iteratively diagonalize **A**. These matrices are then concatenated to create **R**. The problem with this method is that it is not always guaranteed to converge to a solution. An alternative is the Householder–QR/QL method, which similarly computes a series of matrices, but this time the end result is a tridiagonal matrix. From this we can perform a series of steps that factor the matrix into an orthogonal matrix **Q** and upper triangular matrix **R** (or an orthogonal matrix **Q** and a lower triangular matrix **L**). This will eventually diagonalize the matrix, again allowing us to compute the eigenvectors and eigenvalues. This can take more steps than the Jacobi method, but is guaranteed to complete in a fixed amount of time.

For 3×3 real symmetric matrices, Eberly [38] has a method that solves for the roots of the characteristic equation. This is considerably more efficient than the Householder method, and is relatively straightforward to compute.

3.8   Chapter Summary

In this chapter, we've discussed the general properties of linear transformations and how they are represented and performed by matrices. Matrices also can be used to compute solutions to linear systems of equations by using either Gaussian elimination or similar methods. We covered some basic matrix properties, the concepts of matrix identity and inverse (and various methods for calculating the latter), and the meaning and calculation of the determinant. This lays the foundation for what we'll be discussing in Chapter 4: using matrix transformations to manipulate models in a 3D world.

For those who are interested in reading further, Anton and Rorres [6] is a standard reference for many first courses in linear algebra. Other texts with slightly different approaches include Axler [8] and Friedberg et al. [51]. More information on Gaussian elimination and its extensions, such as 'lower upper' or LU decomposition, can be found in Anton and Rorres [6] as well as in the *Numerical Recipes* series [126]. Finally, Blinn has an excellent article in his collection *Notation, Notation, Notation* [13] on the geometry underlying 2×2 matrix operations.

4 Affine Transformations

4.1 Introduction

Now that we've chosen a mathematically sound basis for representing geometry in our game and discussed some aspects of matrix arithmetic, we need to combine them into an efficient method for placing and moving virtual objects or models. There are a few reasons we seek this efficiency. Suppose we wish to build a core level in our game space, say the office of a computer company. We could build all of our geometry in place and hard-code all of the locations. However, if we have a number of objects that are duplicated throughout the space—computers, desks, and chairs, for example—it would be more memory-efficient to create one master copy of the geometry for each type of object. Then, for each instance of a particular object, we can specify just a position and orientation and let the rendering and simulation engine handle the placement.

Another, more obvious reason is that objects in games generally move so that setting them at a fixed location is not practical. We will need to have some means to specify, for a model as a whole, its position and orientation in space.

There are a few characteristics we desire in our method. We want it to be fast and work well with our existing data and math library. We want to be able to concatenate a series of operations so we can perform them with a single operation, just as we did with linear transformations. Since our objects consist of collections of points, we need our method to work on points in an affine space, but we'll still need to transform vectors as well. The specific method we will use is called an *affine transformation*.

4.2   Affine Transformations

4.2.1   Matrix Definition

In Chapter 3 we discussed linear transformations, which map from one vector space to another. We can apply such transformations to vectors using matrix operations. There is a nearly equivalent set of transformations that map between affine spaces, which we can apply to points and vectors in an affine space. These are known as *affine transformations*, and they too can be applied using matrix operations, albeit in a slightly different form.

In the simplest terms, an affine transformation on a point can be represented by a matrix multiplication followed by a vector add, or

$$\mathbf{A}\mathbf{x} + \mathbf{y}$$

where the matrix $\mathbf{A}$ is an $m \times n$ matrix, $\mathbf{y}$ is an m-vector, and $\mathbf{x}$ consists of the point coordinates $(x_0, \ldots, x_{n-1})$.

We can represent this process of transformation by using block matrices:

$$\begin{bmatrix} \mathbf{A} & \mathbf{y} \\ \mathbf{0}^T & 1 \end{bmatrix} \begin{bmatrix} \mathbf{x} \\ 1 \end{bmatrix} = \begin{bmatrix} \mathbf{A}\mathbf{x} + \mathbf{y} \\ 1 \end{bmatrix} \tag{4.1}$$

As we can see, in order to allow the multiplication to proceed, we'll represent our point with a trailing 1 component. However, for the purposes of computation, the vector $\mathbf{0}^T$, the 1 in the lower right-hand corner of the matrix, and the trailing 1s in the points are unnecessary. They take up memory and using the full matrix takes additional instructions to multiply by constant values. Because of this, an affine transformation matrix is sometimes represented in a form where these constant terms are implied, either as an $m \times (n+1)$ matrix or as the matrix multiplication plus vector add form above.

If we subtract two points in an affine space, we get a vector:

$$\mathbf{v} = P_0 - P_1$$

$$= \begin{bmatrix} \mathbf{x}_0 \\ 1 \end{bmatrix} - \begin{bmatrix} \mathbf{x}_1 \\ 1 \end{bmatrix}$$

$$= \begin{bmatrix} \mathbf{x}_0 - \mathbf{x}_1 \\ 0 \end{bmatrix}$$

As we can see, a vector is represented in an affine space with a trailing 0. As previously noted in Chapter 2, this provides justification for some math libraries to use the trailing 1 on points and trailing 0 on vectors. If we multiply a vector using this representation by our $(m+1) \times (n+1)$ matrix,

$$\begin{bmatrix} \mathbf{A} & \mathbf{y} \\ \mathbf{0}^T & 1 \end{bmatrix} \begin{bmatrix} \mathbf{v} \\ 0 \end{bmatrix} = \begin{bmatrix} \mathbf{A}\mathbf{v} \\ 0 \end{bmatrix}$$

we see that the vector is affected by the upper left $m \times n$ matrix $\mathbf{A}$, but not the vector $\mathbf{y}$. This has the same effect on the first n elements of $\mathbf{v}$ as multiplying an n-dimensional vector by $\mathbf{A}$, which is a linear transformation. So, this representation allows us to use affine transformation matrices to apply linear transformations on vectors in an affine space.

Suppose we wish to concatenate two affine transformations $\mathcal{S}$ and $\mathcal{T}$, where the matrix representing $\mathcal{S}$ is

$$\begin{bmatrix} \mathbf{A} & \mathbf{y} \\ \mathbf{0}^T & 1 \end{bmatrix}$$

and the matrix representing $\mathcal{T}$ is

$$\begin{bmatrix} \mathbf{B} & \mathbf{z} \\ \mathbf{0}^T & 1 \end{bmatrix}$$

As with linear transformations, to find the matrix that represents the composition of $\mathcal{S}$ and $\mathcal{T}$, we multiply the matrices together. This gives

$$\begin{bmatrix} \mathbf{A} & \mathbf{y} \\ \mathbf{0}^T & 1 \end{bmatrix} \begin{bmatrix} \mathbf{B} & \mathbf{z} \\ \mathbf{0}^T & 1 \end{bmatrix} = \begin{bmatrix} \mathbf{AB} & \mathbf{Az}+\mathbf{y} \\ \mathbf{0}^T & 1 \end{bmatrix} \tag{4.2}$$

Finding the inverse for an affine transformation is equally as straightforward. Again, we can use a process similar to the one we used with linear transformation matrices. Starting with

$$\begin{bmatrix} \mathbf{A} & \mathbf{y} \\ \mathbf{0}^T & 1 \end{bmatrix} \begin{bmatrix} \mathbf{A} & \mathbf{y} \\ \mathbf{0}^T & 1 \end{bmatrix}^{-1} = \begin{bmatrix} \mathbf{I} & 0 \\ \mathbf{0}^T & 1 \end{bmatrix}$$

we multiply by both sides to remove the $\mathbf{y}$ component from the leftmost matrix:

$$\begin{bmatrix} \mathbf{I} & -\mathbf{y} \\ \mathbf{0}^T & 1 \end{bmatrix} \begin{bmatrix} \mathbf{A} & \mathbf{y} \\ \mathbf{0}^T & 1 \end{bmatrix} \begin{bmatrix} \mathbf{A} & \mathbf{y} \\ \mathbf{0}^T & 1 \end{bmatrix}^{-1} = \begin{bmatrix} \mathbf{I} & -\mathbf{y} \\ \mathbf{0}^T & 1 \end{bmatrix} \begin{bmatrix} \mathbf{I} & 0 \\ \mathbf{0}^T & 1 \end{bmatrix}$$

$$\begin{bmatrix} \mathbf{A} & 0 \\ \mathbf{0}^T & 1 \end{bmatrix} \begin{bmatrix} \mathbf{A} & \mathbf{y} \\ \mathbf{0}^T & 1 \end{bmatrix}^{-1} = \begin{bmatrix} \mathbf{I} & -\mathbf{y} \\ \mathbf{0}^T & 1 \end{bmatrix}$$

We then multiply by both sides to change the leftmost matrix to the identity:

$$\begin{bmatrix} \mathbf{A}^{-1} & 0 \\ \mathbf{0}^T & 1 \end{bmatrix} \begin{bmatrix} \mathbf{A} & 0 \\ \mathbf{0}^T & 1 \end{bmatrix} \begin{bmatrix} \mathbf{A} & \mathbf{y} \\ \mathbf{0}^T & 1 \end{bmatrix}^{-1} = \begin{bmatrix} \mathbf{A}^{-1} & 0 \\ \mathbf{0}^T & 1 \end{bmatrix} \begin{bmatrix} \mathbf{I} & -\mathbf{y} \\ \mathbf{0}^T & 1 \end{bmatrix} \tag{4.3}$$

$$\begin{bmatrix} \mathbf{A} & \mathbf{y} \\ \mathbf{0}^T & 1 \end{bmatrix}^{-1} = \begin{bmatrix} \mathbf{A}^{-1} & -\mathbf{A}^{-1}\mathbf{y} \\ \mathbf{0}^T & 1 \end{bmatrix}$$

thereby giving us the inverse on the right-hand side.

When we're working in $\mathbb{R}^3$, $\mathbf{A}$ will be a 3×3 matrix and $\mathbf{y}$ will be a 3-vector; hence, the full affine matrix will be a 4×4 matrix. Most graphics libraries expect transformations to be in the 4×4 matrix form, so if we do use the more compact forms in our math library to save memory, we will still have to expand them before rendering our objects. Because of this, we will use the 4×4 form for our following discussions, with the understanding that in our ultimate implementation we may choose one of the other forms for efficiency's sake.

4.2.2 Formal Definition

While the definition above will work for most practical purposes, to truly understand what our matrix form does requires some further explanation. We'll begin by formally defining an affine transformation. Recall that linear transformations preserve the linear operations of vector addition and scalar multiplication. In other words, linear transformations map from one vector space to another and preserve linear combinations. Thus, for a given linear transformation $\mathcal{S}$:

$$\mathcal{S}(a_0\mathbf{v}_0 + a_1\mathbf{v}_1 + \cdots + a_{n-1}\mathbf{v}_{n-1}) = a_0\mathcal{S}(\mathbf{v}_0) + a_1\mathcal{S}(\mathbf{v}_1) + \cdots + a_{n-1}\mathcal{S}(\mathbf{v}_{n-1})$$

Correspondingly, an affine transformation $\mathcal{T}$ maps between two affine spaces A and B and preserves affine combinations. For scalars $a_0, \ldots, a_{n-1}$ and points $P_0, \ldots, P_{n-1}$ in A:

$$\mathcal{T}(a_0 P_0 + \cdots + a_{n-1}P_{n-1}) = a_0\mathcal{T}(P_0) + \cdots + a_{n-1}\mathcal{T}(P_{n-1})$$

where $a_0 + \cdots + a_{n-1} = 1$.

As with our test for linear transformations, to determine whether a given transformation $\mathcal{T}$ is an affine transformation, it is sufficient to test a single affine combination:

$$\mathcal{T}(a_0 P_0 + a_1 P_1) = a_0\mathcal{T}(P_0) + a_1\mathcal{T}(P_1)$$

where $a_0 + a_1 = 1$.

Affine transformations are particularly useful to us because they preserve certain properties of geometry. First, they maintain collinearity, so points on a line will remain collinear and points on a plane will remain coplanar when transformed.

If we transform a line:

$$L(t) = (1-t)P_0 + tP_1$$
$$\mathcal{T}(L(t)) = \mathcal{T}((1-t)P_0 + tP_1)$$
$$= (1-t)\mathcal{T}(P_0) + t\mathcal{T}(P_1)$$

the result is clearly still a line (assuming $\mathcal{T}(P_0)$ and $\mathcal{T}(P_1)$ aren't coincident). Similarly, if we transform a plane:

$$P(t) = (1-s-t)P_0 + sP_1 + tP_2$$
$$\mathcal{T}(P(t)) = \mathcal{T}((1-s-t)P_0 + sP_1 + tP_2)$$
$$= (1-s-t)\mathcal{T}(P_0) + s\,\mathcal{T}(P_1) + t\,\mathcal{T}(P_2)$$

the result is clearly a plane (assuming $\mathcal{T}(P_0)$, $\mathcal{T}(P_1)$, and $\mathcal{T}(P_2)$ aren't collinear).

The second property of affine transformations is that they preserve relative proportions. The point that lies at t distance between P_0 and P_1 on the original line will map to the point that lies at t distance between $\mathcal{T}(P_0)$ and $\mathcal{T}(P_1)$ on the transformed line.

Note that while ratios of distances remain constant, angles and exact distances don't necessarily stay the same. The specific subset of affine transformations that preserve these features is called *rigid transformations*; those that don't are called *deformations*. It should be no surprise that we find rigid transformations useful. When transforming our models, in most cases we don't want them distorted unrecognizably. A bottle should maintain its size and shape—it should look like a bottle no matter where we place it in space. However, the deformations have their use as well. On occasion we may want to make an object larger or smaller or reflect it across a plane, as in a mirror.

To apply an affine transformation to a vector in an affine space, we can apply it to the difference of two points that equal the vector, or

$$T(\mathbf{v}) = T(P - Q) = T(P) - T(Q)$$

So, as we've seen above, an affine transformation that is applied to a vector performs a linear transformation.

4.2.3 Formal Representation

Suppose we have an affine transformation that maps from affine space A to affine space B, where the frame for A has basis vectors $(\mathbf{v}_0, \ldots, \mathbf{v}_{n-1})$ and origin O_A, and the frame for B has basis vectors $(\mathbf{w}_0, \ldots, \mathbf{w}_{m-1})$ and origin O_B. If we apply an affine transformation to a point $P = (x_0, \ldots, x_{n-1})$ in A, this gives

$$\begin{aligned} T(P) &= T(x_0\mathbf{v}_0 + \cdots + x_{n-1}\mathbf{v}_{n-1} + O_A) \\ &= x_0 T(\mathbf{v}_0) + \cdots + x_{n-1} T(\mathbf{v}_{n-1}) + T(O_A) \end{aligned}$$

As we did with linear transformations, we can express a given $T(\mathbf{v})$ in terms of B's frame:

$$T(\mathbf{v}_j) = a_{0,j}\mathbf{w}_0 + a_{1,j}\mathbf{w}_1 + \cdots + a_{m-1,j}\mathbf{w}_{m-1}$$

Similarly, we can express $T(O_A)$ in terms of B's frame:

$$T(O_A) = y_0\mathbf{w}_0 + y_1\mathbf{w}_1 + \cdots + y_{m-1}\mathbf{w}_{m-1} + O_B$$

Again, as we did with linear transformations, we can rewrite this as a matrix product. However, unlike linear transformations, we write a mapping from an n-dimensional affine space to an m-dimensional affine space as an $(m+1) \times (n+1)$ matrix:

$$\begin{bmatrix} a_{0,0}\mathbf{w}_0 & a_{0,1}\mathbf{w}_0 & \cdots & a_{0,n-1}\mathbf{w}_0 & y_0\mathbf{w}_0 \\ a_{1,0}\mathbf{w}_1 & a_{1,1}\mathbf{w}_1 & \cdots & a_{1,n-1}\mathbf{w}_1 & y_1\mathbf{w}_1 \\ \vdots & \vdots & \ddots & \vdots & \vdots \\ a_{m-1,0}\mathbf{w}_{m-1} & a_{m-1,1}\mathbf{w}_{m-1} & \cdots & a_{m-1,n-1}\mathbf{w}_{m-1} & y_{m-1}\mathbf{w}_{m-1} \\ 0 & 0 & \cdots & 0 & O_B \end{bmatrix} \begin{bmatrix} x_0 \\ x_1 \\ \vdots \\ x_{n-1} \\ 1 \end{bmatrix}$$

The dimensions of our matrix now make sense. The $n+1$ columns represent the n transformed basis vectors plus the transformed origin. We need $m+1$ rows since the frame of B has m basis vectors plus the origin O_B.

We can pull out the frame terms to get

$$\begin{bmatrix} \mathbf{w}_0 & \mathbf{w}_1 & \cdots & \mathbf{w}_{m-1} & O_B \end{bmatrix} \begin{bmatrix} a_{0,0} & a_{0,1} & \cdots & a_{0,n-1} & y_0 \\ a_{1,0} & a_{1,1} & \cdots & a_{1,n-1} & y_1 \\ \vdots & \vdots & \ddots & \vdots & \vdots \\ a_{m-1,0} & a_{m-1,1} & \cdots & a_{m-1,n-1} & y_{m-1} \\ 0 & 0 & \cdots & 0 & 1 \end{bmatrix} \begin{bmatrix} x_0 \\ x_1 \\ \vdots \\ x_{n-1} \\ 1 \end{bmatrix}$$

So, similar to linear transformations, if we know how the affine transformation affects the frame for A, we can copy the transformed frame *in terms of the frame for B* into the columns of a matrix and use matrix multiplication to apply the affine transformation to an arbitrary point.

4.3   Standard Affine Transformations

Now that we've defined affine transformations in general, we can discuss some specific affine transformations that will prove useful when manipulating objects in our game. We'll cover these in terms of transformations from $\mathbb{R}^3$ to $\mathbb{R}^3$, since they will be the most common uses. However, we can apply similar principles to find transformations from $\mathbb{R}^2$ to $\mathbb{R}^2$ or even $\mathbb{R}^4$ to $\mathbb{R}^4$ if we desire.

Since affine spaces A and B are the same in this case, to simplify things we'll use the same frame for each one: the standard Cartesian frame of $(\mathbf{i}, \mathbf{j}, \mathbf{k}, O)$.

4.3.1   Translation

The most basic affine transformation is *translation*. For a single point, it's the same as adding a vector $\mathbf{t}$ to it, and when applied to an entire set of points it has the effect of moving them rigidly through space (Figure 4.1). Since all the points are shifted equally in space, the size and shape of the object will not change, so this is a rigid transformation.

We can determine the matrix for a translation by computing the transformation for each of the frame elements. For the origin O, this is

$$\begin{aligned} \mathcal{T}(O) &= \mathbf{t} + O \\ &= t_x \mathbf{i} + t_y \mathbf{j} + t_z \mathbf{k} + O \end{aligned}$$

For a given basis vector, we can find two points P and Q that define the vector and compute the transformation of their difference. For example, for $\mathbf{i}$:

$$\begin{aligned} \mathcal{T}(\mathbf{i}) &= \mathcal{T}(P - Q) \\ &= \mathcal{T}(P) - \mathcal{T}(Q) \\ &= (\mathbf{t} + P) - (\mathbf{t} + Q) \\ &= P - Q \\ &= \mathbf{i} \end{aligned}$$

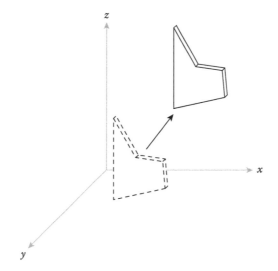

Figure 4.1. Translation.

The same holds true for **j** and **k**, so translation has no effect on the basis vectors in our frame. We end up with a 4×4 matrix:

$$\begin{bmatrix} 1 & 0 & 0 & t_x \\ 0 & 1 & 0 & t_y \\ 0 & 0 & 1 & t_z \\ 0 & 0 & 0 & 1 \end{bmatrix}$$

Or, in block form:

$$\mathbf{T_t} = \begin{bmatrix} \mathbf{I} & \mathbf{t} \\ \mathbf{0}^T & 1 \end{bmatrix}$$

Translation only affects points. To see why, suppose we have a vector **v**, which equals the displacement between two points P and Q, that is, $\mathbf{v} = P - Q$. If we translate $P - Q$, we get

$$\begin{aligned} \text{trans}(P - Q) &= (P + \mathbf{t}) - (Q + \mathbf{t}) \\ &= (P - Q) + (\mathbf{t} - \mathbf{t}) \\ &= \mathbf{v} \end{aligned}$$

This fits with our geometric notion that points have position and hence can be translated in space, while vectors do not and cannot.

We can use Equation 4.3 to compute the inverse translation transformation:

$$\mathbf{T_t}^{-1} = \begin{bmatrix} \mathbf{I}^{-1} & -\mathbf{I}^{-1}\mathbf{t} \\ \mathbf{0}^T & 1 \end{bmatrix} \tag{4.4}$$

$$= \begin{bmatrix} \mathbf{I} & -\mathbf{t} \\ \mathbf{0}^T & 1 \end{bmatrix} \tag{4.5}$$

$$= \mathbf{T_{-t}} \tag{4.6}$$

So, the inverse of a given translation negates the original translation vector to displace the point back to its original position.

4.3.2   Rotation

The other common rigid transformation is *rotation*. If we consider the rotation of a vector, we are rigidly changing its direction around an axis without changing its length. In $\mathbb{R}^2$, this is the same as replacing a vector with the one that's θ degrees counterclockwise (Figure 4.2).

In $\mathbb{R}^3$, we usually talk about an *axis of rotation*. In his rotation theorem, Euler showed that when applying a rotation in three-dimensional (3D) space, there is a linear set of points (i.e., a line) that does not change. This is called the axis of rotation, and the amount we rotate around this axis is the *angle of rotation*. A helpful mnemonic is the right-hand rule: if you point your right thumb in the direction of the axis vector, the curl of your fingers represents the direction of positive rotation (Figure 4.3).

For a given point, we rotate it by moving it along a planar arc a constant distance from another point, known as the *center of rotation* (Figure 4.4). This center of rotation is commonly defined as the origin of the current frame (we'll refer to this as a *pure rotation*) but can be any arbitrary point. We can think of this as defining a vector **v** from the center of rotation to the point to be rotated, rotating **v**, and then adding the result to the center of rotation to compute the new position of the point. For now we'll only cover pure rotations; applying general affine transformations about an arbitrary center will be discussed later.

To keep things simple, we'll begin with rotations around one of the three frame axes, with a center of rotation equal to the origin. The following system of equations rotates a

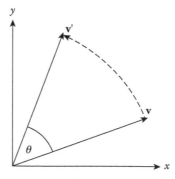

Figure 4.2. Rotation of vector in $\mathbb{R}^2$.

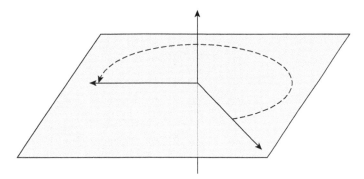

Figure 4.3. Axis and plane of rotation.

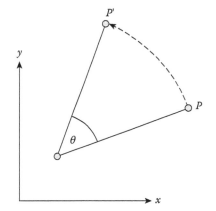

Figure 4.4. Rotation of point in $\mathbb{R}^2$.

vector or point counterclockwise (assuming the axis is pointing at us) around **k**, or the z-axis (Figure 4.5c):

$$x' = x \cos \theta - y \sin \theta$$
$$y' = x \sin \theta + y \cos \theta \qquad (4.7)$$
$$z' = z$$

Figure 4.6 shows why this works. Since we're rotating around the z-axis, no z values will change, so we will consider only how the rotation affects the xy values of the points. The starting position of the point is (x, y), and we want to rotate that θ degrees counterclockwise. Handling this in Cartesian coordinates can be problematic, but this is one case where polar coordinates are useful.

Recall that a point P in polar coordinates has representation (r, ϕ), where r is the distance from the origin and ϕ^1 is the counterclockwise angle from the x-axis. We can think of this

---

[1] We're using ϕ for polar coordinates in this case to distinguish it from the rotation angle θ.

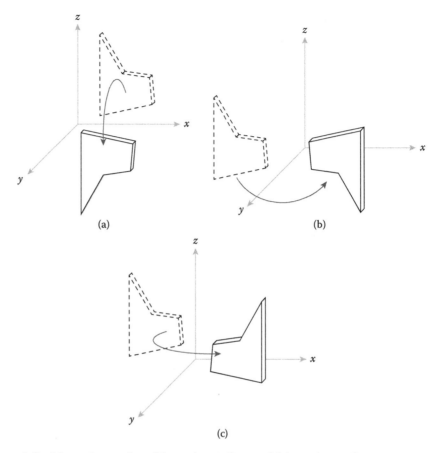

Figure 4.5. (a) *x*-axis rotation, (b) *y*-axis rotation, and (c) *z*-axis rotation.

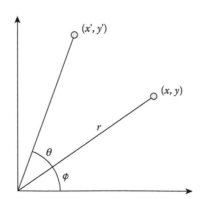

Figure 4.6. Rotation in *xy* plane.

as rotating an r length radius lying along the x-axis by ϕ degrees. If we rotate this a further θ degrees, the end of the radius will be at $(r, \phi + \theta)$ (in polar coordinates). Converting to Cartesian coordinates, the final point will lie at

$$x' = r\cos(\phi + \theta)$$
$$y' = r\sin(\phi + \theta)$$

Using trigonometric identities, this becomes

$$x' = r\cos\phi\cos\theta - r\sin\phi\sin\theta$$
$$y' = r\cos\phi\sin\theta + r\sin\phi\cos\theta$$

But $r\cos\phi = x$, and $r\sin\phi = y$, so we can substitute and get

$$x' = x\cos\theta - y\sin\theta$$
$$y' = x\sin\theta + y\cos\theta$$

We can derive similar equations for rotation around the x-axis (Figure 4.5a):

$$x' = x$$
$$y' = y\cos\theta - z\sin\theta$$
$$z' = y\sin\theta + z\cos\theta$$

and rotation around the y-axis (Figure 4.5b):

$$x' = z\sin\theta + x\cos\theta$$
$$y' = y$$
$$z' = z\cos\theta - x\sin\theta$$

To create the corresponding transformation, we need to determine how the frame elements are transformed. The frame's origin will not change since it's our center of rotation, so $\mathbf{y} = \mathbf{0}$. Therefore, our primary concern will be the contents of the 3×3 matrix $\mathbf{A}$.

For this matrix, we need to compute where $\mathbf{i}, \mathbf{j}$, and $\mathbf{k}$ will go. For example, for rotations around the z-axis we can transform $\mathbf{i}$ to get

$$x' = (1)\cos\theta - (0)\sin\theta = \cos\theta$$
$$y' = (1)\sin\theta + (0)\cos\theta = \sin\theta$$
$$z' = 0$$

Transforming $\mathbf{j}$ and $\mathbf{k}$ similarly and copying the results into the columns of a 3×3 matrix gives

$$\mathbf{R}_z = \begin{bmatrix} \cos\theta & -\sin\theta & 0 \\ \sin\theta & \cos\theta & 0 \\ 0 & 0 & 1 \end{bmatrix}$$

Similar matrices can be created for rotation around the x-axis:

$$\mathbf{R}_x = \begin{bmatrix} 1 & 0 & 0 \\ 0 & \cos\theta & -\sin\theta \\ 0 & \sin\theta & \cos\theta \end{bmatrix}$$

and around the y-axis:

$$\mathbf{R}_y = \begin{bmatrix} \cos\theta & 0 & \sin\theta \\ 0 & 1 & 0 \\ -\sin\theta & 0 & \cos\theta \end{bmatrix}$$

One thing to note about these matrices is that their determinants are equal to 1, and they are all orthogonal. For example, look at the component 3-vectors of the z-axis rotation matrix. We have $(\cos\theta, \sin\theta, 0)$, $(-\sin\theta, \cos\theta, 0)$, and $(0, 0, 1)$. The first two lie on the xy plane and so are perpendicular to the third, and they are perpendicular to each other. All three are unit length and so form an orthonormal basis.

The product of two orthogonal matrices is also an orthogonal matrix; thus, the product of a series of pure rotation matrices is also a rotation matrix. For example, by concatenating matrices that rotate around the z-axis, then the y-axis, and then the x-axis, we can create one form of a generalized rotation matrix:

$$\mathbf{R}_x\mathbf{R}_y\mathbf{R}_z = \begin{bmatrix} CyCz & -CySz & Sy \\ SxSyCz+CxSz & -SxSySz+CxCz & -SxCy \\ -CxSyCz+SxSz & CxSySz+SxCz & CxCy \end{bmatrix} \tag{4.8}$$

where

$$\begin{aligned} Cx &= \cos\theta_x & Sx &= \sin\theta_x \\ Cy &= \cos\theta_y & Sy &= \sin\theta_y \\ Cz &= \cos\theta_z & Sz &= \sin\theta_z \end{aligned}$$

Recall that the inverse of an orthogonal matrix is its transpose. Because pure rotation matrices are orthogonal, the inverse of any rotation matrix is also its transpose. Therefore, the inverse of the z-axis rotation, centered on the origin, is

$$\mathbf{R}_z^{-1} = \begin{bmatrix} \cos\theta & \sin\theta & 0 \\ -\sin\theta & \cos\theta & 0 \\ 0 & 0 & 1 \end{bmatrix}$$

This follows if we think of the inverse transformation as "undoing" the original transformation. If you substitute $-\theta$ for θ in the original matrix and replace $\cos(-\theta)$ with $\cos\theta$ and $\sin(-\theta)$ with $-\sin\theta$, then we have

$$\begin{bmatrix} \cos(-\theta) & -\sin(-\theta) & 0 \\ \sin(-\theta) & \cos(-\theta) & 0 \\ 0 & 0 & 1 \end{bmatrix} = \begin{bmatrix} \cos\theta & \sin\theta & 0 \\ -\sin\theta & \cos\theta & 0 \\ 0 & 0 & 1 \end{bmatrix}$$

which, as we can see, results in the immediately preceding inverse matrix.

Now that we have looked at rotations around the coordinate axes, we will consider rotations around an arbitrary axis. The formula for a rotation of a vector $\mathbf{v}$ by an angle θ around a general axis $\hat{\mathbf{r}}$ is derived as follows. We begin by breaking $\mathbf{v}$ into two parts: the part parallel with $\hat{\mathbf{r}}$ and the part perpendicular to it, which lies on the plane of rotation (Figure 4.7a). Recall from Chapter 1 that the parallel part $\mathbf{v}_\parallel$ is the projection of $\mathbf{v}$ onto $\hat{\mathbf{r}}$, or

$$\mathbf{v}_\parallel = (\mathbf{v} \cdot \hat{\mathbf{r}})\hat{\mathbf{r}}$$

The perpendicular part is what remains of $\mathbf{v}$ after we subtract the parallel part, or

$$\mathbf{v}_\perp = \mathbf{v} - (\mathbf{v} \cdot \hat{\mathbf{r}})\hat{\mathbf{r}}$$

To properly compute the effect of rotation, we need to create a two-dimensional (2D) basis on the plane of rotation (Figure 4.7b). We'll use $\mathbf{v}_\perp$ as our first basis vector, and we'll need a vector $\mathbf{w}$ perpendicular to it for our second basis vector. We can take the cross product with $\hat{\mathbf{r}}$ for this:

$$\mathbf{w} = \hat{\mathbf{r}} \times \mathbf{v}_\perp = \hat{\mathbf{r}} \times \mathbf{v}$$

In the standard basis for $\mathbb{R}^2$, if we rotate the vector $\mathbf{i} = (1, 0)$ by θ, we get the vector $(\cos\theta, \sin\theta)$. Equivalently,

$$R\mathbf{i} = (\cos\theta)\mathbf{i} + (\sin\theta)\mathbf{j}$$

If we use $\mathbf{v}_\perp$ and $\mathbf{w}$ as the 2D basis for the rotation plane, we can find the rotation of $\mathbf{v}_\perp$ by θ in a similar manner:

$$R\mathbf{v}_\perp = (\cos\theta)\mathbf{v}_\perp + (\sin\theta)\mathbf{w}$$

The parallel part of $\mathbf{v}$ doesn't change with the rotation, so the final result of rotating $\mathbf{v}$ around $\hat{\mathbf{r}}$ by θ is

$$\begin{aligned} R\mathbf{v} &= R\mathbf{v}_\parallel + R\mathbf{v}_\perp \\ &= R\mathbf{v}_\parallel + (\cos\theta)\mathbf{v}_\perp + (\sin\theta)\mathbf{w} \\ &= (\mathbf{v} \cdot \hat{\mathbf{r}})\hat{\mathbf{r}} + \cos\theta[\mathbf{v} - (\mathbf{v} \cdot \hat{\mathbf{r}})\hat{\mathbf{r}}] + \sin\theta(\hat{\mathbf{r}} \times \mathbf{v}) \\ &= \cos\theta\mathbf{v} + [1 - \cos\theta](\mathbf{v} \cdot \hat{\mathbf{r}})\hat{\mathbf{r}} + \sin\theta(\hat{\mathbf{r}} \times \mathbf{v}) \end{aligned} \tag{4.9}$$

This is one form of what is known as the *Rodrigues formula*.

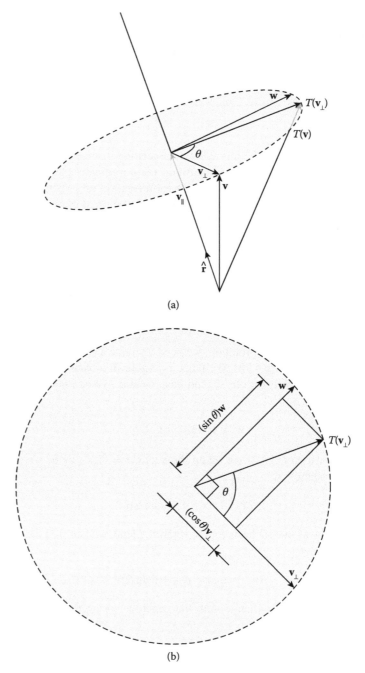

Figure 4.7. General rotation: (a) showing axis of rotation and rotation plane and (b) showing vectors on rotation plane.

The projection $(\mathbf{v} \cdot \hat{\mathbf{r}})\hat{\mathbf{r}}$ can be replaced by the tensor product $(\hat{\mathbf{r}} \otimes \hat{\mathbf{r}})\mathbf{v}$. Similarly, the cross product $\hat{\mathbf{r}} \times \mathbf{v}$ can be replaced by a multiplication by a skew symmetric matrix $\tilde{\mathbf{r}}\mathbf{v}$. This gives

$$
\begin{aligned}
R\mathbf{v} &= \cos\theta \mathbf{v} + (1 - \cos\theta)(\hat{\mathbf{r}} \otimes \hat{\mathbf{r}})\mathbf{v} + \sin\theta \tilde{\mathbf{r}}\mathbf{v} \\
&= [\cos\theta \mathbf{I} + (1 - \cos\theta)(\hat{\mathbf{r}} \otimes \hat{\mathbf{r}}) + \sin\theta \tilde{\mathbf{r}}]\mathbf{v}
\end{aligned}
$$

Expanding the terms, we end up with a matrix:

$$
\mathbf{R}_{\hat{\mathbf{r}}\theta} = \begin{bmatrix} tx^2 + c & txy - sz & txz + sy \\ txy + sz & ty^2 + c & tyz - sx \\ txz - sy & tyz + sx & tz^2 + c \end{bmatrix}
$$

where

$$
\begin{aligned}
\hat{\mathbf{r}} &= (x, y, z) \\
c &= \cos\theta \\
s &= \sin\theta \\
t &= 1 - \cos\theta
\end{aligned}
$$

As we can see, there is a wide variety of choices for the 3×3 matrix $\mathbf{A}$, depending on what sort of rotation we wish to perform. The full affine matrix for rotation around the origin is

$$
\begin{bmatrix} \mathbf{R} & \mathbf{0} \\ \mathbf{0}^T & 1 \end{bmatrix}
$$

where $\mathbf{R}$ is one of the rotation matrices just given. For example, the affine matrix for rotation around the x-axis is

$$
\begin{bmatrix} \mathbf{R}_x & \mathbf{0} \\ \mathbf{0}^T & 1 \end{bmatrix} = \begin{bmatrix} 1 & 0 & 0 & 0 \\ 0 & \cos\theta & -\sin\theta & 0 \\ 0 & \sin\theta & \cos\theta & 0 \\ 0 & 0 & 0 & 1 \end{bmatrix}
$$

This is also an orthogonal matrix and its inverse is the transpose, as before.

Finally, when discussing rotations one has to be careful to distinguish rotation from orientation, which is to rotation as position is to translation. If we consider the representation of a point in an affine space,

$$
P = \mathbf{v} + O
$$

then we can think of the origin as a reference position and the vector $\mathbf{v}$ as a translation that relates our position to the reference. We can represent our position as just the components of the translation. Similarly, we can define a reference orientation Ω_0, and any orientation Ω is related to it by a rotation, or

$$\Omega = \mathbf{R}_0 \Omega_0$$

Just as we might use the components of the vector $\mathbf{v}$ to represent our position, we can use the rotation $\mathbf{R}_0$ to represent our orientation. To change our orientation, we apply an additional rotation, just as we might add a translation vector to change our position:

$$\Omega' = \mathbf{R}_1 \Omega$$

In this case, our final orientation, using the rotation component, is

$$\mathbf{R}_1 \mathbf{R}_0$$

Remember that the order of concatenation matters, because matrix multiplication—particularly for rotation matrices—is not a commutative operation.

4.3.3   Scaling

The remaining affine transformations that we will cover are *deformations*, since they don't preserve exact lengths or angles. The first is *scaling*, which can be thought of as corresponding to our other basic vector operation, scalar multiplication; however, it is not quite the same. Scalar multiplication of a vector has only one multiplicative factor and changes a vector's length equally in all directions. We can also multiply a vector by a negative scalar. In comparison, scaling as it is commonly used in computer graphics applies a possibly different but positive factor to each basis vector in our frame.[2] If all the factors are equal, then it is called uniform scaling and is—for vectors in the affine space—equivalent to scalar multiplication by a single positive scalar. Otherwise, it is called nonuniform scaling. Full nonuniform scaling can be applied differently in each axis direction, so we can scale by 2 in z to make an object twice as tall, but $1/2$ in x and y to make it half as wide.

A point doesn't have a length per se, so instead we change its relative distance from another point C_s, known as the center of scaling. We can consider this as scaling the vector from the center of scaling to our point P. For a set of points, this will end up scaling their distance relative to each other, but still maintaining the same relative shape (Figure 4.8).

For now we'll consider only scaling around the origin, so $C_s = O$ and $\mathbf{y} = \mathbf{0}$. For the upper 3×3 matrix $\mathbf{A}$, we again need to determine how the frame basis vectors change, which is defined as

$$T(\mathbf{i}) = a\mathbf{i}$$
$$T(\mathbf{j}) = b\mathbf{j}$$
$$T(\mathbf{k}) = c\mathbf{k}$$

---

[2] We'll consider negative factors when we discuss reflections in the following section.

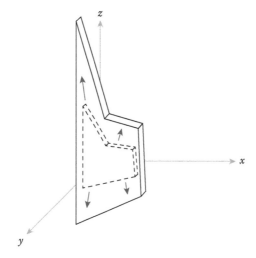

Figure 4.8. Nonuniform scaling.

where a, b, $c > 0$ and are the scale factors in the x, y, z directions, respectively. Writing these transformed basis vectors as the columns of $\mathbf{A}$, we get an affine matrix of

$$\mathbf{S}_{abc} = \begin{bmatrix} a & 0 & 0 & 0 \\ 0 & b & 0 & 0 \\ 0 & 0 & c & 0 \\ 0 & 0 & 0 & 1 \end{bmatrix}$$

This is a diagonal matrix, with the positive scale factors lying along the diagonal, so the inverse is

$$\mathbf{S}_{abc}^{-1} = \mathbf{S}_{\frac{1}{a}\frac{1}{b}\frac{1}{c}} = \begin{bmatrix} 1/a & 0 & 0 & 0 \\ 0 & 1/b & 0 & 0 \\ 0 & 0 & 1/c & 0 \\ 0 & 0 & 0 & 1 \end{bmatrix}$$

4.3.4 Reflection

The reflection transformation symmetrically maps an object across a plane or through a point. One possible reflection is (Figure 4.9a)

$$x' = -x$$
$$y' = y$$
$$z' = z$$

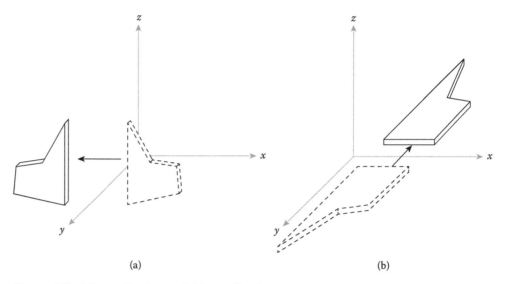

Figure 4.9. (a) yz reflection and (b) xz reflection.

This reflects across the yz plane and gives an effect like a standard mirror (mirrors don't swap left to right, they swap front to back). If we want to reflect across the xz plane instead, we would use (Figure 4.9b)

$$x' = x$$
$$y' = -y$$
$$z' = z$$

As one might expect, we can create a planar reflection that reflects across a general plane, defined by a normal $\hat{\mathbf{n}}$ and a point on the plane P_0. For now we'll consider only planes that pass through the origin. If we have a vector $\mathbf{v}$ in our affine space, we can break it into two parts relative to the plane normal: the orthogonal part $\mathbf{v}_\perp$, which will remain unchanged, and parallel part $\mathbf{v}_\parallel$, which will be reflected to the other side of the plane to become $-\mathbf{v}_\parallel$. The transformed vector will be the sum of $\mathbf{v}_\perp$ and the reflected $-\mathbf{v}_\parallel$ (Figure 4.10).

To compute $\mathbf{v}_\parallel$, we merely have to take the projection of $\mathbf{v}$ against the plane normal $\hat{\mathbf{n}}$, or

$$\mathbf{v}_\parallel = (\mathbf{v} \cdot \hat{\mathbf{n}})\hat{\mathbf{n}} \tag{4.10}$$

Subtracting this from $\mathbf{v}$, we can compute $\mathbf{v}_\perp$:

$$\mathbf{v}_\perp = \mathbf{v} - \mathbf{v}_\parallel \tag{4.11}$$

We know that the transformed vector will be $\mathbf{v}_\perp - \mathbf{v}_\parallel$. Substituting Equations 4.10 and 4.11 into this gives us

$$\begin{aligned}
\mathcal{T}(\mathbf{v}) &= \mathbf{v}_\perp - \mathbf{v}_\parallel \\
&= \mathbf{v} - 2\mathbf{v}_\parallel \\
&= \mathbf{v} - 2(\mathbf{v} \cdot \hat{\mathbf{n}})\hat{\mathbf{n}}
\end{aligned}$$

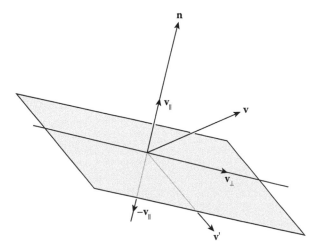

Figure 4.10. General reflection.

From Chapter 2, we know that we can perform the projection of $\mathbf{v}$ on $\hat{\mathbf{n}}$ by multiplying by the tensor product matrix $\hat{\mathbf{n}} \otimes \hat{\mathbf{n}}$, so this becomes

$$\mathcal{T}(\mathbf{v}) = \mathbf{v} - 2(\hat{\mathbf{n}} \otimes \hat{\mathbf{n}})\mathbf{v}$$
$$= [\mathbf{I} - 2(\hat{\mathbf{n}} \otimes \hat{\mathbf{n}})]\mathbf{v}$$

Thus, the linear transformation part $\mathbf{A}$ of our affine transformation is $[\mathbf{I} - 2(\hat{\mathbf{n}} \otimes \hat{\mathbf{n}})]$. Writing this as a block matrix, we get

$$\mathbf{F_n} = \left[\begin{array}{cc} \mathbf{I} - 2(\hat{\mathbf{n}} \otimes \hat{\mathbf{n}}) & \mathbf{0} \\ \mathbf{0}^T & 1 \end{array} \right]$$

While in the real world we usually see planar reflections, in our virtual world we can also compute a reflection through a point. The following performs a reflection through the origin (Figure 4.11):

$$x' = -x$$
$$y' = -y$$
$$z' = -z$$

The corresponding block matrix is

$$\mathbf{F}_O = \left[\begin{array}{cc} -\mathbf{I} & \mathbf{0} \\ \mathbf{0}^T & 1 \end{array} \right]$$

Reflections are a symmetric operation; that is, the reflection of a reflection returns the original point or vector. Because of this, the inverse of a reflection matrix is the matrix itself.

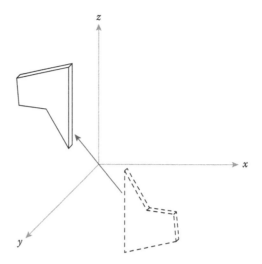

Figure 4.11. Point reflection.

As an aside, we would (incorrectly) expect that if we can reflect through a plane and a point, we can reflect through a line. The system

$$x' = -x$$
$$y' = -y$$
$$z' = z$$

appears to reflect through the z-axis, giving a "funhouse mirror" effect, where right and left are swapped (if y is left, it becomes $-y$ in the reflection, and so ends up on the right side). However, if we examine the transformation closely, we see that while it does perform the desired effect, this is actually a rotation of 180 degrees around the z-axis. While both pure rotations and pure reflections through the origin are orthogonal matrices, we can distinguish between them by noting that reflection matrices have a determinant of -1, while rotation matrices have a determinant of 1.

4.3.5 Shear

The final affine transformation that we will cover is shear (Figure 4.12). Because it affects the angles of objects it is not used all that often, but we will use it particularly when discussing oblique projections in Chapter 7. An axis-aligned shear provides a shift in one or two axes proportional to the component in a third axis. One often sees shear in buildings that have been affected by an earthquake—the bottom of the building remains fixed, but the sides have shifted so that they're now diagonal to the ground. Transforming a square to a rhombus or a cube to a rhomboid solid is a shear transformation.

There are a number of ways of specifying shear [1, 133]. In our case, we will define a shear plane, with normal $\hat{\mathbf{n}}$, that does not change due to the transformation. We define an orthogonal shear vector **s**, which indicates how planes parallel to the shear plane will be transformed. Points on the plane 1 unit of distance from the shear plane, in the direction

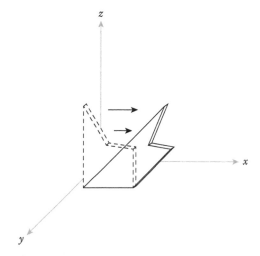

Figure 4.12. z-shear on shape.

of the plane normal, will be displaced by **s**. Points on the plane 2 unit of distance from the shear plane will be displaced by 2**s**, and so on. In general, if we take a point P and define it as $P_0 + \mathbf{v}$, where P_0 is a point on the shear plane, then P will be displaced by $(\hat{\mathbf{n}} \bullet \mathbf{v})\mathbf{s}$.

The simplest case is when we apply shear perpendicular to one of the main coordinate axes. For example, if we take the yz plane as our shear plane, our normal is **i** and the shear plane passes through the origin O. We know from this that O will not change with the transformation, so our translation vector **y** is **0**. As before, to find **A** we need to figure out how the transformation affects our basis vectors. If we define **j** as $P_1 - O$, then

$$\mathcal{T}(\mathbf{j}) = \mathcal{T}(P_1) - \mathcal{T}(O)$$

But P_1 and O lie on the shear plane, so

$$\mathcal{T}(\mathbf{j}) = P_1 - O$$
$$= \mathbf{j}$$

The same is true for the basis vector **k**. For **i**, we can define it as $P_0 - O$. We know that P_0 is distance 1 from the shear plane, so it will become $P_0 + \mathbf{s}$, so

$$\mathcal{T}(\mathbf{i}) = \mathcal{T}(P_0) - \mathcal{T}(O)$$
$$= P_0 + \mathbf{s} - O$$
$$= \mathbf{i} + \mathbf{s}$$

The vector **s** in this case is orthogonal to **i**; therefore, it is of the form $(0, a, b)$, so our transformed basis vector will be $(1, a, b)$. Our final matrix **A** is

$$\mathbf{H}_x = \begin{bmatrix} 1 & 0 & 0 \\ a & 1 & 0 \\ b & 0 & 1 \end{bmatrix}$$

We can go through a similar process to get shear by the y-axis:

$$\mathbf{H}_y = \begin{bmatrix} 1 & c & 0 \\ 0 & 1 & 0 \\ 0 & d & 1 \end{bmatrix}$$

and shear by the z-axis:

$$\mathbf{H}_z = \begin{bmatrix} 1 & 0 & e \\ 0 & 1 & f \\ 0 & 0 & 1 \end{bmatrix}$$

For shearing by a general plane through the origin, we already have the formula for the displacement: $(\hat{\mathbf{n}} \bullet \mathbf{v})\mathbf{s}$. We can rewrite this as a tensor product to get $(\mathbf{s} \otimes \hat{\mathbf{n}})\mathbf{v}$. Because this is merely the displacement, we need to include the original point, and thus our origin-centered general shear matrix is simply $\mathbf{I} + \mathbf{s} \otimes \hat{\mathbf{n}}$. Our final shear matrix is

$$\mathbf{H}_{\hat{\mathbf{n}},\mathbf{s}} = \begin{bmatrix} \mathbf{I} + \mathbf{s} \otimes \hat{\mathbf{n}} & \mathbf{0} \\ \mathbf{0}^T & 1 \end{bmatrix}$$

The inverse shear transformation is shear in the opposite direction, so the corresponding matrix is

$$\mathbf{H}_{\hat{\mathbf{n}},\mathbf{s}}^{-1} = \begin{bmatrix} \mathbf{I} - \mathbf{s} \otimes \hat{\mathbf{n}} & \mathbf{0} \\ \mathbf{0}^T & 1 \end{bmatrix} = \mathbf{H}_{\hat{\mathbf{n}},-\mathbf{s}}$$

4.3.6 Applying an Affine Transformation around an Arbitrary Point

Up to this point, we have been assuming that our affine transformations are applied around the origin of the frame. For example, when discussing rotation we treated the origin as our center of rotation. Similarly, our shear planes were assumed to pass through the origin. This doesn't necessarily have to be the case.

Since we're no longer transforming around the origin, we need to consider how it is affected by our transformation. Let's look at a particular example—the rotation of a point around an arbitrary center of rotation C. If we look at Figure 4.13, we see the situation. We have a point C and our origin O. We want to rotate the difference vector $\mathbf{v} = O - C$ between the two points by matrix $\mathbf{R}$ and determine where the resulting point $\mathcal{T}(O)$, or $C + \mathcal{T}(\mathbf{v})$, will be. From that we can compute the difference vector $\mathbf{y} = \mathcal{T}(O) - O$. From Figure 4.13, we can see that $\mathbf{y} = \mathcal{T}(\mathbf{v}) - \mathbf{v}$, so we can reduce this as follows:

$$\begin{aligned} \mathbf{y} &= \mathcal{T}(\mathbf{v}) - \mathbf{v} \\ &= \mathbf{R}\mathbf{v} - \mathbf{v} \\ &= (\mathbf{R} - \mathbf{I})\mathbf{v} \end{aligned}$$

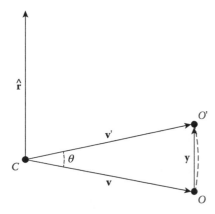

Figure 4.13. Rotation of origin around arbitrary center.

It's usually more convenient to write this in terms of the vector dual to C, which is $\mathbf{x} = C - O = -\mathbf{v}$, so this becomes

$$\mathbf{y} = -(\mathbf{R} - \mathbf{I})\mathbf{x}$$
$$= (\mathbf{I} - \mathbf{R})\mathbf{x}$$

We can achieve the same result by translating our center C to the frame origin by $-\mathbf{x}$, performing our origin-centered rotation, and then translating back by $\mathbf{x}$:

$$\mathbf{M}_c = \begin{bmatrix} \mathbf{I} & \mathbf{x} \\ \mathbf{0}^T & 1 \end{bmatrix} \begin{bmatrix} \mathbf{R} & \mathbf{0} \\ \mathbf{0}^T & 1 \end{bmatrix} \begin{bmatrix} \mathbf{I} & -\mathbf{x} \\ \mathbf{0}^T & 1 \end{bmatrix}$$

$$= \begin{bmatrix} \mathbf{R} & \mathbf{x} \\ \mathbf{0}^T & 1 \end{bmatrix} \begin{bmatrix} \mathbf{I} & -\mathbf{x} \\ \mathbf{0}^T & 1 \end{bmatrix}$$

$$= \begin{bmatrix} \mathbf{R} & (\mathbf{I} - \mathbf{R})\mathbf{x} \\ \mathbf{0}^T & 1 \end{bmatrix}$$

Notice that the upper left-hand block $\mathbf{R}$ is not affected by this process.

The same construction can be used for all affine transformations that use a center of transformation: rotation, scale, reflection, and shear. The exception is translation, since such an operation has no effect: $P - \mathbf{x} + \mathbf{t} + \mathbf{x} = P + \mathbf{t}$. But for the others, using a point $C = (\mathbf{x}, 1)$ as our arbitrary center of transformation gives

$$\mathbf{M}_c = \begin{bmatrix} \mathbf{A} & (\mathbf{I} - \mathbf{A})\mathbf{x} \\ \mathbf{0}^T & 1 \end{bmatrix}$$

where $\mathbf{A}$ is the upper 3×3 matrix of an origin-centered transformation. The corresponding inverse is

$$\mathbf{M}_c^{-1} = \begin{bmatrix} \mathbf{A}^{-1} & (\mathbf{I} - \mathbf{A}^{-1})\mathbf{x} \\ \mathbf{0}^T & 1 \end{bmatrix}$$

4.3.7 Transforming Plane Normals

As we saw in the previous section, if we want to transform a line or plane represented in parametric form, we transform the points in the affine combination. For example,

$$T(P(t)) = (1 - s - t)T(P_0) + sT(P_1) + tT(P_2)$$

But suppose we have a plane represented using the generalized plane equation. One way of considering this is as a plane normal (a, b, c) and a point on the plane P_0. We could transform these and try to use the resulting vector and point to build the new plane. However, if we apply an affine transform to the plane normal (a, b, c) directly, we may end up performing a deformation. Since angles aren't preserved under deformations, the resulting normal may no longer be orthogonal to the points in the plane.

The correct approach is as follows. We can represent the generalized plane equation as the product of a row matrix and column matrix, or

$$ax + by + cz + d = \begin{bmatrix} a & b & c & d \end{bmatrix} \begin{bmatrix} x \\ y \\ z \\ 1 \end{bmatrix}$$

$$= \mathbf{n}^T P$$

Now P is clearly a point, and $\mathbf{n}$ is the vector of coefficients for the plane. For points that lie on the plane,

$$\mathbf{n}^T P = 0$$

If we transform all the points on the plane by some matrix $\mathbf{M}$, then to maintain the relationship between $\mathbf{n}^T$ and P, we'll have to transform $\mathbf{n}$ by some unknown matrix $\mathbf{Q}$, or

$$(\mathbf{Q}\mathbf{n})^T (\mathbf{M}P) = 0$$

This can be rewritten as

$$\mathbf{n}^T \mathbf{Q}^T \mathbf{M} P = 0$$

One possible solution for this is if

$$\mathbf{I} = \mathbf{Q}^T \mathbf{M}$$

Solving for $\mathbf{Q}$ gives

$$\mathbf{Q} = \left(\mathbf{M}^{-1} \right)^T$$

So, the transformed plane coefficients become

$$\mathbf{n}' = \left(\mathbf{M}^{-1} \right)^T \mathbf{n}$$

The same approach will work if we're transforming the plane normal and point as described earlier. We transform the point P_0 by $\mathbf{M}$ and the normal by $(\mathbf{M}^{-1})^T$. Note that we've picked one particular solution out of many, so our transformed normal will be orthogonal to the transformed plane but likely no longer normalized. If the result is required to be unit length, we must be sure to normalize it afterwards.

In many cases the inverse matrix $\mathbf{M}^{-1}$ may not exist. So, if we're just transforming a normal vector (a, b, c), we can use a different method. Instead of $\mathbf{M}^{-1}$, we use the adjoint matrix from Cramer's rule. Normally we couldn't proceed at this point: if the inverse doesn't exist, we end up dividing by a zero determinant. However, even when the inverse exists, the division by the determinant is a scale factor. So, we can ignore it in all cases and just use the adjoint matrix directly, because we're going to normalize the resulting vector anyway.

4.4 Using Affine Transformations

4.4.1 Manipulation of Game Objects

The primary use of affine transformations is for the manipulation of objects in our game world. Suppose, from our earlier hypothetical example, we have an office environment that is acting as our game space. The artists could build the basic level—the walls, the floor, the ceilings, and so forth—as a single set of triangles with coordinates defined to place them exactly where we might want them in the world. However, suppose we have a single desk model that we want to duplicate and place in various locations in the level. The artist could build a new version of the desk for each location in the core-level geometry, but that would involve unnecessarily duplicating all the memory needed for the model. Instead, we could have one version, or *master*, of the desk model and then set a series of transformations that indicate where in the level each copy, or *instance*, of the desk should be placed [142].

Before we can begin to discuss how we specify these transformations and what they might mean, we need to define the two different coordinate frames we are working in: the local coordinate frame and the world coordinate frame.

4.4.1.1 Local and World Coordinate Frames

When artists create an object or we create an object directly in a program, the coordinates of the points that make up that object are defined in that particular object's *local frame*. This is also commonly known as *local space*. In addition, often the frame is named after the object itself, so you might also see terms like *model space* or *camera space*.

The orientation of the basis vectors in the local frame is usually set so that the engineers know which part of the object is the front, which is the top, and which is the side. This allows us to orient the object correctly relative to the rest of the world and to translate it in the correct direction if we want to move it forward. The convention that we will be using in this book is one where the x-axis points along the forward direction of the object, the y-axis points toward the left of the object, and the z-axis points out the top of the object (Figure 4.14). Another common convention is to use the y-axis for up, the z-axis for forward, and the x-axis for out to either the left or the right, depending on whether we want to work in a right- or left-handed frame.

Typically, the origin of the frame is placed in a position convenient for the game, either at the center or at the bottom of the object. The first is useful when we want to rotate objects around their centers, the second for placement on the ground.

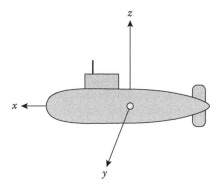

Figure 4.14. Local object frame.

When constructing our world, we define a specific coordinate frame, or *world frame*, also known as *world space*. The world frame acts as a common reference among all the objects, much as the origin acts as a common reference among points. Ultimately, in order to render, simulate, or otherwise interact with objects, we will need to transform their local coordinates into the world frame.

When an artist builds the level geometry, the coordinates are usually set in the world frame. Orientation of the level relative to our world frame is set by convention. Knowing which direction is "up" is important in a 3D game; in our case, we'll be using the z-axis, but the y-axis is also commonly used. Aligning the level to the other two axes (in our case, x and y) is arbitrary, but if our level is either gridlike or box shaped, it is usually convenient to orient the grid lines or box sides to these remaining axes.

Positioning the level relative to the origin of the frame is also arbitrary but is usually set so that the origin lies in the center of a box defining our maximum play area. This helps avoid precision problems, since floating-point precision is centered around 0 (see Chapter 1). For example, we might have a 300 m by 300 m play area, so that in the xy directions the origin will lie directly in the center. While we can set things so that the origin is centered in z as well, we may want to adjust that depending on our application. If our game mainly takes place on a flat play area, such as in an arena fighting game, we might set the floor so that it lies at the origin; this will make it simple to place objects and characters exactly at floor level. In a submarine game, we might place sea level at the origin; negative z lies under the waterline and positive z above.

4.4.1.2 Placing Objects

If we were to use the objects' local coordinates directly in the world frame, they would end up interpenetrating and centered around the world origin. To avoid that situation, we apply affine transformations to each object to place them at their own specific position and orientation in the world. For each object, this is known as its particular *local-to-world transformation*. We often display the relative position and orientation of a particular object in the world by drawing its frame relative to the world frame (Figure 4.15). The local-to-world transformation, or world transformation for short, describes this relative relationship: the column vectors of the local-to-world matrix **A** describe where the local

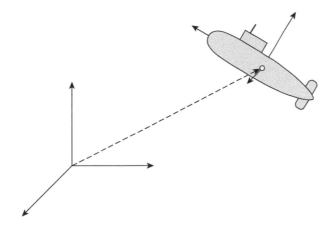

Figure 4.15. Local-to-world transformation.

frame's basis vectors will lie relative to the world space basis, and the vector **y** describes where the local frame's origin lies relative to the world origin.

The most commonly used affine transformations for object placement are translation, rotation, and scaling. Translation and rotation are convenient for two reasons. First, they correspond naturally to two of the characteristics we want to control in our objects: position and orientation. Second, they are rigid transformations, meaning they don't affect the size or shape of our object, which is generally the desired effect. Scaling is a deformation but is commonly useful to change the size of objects. For example, if two artists build two objects but fail to agree on a relative measure of size, you might end up with a table bigger than a room, if placed directly in the level. Rather than have the artist redo the model, we can use scaling to make it appear smaller. Scaling is also useful in fantastical games to either shrink a character to fit in a small space or grow a character to be more imposing. However, for most games you can actually get away with not using scaling at all. Objects are often modeled at world space scale; for example, in our office environment example, desks may be placed at different orientations and locations, but they are all the same size. A meter in object space equals a meter in world space.

To create the final world transformation, we'll be concatenating a sequence of these translation, rotation, and scaling transformations together. However, remember that concatenation of transformations is not commutative. So, the order in which we apply our transformations affects the final result, sometimes in surprising ways. One basic example is transforming the point $(0, 0, 0)$. A pure rotation around the origin has no effect on $(0, 0, 0)$, so rotating by 90 degrees around z and then translating by (t_x, t_y, t_z) will just act as a translation, and we end up with (t_x, t_y, t_z). Translating the point first will transform it to (t_x, t_y, t_z), so in this case a subsequent rotation of 90 degrees around z will have an effect, with the final result of $(-t_y, t_x, t_z)$. As another example, look at Figure 4.16a, which shows a rotation and translation. Figure 4.16b shows the equivalent translation and rotation.

Nonuniform scaling and rotation are also noncommutative. If we first scale $(1, 0, 0)$ by (s_x, s_y, s_z), we get the point $(s_x, 0, 0)$. Rotating this by 90 degrees around z, we end up with $(0, s_x, 0)$. Reversing the transformation order, if we rotate $(1, 0, 0)$ by 90 degrees around z, we get the point $(0, 1, 0)$. Scaling this by (s_x, s_y, s_z), we get the point $(0, s_y, 0)$. Note that in

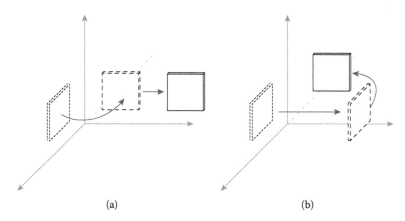

Figure 4.16. (a) Rotation, then translation and (b) translation, then rotation.

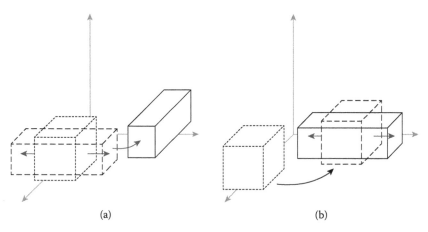

Figure 4.17. (a) Scale, then rotation and (b) rotation, then scale.

the second case we rotated our object so that our original x-axis lies along the y-axis and then applied our scale, giving us the unexpected result. Figure 4.17a and b shows another example of this applied to an object.

The final combination is scaling and translation. Again, this is not commutative. Remember that pure scaling is applied from the origin of the frame. If we translate an object from the origin and then scale, there will be additional scaling done to the translation of the object. So, for example, if we scale $(1, 1, 1)$ by (s_x, s_y, s_z) and then translate by (t_x, t_y, t_z), we end up with $(t_x + s_x, t_y + s_y, t_z + s_z)$. If instead we translate first, we get $(t_x + 1, t_y + 1, t_z + 1)$, and then scaling gives us $(s_x t_x + s_x, s_y t_y + s_y, s_z t_z + s_z)$. Another example can be seen in Figure 4.18a and b.

Generally, the desired order we wish to use for these transforms is to scale first, then rotate, then translate. Scaling first gives us the scaling along the axes we expect. We can

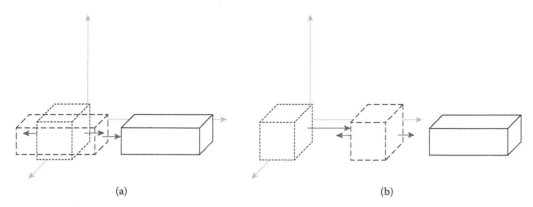

Figure 4.18. (a) Scale, then translation and (b) translation, then scale.

then rotate around the origin of the frame, and then translate it into place. This gives us the following multiplication order:

$$\mathbf{M = TRS}$$

4.4.2   Matrix Decomposition

It is sometimes useful to break an affine transformation matrix into its component basic affine transformations. This is called *matrix decomposition*. We performed one such decomposition when we pulled the translation information out of the matrix, effectively representing our transformation as the product of two matrices:

$$\begin{bmatrix} \mathbf{A} & \mathbf{y} \\ \mathbf{0}^T & 1 \end{bmatrix} = \begin{bmatrix} \mathbf{I} & \mathbf{y} \\ \mathbf{0}^T & 1 \end{bmatrix} \begin{bmatrix} \mathbf{A} & \mathbf{0} \\ \mathbf{0}^T & 1 \end{bmatrix}$$

Suppose we continue the process and break down $\mathbf{A}$ into the product of more basic affine transformations. For example, if we're using only scaling, rotation, and translation, it would be ideal if we could break $\mathbf{A}$ into the product of a scaling and rotation matrix. If we know for a fact that $\mathbf{A}$ is the product of only a scaling and rotation matrix, in the order $\mathbf{RS}$, we can multiply it out to get

$$\begin{bmatrix} r_{11} & r_{12} & r_{13} & 0 \\ r_{21} & r_{22} & r_{23} & 0 \\ r_{31} & r_{32} & r_{33} & 0 \\ 0 & 0 & 0 & 1 \end{bmatrix} \begin{bmatrix} s_x & 0 & 0 & 0 \\ 0 & s_y & 0 & 0 \\ 0 & 0 & s_z & 0 \\ 0 & 0 & 0 & 1 \end{bmatrix} = \begin{bmatrix} s_x r_{11} & s_y r_{12} & s_z r_{13} & 0 \\ s_x r_{21} & s_y r_{22} & s_z r_{23} & 0 \\ s_x r_{31} & s_y r_{32} & s_z r_{33} & 0 \\ 0 & 0 & 0 & 1 \end{bmatrix}$$

In this case, the lengths of the first three column vectors will give our three scale factors, s_x, s_y, and s_z. To get the rotation matrix, all we need to do is normalize those three vectors.

Unfortunately, it isn't always that simple. As we'll see in Section 4.5, often we'll be concatenating a series of **TRS** transformations to get something like

$$\mathbf{M} = \mathbf{T}_n\mathbf{R}_n\mathbf{S}_n \cdots \mathbf{T}_1\mathbf{R}_1\mathbf{S}_1\mathbf{T}_0\mathbf{R}_0\mathbf{S}_0$$

In this case, even ignoring the translations, it is impossible to decompose **M** into the form **RS**. As a quick example, suppose that all these transformations with the exception of $\mathbf{S}_1$ and $\mathbf{R}_0$ are the identity transformation. This simplifies to

$$\mathbf{M} = \mathbf{S}_1\mathbf{R}_0$$

Now suppose $\mathbf{S}_1$ scales by 2 along y and by 1 along x and z, and $\mathbf{R}_0$ rotates by 60 degrees around z. Figure 4.19 shows how this affects a square on the xy plane. The sides of the transformed square are no longer perpendicular. Somehow, we have ended up applying a shear within our transformation, and clearly we cannot represent this by a simple concatenation **RS**.

One solution is to decompose the matrix using a technique known as *singular value decomposition*, or simply SVD. Assuming no translation, the matrix **M** can be represented by three matrices **L**, **D**, and **R**, where **L** and **R** are orthogonal matrices, **D** is a diagonal matrix with nonnegative entries, and

$$\mathbf{M} = \mathbf{LDR}$$

An alternative formulation to this is *polar decomposition*, which breaks the nontranslational part of the matrix into two pieces: an orthogonal matrix **Q** and a stretch matrix **S**, where

$$\mathbf{S} = \mathbf{U}^T\mathbf{KU}$$

Matrix **U** in this case is another orthogonal matrix, and **K** is a diagonal matrix. The stretch matrix combines the scale-plus-shear effect we saw in our example: it rotates the frame

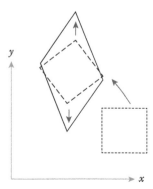

Figure 4.19. Effect of rotation, then scale.

to an orientation, scales along the axes, and then rotates back. Using this, a general affine matrix can be broken into four transformations:

$$\mathbf{M} = \mathbf{TRNS}$$

where $\mathbf{T}$ is a translation matrix, $\mathbf{Q}$ has been separated into a rotation matrix $\mathbf{R}$ and a reflection matrix $\mathbf{N} = \pm\mathbf{I}$, and $\mathbf{S}$ is the preceding stretch matrix.

Performing either SVD or polar decomposition is out of the purview of this text. As we'll see, there are ways to avoid matrix decomposition at the cost of some conversion before we send our models down the graphics pipeline. However, at times we may get a matrix of unknown structure from a library module that we don't control. For example, we could be using a commercial physics engine or writing a plug-in for a 3D modeling package such as Max or Maya. Most of the time a function is provided that will decompose such matrices for us, but this isn't always the case. For those times and for those who are interested in pursuing this topic, more information on decompositions can be found in Goldman [56], Golub and Van Loan [57], and Shoemake and Duff [138].

4.4.3 Avoiding Matrix Decomposition

In the preceding section, we made no assumptions about the values for our scaling factors. Now let's assume that each scaling matrix performs a uniform scale. Looking at just the rotation and scaling transformations, we have

Source Code
Demo
Centered

$$\mathbf{M} = \mathbf{R}_n\mathbf{S}_n \cdots \mathbf{R}_1\mathbf{S}_1\mathbf{R}_0\mathbf{S}_0$$

Since each scaling transformation is uniformly scaling, we can simplify this to

$$\mathbf{M} = \mathbf{R}_n\sigma_n \cdots \mathbf{R}_1\sigma_1\mathbf{R}_0\sigma_0$$

Using matrix algebra, we can shuffle terms to get

$$\mathbf{M} = \mathbf{R}_n \cdots \mathbf{R}_1\mathbf{R}_0\sigma_n \cdots \sigma_1\sigma_0$$
$$= \mathbf{R}\sigma$$
$$= \mathbf{RS}$$

where $\mathbf{R}$ is a rotation matrix and $\mathbf{S}$ is a uniform scaling matrix. So, if we use uniform scaling, we can in fact decompose our matrix into a rotation and scaling matrix, as we just did.

However, even in this case, the decomposition takes three square roots and nine scaling operations to perform. This leads to an alternate approach to handling transformations. Instead of storing transformations for our objects as a single 4×4 or even 3×4 matrix, we will break out the individual parts: a scale factor s, a 3×3 rotation matrix $\mathbf{R}$, and a translation vector $\mathbf{t}$. To apply this transformation to a point P, we use

Source Code
Demo
Separate

$$\mathcal{T}(P) = \left[\begin{array}{c} s\mathbf{R}\mathbf{x} + \mathbf{t} \\ 1 \end{array} \right]$$

Note the similarity to Equation 4.1. We've replaced $\mathbf{A}$ with $s\mathbf{R}$ and $\mathbf{y}$ with $\mathbf{t}$. In practice we ignore the trailing 1.

Concatenating transformations in matrix format is as simple as performing a multiplication. Concatenating in our alternate format is a little less straightforward but is not difficult and actually takes fewer operations on a standard floating-point processor:

$$\begin{aligned} s' &= s_1 s_0 \\ \mathbf{R}' &= \mathbf{R}_1 \mathbf{R}_0 \\ \mathbf{t}' &= \mathbf{t}_1 + s_1 \mathbf{R}_1 \mathbf{t}_0 \end{aligned} \qquad (4.12)$$

Computing the new scale and rotation makes a certain amount of sense, but it may not be clear why we don't add the two translations together to get the new translation. If we multiply the two transforms in matrix format, we have the following order:

$$\mathbf{M} = \mathbf{T}_1 \mathbf{R}_1 \mathbf{S}_1 \mathbf{T}_0 \mathbf{R}_0 \mathbf{S}_0$$

But since $\mathbf{T}_0$ is applied after $\mathbf{R}_0$ and $\mathbf{S}_0$, they have no effect on it. So, if we want to find how the translation changes, we drop them:

$$\mathbf{M}' = \mathbf{T}_1 \mathbf{R}_1 \mathbf{S}_1 \mathbf{T}_0$$

Multiplying this out in block format gives us

$$\begin{aligned} \mathbf{M}' &= \begin{bmatrix} \mathbf{I} & \mathbf{t}_1 \\ \mathbf{0}^T & 1 \end{bmatrix} \begin{bmatrix} \mathbf{R}_1 & \mathbf{0} \\ \mathbf{0}^T & 1 \end{bmatrix} \begin{bmatrix} s_1\mathbf{I} & \mathbf{0} \\ \mathbf{0}^T & 1 \end{bmatrix} \begin{bmatrix} \mathbf{I} & \mathbf{t}_0 \\ \mathbf{0}^T & 1 \end{bmatrix} \\ &= \begin{bmatrix} \mathbf{R}_1 & \mathbf{t}_1 \\ \mathbf{0}^T & 1 \end{bmatrix} \begin{bmatrix} s_1\mathbf{I} & s_1\mathbf{t}_0 \\ \mathbf{0}^T & 1 \end{bmatrix} \\ &= \begin{bmatrix} s_1\mathbf{R}_1 & s_1\mathbf{R}_1\mathbf{t}_0 + \mathbf{t}_1 \\ \mathbf{0}^T & 1 \end{bmatrix} \end{aligned}$$

We can see that the right-hand column vector $\mathbf{y}$ is equal to Equation 4.12. To get the final translation, we need to apply the second scale and rotation before adding the second translation. Another way of thinking of this is that we need to scale and rotate the first translation vector into the frame of the second translation vector before they can be combined together.

There are a few advantages to this alternate format. First of all, it's clear what each part does—the scale and rotation aren't combined into a single 3×3 matrix. Because of this, it's also easier to change individual elements. We can update rotation, scale through a simple multiplication, or even just set them directly. Surprisingly, on a serial processor concatenation is also cheaper. It takes 48 multiplications and 32 adds to do a traditional matrix multiplication, but only 40 multiplications and 27 adds to perform our alternate concatenation. This advantage disappears when using vector processor operations, however. In that case, it's much easier to parallelize the matrix multiplication (16 operations on some systems), and the cost of scaling and rotating the translation vector becomes more of an issue.

Even with serial processors our alternate format does have one main disadvantage, which is that we need to create a 4×4 matrix to be sent to the graphics application programming interface (API). Based on our previous explorations of the transformation matrix, we can create a matrix from our alternate format quite quickly, scale the three columns of the rotation matrix, and then copy it and the translation vector into our 4×4:

$$\begin{bmatrix} sr_{0,0} & sr_{0,1} & sr_{0,2} & t_x \\ sr_{1,0} & sr_{1,1} & sr_{1,2} & t_y \\ sr_{2,0} & sr_{2,1} & sr_{2,2} & t_z \\ 0 & 0 & 0 & 1 \end{bmatrix}$$

Which representation is better? It depends on your application. If all you wish to do is an initial scale and then apply sequences of rotations and translations, the 4×4 matrix format works fine and will be faster on a vector processor. If, on the other hand, you wish to make changes to scale as well, using the alternate format should at least be considered. And, as we'll see, if we wish to use a rotation representation other than a matrix, the alternate formation is almost certainly the way to go.

4.5 Object Hierarchies

Source Code
Demo
Hierarchy

In describing object transformations, we have considered them as transforming from the object's local frame (or local space) to a world frame (or world space). However, it is possible to define an object's transformation as being relative to another object's space instead. We could carry this out for a number of steps, thereby creating a hierarchy of objects, with world space as the root and each object's local space as a node in a tree.

For example, suppose we wish to attach an arm to a body. The body is built with its origin relative to its center. The arm has its origin at the shoulder joint location because that will be our center of rotation. If we were to place them in the world using the same transformation, the arm would end up inside the body instead of at the shoulder. We want to find the transformation that modifies the arm's world transformation so that it matches the movement of the body and still remains at the shoulder. The way to do this is to define a transformation for the arm relative to the body's local space. If we combine this with the transformation for the body, this should place the arm in the correct place in world space relative to the body, no matter its position and orientation.

The idea is to transform the arm to body space (Figure 4.20a) and then continue the transform into world space (Figure 4.20b). In this case, for each stage of transformation we perform the order as scale, rotate, and then translate. In matrix format the world transformation for the arm would be

$$\mathbf{W} = \mathbf{T}_{body}\mathbf{R}_{body}\mathbf{S}_{body}\mathbf{T}_{arm}\mathbf{R}_{arm}\mathbf{S}_{arm}$$

As we've indicated, the body and arm are treated as two separate objects, each with its own transformations, placed in a hierarchy. The body transformation is relative to world space, and the arm transformation is relative to the body's space. When rendering, for example, we begin by drawing the body with its world transformation and then drawing the arm with the concatenation of the body's transformation and the arm's transformation. By doing this, we can change them independently—rotating the arm around the shoulder, for example,

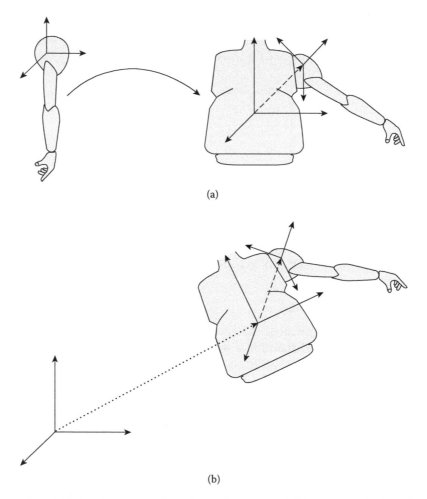

Figure 4.20. (a) Mapping arm to body's local space and (b) mapping body and arm to world space.

without affecting the body at all. Similar techniques can be used to create deeper hierarchies, for example, a turret that rotates on top of a tank chassis, with a gun barrel that elevates up and down relative to the turret.

One way of coding this is to create separate objects, each of which handles all the work of grabbing the transformation from the parent objects and combining to get the final display transform. The problem with this approach is that it generates a lot of duplicated code. Using the tank example, the code necessary for handling the hierarchy for the turret is going to be almost identical to that for the barrel. What is usually done is to design a data structure that handles the generalized case of a hierarchy of frames and use that to manage our hierarchical objects. We've implemented an example using one such data structure.

There are a few caveats to be aware of when managing hierarchies. One common approach is to store the local and world transformations as members of a base class, and then derive subclasses from that, stored in a tree. A further addition is to store a dirty bit—when you change a local transformation, you set the bit. When it's time to update the world transformations and any bounding information (see Chapter 12), you only update those below a node that's been marked dirty.

This all seems very reasonable on the surface, but it wreaks havoc with modern processors. First of all, to check the dirty bit you must load it into memory and then branch on the result. Most of the time the dirty bit will not be set, and due to branch misprediction the processor will stall. In this case it can be faster to not branch at all and just always update. Secondly, because our subclass will probably have additional members, the specific data we're focused on accessing and updating—all the local and world transformations—will probably not be located near each other in memory. Add in the fact that we are using nodes in a tree that are most likely allocated from a heap, and data locality becomes even more unlikely. This means that our data cache will be spending a good part of time loading memory we won't even use, and will have to access memory frequently to get the data we want. If our interest is speed (and it usually is), it is far better to store the transformations and any bounding information separately, and run our update step on just that data. See [2] or our example for more details.

4.6 Chapter Summary

In this chapter we've discussed the general properties of affine transformations, how they map between affine spaces, and how they can be represented and performed by matrices at one dimension higher than the affine spaces involved. We've covered the basic affine transformations as used in interactive applications and how to combine three of them— scaling, rotation, and translation—to manipulate our objects within our world. While it may be desirable to separate a given affine transformation back into scaling, rotation, and translation components, we have seen that it is not always possible when using nonuniform scaling. Separating components in this manner may not be efficient, so we have presented an alternative affine transformation representation with the three components separated. Finally, we have discussed how to construct transformations relative to other objects, which allows us to create jointed, hierarchical structures.

For those interested in reading further, information on affine algebra can be found in deRose [32], as well as in Schneider and Eberly [133]. And the standard affine transformations are described in most graphics textbooks, such as Möller et al. [1] and Hughes et al. [82].

5 Orientation Representation

5.1 Introduction

In Chapter 4 we discussed various types of affine transformations in $\mathbb{R}^3$ and how they can be represented by a matrix. In this chapter we will focus specifically on orientation and the rotation transformation. We'll look at four different orientation formats and compare them on the basis of the following criteria:

- Represents orientation/rotation with a small number of values

- Can be concatenated efficiently to form new orientations/rotations

- Rotates points and vectors efficiently

The first item is important if memory usage is an issue, either because we are working with a memory-limited machine such as a mobile device, or because we want to store a large number of transformations, such as in animation data. In either case, any reduction in representation size means that we have freed-up memory that can be used for more animations, for more animation frames (leading to a smoother result), or for some other aspect of the game. Rotating points and vectors efficiently may seem like an obvious requirement, but one that merits mentioning; not all representations are good at this. Similarly, for some representations concatenation is not possible.

There are two other criteria we might consider for an orientation format that we will not discuss here: how well the representation can be interpolated and how suitable it is for numeric integration in physics. Both of these topics will be discussed in Chapters 6 and 13, respectively.

As we'll see, there is no one choice that meets all of our requirements; each has its strengths and weaknesses in each area, depending on our implementation needs.

5.2   Rotation Matrices

Since we have been using matrices as our primary orientation/rotation representation, it is natural to begin our discussion with them.

For our first desired property, memory usage, matrices do not fare well. Euler's rotation theorem states that the minimum number of values needed to represent a rotation in three dimensions is three. The smallest possible rotation matrix requires nine values, or three orthonormal basis vectors. It is possible to compress a rotation matrix, but in most cases this is not done unless we're sending data across a network. Even then it is better to convert to one of the more compact representations that we present in the following sections, rather than compress the matrix.

However, for the second two properties, matrices do quite well. Concatenation is done through a matrix–matrix multiplication, and rotating a vector is done through a matrix–vector multiplication. Both of these are reasonably efficient on a standard floating-point processor. But on a processor that supports SSE or NEON instructions, which can perform matrix and vector operations in parallel, both of these operations can be performed even faster. Most graphics hardware has built-in circuitry that performs similarly. And as we've seen, 4×4 matrices can be useful for more than just rotation. Because of all these reasons, matrices continue to be useful despite their memory footprint.

5.3   Euler Angles

5.3.1   Definition

We've just stated that the minimum number of values needed to represent a rotation in three-dimensional (3D) space is three. As it happens, these three values can be the angles of three sequential rotations around a set of orthogonal axes. In Chapter 4, we used this as one means of building a generalized rotation matrix. Our chosen sequence of axes in this case was z-y-x, so, for example, the values $(0, \pi/4, \pi/2)$ represent a rotation of 0 radians around the z-axis, followed by a rotation of $\pi/4$ radians (or 45 degrees) around the y-axis, and concluding with a rotation of $\pi/2$ radians (90 degrees) around the x-axis. Angles can be less than 0 or greater than 2π, to represent reversed rotations and multiple rotations around a given axis. Note that we are using radians rather than degrees to represent our angles; either convention is acceptable, but the trigonometric functions used in C or C++ expect radians.

The order we've given is somewhat arbitrary, as there is no standard order that is used for the three axes. We could have used the sequence x-y-z or z-x-y just as well. We can even use the same axis for the first and third rotations, so y-z-y is a valid sequence. However, an axis rotation sequence such as z-y-y is not permitted, because duplicating an axis in sequence is redundant and doesn't add an additional degree of freedom. If the three axes are different, they can also be referred to as *Tait–Bryan angles*, whereas if an axis is repeated, they are referred to as *classic Euler angles*.

These rotations are performed around either the world axes or the object's model axes. When the angles represent world axis rotations, they are usually called *fixed angles* or *extrinsic rotations* (Figure 5.1). The most convenient way to use extrinsic rotations is to

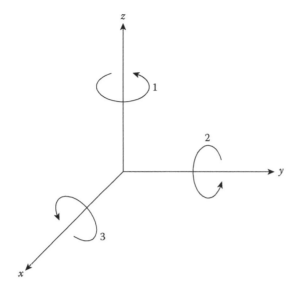

Figure 5.1. Order and direction of rotation for *z-y-x* extrinsic Euler angles.

create an *x*, *y*, or *z* rotation matrix for each angle and apply it in turn to our set of vertices. So an *y-z-x* extrinsic representation can be concatenated into a single matrix $\mathbf{R} = \mathbf{R}_x \mathbf{R}_z \mathbf{R}_y$ in matrix form.

A sequence of model axis rotations, in turn, is said to consist of *intrinsic rotations*. Intrinsic Tait–Bryan angles are commonly known as *roll*, *pitch*, and *heading*, after the three axes in a ship or an airplane. Heading is also sometimes referred to as *yaw*. Roll represents rotation around the forward axis, pitch rotation around a side axis, and heading rotation around the up axis (Figure 5.2). Whether a given roll, pitch, or heading rotation is around *x*, *y*, or *z* depends on how we've defined our coordinate frame. Suppose we are using a coordinate system where the *z*-axis represents up, the *x*-axis represents forward, and the *y*-axis represents left. Then heading is rotation around the *z*-axis, pitch is rotation around the *y*-axis, and roll is rotation around the *x*-axis. They are commonly applied in the order roll-pitch-heading, so the corresponding intrinsic rotations for our case are *x-y-z*.

To create a rotation matrix that applies intrinsic rotations, we concatenate in the reverse order of extrinsic rotations. To see why, let's take our set of *x-y-z* intrinsic rotations. We begin by applying the $\mathbf{R}_x$ matrix, to give us a rotation around *x*. We then want to apply a rotation around the object's initial model *y*-axis. However, because of the *x* rotation, the *y*-axis has been transformed to a new orientation. So, if we concatenate as we normally would, our rotation will be about the transformed *y*-axis, which is not what we want. To avoid this, we transform by $\mathbf{R}_y$ first, then by $\mathbf{R}_x$, giving $\mathbf{R}_x \mathbf{R}_y$. The same is true for the *z* rotation: We need to rotate around *z* first to ensure we rotate around the original model *z*-axis, not the transformed one. The resulting matrix is

$$\mathbf{R}_{intrinsic} = \mathbf{R}_x \mathbf{R}_y \mathbf{R}_z$$

So *x-y-z* intrinsic angles are the same as *z-y-x* extrinsic angles.

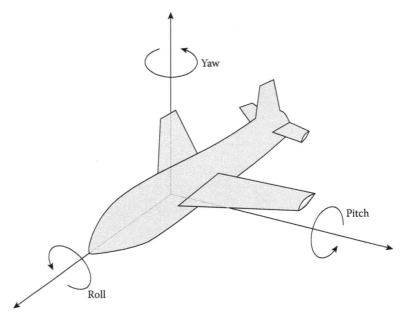

Figure 5.2. Roll, pitch, and rotations relative to model coordinate axes.

5.3.2 Format Conversion

By concatenating three general axis rotation matrices and expanding out the terms, we can create a generalized rotation matrix. The particular matrix will depend on which axis rotations we're using and whether they are extrinsic or intrinsic. For *z-y-x* extrinsic rotations or *x-y-z* intrinsic rotations, the matrix looks like

$$\mathbf{R} = \mathbf{R}_x\mathbf{R}_y\mathbf{R}_z = \begin{bmatrix} CyCz & -CySz & Sy \\ SxSyCz + CxSz & -SxSySz + CxCz & -SxCy \\ -CxSyCz + SxSz & CxSySz + SxCz & CxCy \end{bmatrix}$$

where

$$Cx = \cos\theta_x \quad Sx = \sin\theta_x$$
$$Cy = \cos\theta_y \quad Sy = \sin\theta_y$$
$$Cz = \cos\theta_z \quad Sz = \sin\theta_z$$

This should look familiar from Chapter 4.

When possible, we can save some instructions by computing each sine and cosine using a single `sincos()` call. This function is not supported on all processors, or even in all math libraries, so we have provided a wrapper function `IvSinCosf()` (accessible by including `IvMath.h`) that will calculate it depending on the platform.

We can convert from a matrix back to a possible set of extrinsic rotations by inverting this process. Note that since we'll be using inverse trigonometric functions there are multiple resulting angles. We'll also be taking a square root, the result of which could be positive or negative. Hence, there are multiple possibilities of intrinsic or extrinsic rotations for a given matrix—the best we can do is find one. Assuming we're using z-y-x extrinsic rotations, we can see that $\sin\theta_y$ is equal to $\mathbf{R}_{02}$. Finding $\cos\theta_y$ can be done by using the identity $\cos\theta_y = \sqrt{1 - \sin^2\theta_y}$. The rest falls out from dividing quantities out of the first row and last column of the matrix, so

$$\sin\theta_y = \mathbf{R}_{02}$$
$$\cos\theta_y = \sqrt{1 - \sin^2\theta_y}$$
$$\sin\theta_x = -\mathbf{R}_{12}/\cos\theta_y$$
$$\cos\theta_x = \mathbf{R}_{22}/\cos\theta_y$$
$$\sin\theta_z = -\mathbf{R}_{01}/\cos\theta_y$$
$$\cos\theta_z = \mathbf{R}_{00}/\cos\theta_y$$

Note that we have no idea whether $\cos\theta_y$ should be positive or negative, so we assume that it's positive. Also, if $\cos\theta_y = 0$, then the x- and z-axes have become aligned (see Section 5.3.5) and we can't distinguish between rotations around x and rotations around z. One possibility is to assume that rotation around z is 0, so

$$\sin\theta_z = 0$$
$$\cos\theta_z = 1$$
$$\sin\theta_x = \mathbf{R}_{21}$$
$$\cos\theta_x = \mathbf{R}_{11}$$

Calling `arctan2()` for each sin–cos pair will return a possible angle in radians, generally in the range $[-\pi, \pi]$. Note that we have lost one of the few benefits of Euler angles, which is that they can represent multiple rotations around an axis by using angles greater than 2π radians, or 360 degrees. We have also lost any notion of "negative" rotation.

5.3.3 Concatenation

Clearly, Euler angles meet our first criterion for a good orientation representation: they use the minimum number of values. However, they don't really meet the remainder of our requirements. First of all, they don't concatenate well. Adding angles doesn't work: Applying $(\pi/2, \pi/2, \pi/2)$ twice doesn't end up at the same orientation as (π, π, π). The most straightforward method for concatenating two Euler angle triples is to convert each sequence of angles to a matrix, concatenate the matrix, and then convert the matrix back to Euler angles. This will take a large number of operations, and will only give an approximate result, due to the ill-formed nature of the matrix to Euler conversion.

5.3.4 Vector Rotation

Euler angles also aren't the most efficient method for rotating vectors. Recall that to rotate a vector around z uses the formula

$$R_z(x, y, \theta) = (x\cos\theta - y\sin\theta, x\sin\theta + y\cos\theta)$$

Using the angles directly means that for each axis, we compute a sine and cosine and then apply the preceding formula. Even if we cache the sine and cosine values for a set of vectors, this ends up being more expensive than the cost of a matrix multiplication. Therefore, when rotating multiple vectors (in general the break-even point is five vectors), it's more efficient to convert to matrix format.

5.3.5   Other Issues

As if all of these disadvantages are not enough, the fatal blow is that in certain cases Euler angles can lose one degree of freedom. We can think of this as a mathematical form of *gimbal lock*. In aeronautic navigational systems, there is often a set of gyroscopes, or gimbals, that control the orientation of an airplane or rocket. Gimbal lock is a mechanical failure where one gimbal is rotated to the end of its physical range and it can't be rotated any further, thereby losing one degree of freedom. While in the virtual world, we don't have mechanical gyroscopes to worry about, a similar situation can arise.

Suppose we are using x-y-z extrinsic rotations and we consider the case where, no matter what we use for the x and z angles, we will always rotate around the y-axis by 90 degrees. This rotates the original world x-axis—the axis we first rotate around—to be aligned with the world negative z-axis (Figure 5.3). Now any rotation we do with θ_z will subtract from any rotation to which we have applied θ_x. The combination of x and z rotations can be represented by one value $\theta_x - \theta_z$, applied as the initial x-axis rotation. Instead of using $(\theta_x, \pi/2, \theta_z)$, we could just as well use $(\theta_x - \theta_z, \pi/2, 0)$ or $(0, \pi/2, \theta_z - \theta_x)$. Another way to think of this is: were this in matrix form we would not be able to extract unique values for θ_x and θ_z. We have effectively lost one degree of freedom.

To try this for yourself, take an object whose orientation can be clearly distinguished, like a book.[1] From your point of view, rotate the object clockwise 90 degrees around

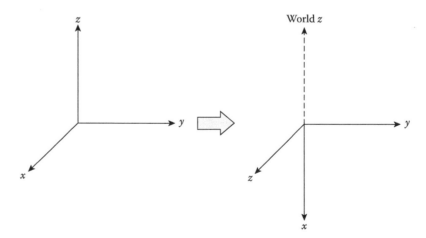

Figure 5.3. Demonstration of mathematical gimbal lock. A rotation of 90 degrees around y will lead to the local x-axis aligning with the $-z$ world axis, and a loss of a degree of freedom.

---

[1] Or a close friend.

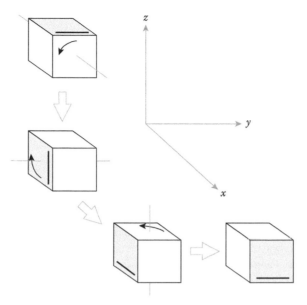

Figure 5.4. Effect of gimbal lock. Rotating the box around the world x-axis, then the world y-axis, then the world z-axis ends up having the same effect as rotating the box around just the y-axis.

an axis pointing forward (roll). Now rotate the new top of the object towards you by 90 degrees (pitch). Now rotate the object counterclockwise 90 degrees around an axis pointing up (heading). The result is the same as pitching the object downward 90 degrees (see Figure 5.4).

Still, in some cases Euler angles do provide an intuitive representation for orientation. For example, in a hierarchical system it is very intuitive to define rotations at each joint as a set of Euler angles and to constrain certain axes to remain fixed. An elbow or knee joint, for instance, could be considered a set of Euler angles with two constraints and only one axis available for applying rotation. It's also easy to set a range of angles so that the joint doesn't bend too far one way or the other. However, these limited advantages are not enough to outweigh the problems with Euler angles. So in most cases, Euler angles are used as a means to semi-intuitively set other representations (being aware of the dangers of gimbal lock, of course), and our library will be no exception.

5.4   Axis–Angle Representation

5.4.1   Definition

Recall from Chapter 4 that we can represent a general rotation in $\mathbb{R}^3$ by an axis of rotation, and the amount we rotate around this axis by an angle of rotation. Therefore, we can represent rotations in two parts: a 3-vector $\mathbf{r}$ that lies along the axis of rotation, and a scalar θ that corresponds to a counterclockwise rotation around the axis, if the axis is pointing toward us. Usually, a normalized vector $\hat{\mathbf{r}}$ is used instead, which constrains the

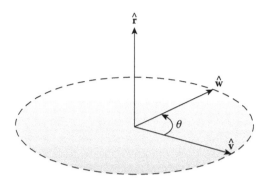

Figure 5.5. Axis–angle representation. Rotation by **r** by angle θ rotates **v** into **w**.

four values to three degrees of freedom, corresponding to the three degrees of freedom necessary for 3D rotations.

Generating the axis–angle rotation that takes us from one normalized vector $\hat{\mathbf{v}}$ to another vector $\hat{\mathbf{w}}$ is straightforward (Figure 5.5). The angle of rotation is the angle between the two vectors:

$$\theta = \arccos(\hat{\mathbf{v}} \cdot \hat{\mathbf{w}}) \tag{5.1}$$

The two vectors lie in the plane of rotation, and so the axis of rotation is perpendicular to both of them:

$$\mathbf{r} = \hat{\mathbf{v}} \times \hat{\mathbf{w}} \tag{5.2}$$

Normalizing **r** gives us $\hat{\mathbf{r}}$. Near-parallel vectors may cause us some problems either because the dot product is near 0, or normalizing the cross product ends up dividing by a near-zero value. In those cases, we set θ to 0 and $\hat{\mathbf{r}}$ to any arbitrary, normalized vector.

5.4.2 Format Conversion

To convert an axis–angle representation to a matrix, we can use the derivation from Chapter 4:

$$\mathbf{R}_{\hat{\mathbf{r}}\theta} = \begin{bmatrix} tx^2 + c & txy - sz & txz + sy \\ txy + sz & ty^2 + c & tyz - sx \\ txz - sy & tyz + sx & tz^2 + c \end{bmatrix} \tag{5.3}$$

where

$$\hat{\mathbf{r}} = (x, y, z)$$
$$c = \cos\theta$$
$$s = \sin\theta$$
$$t = 1 - \cos\theta$$

Converting from a matrix to the axis–angle format has similar issues as the Euler angle format, since opposing vectors $\hat{\mathbf{r}}$ and $-\hat{\mathbf{r}}$ can be used to generate the same rotation by rotating in opposite directions, and multiple angles (0 and 2π, for example) applied to the same axis can rotate to the same orientation. The following method is from Eberly [36].

We begin by computing the angle. The sum of the diagonal elements, or *trace* of a rotation matrix $\mathbf{R}$, is equal to $2\cos\theta + 1$, where θ is our angle of rotation. This gives us an easy method for computing θ:

$$\theta = \arccos\left(\frac{1}{2}(\text{trace}(\mathbf{R}) - 1)\right)$$

There are three possibilities for θ. If θ is 0, then we can use any arbitrary unit vector as our axis. If θ lies in the range $(0, \pi)$, then we can compute the axis by using the formula

$$\mathbf{R} - \mathbf{R}^T = 2\sin\theta\mathbf{S} \tag{5.4}$$

where $\mathbf{S}$ is a skew symmetric matrix of the form

$$\mathbf{S} = \begin{bmatrix} 0 & -z & y \\ z & 0 & -x \\ -y & x & 0 \end{bmatrix}$$

The values x, y, and z in this case are the components of our axis vector $\hat{\mathbf{r}}$. We can compute $\mathbf{r}$ as $(R_{21} - R_{12}, R_{02} - R_{20}, R_{10} - R_{01})$, and normalize to get $\hat{\mathbf{r}}$.

If θ equals π, then $\mathbf{R} - \mathbf{R}^T = \mathbf{0}$, which doesn't help us at all. In this case, we can use another formulation for the rotation matrix, which only holds if $\theta = \pi$:

$$\mathbf{R} = \mathbf{I} + 2\mathbf{S}^2 = \begin{bmatrix} 1 - 2y^2 - 2z^2 & 2xy & 2xz \\ 2xy & 1 - 2x^2 - 2z^2 & 2yz \\ 2xz & 2yz & 1 - 2x^2 - 2y^2 \end{bmatrix}$$

The idea is that we can use the diagonal elements to compute the three axis values. By subtracting appropriately, we can solve for one term, and then use that value to solve for the other two. For example, $R_{00} - R_{11} - R_{22} + 1$ expands to

$$R_{00} - R_{11} - R_{22} + 1 = 1 - 2y^2 - 2z^2 - 1 + 2x^2 + 2z^2 - 1 + 2x^2 + 2y^2 + 1$$
$$= 4x^2$$

So,

$$x = \frac{1}{2}\sqrt{R_{00} - R_{11} - R_{22} + 1} \tag{5.5}$$

and consequently,

$$y = \frac{R_{01}}{2x}$$

$$z = \frac{R_{02}}{2x}$$

To avoid problems with numeric precision and square roots of negative numbers, we'll choose the largest diagonal element as the term that we'll solve for. So, if R_{00} is the largest diagonal element, we'll use the preceding equations. If R_{11} is the largest, then

$$y = \frac{1}{2}\sqrt{R_{11} - R_{00} - R_{22} + 1}$$

$$x = \frac{R_{01}}{2y}$$

$$z = \frac{R_{12}}{2y}$$

Finally, if R_{22} is the largest element we use

$$z = \frac{1}{2}\sqrt{R_{22} - R_{00} - R_{11} + 1}$$

$$x = \frac{R_{02}}{2z}$$

$$y = \frac{R_{12}}{2z}$$

5.4.3 Concatenation

Concatenating two axis–angle representations is not straightforward. One method is to convert them to two matrices or two quaternions (see below), multiply, and then convert back to the axis–angle format. As one can easily see, this is more expensive than just concatenating two matrices. Because of this, one doesn't often perform this operation on axis–angle representations.

5.4.4 Vector Rotation

For the rotation of a vector $\mathbf{v}$ by the axis–angle representation $(\hat{\mathbf{r}}, \theta)$, we can use the Rodrigues formula that we derived in Chapter 4:

$$R\mathbf{v} = \cos\theta\,\mathbf{v} + [1 - \cos\theta](\mathbf{v} \cdot \hat{\mathbf{r}})\hat{\mathbf{r}} + \sin\theta(\hat{\mathbf{r}} \times \mathbf{v})$$

If we precompute $\cos\theta$ and $\sin\theta$ and reuse intermediary values, we can compute this relatively efficiently. We can improve this slightly by using the identity

$$\hat{\mathbf{r}} \times (\hat{\mathbf{r}} \times \mathbf{v}) = (\mathbf{v} \cdot \hat{\mathbf{r}})\hat{\mathbf{r}} - (\hat{\mathbf{r}} \cdot \hat{\mathbf{r}})\mathbf{v}$$
$$= (\mathbf{v} \cdot \hat{\mathbf{r}})\hat{\mathbf{r}} - \mathbf{v}$$

and substituting to get an alternate Rodrigues formula:

$$R\mathbf{v} = \mathbf{v} + (1 - \cos\theta)[\hat{\mathbf{r}} \times (\hat{\mathbf{r}} \times \mathbf{v})] + \sin\theta(\hat{\mathbf{r}} \times \mathbf{v})$$

In both these cases, the trade-off is whether to store the results of the transcendental functions and thereby use more memory, or compute them every time and lose speed. The answer will depend on the needs of the implementation.

When rotating two or more vectors, it is more efficient on a serial processor to convert the axis–angle format to a matrix and then multiply. So if you're only transforming one vector, don't bother converting; otherwise, use a matrix.

5.4.5 Axis-Angle Summary

While being a useful way of thinking about rotation, the axis–angle format still has some problems. Concatenating two axis–angle representations is extremely expensive. And unless we store two additional values, rotating vectors requires computing transcendental functions, which is not very efficient either. Our next representation encapsulates some of the useful properties of the axis–angle format, while providing a more efficient method for concatenation. It precomputes the transcendental functions and uses them to rotate vectors in nearly equivalent time to the axis–angle method. Because of this, we have not explicitly provided an implementation in our library for the axis–angle format.

5.5 Quaternions

5.5.1 Definition

Source Code
Library
IvMath
Filename
IvQuat

The final orientation representation we'll consider could be considered a variant of the axis–angle representation, and in fact when using it for rotation, it's often simplest to think of it that way. It is called the *quaternion* and was created by the Irish mathematician Sir William Hamilton [69] in the nineteenth century and introduced to computer graphics by Ken Shoemake [136] in the 1980s. Quaternions require only four values, they don't have problems of gimbal lock, the mathematics for concatenation are relatively simple, and if properly constructed, they can be used to rotate vectors in a reasonably efficient manner.

Hamilton's original purpose for creating quaternions was to extend complex numbers beyond two dimensions. Complex numbers have the form

$$x = a + bi$$

where i represents the square root of negative 1. The value a is called the real part, and the value b the imaginary part. We can treat $\{1, i\}$ as a 2D basis, and thereby map a complex number to a 2D vector (a, b).

Hamilton tried to extend this to three dimensions, but couldn't create an algebra (known as a division algebra) where every value had a multiplicative inverse. During a walk along the Grand Canal in Dublin, he realized that it was possible if he extended it to four dimensions rather than three, and so quaternions were discovered.

Hamilton's general formula for a quaternion $\mathbf{q}$ is as follows:

$$\mathbf{q} = w + xi + yj + zk$$

The quantities 1, i, j, and k can be thought of as the standard basis for all quaternions, so it is common to write a quaternion as just

$$\mathbf{q} = (w, x, y, z)$$

The $xi + yj + zk$ part of the quaternion is akin to a vector in $\mathbb{R}^3$, so a quaternion also can be written as

$$\mathbf{q} = (w, \mathbf{v})$$

where w is called the scalar part and $\mathbf{v}$ is called the vector part.

Frequently, we'll want to use vectors in combination with quaternions. To do so, we'll zero out the scalar part and set the vector part equal to our original vector. So, the quaternion corresponding to a vector $\mathbf{u}$ is

$$\mathbf{q_u} = (0, \mathbf{u})$$

Other than terminology, we aren't that concerned about Hamilton's intentions for generalized quaternions, because we are only going to consider a specialized case discovered by Arthur Cayley [23]. In particular, he showed that quaternions can be used to describe pure rotations. Later on, Courant and Hilbert [27] determined the relationship between normalized quaternions and the axis–angle representation.

5.5.2 Quaternions as Rotations

While any quaternion can be used to represent rotation (as we will see later), we will be primarily using unit quaternions, where

$$w^2 + \mathbf{v} \bullet \mathbf{v} = 1$$

There are three reasons for this. First of all, it makes the calculations for rotation and conversions more efficient. Secondly, it manages floating-point error. By normalizing, our data will lie in the range $[-1, 1]$, and floating-point values in that range have a high degree of relative precision. Finally, it provides a natural correspondence between an axis–angle rotation and a quaternion. In a unit quaternion, w can be thought of as representing the angle of rotation θ. More specifically, $w = \cos(\theta/2)$. The vector $\mathbf{v}$ represents the axis of rotation, but normalized and scaled by $\sin(\theta/2)$. So, $\mathbf{v} = \sin(\theta/2)\hat{\mathbf{r}}$. For example, suppose we wanted to rotate by 90 degrees around the z-axis. Our axis is $(0, 0, 1)$ and half our angle is $\pi/4$ (in radians). The corresponding quaternion components are

$$w = \cos\left(\frac{\pi}{4}\right) = \frac{\sqrt{2}}{2}$$

$$x = 0 \cdot \sin\left(\frac{\pi}{4}\right) = 0$$

$$y = 0 \cdot \sin\left(\frac{\pi}{4}\right) = 0$$

$$z = 1 \cdot \sin\left(\frac{\pi}{4}\right) = \frac{\sqrt{2}}{2}$$

giving us a final quaternion of

$$\mathbf{q} = \left(\frac{\sqrt{2}}{2}, 0, 0, \frac{\sqrt{2}}{2}\right)$$

So, why reformat our previously simple axis and angle to this somewhat strange representation, particularly using only half the angle? As we'll see, precooking the data in this way allows us to rotate vectors and concatenate with ease.

Our class implementation for quaternions looks like

```
class IvQuat
{
public:
    // constructor/destructor
    inline IvQuat() {}
    inline IvQuat( float_w, float _x, float _y, float _z )  :
        w(_w), x(_x), y(_y), z(_z)
    {
    }
    IvQuat(const IvVector3& axis, float angle);
    explicit IvQuat(const IvVector3& vector);
    inline ~IvQuat() {}

    // member variables
    float x, y, z, w;
};
```

Much of this follows from what we've already discussed. We can set our quaternion values directly, use an axis–angle format, or explicitly use a vector. Recall that in this last case, we use the vector to set our x, y, and z terms, and set w to 0.

5.5.3 Addition and Scalar Multiplication
Like vectors, quaternions can be scaled and added componentwise. For both operations a quaternion acts just like a 4-vector, so

$$(w_1, x_1, y_1, z_1) + (w_2, x_2, y_2, z_2) = (w_1 + w_2, x_1 + x_2, y_1 + y_2, z_1 + z_2)$$
$$a(w, x, y, z) = (aw, ax, ay, az)$$

The algebraic rules for addition and scalar multiplication that apply to vectors and matrices apply here, so like them, the set of all quaternions is also a vector space. However, the set of unit quaternions is not, since neither operation maintains unit length. Therefore, if we use one of these operations, we'll need to normalize afterwards. In general, however, we will not be using these operations except in special cases.

5.5.4 Negation
Negation is a subset of scale, but it's worth discussing separately. One would expect that negating a quaternion would produce a quaternion that applies a rotation in the opposite direction—that is, that it would be the inverse. However, while it does rotate in the opposite direction, it also rotates around the negative axis. The end result is that a vector rotated by either quaternion ends up in the same place, but if one quaternion rotates by θ radians around $\hat{\mathbf{r}}$, its negation rotates $2\pi - \theta$ radians around $-\hat{\mathbf{r}}$. Figure 5.6 shows what this looks like on the rotation plane. The negated quaternion can be thought of as "taking the other way around," but both quaternions rotate the vector to the same orientation. This will

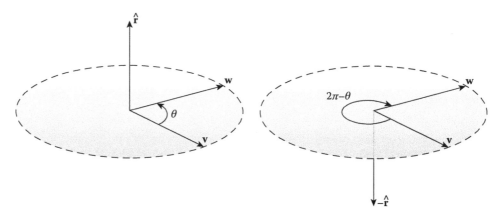

Figure 5.6. Comparing rotation performed by a normalized quaternion (left) with its negation (right).

cause some issues when blending between quaternions but can be handled by adjusting our values appropriately, which we'll discuss in Chapter 6. Otherwise, we can use $\mathbf{q}$ and $-\mathbf{q}$ interchangeably.

5.5.5 Magnitude and Normalization

As we've implied, we will be normalizing quaternions, and will do so as if we were using 4-vectors. The magnitude of a quaternion is therefore as follows:

$$\|\mathbf{q}\| = \sqrt{(w^2 + x^2 + y^2 + z^2)}$$

A normalized quaternion $\hat{\mathbf{q}}$ is

$$\hat{\mathbf{q}} = \frac{\mathbf{q}}{\|\mathbf{q}\|}$$

Since we're assuming that our quaternions are normalized, we'll forgo the use of the notation $\hat{\mathbf{q}}$ to keep our equations from being too cluttered.

5.5.6 Dot Product

The dot product of two quaternions should also look familiar:

$$\mathbf{q}_1 \cdot \mathbf{q}_2 = w_1 w_2 + x_1 x_2 + y_1 y_2 + z_1 z_2$$

As with vectors, this is still equal to the cosine of the angle between the quaternions, except that our angle is in four dimensions instead of the usual three. What this gives us is a way of measuring how different two quaternions are. If $\mathbf{q}_1 \cdot \mathbf{q}_2$ is close to 1 (assuming that they're normalized), then they apply very similar rotations. Also, since we know that the negation of a quaternion performs the same rotation as the original, if the dot product is close to -1, the two still apply very similar rotations. So parallel normalized quaternions ($|\mathbf{q}_1 \cdot \mathbf{q}_2| \approx 1$) are similar. Correspondingly, orthogonal normalized quaternions ($\mathbf{q}_1 \cdot \mathbf{q}_2 = 0$) produce extremely different rotations.

5.5.7 Format Conversion

Converting from axis–angle format to a quaternion requires multiplying the angle by one-half, computing the sine and cosine of that result, and scaling the normalized axis vector by the sine. To convert back, we take the arccos of w to get half the angle, and then use $\sqrt{1 - w^2}$ to get the length of $\mathbf{v}$ so we can normalize it. The full conversion is

$$\theta = 2 \arccos(w)$$
$$\|\mathbf{v}\| = \sqrt{1 - w^2}$$
$$\hat{\mathbf{r}} = \mathbf{v}/\|\mathbf{v}\|$$

Converting a normalized quaternion to a 3×3 rotation matrix takes the following form:

$$\mathbf{M_q} = \begin{bmatrix} 1 - 2y^2 - 2z^2 & 2xy - 2wz & 2xz + 2wy \\ 2xy + 2wz & 1 - 2x^2 - 2z^2 & 2yz - 2wx \\ 2xz - 2wy & 2yz + 2wx & 1 - 2x^2 - 2y^2 \end{bmatrix} \tag{5.6}$$

If the quaternion is not normalized, it's easiest to normalize the quaternion first and then create the matrix.

To compute this on a serial processor we can make use of the fact that there are a lot of duplicated terms. The following is derived from Shoemake [137]:

```
IvMatrix33&
IvMatrix33::Rotation( const IvQuat& q )
{
    float s, xs, ys, zs, wx, wy, wz, xx, xy, xz, yy, yz, zz;

    // if q is normalized, s = 2.0f
    s = 2.0f/( q.x*q.x + q.y*q.y + q.z*q.z + q.w*q.w );

    xs = s*q.x;      ys = s*q.y;      zs = s*q.z;
    wx = q.w*xs;     wy = q.w*ys;     wz = q.w*zs;
    xx = q.x*xs;     xy = q.x*ys;     xz = q.x*zs;
    yy = q.y*ys;     yz = q.y*zs;     zz = q.z*zs;

    mV[0] = 1.0f - (yy + zz);
    mV[3] = xy - wz;
    mV[6] = xz + wy;

    mV[1] = xy + wz;
    mV[4] = 1.0f - (xx + zz);
    mV[7] = yz - wx;

    mV[2] = xz - wy;
    mV[5] = yz + wx;
    mV[8] = 1.0f - (xx + yy);

    return *this;

}    // End of Rotation()
```

If we have a parallel vector processor that can perform fast matrix multiplication, another way of doing this is to generate two 4×4 matrices and multiply them together:

$$
\mathbf{M_q} =
\begin{bmatrix}
w & -z & y & x \\
z & w & -x & y \\
-y & x & w & z \\
-x & -y & -z & w
\end{bmatrix}
\begin{bmatrix}
w & -z & y & -x \\
z & w & -x & -y \\
-y & x & w & -z \\
x & y & z & w
\end{bmatrix}
$$

If the quaternion is normalized, the product will be the affine rotation matrix corresponding to the quaternion.

To convert a matrix to a quaternion, we can use an approach that is similar to converting from a rotation matrix to axis–angle format. Recall that the trace of a rotation matrix is $2\cos\theta + 1$, where θ is our angle of rotation. Also, from Equation 5.4, we know that the vector $\mathbf{r} = (R_{21} - R_{12}, R_{02} - R_{20}, R_{10} - R_{01})$ will have length $2\sin\theta$. If we add 1 to the trace and use these as the scalar and vector parts, respectively, of a quaternion, we get

$$
\hat{\mathbf{q}} = (2\cos\theta + 2, 2\sin\theta\,\hat{\mathbf{r}}) \tag{5.7}
$$

Surprisingly, all we need to do now is normalize to get the final result. To see why, suppose we started with a quaternion

$$
\hat{\mathbf{q}}_1 = (\cos\theta, \sin\theta\,\hat{\mathbf{r}})
$$

This is close to what we need, which is

$$
\hat{\mathbf{q}}_h = \left(\cos\frac{\theta}{2}, \sin\frac{\theta}{2}\,\hat{\mathbf{r}}\right)
$$

To get from $\hat{\mathbf{q}}_1$ to $\hat{\mathbf{q}}_h$, let's consider two vectors. If we have a vector $\mathbf{w}_0$ and a vector $\mathbf{w}_1$ rotated θ degrees from $\mathbf{w}_0$, then to find the vector $\mathbf{v}_h$ that lies between them on the rotation plane (i.e., the vector rotated $\theta/2$ degrees from $\mathbf{w}_0$), we just need to compute $(\mathbf{w}_1 + \mathbf{w}_2)/2$. If we want a normalized vector, we can skip the division by two and just do the normalize step.

So to do the same with quaternions, we take as our $\mathbf{q}_0$ the quaternion $(1, \mathbf{0})$, which represents no rotation. If we add that to $\mathbf{q}_1$ and normalize, that will give us our desired result. That boils down to adding 1 to w and normalizing. Equation 5.7 is just that scaled by 2; the scaling factor drops out nicely when we normalize.

If the trace of the matrix is less than 0, then this will not work. We'll need to use an approach similar to when we extracted the axis from a rotation matrix. By taking the largest diagonal element and subtracting the elements from it, we can derive an equation to solve for a single-axis component (e.g., Equation 5.5). Using that value as before, we can then compute the other quaternion components from the elements of the matrix.

So, if the largest diagonal element is R_{00}, then

$$x = \frac{1}{2}\sqrt{R_{00} - R_{11} - R_{22} + 1}$$

$$y = \frac{R_{01} + R_{10}}{4x}$$

$$z = \frac{R_{02} + R_{20}}{4x}$$

$$w = \frac{R_{21} - R_{12}}{4x}$$

We can simplify this by noting that

$$4x^2 = R_{00} - R_{11} - R_{22} + 1$$

$$\frac{4x^2}{4x} = \frac{R_{00} - R_{11} - R_{22} + 1}{4x}$$

$$x = \frac{R_{00} - R_{11} - R_{22} + 1}{4x}$$

Substituting this formula for x, we now see that all of the components are scaled by $1/4x$. We can accomplish the same thing by taking the numerators

$$x = R_{00} - R_{11} - R_{22} + 1$$
$$y = R_{01} + R_{10}$$
$$z = R_{02} + R_{20}$$
$$w = R_{21} - R_{12}$$

and normalizing.

Similarly, if the largest diagonal element is R_{11}, we start with

$$y = R_{11} - R_{00} - R_{22} + 1$$
$$x = R_{01} + R_{10}$$
$$z = R_{12} + R_{21}$$
$$w = R_{02} - R_{20}$$

and normalize.

And, if the largest diagonal element is R_{22}, we take

$$z = R_{22} - R_{00} - R_{11} + 1$$
$$x = R_{02} + R_{20}$$
$$y = R_{21} + R_{12}$$
$$w = R_{10} - R_{01}$$

and normalize.

Converting from an extrinsic Euler angle format to a quaternion requires creating a quaternion for each rotation around a coordinate axis, and then concatenating them together. For *z-y-x* extrinsic rotations, the result is

$$w = \cos\frac{\theta_x}{2}\cos\frac{\theta_y}{2}\cos\frac{\theta_z}{2} - \sin\frac{\theta_x}{2}\sin\frac{\theta_y}{2}\sin\frac{\theta_z}{2}$$

$$x = \sin\frac{\theta_x}{2}\cos\frac{\theta_y}{2}\cos\frac{\theta_z}{2} + \cos\frac{\theta_x}{2}\sin\frac{\theta_y}{2}\sin\frac{\theta_z}{2}$$

$$y = \cos\frac{\theta_x}{2}\sin\frac{\theta_y}{2}\cos\frac{\theta_z}{2} - \sin\frac{\theta_x}{2}\cos\frac{\theta_y}{2}\sin\frac{\theta_z}{2}$$

$$z = \cos\frac{\theta_x}{2}\cos\frac{\theta_y}{2}\sin\frac{\theta_z}{2} + \sin\frac{\theta_x}{2}\sin\frac{\theta_y}{2}\cos\frac{\theta_z}{2}$$

Converting a quaternion to Euler angles is, quite frankly, an awful thing to do. If it's truly necessary (e.g., for an interface), the simplest method is to convert the quaternion to a matrix, and extract the Euler angles from the matrix.

5.5.8   Concatenation

As with matrices, if we wish to concatenate the transformations performed by two quaternions, we multiply them together to get a new quaternion. If we consider the simpler case of complex numbers, after multiplying the result is

$$\begin{aligned}(a + b\boldsymbol{i})(c + d\boldsymbol{i}) &= ac + (ad + bc)\boldsymbol{i} + bd\boldsymbol{i}^2 \\ &= ac + (ad + bc)\boldsymbol{i} + bd(-1) \\ &= (ac - bd) + (ad + bc)\boldsymbol{i}\end{aligned} \tag{5.8}$$

Multiplying complex numbers is commutative, so order does not matter.

For quaternions, expanding out the terms of the multiplication produces the following result:

$$\begin{aligned}(w_2 + x_2\boldsymbol{i} &+ y_2\boldsymbol{j} + z_2\boldsymbol{k})(w_1 + x_1\boldsymbol{i} + y_1\boldsymbol{j} + z_1\boldsymbol{k}) \\ &= w_2w_1 + w_2x_1\boldsymbol{i} + w_2y_1\boldsymbol{j} + w_2z_1\boldsymbol{k} \\ &\quad + x_2w_1\boldsymbol{i} + x_2x_1\boldsymbol{i}^2 + x_2y_1\boldsymbol{ij} + x_2z_1\boldsymbol{ik} \\ &\quad + y_2w_1\boldsymbol{j} + y_2x_1\boldsymbol{ji} + y_2y_1\boldsymbol{j}^2 + y_2z_1\boldsymbol{jk} \\ &\quad + z_2w_1\boldsymbol{k} + z_2x_1\boldsymbol{ki} + z_2y_1\boldsymbol{kj} + z_2z_1\boldsymbol{k}^2\end{aligned} \tag{5.9}$$

We define the products of the $\boldsymbol{i}, \boldsymbol{j}$, and $\boldsymbol{k}$ quantities as follows:

$$\begin{aligned}\boldsymbol{ij} &= \boldsymbol{k} & \boldsymbol{jk} &= \boldsymbol{i} & \boldsymbol{ki} &= \boldsymbol{j} \\ \boldsymbol{ji} &= -\boldsymbol{k} & \boldsymbol{kj} &= -\boldsymbol{i} & \boldsymbol{ik} &= -\boldsymbol{j}\end{aligned}$$

and

$$i^2 = j^2 = k^2 = i \quad jk = -1$$

Note that multiplying these terms is anticommutative, so here order does matter.

We can use these properties and well-known vector operations to simplify the product to

$$\mathbf{q}_2 \cdot \mathbf{q}_1 = (w_1 w_2 - \mathbf{v}_1 \bullet \mathbf{v}_2, w_1 \mathbf{v}_2 + w_2 \mathbf{v}_1 + \mathbf{v}_2 \times \mathbf{v}_1)$$

We've expressed this in a right-to-left order, like our matrices. This is because the rotation defined by $\mathbf{q}_1$ will be applied first, followed by the rotation defined by $\mathbf{q}_2$. We'll see this more clearly when we look at how we use quaternions to transform vectors. Also note the cross product; due to this, quaternion multiplication is also not commutative. This is what we expect with rotations; applying two rotations in one order does not necessarily provide the same result as applying them in the reverse order. Finally, other than the cross product, this has a pleasing similarity to Equation 5.8.

Multiplying two normalized quaternions does produce a normalized quaternion. However, due to floating-point error, it is wise to renormalize the result—if not after every multiplication, at least often and definitely before using the quaternion to rotate vectors.

A straightforward implementation of quaternion multiplication might look like

```
IvQuat operator*(IvQuat q2, IvQuat q1)
{
    IvVector3 v1(q1.x, q1.y, q1.z);
    IvVector3 v2(q2.x, q2.y, q2.z);

    float w = q1.w*q2.w - v1.Dot(v2);
    IvVector3 v = q1.w*v2 + q2.w*v1 + v2.Cross(v1);
    IvQuat q(w, v);

    return q;
}
```

Alternatively, we can unroll the operations to get

```
IvQuat operator*(IvQuat q2, IvQuat q1)
{
    w = q2.w*q1.w - q2.x*q1.x
        - q2.y*q1.y - q2.z*q1.z;
    x = q2.y* q1.z - q2.z*q1.y
        + q2.w*q1.x + q1.w*q2.x;
    y = q2.z*q1.x - q2.x*q1.z
        + q2.w*q1.y + q1.w*q2.y;
    z = q2.x*q1.y - q2.y*q1.x
        + q2.w*q1.z + q1.w*q2.z;
    return IvQuat(w,x,y,z);
}
```

Note that on a scalar processor that concatenating two quaternions can actually be faster than multiplying two matrices together.

An example of concatenating quaternions is the conversion from z-y-x extrinsic Euler angles to a quaternion. The corresponding quaternions for each axis are

$$\mathbf{q}_z = \left(\cos\frac{\theta_z}{2}, 0, 0, \sin\frac{\theta_z}{2}\right)$$

$$\mathbf{q}_y = \left(\cos\frac{\theta_y}{2}, 0, \sin\frac{\theta_y}{2}, 0\right)$$

$$\mathbf{q}_x = \left(\cos\frac{\theta_x}{2}, \sin\frac{\theta_x}{2}, 0, 0\right)$$

Multiplying these together in the order $\mathbf{q}_x\mathbf{q}_y\mathbf{q}_z$ gives the result in Section 5.5.7.

5.5.9   Identity and Inverse

As with matrix products, there is an identity quaternion and, subsequently, there are multiplicative inverses. As we've mentioned, the identity quaternion is $(1, 0, 0, 0)$, or $(1, \mathbf{0})$. Multiplying this by any quaternion $\mathbf{q} = (w, \mathbf{v})$ gives

$$\mathbf{q} \cdot (1, \mathbf{0}) = (1 \cdot w - \mathbf{0} \bullet \mathbf{v}, 1\mathbf{v} + w\mathbf{0} + \mathbf{v} \times \mathbf{0})$$
$$= (w, \mathbf{v})$$

In this case, multiplication is commutative, so $\mathbf{q} \cdot (1, \mathbf{0}) = (1, \mathbf{0}) \cdot \mathbf{q} = \mathbf{q}$.

As with matrices, the inverse $\mathbf{q}^{-1}$ of a quaternion $\mathbf{q}$ is one such that $\mathbf{q}^{-1}\mathbf{q} = \mathbf{q}\mathbf{q}^{-1} = (1, \mathbf{0})$. If we consider a quaternion as rotating θ degrees counterclockwise around an axis $\hat{\mathbf{r}}$, then to undo the rotation we should rotate θ degrees clockwise around the same axis. This is the same as rotating $-\theta$ degrees counterclockwise: to create the inverse we negate the angle (Figure 5.7a). So, if

$$(w, \mathbf{v}) = \left(\cos\left(\frac{\theta}{2}\right), \hat{\mathbf{r}}\sin\left(\frac{\theta}{2}\right)\right)$$

then

$$(w, \mathbf{v})^{-1} = \left(\cos\left(-\frac{\theta}{2}\right), \hat{\mathbf{r}}\sin\left(-\frac{\theta}{2}\right)\right)$$
$$= \left(\cos\left(\frac{\theta}{2}\right), -\hat{\mathbf{r}}\sin\left(\frac{\theta}{2}\right)\right) \tag{5.10}$$
$$(w, \mathbf{v})^{-1} = (w, -\mathbf{v})$$

At first glance, negating the vector part of the quaternion (also known as the *conjugate*) to reverse the rotation is counterintuitive. But after some thought this still makes sense geometrically. A clockwise rotation around an axis turns in the same direction as a counterclockwise rotation around the negative of the axis (Figure 5.7b).

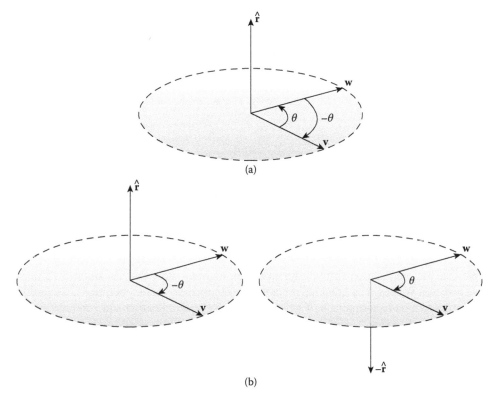

Figure 5.7. (a) Relationship between quaternion and its inverse. Inverse rotates around the same axis but negative angle. (b) Rotation direction around axis by negative angle is the same as rotation direction around negative axis by positive angle.

Equation 5.10 only holds if our quaternion is normalized. While in most cases it should be since we're trying to maintain unit quaternions, if it is not then we need to scale by 1 over the length squared, or

$$\mathbf{q}^{-1} = \frac{1}{\|\mathbf{q}\|^2}(w, -\mathbf{v}) \tag{5.11}$$

Avoiding the floating-point divide in this case is another good reason to keep our quaternions normalized.

It bears repeating that the negative of a quaternion, where both w and $\mathbf{v}$ are negated, is not the same as the inverse. When applied to vectors, the negative rotates the vector to the same orientation, but going the other way around the axis.

5.5.10 Vector Rotation

We can use complex numbers to rotate 2D vectors simply by defining one complex number as $\cos\theta + \sin\theta i$ and the other as our 2D vector $x + yi$, and multiplying to get

$$(\cos\theta + \sin\theta i)(x + yi) = (x\cos\theta - y\sin\theta) + (x\sin\theta + y\cos\theta)i$$

If we consider only the real and imaginary parts, this is exactly the same as Equation 4.7, restricted to the 2D plane. It follows that quaternions might behave similarly.

If **qr** is used to concatenate two quaternions **q** and **r**, then for a vector **p** we might expect **qp** to rotate the vector by the quaternion, just as it does for complex numbers or a matrix. Unfortunately for intuition, this is not the case. For one thing, the result of this multiplication is not a vector (w will not be 0). The actual formula for rotating a vector by a quaternion is

$$R_{\mathbf{q}}\mathbf{p} = \mathbf{qpq}^{-1} \tag{5.12}$$

It may look like the effect of the operation is to perform the rotation and then undo it, but this is not the case. Remember that quaternion multiplication is not commutative, so if **q** is not the identity,

$$\mathbf{qpq}^{-1} \neq \mathbf{qq}^{-1}\mathbf{p} = \mathbf{p}$$

We can use our rotation formula for axis and angle to show that Equation 5.12 does rotate a vector. We begin by breaking it out into its component vector operations. Assuming that our quaternion is normalized, if we expand the full multiplication and combine terms, we get

$$R_{\mathbf{q}}\mathbf{p} = (2w^2 - 1)\mathbf{p} + 2(\mathbf{v} \cdot \mathbf{p})\mathbf{v} + 2w(\mathbf{v} \times \mathbf{p}) \tag{5.13}$$

Substituting $\cos(\theta/2)$ for w, and $\hat{\mathbf{r}} \sin(\theta/2)$ for **v**, we get

$$R_{\mathbf{q}}(\mathbf{p}) = \left(2\cos^2\left(\frac{\theta}{2}\right) - 1\right)\mathbf{p} + \left(\hat{\mathbf{r}}\sin\left(\frac{\theta}{2}\right) \cdot \mathbf{p}\right)\hat{\mathbf{r}}\sin\left(\frac{\theta}{2}\right)$$
$$+ 2\cos\left(\frac{\theta}{2}\right)\left(\hat{\mathbf{r}}\sin\left(\frac{\theta}{2}\right) \times \mathbf{p}\right)$$

Reducing terms and using the appropriate trigonometric identities, we end up with

$$R_{\mathbf{q}}(\mathbf{p}) = \left(\cos^2\left(\frac{\theta}{2}\right) - \sin^2\left(\frac{\theta}{2}\right)\right)\mathbf{p} + 2\sin^2\left(\frac{\theta}{2}\right)(\hat{\mathbf{r}} \cdot \mathbf{p})\hat{\mathbf{r}}$$
$$+ 2\cos\left(\frac{\theta}{2}\right)\sin\left(\frac{\theta}{2}\right)(\hat{\mathbf{r}} \times \mathbf{p}) \tag{5.14}$$
$$= \cos\theta\mathbf{p} + [1 - \cos\theta](\hat{\mathbf{r}} \cdot \mathbf{p})\hat{\mathbf{r}} + \sin\theta(\hat{\mathbf{r}} \times \mathbf{p})$$

We see that Equation 5.14 is equal to Equation 4.9, so our quaternion multiplication—odd as it may look—does rotate a vector around an axis by a given angle.

But what exactly is going on here? The explanation lies in how quaternions actually rotate. Multiplying by a complex number performs a single rotation in the 2D plane. However, multiplying by a quaternion performs two simultaneous rotations in 4D space through two orthogonal planes. Since the planes are orthogonal, the two rotations are independent. If we multiply the quaternion on the left (e.g., **qp** in Equation 5.12), then both rotations are in the same direction (e.g., counterclockwise relative to the two plane normals). Multiplying the quaternion on the right (e.g., $\mathbf{pq}^{-1}$ in Equation 5.12) reverses one of the rotations, but not the other. We want to use quaternions to rotate a vector in $\mathbb{R}^3$, and that only occurs in one plane; therefore, we need to cancel out one of the 4D rotations. So rotating by an angle θ around an axis $\hat{\mathbf{r}}$ becomes a three-step process:

1. Create a quaternion that rotates by $\theta/2$, with one plane orthogonal to $\hat{\mathbf{r}}$ and a second plane orthogonal to the first.

2. Multiply on the left to rotate through both planes by $\theta/2$.

3. Multiply by the inverse on the right to rotate through the desired plane by $\theta/2$, and rotate through the other plane by $-\theta/2$.

Effectively, the second multiplication doubles one of the rotations and eliminates the other. And this explains why we need to use the half angle when creating the quaternion—by doing so we end up with the correct rotation. For more details on this, as well as using this "sandwiching" technique with quaternions to perform reflections and even perspective transformations (covered in Chapter 7), see [55].

In our code, we won't want to use the $\mathbf{qpq}^{-1}$ form, since performing both quaternion multiplications isn't very efficient. Instead, we'll use Equation 5.13:

```
IvVector3
IvQuat::Rotate( const IvVector3& vector ) const
{
    ASSERT( IsUnit() );

    float vMult = 2.0f*(x*vector.x + y*vector.y + z*vector.z);
    float crossMult = 2.0f*w;
    float pMult = crossMult*w - 1.0f;

    return IvVector3( pMult*vector.x + vMult*x + crossMult*(y*vector.z - z*vector.y),
                      pMult*vector.y + vMult*y + crossMult*(z*vector.x - x*vector.z),
                      pMult*vector.z + vMult*z + crossMult*(x*vector.y - y*vector.x) );

}   // End of IvQuat::Rotate()
```

The operation count is more than that of matrix multiplication, but comparable to the Rodrigues formula for axis–angle representation.

An alternate version,

$$R_{\mathbf{q}}\mathbf{p} = (\mathbf{v} \bullet \mathbf{p})\mathbf{v} + w^2\mathbf{p} + 2w(\mathbf{v} \times \mathbf{p}) + \mathbf{v} \times (\mathbf{v} \times \mathbf{p})$$

is useful for processors that have fast cross product operations.

Neither of these formulas is as efficient as matrix multiplication, but for a single vector it is more efficient to perform these operations rather than convert the quaternion to a matrix and then multiply. However, if we need to rotate multiple vectors by the same quaternion, matrix conversion becomes worthwhile.

To see how concatenation of rotations works, suppose we apply a rotation from one quaternion followed by a second rotation from another quaternion. We can rearrange parentheses to get

$$\mathbf{q}(\mathbf{rpr}^{-1})\mathbf{q}^{-1} = (\mathbf{qr})\mathbf{p}(\mathbf{qr})^{-1}$$

As we see, concatenated quaternions will apply their rotation, one after the other. The order is right to left, as we have stated.

If we substitute $-\mathbf{q}$ in place of $\mathbf{q}$ in Equation 5.12, we can see in another way how negating the quaternion doesn't affect rotation. By Equation 5.11, $(-\mathbf{q})^{-1} = -\mathbf{q}^{-1}$, so

$$R_{-\mathbf{q}}(\mathbf{p}) = -\mathbf{q}\mathbf{p}(-\mathbf{q})^{-1}$$
$$= \mathbf{q}\mathbf{p}\mathbf{q}^{-1}$$

The two negatives cancel, and we're back with our familiar result.

Similarly, if $\mathbf{q}$ is a nonunit quaternion, we can show that the same result occurs as if the quaternion were normalized:

$$(s\hat{\mathbf{q}})\mathbf{p}(s\hat{\mathbf{q}})^{-1} = (s\hat{\mathbf{q}})\mathbf{p}\left(\frac{1}{s}\hat{\mathbf{q}}^{-1}\right)$$
$$= s\frac{1}{s}\hat{\mathbf{q}}\mathbf{p}\hat{\mathbf{q}}^{-1}$$
$$= \hat{\mathbf{q}}\mathbf{p}\hat{\mathbf{q}}^{-1}$$

5.5.11 Shortest Path of Rotation

As with the axis–angle format, it is often useful to create a quaternion that rotates a vector $\mathbf{v}_1$ into another vector $\mathbf{v}_2$, although in this case we'll use a different approach discussed by Baker and Norel [9] that also avoids some issues with numerical error when $\mathbf{v}_1$ and $\mathbf{v}_2$ are nearly collinear.

We begin by taking the dot product and cross product of the two vectors:

$$\mathbf{v}_1 \bullet \mathbf{v}_2 = \|\mathbf{v}_1\|\|\mathbf{v}_2\| \cos\theta$$
$$\mathbf{v}_1 \times \mathbf{v}_2 = \|\mathbf{v}_1\|\|\mathbf{v}_2\| \sin\theta\hat{\mathbf{r}}$$

where $\hat{\mathbf{r}}$ is our normalized rotation axis. Using these as the scalar and vector parts, respectively, of a quaternion and normalizing gives us

$$\hat{\mathbf{q}}_1 = (\cos\theta, \sin\theta\hat{\mathbf{r}})$$

This should look familiar from our previous discussion of matrix-to-quaternion conversion. As before, if we add 1 to w,

$$\hat{\mathbf{q}}_h = (\cos\theta + 1, \sin\theta\hat{\mathbf{r}})$$

and normalize, we get

$$\hat{\mathbf{q}} = \left(\cos\frac{\theta}{2}, \sin\frac{\theta}{2}\hat{\mathbf{r}}\right)$$

Note that we haven't handled the case where the two vectors are parallel. In this case, there are an infinite number of possible rotation axes, and hence an infinite number of possible quaternions. A stop-gap solution is to pick one by taking the cross product between one of the vectors and a known vector such as **i** or **j**. While this will work, it may lead to discontinuities—something we'll discuss in Chapter 6 when we cover interpolation.

5.5.12 Quaternions and Transformations

While quaternions are good for rotations, they don't help us much when performing translation and scale. Fortunately, we already have a transformation format that quaternions fit
right into. Recall that in Chapter 4, instead of using a generalized 4×4 matrix for affine
transformations, we used a single scale factor s, a 3×3 rotation matrix **R**, and a translation
vector **t**. Our formula for transformation was

$$\mathbf{p}' = \mathbf{R}(s\mathbf{p}) + \mathbf{t}$$

We can easily replace our matrix **R** with an equivalent quaternion **r**, which gives us

$$\mathbf{p}' = \mathbf{r}(s\mathbf{p})\mathbf{r}^{-1} + \mathbf{t}$$

Concatenation using the quaternion is similar to concatenation with our original separated format, except that we replace multiplication by the rotation matrix with quaternion
operations:

$$s' = s_1 s_0$$
$$\mathbf{r}' = \mathbf{r}_1 \mathbf{r}_0$$
$$\mathbf{t}' = \mathbf{t}_1 + \mathbf{r}_1(s_1 \mathbf{t}_0)\mathbf{r}_1^{-1}$$

Again, to add the translations, we first need to scale $\mathbf{t}_0$ by s_1 and then rotate by the quaternion $\mathbf{r}_1$.

As with lone quaternions, concatenation on a serial processor can be much cheaper in
this format than using a 4×4 matrix. However, transformation of points is more expensive.
As was the case with simple rotation, for multiple points it will be better to convert the
quaternion to a matrix and transform them that way.

5.6 Chapter Summary

In this chapter we've discussed four different representations for orientation and rotation:
matrices, Euler angles, axis and angle, and quaternions. In the introduction we gave three
criteria for our format: it may be informative to compare them along with their usefulness
in interpolation.

As far as size, matrices are the worst at nine values, and Euler angles are the best at
three values. However, quaternions and axis–angle representation are close to Euler angles
at four values, and they avoid the problems engendered by gimbal lock.

For concatenation, quaternions take the fewest number of operations, followed closely by
matrices, and then by axis–angle and the Euler representations. The last two are hampered

by not having low-cost methods for direct concatenation, and so the majority of their expense is tied up in converting to a more favorable format.

When transforming vectors, matrices are the clear winner. Assuming precached sine and cosine data, Euler angles are close behind, while axis–angle representation and quaternions take a bit longer. However, if we don't precache our data, the sine and cosine computations will probably take longer, and quaternions come in second.

Finally, it is worth noting that due to floating-point error, the numbers representing our orientation may drift. The axis–angle and Euler angle formats do not provide an intuitive method for correcting for this. On the other hand, matrices can use Gram–Schmidt orthonormalization and quaternions can perform a normalization step. Quaternions are a clear winner here, as normalizing four values is a relatively inexpensive operation.

For further reading about quaternions, the best place to start is with the writings of Shoemake, in particular [136]. Hamilton's original series of articles on quaternions [69] are in the public domain and can be found by searching online. Courant and Hilbert [27] cover applications of quaternions, in particular to represent rotations. Finally, we recommend two books that provide further insights into quaternions, by Hanson [71] and Goldman [55].

6 Interpolation

6.1 Introduction

Up to this point, we have considered only motions (more specifically, transformations) that have been created programmatically. In order to create a particular motion (e.g., a submarine moving through the world), we have to write a specific program to generate the appropriate sequence of transformations for our model. However, this takes time and it can be quite tedious to move objects in this fashion. It would be much more convenient to predefine our transformation set in a tool and then somehow regenerate it within our game. An artist could create the sequence using a modeling package, and then a programmer would just write the code to play it back, much as a projector plays back a strip of film. This process of pregenerating a set of data and then playing it back is known as *animation*.

The best way to understand animation is to look at the art form in which it has primarily been used: motion pictures. In this case, the illusion of motion is created by drawing or otherwise recording a series of images on film and then projecting them at 24 or 30 frames per second (for film and video, respectively). The illusion is maintained by a property of the eye–brain combination known as persistence of motion: the eye–brain system sees two frames and invisibly (to our perception) fills in the gaps between them, thus giving us the notion of smooth motion.

We could do something similar in our game. Suppose we had a character that we want to move around the world. The artist could generate various animation sets at 60 frames per second (f.p.s.), and then when we want the character to run, we play the appropriate running animation. When we want the character to walk, we switch to the walking animation. The same process can be used for all the possible motions in the game.

However, there are a number of problems with this. First, by setting the animation set to a rate of 60 f.p.s. and then playing it back directly, we have effectively locked the frame rate for the game at 60 f.p.s. as well. Many monitors and televisions can run at 120 Hz, and when running in windowed mode, the graphics can be updated much faster than that. It would be much better if we could find some way to generate 120 f.p.s. or more from a 60 f.p.s. dataset. In other words, we need to take our initial dataset and generate a new one at a different rate. This is known as *resampling*.

This brings us to our second problem. Storing 60 f.p.s. per animation adds up to a lot of data. As an example, if we have 10 data points per model that we're storing, with 16 floats per point (i.e., a 4×4 matrix), that adds up to about 38 KB/s of animation. A minute of animation adds up to over 2 MB of data, which can be a serious hit, particularly if we're running on a low-memory platform such as a mobile phone. It would be better if we could generate our data at a lower rate, say 10 or 15 f.p.s., and then resample up to the speed we need. This is essentially the same problem as our first one—it's just that our initial dataset has fewer samples.

Alternately, we could take another cue from movie animation. The primary animators on a film draw only the important, infrequent "key" frames that capture the essential flow of an animation. The work of generating the remaining "in-between" frames is left to secondary animators, who generate these intermediate frames from the supplied key frames. These artists are known as 'tweeners. In our case, we could store key frames that represent the essential positions of our motion. These key frames would not have to be separated by a constant time interval, rather at smaller intervals when the positions are changing quickly, and at larger intervals when the positions change very slowly. The resampling function would act as our 'tweener for this key frame data.

Fortunately, we have already been introduced to one technique for doing all of this, albeit in another form. This method is known as *interpolation*, and we first saw it when generating a line from two points. Interpolation takes a set of discrete sample points at given time intervals and generates a continuous function that passes through the points. Using this, we can pick any time along the domain of the function and generate a new point so that we might fill in the gaps. We're using the interpolation function to sample at a different rate.

An alternative is approximation, which uses the points to guide the resulting function. In this case, the function does not pass through the points. This may seem odd, but it can help us better control the shape of the function. However, the same principle applies: we generate a function based on the initial sample data and resample later at a different frame rate.

In this chapter, we'll be relying heavily on concepts from calculus—in particular limits, derivatives, and integrals—so it may be worthwhile reviewing them before proceeding (we have a review article on our web site, www.essentialmath.com). We'll be breaking our discussion of interpolation and approximation into three parts. First, we'll look at some techniques for interpolating and approximating position. Next, we'll look at how we can extend those techniques for orientation. Finally, we'll look at some applications, in particular, the motion of a constrained camera.

6.2 Interpolation of Position

6.2.1 General Definitions

The general class of functions we'll be using for both interpolating and approximating are called *parametric curves*. We can think of a curve as a squiggle in space, where the parameter controls where we are in the squiggle. The simplest example of a parametric curve is our old line equation,

$$L(t) = P_0 + (P_1 - P_0)t$$

Here t controls where we are on the line, relative to P_0 and P_1.

When curves are used for animation, our parameter is usually represented by u or t. We can think of this as representing time, although the units used don't necessarily have any relationship to seconds. In our discussion we will use u as the parameter to a uniform curve Q such that $Q(0)$ is the start of the curve and $Q(1)$ is the end. When we want to use a general parameterization, we will use t. In this case, we usually set a time value t_i for each point P_i; we expect to end up at position P_i in space at time t_i. The sequence $t_0, t_1, \ldots, t_n$ is sorted (as are the corresponding points) so that it is monotonically increasing.

We can formally define a parametric curve as a function $Q(u)$ that maps an interval of real values (represented by the parameter u, as above) to a continuous set of points. When mapping to $\mathbb{R}^3$, we commonly use a parametric curve broken into three separate functions, one for each coordinate: $Q(u) = (x(u), y(u), z(u))$. This is also known as a *space curve*.

The term *continuous* in our definition is a difficult one to grasp mathematically. Informally, we can think of a continuous function as one that we can draw without ever lifting the pen from the page—there are no gaps or jumps in the function. Formally, we represent that by saying that a function f is continuous at a value x_0 if

$$\lim_{x \to x_0} f(x) = f(x_0)$$

In addition, we say that a function $f(x)$ is continuous over an interval (a, b) if it is continuous for every value x in the interval. We can also say that the function has positional, or C^0, continuity over the interval (a, b).

This can be taken further: a function $f(x)$ has tangential, or C^1, continuity across an interval (a, b) if the first derivative $f'(x)$ of the function is continuous across the interval. In our case, the derivative $\mathbf{Q}'(u)$ for parameter u is a tangent vector to the curve at location $Q(u)$. Correspondingly, the derivative of a space curve is $\mathbf{Q}'(u) = (x'(u), y'(u), z'(u))$.

Occasionally, we may be concerned with C^2 continuity, also known as curvature continuity. A function $f(x)$ has C^2 continuity across an interval (a, b) if the second derivative $f''(x)$ of the function is continuous across the interval. Higher orders of continuity are possible, but they are not relevant to the discussion that follows.

A few more definitions will be useful to us. The average speed r we travel along a curve is related to the distance d traveled along the curve and the time it takes to travel that distance, namely,

$$r = d/u$$

The instantaneous speed at a particular parameter u is the length of the derivative vector $\mathbf{Q}'(u)$.

A parametric curve $Q(u)$ is defined as *smooth* on an interval $[a, b]$ if it has a continuous derivative on $[a, b]$ and $Q'(u) \neq 0$ for all u in (a, b). For a given point P on a smooth curve $Q(u)$, we define a circle with first and second derivative vectors equal to those at P as the osculating[1] circle. If the radius of the osculating circle is ρ, the *curvature κ* at P is $1/\rho$. The curvature at any point is always nonnegative. The higher the curvature, the more the curve bends at that point; the curvature of a straight line is 0.

In general, it is not practical to construct a single, closed-form polynomial that uses all of the sample points—most of the curves we will discuss use at most four points as their geometric foundation. Instead, we will create a *piecewise curve*. This consists of curve segments that each apply over a sequential subset of the points and are joined together to create a function across the entire domain. How we create this joint determines the type of continuity we will have in our function as whole. We can achieve C^0 continuity by ensuring that the endpoint of one curve segment is equal to the start point of the next segment. In general, this is desirable.

We can achieve C^1 continuity over the entire piecewise curve by guaranteeing that tangent vectors are equal at the end of one segment and the start of the next segment. A related form of continuity in this case is G^1 continuity, where the tangents at a pair of segment endpoints are not necessarily equal but point in the same direction. In many cases G^1 continuity is good enough for our purposes. And as one might expect, we can achieve C^2 continuity by guaranteeing that the second derivative vectors are equal at the end of one segment and the start of the next segment.

6.2.2 Linear Interpolation

6.2.2.1 Definition

The most basic parametric curve is our example above: a line passing through two points. By using the parameterized line equation based on the two points, we can generate any point along the line. This is known as *linear interpolation* and is the most commonly used form of interpolation in game programming, mainly because it is the fastest. From our familiar line equation,

$$Q(u) = P_0 + u(P_1 - P_0)$$

we can rearrange to get

$$Q(u) = (1-u)P_0 + uP_1$$

The value u is the factor we use to control our interpolation, or parameter. Recall that if u is 0, $Q(u)$ returns our starting point P_0, and if u is 1, then $Q(u)$ returns P_1, our endpoint. Values of u between 0 and 1 will return a point along the line segment $\overline{P_0P_1}$. When interpolating, we usually care only about values of u within the interval $[0, 1]$ and, in fact, state that the interpolation is undefined outside of this interval.

It is common when creating parametric curves to represent them as matrix equations. As we'll see later, it makes it simple to set certain conditions for a curve and then solve for the equation we want. The standard matrix form is

$$Q(u) = \mathbf{U} \cdot \mathbf{M} \cdot \mathbf{G}$$

where $\mathbf{U}$ is a row matrix containing the polynomial interpolants we're using: $1, u, u^2, u^3$, and so on; $\mathbf{M}$ is a matrix containing the coefficients necessary for the parametric curve; and

---

[1] So called because it "kisses" up to the point.

G is a matrix containing the coordinates of the geometry that defines the curve. In the case of linear interpolation,

$$\mathbf{U} = \begin{bmatrix} u & 1 \end{bmatrix}$$

$$\mathbf{M} = \begin{bmatrix} -1 & 1 \\ 1 & 0 \end{bmatrix}$$

$$\mathbf{G} = \begin{bmatrix} x_0 & y_0 & z_0 \\ x_1 & y_1 & z_1 \end{bmatrix}$$

Note that the columns of **M** are the $(u, 1)$ coefficients for P_0 and P_1, respectively.

With this formulation, the result **UMG** will be a 1×3 matrix:

$$\mathbf{UMG} = \begin{bmatrix} x(u) & y(u) & z(u) \end{bmatrix}$$

$$= \begin{bmatrix} (1-u)x_0 + ux_1 & (1-u)y_0 + uy_1 & (1-u)z_0 + uz_1 \end{bmatrix}$$

This is counter to our standard convention of using column vectors. However, rather than write out **G** as individual coordinates, we can write **G** as a column matrix of n points, where for linear interpolation this is

$$\mathbf{G} = \begin{bmatrix} P_0 \\ P_1 \end{bmatrix}$$

Then, using block matrix multiplication, the result **UMG** becomes

$$\mathbf{UMG} = (1-u)P_0 + uP_1$$

This form allows us to use a convenient shorthand to represent a general parameterized curve without having to expand into three essentially similar functions.

Recall that in most cases we are given time values t_0 and t_1 that are associated with points P_0 and P_1, respectively. In other words, we want to start at point P_0 at time t_0 and end up at point P_1 at time t_1. These times are not necessarily 0 and 1, so we'll need to remap our time value t in the interval $[t_0, t_1]$ to a parameter u in the interval $[0, 1]$, which we'll use in our original interpolation equation. If we want the percentage u that a time value t lies between t_0 and t_1, we can use the formula

$$u = \frac{t - t_0}{t_1 - t_0} \tag{6.1}$$

Using this parameter u with the linear interpolation will give us the effect we desire. We can use this approach to change any curve valid over the interval $[0, 1]$ and using u as a parameter to be valid over $[t_0, t_1]$ and using t as a parameter.

6.2.2.2   Piecewise Linear Interpolation

Pure linear interpolation works fine if we have only two values, but in most cases, we will have many more than two. How do we interpolate among multiple points? The simplest method is to use piecewise curves; that is, we linearly interpolate from the first point to the second, then from the second point to the third, and so on, until we get to the end. For each pair of points P_i and P_{i+1}, we use Equation 6.1 to adjust the time range $[t_i, t_{i+1}]$ to $[0, 1]$ so we can interpolate properly.

For a given time value t, we need to find the stored time values t_i and t_{i+1} such that $t_i \leq t \leq t_{i+1}$. From there we look up their corresponding P_i and P_{i+1} values and interpolate. If we start with $n+1$ points, we will end up with a series of n segments labeled $Q_0, Q_1, \ldots, Q_{n-1}$. Each Q_i is defined by points P_i and P_{i+1} where

$$Q_i(u) = (1 - u)P_i + uP_{i+1}$$

and $Q_i(1) = Q_{i+1}(0)$. This last condition guarantees C^0 continuity. This is expressed as code as follows:

```
IvVector3 EvaluatePiecewiseLinear( float t, unsigned int count,
                                   const IvVector3* positions,
                                   const float* times)
{
    // handle boundary conditions
    if ( t <= times[0] )
        return positions[0];
    else if ( t >= times[count-1] )
        return positions[count-1];

    // find segment and parameter
    unsigned int i;
    for ( i = 0; i < count-1; ++i )
    {
        if ( t < times[i+1] )
            break;
    }
    float t0 = times[i];
    float t1 = times[i+1];
    float u = (t - t0)/(t1 - t0);

    //evaluate
    return (1-u)*positions[i] + u*positions[i+1];
}
```

In the pseudocode we found the subcurve by using a straight linear search. For large sets of points, using a binary search will be more efficient since we'll be storing the values in sorted order. We can also use temporal coherence: since our time values won't be varying wildly and most likely will be increasing in value, we can first check whether we lie in the interval $[t_i, t_{i+1}]$ from the last frame and then check subsequent intervals.

This works reasonably well and is quite fast, but as Figure 6.1 demonstrates, will lead to sharp changes in direction. If we treat the piecewise interpolation of $n+1$ points as a single function $f(t)$ over $[t_0, t_n]$, we find that the derivative $f'(t)$ is discontinuous at the sample

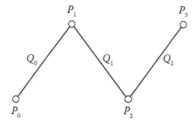

Figure 6.1. Piecewise linear interpolation.

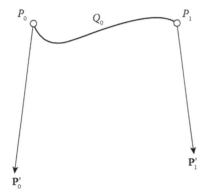

Figure 6.2. Hermite curve.

points, so $f(t)$ is not C^1 continuous. In animation this expresses itself as sudden changes in the speed and direction of motion, which may not be desirable. Despite this, because of its speed, piecewise linear interpolation is a reasonable choice if the slopes of the piecewise line segments are relatively close. If not, or if smoother motion is desired, other methods using higher-order polynomials are necessary.

6.2.3 Hermite Curves

6.2.3.1 Definition

The standard method of improving on piecewise linear equations is to use piecewise cubic curves. If we control the curve properly at each point, then we can smoothly transition from one point to the next, avoiding the obvious discontinuities. In particular, what we want to do is to set up our piecewise curves so that the tangent at the end of one curve matches the tangent at the start of the next curve. This will remove the first-order discontinuity at each point—the derivative will be continuous over the entire time interval that we are concerned with.

Why a cubic curve and not a quadratic curve? Take a look at Figure 6.2. We have set two positions P_0 and P_1, and two tangents $\mathbf{P}_0'$ and $\mathbf{P}_1'$. Clearly, a line won't pass through the two points and also have a derivative at each point that matches its corresponding tangent vectors. The same is true for a parabola. The next order curve is cubic, which will satisfy these conditions. Intuitively, this makes sense. A line is constrained by two points, or one

Source Code

Hermite

point and a vector; a parabola can be defined by three points, or by two points and a tangent; and a cubic curve can be defined by four points, or two points and two tangents.

Using our given constraints, or *boundary conditions*, let's derive our cubic equation. A generalized cubic function and corresponding derivative are

$$Q(u) = \mathbf{a}u^3 + \mathbf{b}u^2 + \mathbf{c}u + D \tag{6.2}$$

$$\mathbf{Q}'(u) = 3\mathbf{a}u^2 + 2\mathbf{b}u + \mathbf{c} \tag{6.3}$$

We'll solve for our four unknowns $\mathbf{a}$, $\mathbf{b}$, $\mathbf{c}$, and D by using our four boundary conditions. We'll assume that when $u = 0$, $Q(0) = P_0$ and $\mathbf{Q}'(0) = \mathbf{P}'_0$. Similarly, at $u = 1$, $Q(1) = P_1$ and $\mathbf{Q}'(1) = \mathbf{P}'_1$. Substituting these values into Equations 6.2 and 6.3, we get

$$Q(0) = D = P_0 \tag{6.4}$$

$$Q(1) = \mathbf{a} + \mathbf{b} + \mathbf{c} + D = P_1 \tag{6.5}$$

$$\mathbf{Q}'(0) = \mathbf{c} = \mathbf{P}'_0 \tag{6.6}$$

$$\mathbf{Q}'(1) = 3\mathbf{a} + 2\mathbf{b} + \mathbf{c} = \mathbf{P}'_1 \tag{6.7}$$

We can see that Equations 6.4 and 6.6 already determine that $\mathbf{c}$ and D are $\mathbf{P}'_0$ and P_0, respectively. Substituting these into Equations 6.5 and 6.7 and solving for $\mathbf{a}$ and $\mathbf{b}$ gives

$$\mathbf{a} = 2(P_0 - P_1) + \mathbf{P}'_0 + \mathbf{P}'_1$$

$$\mathbf{b} = 3(P_1 - P_0) - 2\mathbf{P}'_0 - \mathbf{P}'_1$$

Substituting our now known values for $\mathbf{a}$, $\mathbf{b}$, $\mathbf{c}$, and D into Equation 6.2 gives

$$Q(u) = \left[2(P_0 - P_1) + \mathbf{P}'_0 + \mathbf{P}'_1\right] u^3 + \left[3(P_1 - P_0) - 2\mathbf{P}'_0 - \mathbf{P}'_1\right] u^2 + \mathbf{P}'_0 u + P_0$$

This can be rearranged in terms of the boundary conditions to produce our final equation:

$$Q(u) = (2u^3 - 3u^2 + 1)P_0 + (-2u^3 + 3u^2)P_1 + (u^3 - 2u^2 + u)\mathbf{P}'_0 + (u^3 - u^2)\mathbf{P}'_1$$

This is known as a *Hermite curve*. We can also represent this as the product of a matrix multiplication, just as we did with linear interpolation. In this case, the matrices are

$$\mathbf{U} = \begin{bmatrix} u^3 & u^2 & u & 1 \end{bmatrix}$$

$$\mathbf{M} = \begin{bmatrix} 2 & -2 & 1 & 1 \\ -3 & 3 & -2 & -1 \\ 0 & 0 & 1 & 0 \\ 1 & 0 & 0 & 0 \end{bmatrix}$$

$$\mathbf{G} = \begin{bmatrix} P_0 \\ P_1 \\ \mathbf{P}'_0 \\ \mathbf{P}'_1 \end{bmatrix}$$

We can use either formulation to build piecewise curves just as we did for linear interpolation. As before, we can think of each segment as a separate function, valid over the interval [0, 1]. Then to create a C^1 continuous curve, two adjoining segments Q_i and Q_{i+1} would have to have matching positions such that

$$Q_i(1) = Q_{i+1}(0)$$

and matching tangent vectors such that

$$\mathbf{Q}'_i(1) = \mathbf{Q}'_{i+1}(0)$$

What we end up with is a set of sample positions $\{P_0, \ldots, P_n\}$, tangent vectors $\{\mathbf{P}'_0, \ldots, \mathbf{P}'_n\}$, and times $\{t_0, \ldots, t_n\}$. At a given point adjoining two curve segments Q_i and Q_{i+1},

$$Q_i(1) = Q_{i+1}(0) = P_{i+1}$$
$$\mathbf{Q}'_i(1) = \mathbf{Q}'_{i+1}(0) = \mathbf{P}'_{i+1}$$

Figure 6.3 shows this situation in the piecewise Hermite curve.

The above assumes that our time values occur at uniform intervals; that is, there is a constant Δt between t_0 and t_1, and t_1 and t_2, and so forth. However, as mentioned under linear interpolation, the difference between time values t_i to t_{i+1} may vary from segment to segment. The solution is to do the same thing we did for linear interpolation: if we know that a given value t lies between t_i and t_{i+1}, we can use Equation 6.1 to normalize our time value to the range $0 \le u \le 1$ and use that as our parameter to curve segment Q_i.

This is equivalent to using nonuniform Hermite splines, where the final parameter value is not necessarily equal to 1. These can be derived similarly to the uniform Hermite splines. Assuming a valid range of $[0, t_f]$, their general formula is

$$Q(t) = \left(\frac{2t^3}{t_f^3} - \frac{3t^2}{t_f^2} + 1\right) P_0 + \left(\frac{-2t^3}{t_f^3} + \frac{3t^2}{t_f^2}\right) P_1$$

$$+ \left(\frac{t^3}{t_f^3} - \frac{2t^2}{t_f^2} + \frac{t}{t_f}\right) \mathbf{P}'_0 + \left(\frac{t^3}{t_f^3} - \frac{t^2}{t_f^2}\right) \mathbf{P}'_1$$

In our case, for each (t_i, t_{i+1}) pair, $t_f = t_{i+1} - t_i$.

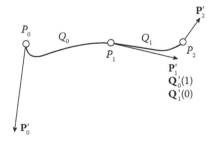

Figure 6.3. Piecewise Hermite curve. Tangents at P_1 match direction and magnitude.

6.2.3.2 Manipulating Tangents

The tangent vectors are used for more than just maintaining first derivative continuity across each sample point. Changing their magnitude also controls the speed at which we move through the point and consequently through the curve. They also affect the shape of the curve. Take a look at Figure 6.4. The longer the vector, the faster we will move and the sharper the curvature. We can create a completely different curve through our sample points, simply by adjusting the tangent vectors.

There is, of course, no reason that the tangents $\mathbf{Q}'_i(1)$ and $\mathbf{Q}'_{i+1}(0)$ have to match. One possibility is to match the tangent directions but not the tangent magnitudes—this gives us G^1 continuity. The resulting function has a discontinuity in its derivative but usually still appears smooth. It also has the advantage that it allows us to control how our curve looks across each segment a little better. For example, it might be that we want to have the appearance of a continuous curve but also be able to have more freedom in how each individual segment is shaped. By maintaining the same direction but allowing for different magnitudes, this function provides for the kind of flexibility we need in this instance (Figure 6.5).

Another possibility is that the tangent directions don't match at all. In this case, we'll end up with a kink, or cusp, in the whole curve (Figure 6.6). While not physically realistic,

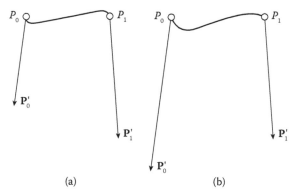

(a)   (b)

Figure 6.4. Hermite curve with (a) small tangent and low curvature and (b) large tangent and higher curvature.

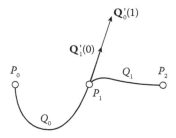

Figure 6.5. Piecewise Hermite curve. Tangents at P_1 have the same direction but differing magnitudes.

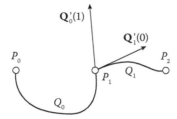

Figure 6.6. Piecewise Hermite curve. Tangents at P_1 have differing directions and magnitudes.

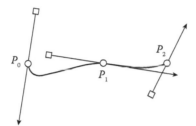

Figure 6.7. Possible interface for Hermite curves, showing in–out tangent vectors.

it does allow for sudden changes in direction. The combination of all the possibilities at each sample point—equal tangents, equal tangent directions with nonequal magnitudes, and nonequal tangent directions—gives us a great deal of flexibility in creating our interpolating function across all the sample points. To allow for this level of control, we need to set two tangents at each internal sample point P_i, which we'll express as $\mathbf{P}'_{i,1}$ (the "incoming" tangent) and $\mathbf{P}'_{i,0}$ (the "outgoing" tangent). Alternatively, we can think of a curve segment as being defined by two points P_i and P_{i+1}, and two tangents $\mathbf{P}'_{i,0}$ and $\mathbf{P}'_{i+1,1}$.

One question remains: How do we generate these tangents? One simple answer is that most existing tools that artists will use, such as Autodesk's Maya and 3D Studio Max, provide ways to set up Hermite curves and their corresponding tangents. When exporting the sample points for subsequent animation, we export the tangents as well. Some tweaking may need to be done to guarantee that the curves generated in internal code match that in the artist program; information on a particular representation is usually available from the manufacturer.

Another common way of generating Hermite data is using in-house tools built for a specific purpose, for example, a tool for managing paths for cameras and other animated objects. In this case, an interface will have to be created to manage construction of the path. One possibility is to click to set the next sample position, and then drag the mouse away from the sample position to set tangent magnitude and direction. A line segment with an arrowhead can be drawn showing the outgoing tangent, and a corresponding line segment with a tail drawn showing the incoming tangent (Figure 6.7).

We will need to modify the tangents so that they can have either different magnitudes or different directions. Many drawing programs control this by allowing three different tangent types. For example, Corel Paint Shop Pro refers to them as symmetric, asymmetric, and cusp. With the symmetric node, clicking and dragging on one of the segment ends rotates both segments and changes their lengths equally, to maintain equal tangents. With an asymmetric node, clicking and dragging will rotate both segments to maintain equal direction but change only the length of the particular tangent clicked on. And with a cusp, clicking and dragging a segment end changes only the length and direction of that tangent. This allows for the full range of possibilities in continuity previously described.

6.2.3.3 Automatic Generation of Hermite Curves

Suppose we don't need the full control of generating tangents for each sample position. Instead, we just want to automatically generate a smooth curve that passes through all the sample points. To do this, we'll need to have a method of creating reasonable tangents for each sample. One solution is to generate a quadratic function using a given sample point and its two neighbors, and then take the derivative of the function to get a tangent value at the sample point. A similar possibility is to take, for a given point P_i, the weighted average of $(P_{i+1} - P_i)$ and $(P_i - P_{i-1})$. However, for both of these it still will be necessary to set a tangent for the two endpoints, since they have only one neighboring point.

Source Code
Demo
AutoHermite

Another method creates tangents that maintain C^2 continuity at the interior sample points. To do this, we'll need to solve a system of linear equations, using our sample points as the known quantities and the tangents as our unknowns. For simplicity's sake, we'll assume we're using uniform curves, and begin by computing the first derivative of the Hermite uniform curve Q:

$$\mathbf{Q}'_i(u) = (6u^2 - 6u)P_i + (-6u^2 + 6u)P_{i+1} + (3u^2 - 4u + 1)\mathbf{P}'_i + (3u^2 - 2u)\mathbf{P}'_{i+1}$$

and from that the second derivative $\mathbf{Q}''$:

$$\mathbf{Q}''_i(u) = (12u - 6)P_i + (-12u + 6)P_{i+1} + (6u - 4)\mathbf{P}'_i + (6u - 2)\mathbf{P}'_{i+1}$$

At a given interior point P_{i+1}, we want the incoming second derivative $\mathbf{P}''_{i+1,1}$ to equal the outgoing second derivative $\mathbf{P}''_{i+1,0}$. We'll assume that each curve segment has a valid parameterization from 0 to 1, so we want

$$\mathbf{Q}''_i(1) = \mathbf{Q}''_{i+1}(0)$$

$$6P_i - 6P_{i+1} + 2\mathbf{P}'_i + 4\mathbf{P}'_{i+1} = -6P_{i+1} + 6P_{i+2} - 4\mathbf{P}'_{i+1} - 2\mathbf{P}'_{i+2}$$

This can be rewritten to place our knowns on one side of the equation and unknowns on the other:

$$2\mathbf{P}'_i + 8\mathbf{P}'_{i+1} + 2\mathbf{P}'_{i+2} = 6[(P_{i+2} - P_{i+1}) + (P_{i+1} - P_i)]$$

This simplifies to

$$\mathbf{P}'_i + 4\mathbf{P}'_{i+1} + \mathbf{P}'_{i+2} = 3(P_{i+2} - P_i)$$

Applying this to all of our sample points $\{P_0, \ldots, P_n\}$ creates $n - 1$ linear equations. This can be written as a matrix product as follows:

$$\begin{bmatrix} 1 & 4 & 1 & & \cdots & 0 & 0 \\ 0 & 1 & 4 & 1 & \cdots & 0 & 0 \\ & & \vdots & & & & \\ 0 & 0 & \cdots & 1 & 4 & 1 & 0 \\ 0 & 0 & \cdots & 0 & 1 & 4 & 1 \end{bmatrix} \begin{bmatrix} \mathbf{P}'_0 \\ \mathbf{P}'_1 \\ \vdots \\ \mathbf{P}'_{n-1} \\ \mathbf{P}'_n \end{bmatrix} = \begin{bmatrix} 3(P_2 - P_0) \\ 3(P_3 - P_1) \\ \vdots \\ 3(P_{n-1} - P_{n-3}) \\ 3(P_n - P_{n-2}) \end{bmatrix}$$

This means we have $n-1$ equations with $n+1$ unknowns. To solve this, we will need two more equations. We have already constrained our interior tangents by ensuring C^2 continuity; what remains is to set our two tangents at each extreme point. One possibility is to set them to given values $\mathbf{v}_0$ and $\mathbf{v}_1$, or

$$\mathbf{Q}'_0(0) = \mathbf{P}'_0 = \mathbf{v}_0 \tag{6.8}$$

$$\mathbf{Q}'_{n-1}(1) = \mathbf{P}'_n = \mathbf{v}_1 \tag{6.9}$$

This is known as a clamped end condition, and the resulting curve is a *clamped cubic spline*. Our final system of equations is

$$\begin{bmatrix} 1 & 0 & 0 & 0 & \cdots & 0 & 0 \\ 1 & 4 & 1 & 0 & \cdots & 0 & 0 \\ 0 & 1 & 4 & 1 & \cdots & 0 & 0 \\ & & \vdots & & & & \\ 0 & 0 & \cdots & 1 & 4 & 1 & 0 \\ 0 & 0 & \cdots & 0 & 1 & 4 & 1 \\ 0 & 0 & \cdots & 0 & 0 & 0 & 1 \end{bmatrix} \begin{bmatrix} \mathbf{P}'_0 \\ \mathbf{P}'_1 \\ \vdots \\ \mathbf{P}'_{n-1} \\ \mathbf{P}'_n \end{bmatrix} = \begin{bmatrix} \mathbf{v}_0 \\ 3(P_2 - P_0) \\ 3(P_3 - P_1) \\ \vdots \\ 3(P_{n-1} - P_{n-3}) \\ 3(P_n - P_{n-2}) \\ \mathbf{v}_1 \end{bmatrix}$$

Solving this system of equations gives us the appropriate tangent vectors. This is not as bad as it might seem. Because this matrix (known as a *tridiagonal matrix*) is sparse and extremely structured, the system is very easy and efficient to solve using a modified version of Gaussian elimination known as the *Thomas algorithm*.

If we express our tridiagonal matrix generally as

$$\begin{bmatrix} b_0 & c_0 & 0 & 0 & \cdots & 0 & 0 \\ a_1 & b_1 & c_1 & 0 & \cdots & 0 & 0 \\ 0 & a_2 & b_2 & c_2 & \cdots & 0 & 0 \\ & & \vdots & & & & \\ 0 & 0 & \cdots & a_{n-2} & b_{n-2} & c_{n-2} & 0 \\ 0 & 0 & \cdots & 0 & a_{n-1} & b_{n-1} & c_{n-1} \\ 0 & 0 & \cdots & 0 & 0 & a_n & b_n \end{bmatrix} \begin{bmatrix} x_0 \\ x_1 \\ \vdots \\ x_{n-1} \\ x_n \end{bmatrix} = \begin{bmatrix} d_0 \\ d_1 \\ d_3 \\ \vdots \\ d_{n-2} \\ d_{n-1} \\ d_n \end{bmatrix}$$

Then we can forward substitute to create array $\mathbf{A}'$ as follows:

$$a_i' = 0$$
$$b_i' = 1$$

$$c_i' = \begin{cases} \frac{c_0}{b_0} & ; \quad i = 0 \\ \frac{c_i}{b_i - c_{i-1}' a_i} & ; \quad 1 \le i \le n - 1 \end{cases}$$

$$d_i' = \begin{cases} \frac{d_0}{b_0} & ; \quad i = 0 \\ \frac{d_i - d_{i-1}' a_i}{b_i - c_{i-1}' a_i} & ; \quad 1 \le i \le n \end{cases}$$

Here $\mathbf{A}'$ and the others represent a modification of their respective counterparts, not a derivative.

We can then solve for $\mathbf{x}$ by using back substitution:

$$x_n = d_n'$$
$$x_i = d_i' - c_i' x_{i+1} \quad ; \quad 0 \le i \le n - 1$$

This is significantly faster than blindly applying Gaussian elimination. In addition to the speed-up, we can also use less space than Gaussian elimination by storing our matrix as three $n + 1$–length arrays: a, b, and c. So the fact that our matrix is tridiagonal leads to a great deal of savings.

6.2.3.4 Natural End Conditions

In the preceding examples, we generated splines assuming that the beginning and end tangents were clamped to values set by the programmer or the user. This may not be convenient; we may want to avoid specifying tangents at all. An alternative approach is to set conditions on the end tangents, just as we did with the internal tangents, to reduce the amount of input needed.

One such possibility is to assume that the second derivative is 0 at the two extremes; that is, $\mathbf{Q}_0''(0) = \mathbf{Q}_{n-1}''(1) = 0$. This is known as a *relaxed* or *natural* end condition, and the spline created is known as a *natural spline*. As the name indicates, this produces a very smooth and natural-looking curve at the endpoints, and in most cases, this is the end condition we would want to use.

With a natural spline, we don't need to specify tangent information at all—we can compute the two unconstrained tangents from the clamped spline using the second derivative condition.

At point P_0, we know that

$$0 = \mathbf{Q}_0''(0)$$
$$= -6P_0 + 6P_1 - 4\mathbf{P}_0' - 2\mathbf{P}_1'$$

As before, we can rewrite this so that the unknowns are on the left side and the knowns on the right:

$$4\mathbf{P}_0' + 2\mathbf{P}_1' = 6P_1 - 6P_0$$

or

$$2\mathbf{P}'_0 + \mathbf{P}'_1 = 3(P_1 - P_0) \tag{6.10}$$

Similarly, at point P_n, we know that

$$0 = \mathbf{Q}''_{n-1}(1)$$
$$= 6P_{n-1} - 6P_n + 2\mathbf{P}'_{n-1} + 4\mathbf{P}'_n$$

This can be rewritten as

$$\mathbf{P}'_{n-1} + 2\mathbf{P}'_n = 3(P_n - P_{n-1}) \tag{6.11}$$

We can substitute Equations 9.12 and 9.13 for our first and last equations in the clamped case, to get the following matrix product:

$$
\begin{bmatrix}
2 & 1 & 0 & 0 & \cdots & 0 & 0 \\
1 & 4 & 1 & 0 & \cdots & 0 & 0 \\
0 & 1 & 4 & 1 & \cdots & 0 & 0 \\
 & & & \vdots & & & \\
0 & 0 & \cdots & 1 & 4 & 1 & 0 \\
0 & 0 & \cdots & 0 & 1 & 4 & 1 \\
0 & 0 & \cdots & 0 & 0 & 1 & 2
\end{bmatrix}
\begin{bmatrix}
\mathbf{P}'_0 \\
\mathbf{P}'_1 \\
\vdots \\
\mathbf{P}'_{n-1} \\
\mathbf{P}'_n
\end{bmatrix}
=
\begin{bmatrix}
3(P_1 - P_0) \\
3(P_2 - P_0) \\
3(P_3 - P_1) \\
\vdots \\
3(P_{n-1} - P_{n-3}) \\
3(P_n - P_{n-2}) \\
3(P_n - P_{n-1})
\end{bmatrix}
$$

Once again, by solving this system of linear equations, we can find the values for our tangents.

6.2.4 Catmull–Rom Splines

An alternative for automatic generation of a parametric curve is the *Catmull–Rom spline*. This takes a similar approach to some of the initial methods we described for Hermite curves (tangent of parabola, weighted average), where tangents are generated based on the positions of the sample points. The standard Catmull–Rom splines create the tangent for a given sample point by taking the neighboring sample points, subtracting to create a vector, and halving the length. So, for sample P_i, the tangent $\mathbf{P}'_i$ is

Source Code
Demo
catmull

$$\mathbf{P}'_i = \frac{1}{2}(P_{i+1} - P_{i-1})$$

If we substitute this into our matrix definition of a Hermite curve between P_i and P_{i+1}, this gives us

$$
Q_i(u) = \begin{bmatrix} u^3 & u^2 & u & 1 \end{bmatrix}
\begin{bmatrix}
2 & -2 & 1 & 1 \\
-3 & 3 & -2 & -1 \\
0 & 0 & 1 & 0 \\
1 & 0 & 0 & 0
\end{bmatrix}
\begin{bmatrix}
P_i \\
P_{i+1} \\
\frac{1}{2}(P_{i+1} - P_{i-1}) \\
\frac{1}{2}(P_{i+2} - P_i)
\end{bmatrix}
$$

We can rewrite this in terms of P_{i-1}, P_i, P_{i+1}, and P_{i+2} to get

$$Q_i(u) = \begin{bmatrix} u^3 & u^2 & u & 1 \end{bmatrix} \frac{1}{2} \begin{bmatrix} -1 & 3 & -3 & 1 \\ 2 & -5 & 4 & -1 \\ -1 & 0 & 1 & 0 \\ 0 & 2 & 0 & 0 \end{bmatrix} \begin{bmatrix} P_{i-1} \\ P_i \\ P_{i+1} \\ P_{i+2} \end{bmatrix}$$

This provides a definition for curve segments Q_1 to Q_{n-2}, so it can be used to generate a C^1 curve from P_1 to P_{n-1}. However, since there is no P_{-1} or P_{n+1}, we once again have the problem that curves Q_0 and Q_{n-1} are not valid due to undefined tangents at the endpoints. And as before, these either can be provided by the artist or programmer, or automatically generated. Parent [117] presents one technique. For P_0, we can take the next two points, P_1 and P_2, and use them to generate a new phantom point, $P_1 + (P_1 - P_2)$. If we subtract P_0 from the phantom point and halve the length, this gives a reasonable tangent for the start of the curve (Figure 6.8). The tangent at P_n can be generated similarly.

Since our knowns for the outer curve segments are two points and a tangent, another possibility is to use a quadratic equation to generate these segments. We can derive this in a similar manner as the Hermite spline equation. The general quadratic equation will have the form

$$Q(u) = \mathbf{a}u^2 + \mathbf{b}u + C \tag{6.12}$$

For the case of Q_0, we know that

$$Q_0(0) = C = P_0$$
$$Q_0(1) = \mathbf{a} + \mathbf{b} + C = P_1$$
$$Q_0'(1) = 2\mathbf{a} + \mathbf{b} = \mathbf{P}_1'$$
$$= \frac{1}{2}(P_2 - P_0)$$

Solving for $\mathbf{a}$, $\mathbf{b}$, and C and substituting into Equation 6.12, we get

$$Q_0(u) = \left(\frac{1}{2}P_0 - P_1 + \frac{1}{2}P_2 \right) u^2 + \left(-\frac{3}{2}P_0 + 2P_1 - \frac{1}{2}P_2 \right) u + P_0$$

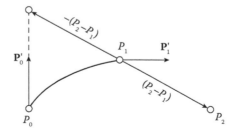

Figure 6.8. Automatic generation of tangent vector at P_1 and P_2.

Rewriting in terms of P_0, P_1, and P_2 gives

$$Q_0(u) = \left(\frac{1}{2}u^2 - \frac{3}{2}u + 1\right)P_0 + \left(-u^2 + 2u\right)P_1 + \left(\frac{1}{2}u^2 - \frac{1}{2}u\right)P_2$$

As before, we can write this in matrix form:

$$Q_0(u) = \begin{bmatrix} u^2 & u & 1 \end{bmatrix} \frac{1}{2} \begin{bmatrix} 1 & -2 & 1 \\ -3 & 4 & -1 \\ 2 & 0 & 0 \end{bmatrix} \begin{bmatrix} P_0 \\ P_1 \\ P_2 \end{bmatrix}$$

A similar process can be used to derive Q_{n-1}:

$$Q_{n-1}(u) = \begin{bmatrix} u^2 & u & 1 \end{bmatrix} \frac{1}{2} \begin{bmatrix} 1 & -2 & 1 \\ -1 & 0 & 1 \\ 0 & 2 & 0 \end{bmatrix} \begin{bmatrix} P_{n-2} \\ P_{n-1} \\ P_n \end{bmatrix}$$

6.2.5 Kochanek–Bartels Splines

Source Code
Demo
Kochanek

An extension of Catmull–Rom splines are Kochanek–Bartels splines [91]. Like Catmull–Rom splines, the tangents are generated based on the positions of the sample points. However, rather than generating a single tangent at each point, Kochanek–Bartels splines separate the incoming and outgoing tangents. In addition, rather than using a fixed function based on the preceding and following points, the tangents are computed from a weighted sum of two vectors: the difference between the following and current point $P_{i+1} - P_i$, and the difference between the current point and the preceding point $P_i - P_{i-1}$.

The weights in this case are based on three parameters: tension (represented as τ), continuity (represented as γ), and bias (represented as β). Because of this, they are also often called TCB splines.

The formulas for the tangents at a sample P_i on a Kochanek–Bartels spline are as follows:

$$\mathbf{P}'_{i,0} = \frac{(1-\tau)(1-\gamma)(1-\beta)}{2}(P_{i+1} - P_i) + \frac{(1-\tau)(1+\gamma)(1+\beta)}{2}(P_i - P_{i-1})$$

$$\mathbf{P}'_{i,1} = \frac{(1-\tau)(1+\gamma)(1-\beta)}{2}(P_{i+1} - P_i) + \frac{(1-\tau)(1-\gamma)(1+\beta)}{2}(P_i - P_{i-1})$$

Note that each of these parameters has a valid range of $[-1, 1]$. Also note that if all are set to 0, then we end up with the formula for a Catmull–Rom spline.

Each parameter has a different effect on the shape of the curve. For example, as the tension at a given control point varies from -1 to 1, the curve passing through the point will change from a very rounded curve to a very tight curve. One can think of it as increasing the influence of the control point on the curve (Figure 6.9a).

Continuity does what one might expect—it varies the continuity at the control point. A continuity setting of 0 means that the curve will have C^1 continuity at that point. As the setting approaches -1 or 1, the curve will end up with a corner at that point; the sign of the continuity controls the direction of the corner (Figure 6.9b).

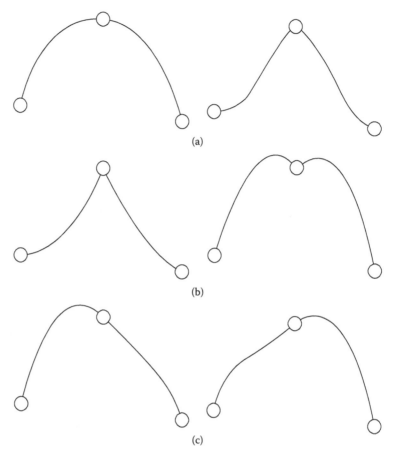

Figure 6.9. Kochanek–Bartels curves. (a) Effect of low versus high tension at central point, (b) effect of low versus high continuity at central point, and (c) effect of low versus high bias at central point.

Bias varies the effect of P_{i+1} and P_{i-1} on the tangents. A bias near -1 means that $P_{i+1} - P_i$ will have the most effect on the tangents; this is called *undershooting*. If the bias is near 1, then $P_i - P_{i-1}$ will have the most effect; this is called *overshooting* (Figure 6.9c).

Note that these splines have the same problem as Catmull–Rom splines with undefined tangents at the endpoints, as there is only one neighboring point. As before, this can be handled by the user setting these tangents by hand or building quadratic curves for the first and last segments. The process for generating these is similar to what we did for Catmull–Rom splines.

Kochanek–Bartels splines are useful because they provide more control over the resulting curve than straight Catmull–Rom splines, and are often used in three-dimensional (3D) packages as an interface to Hermite splines. Because of this, it is useful to be aware of them for use in internal tools and for handling when exporting from commercial software.

6.2.6 Bézier Curves

6.2.6.1 Definition

The previous techniques for generating curves from a set of points meet the functional requirements of controlling curvature and maintaining continuity. However, other than Hermite curves where the tangents are user-specified, they are not so good at providing a means of controlling the shape that is produced. It is not always clear how adjusting the position of a point will change the curve produced, and if we're using a particular type of curve and want to pass through a set of fixed points, there is usually only one possibility.

Bézier curves were created to meet this need. They were devised by Pierre Bézier for modeling car bodies for Renault and further refined by Forrest, Gordon, and Riesenfeld. A cubic Bézier curve uses four *control points*: two endpoints P_0 and P_3 that the curve interpolates, and two points P_1 and P_2 that the curve approximates. Their positions act, as their name suggests, to control the curve. The convex hull, or control polygon, formed by the control points bounds the curve (Figure 6.10). Another way to think of it is that the curve mimics the shape of the control polygon. Note that the four points in this case do not have to be coplanar, which means that the curve generated will not necessarily lie on a plane either.

The tangent vector at point P_0 points in the same direction as the vector $P_1 - P_0$. Similarly, the tangent at P_3 has the same direction as $P_3 - P_2$. As we will see, there is a definite relationship between these vectors and the tangent vectors used in Hermite curves. For now, we can think of the polygon edge between the interpolated endpoint and neighboring control point as giving us an intuitive sense of what the tangent is like at that point.

So far we've only shown cubic Bézier curves, but there is no reason why we couldn't use only three control points to produce a quadratic Bézier curve (Figure 6.11) or more control points to produce higher-order curves. A general Bézier curve is defined by the function

$$Q(u) = \sum_{i=0}^{n} P_i J_{n,i}(u)$$

where the set of P_i are the control points, and

$$J_{n,i}(u) = \binom{n}{i} u^i (1-u)^{n-i}$$

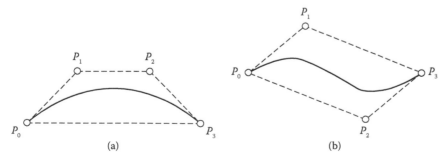

Figure 6.10. Examples of cubic Bézier curve showing convex hull.

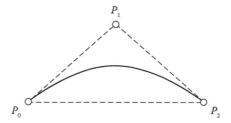

Figure 6.11. Example of quadratic Bézier curve showing convex hull.

where

$$\binom{n}{i} = \frac{n!}{i!(n-i)!}$$

The polynomials generated by $J_{n,i}$ are also known as the Bernstein polynomials, or *Bernstein basis*.

In most cases, however, we will use only cubic Bézier curves. Higher-order curves are more expensive and can lead to odd oscillations in the shape of the curve. Quadratic curves are useful when processing power is limited (a classic example is the game Quake 3) but don't have quite the flexibility of cubic curves. For example, they don't allow for the familiar S shape in Figure 6.10b. To generate something similar with quadratic curves requires two piecewise curves, and hence more data.

The standard representation of an order n Bézier curve is to use an ordered list of points $P_0, \ldots, P_n$ as the control points. Using this representation, we can expand the general definition to get the formula for the cubic Bézier curve:

$$Q(u) = (1-u)^3 P_0 + 3u(1-u)^2 P_1 + 3u^2(1-u)P_2 + u^3 P_3 \tag{6.13}$$

The matrix form is

$$Q(u) = \begin{bmatrix} u^3 & u^2 & u & 1 \end{bmatrix} \begin{bmatrix} -1 & 3 & -3 & 1 \\ 3 & -6 & 3 & 0 \\ -3 & 3 & 0 & 0 \\ 1 & 0 & 0 & 0 \end{bmatrix} \begin{bmatrix} P_0 \\ P_1 \\ P_2 \\ P_3 \end{bmatrix}$$

We can think of the curve as a set of affine combinations of the four points, where the weights are defined by the four basis functions $J_{3,i}$. We can see these basis functions graphed in Figure 6.12. At a given parameter value u, we grab the four basis values and use them to compute the affine combination.

As hinted at, there is a relationship between cubic Bézier curves and Hermite curves. If we set our Hermite tangents to $3(P_1 - P_0)$ and $3(P_3 - P_2)$, substitute those values into our cubic Hermite equation, and simplify, we end up with the cubic Bézier equation.

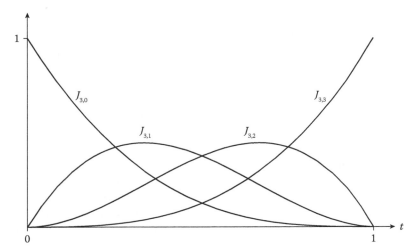

Figure 6.12. Cubic Bézier curve basis functions.

Figure 6.13. Example interface for Bézier curves.

6.2.6.2 Piecewise Bézier Curves

As with linear interpolation and Hermite curves, we can interpolate a curve through more than two points by creating curve segments between each neighboring pair of interpolation points. Many of the same principles apply with Bézier curves as did with Hermite curves. In order to maintain matching direction for our tangents, giving us G^1 continuity, each interpolating point and its neighboring control points need to be collinear. To obtain equal tangents, and therefore C^1 continuity, the control points need to be collinear with and equidistant to the shared interpolating point. Drawing a line segment through the three points gives a three-lobed barbell shape, seen in Figure 6.13.

The barbell makes another very good interface for managing our curves. If we set up our interpolating point as a pivot, then we can grab one neighboring control point and rotate it around to change the direction of the tangent. The other neighboring control point will rotate correspondingly to maintain collinearity and equal distance, and thereby C^1 continuity. If we drag the control point away from our interpolating point, that will increase the length of our tangent. We can leave the other control point at the original distance, if we like, to create different arrival/departure speeds while still maintaining G^1 continuity. Or, we can match its distance from the sample as well, to maintain C^1 continuity. And of course, we can move each neighboring control point independently to create a cusp at that interpolating point.

This seems very similar to our Hermite interface, so the question may be, why use Bézier curves? The main advantage of the Bézier interface over the Hermite interface is that, as mentioned, the control points act to bound the curve, and so give a much better idea of how the shape of the curve will change as we move the control points around. Because of this, many drawing packages use Bézier curves instead of Hermite curves.

While in most cases we will want to make use of user-created data with Bézier curves, it is sometimes convenient to automatically generate them. One possibility is to use the modification of the matrix technique we used with Hermite curves. Alternatively, Parent [117] provides a method for automatically generating Bézier control points from a set of sample positions, as shown in Figure 6.14. Given four points P_{i-1}, P_i, P_{i+1}, and P_{i+2}, we want to compute the two control points between P_i and P_{i+1}. We compute the tangent vector at P_i by computing the difference between P_{i+1} and P_{i-1}. From that we can compute the first control point as $P_i + 1/3(P_{i+1} - P_{i-1})$. The same can be done to create the second control point as $P_{i+1} - 1/3(P_{i+2} - P_i)$. This is very similar to how we created the Catmull–Rom spline, but with tangents twice as large in magnitude.

6.2.7 Other Curve Types

The first set of curves we looked at were interpolating curves, which pass through all the given points. With Bézier curves, the resulting curve interpolates two of the control points, while approximating the others. *B-splines* are a generalization of this—depending on the form of the B-spline, all or none of the points can be interpolated. Because of this, in a B-spline all of the control points can be used as approximating points (Figure 6.15). In fact, B-splines are so flexible they can be used to represent all of the curves we have

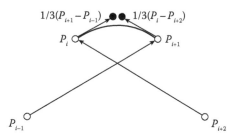

Figure 6.14. Automatic construction of approximating control points with Bézier curve.

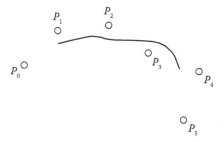

Figure 6.15. B-spline approximating curve.

described so far. However, with flexibility comes a great deal of complexity. Because of this, B-splines are not yet in common usage in games, either for animation or surface construction.

B-splines are computed similarly to Bézier curves. We set up a basis function for each control point in our curve, and then for each parameter value u, we multiply the appropriate basis function by its point and add the results. In general, this can be represented by

$$Q(u) = \sum_{i=0}^{n} P_i B_i(u)$$

where each P_i is a point and B_i is a basis function for that point. The basis functions in this case are far more general than those described for Bézier curves, which gives B-splines their flexibility and their power.

Like our previous piecewise curves, B-splines are broken into smaller segments. The difference is that the number of segments is not necessarily dependent on the number of points, and the intermediary point between each segment is not necessarily one of our control points. These intermediary points are called *knots*. If the knots are spaced equally in time, the curve is known as a *uniform B-spline*; otherwise, it is a *nonuniform B-spline*.

B-splines are not often used for animation; they are more commonly used when building surface representations. A full description of the power and complexity of B-splines is out of the purview of this text, so for those who are interested, more information on B-splines and other curves can be found in Bartels et al. [10], Hughes et al. [82], and Rogers [130].

Another issue is that the curves we have discussed so far have the property that any affine transformation on the set of points (or tangents, in the case of Hermite curves) generating the curve will transform the curve accordingly. So, for example, if we want to transform a Bézier curve from the local frame to the view frame, all we need to do is transform the control points and then generate the curve in the view frame.

However, this will not work for a perspective transformation (see Chapter 7), due to the need for a reciprocal division at each point on the curve. The answer is to apply a process similar to the one we will use when transforming points, by adding an additional parameterized function $w(u)$ that we divide by when generating the points along the curve. This is known as a *rational curve*.

There are a number of uses for rational curves. The first has already been stated: we can use it as a more efficient method for projecting curves. But it also allows us to set weights w_i for the control points so that we can direct the curve to pass closer to one point or another. Another use of rational curves is to create conic section curves, such as circles and ellipses. Nonrational curves, since they are polynomials, can only approximate conic sections.

The most commonly used of the rational curves are nonuniform rational B-splines, or NURBS. Since they can produce conic as well as general curves and surfaces, they are extremely useful in computer-aided design (CAD) systems and modeling for computer animation. Like B-splines, rational curves and particularly NURBS are not yet used much in games because of their relative performance cost and because of concern by artists about lack of control.

6.3   Interpolation of Orientation

So far in our exploration of animation we've considered only interpolation of position. For a coordinate frame, this means only translating the frame in space, without considering rotation. This is fine for moving an object along a path, assuming we wanted it to remain oriented in the same manner as its base frame; however, generally we don't. One possibility is to align the forward vector of the object to the tangent vector of the curve, and use either the second derivative vector or an up vector to build a frame. This will work in general for airplanes and missiles, which tend to orient along their direction of travel. But suppose we want to interpolate a camera so that it travels sideways along a section of curve, or we're trying to model a helicopter, which can face in one direction while moving in another.

Another reason we want to interpolate orientation is for the purpose of animating a character. Usually characters are broken into a scene-graph-like data structure, called the *skeleton*, where each level, or *bone*, is stored at a constant translation from its parent, and only relative rotation is changed to move a particular node (Figure 6.16). So to move a

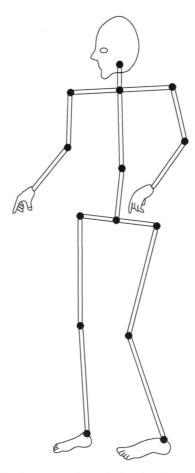

Figure 6.16. Example of skeleton showing relationship between bones.

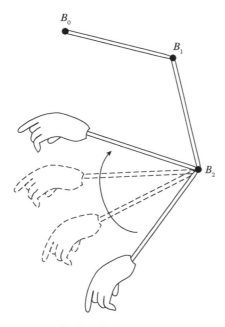

Figure 6.17. Relative bone poses for bending arm.

forearm, for example, we rotate it relative to an upper arm (Figure 6.17). Accordingly, we can generate a set of key frames for an animated character by storing a set of poses generated by setting rotations at each bone. To animate the character, we interpolate from one key frame rotation to another.

As we shall see, when interpolating orientation we can't quite use the same techniques as we did with position. Rotational space doesn't behave in the same way as positional space; we'll be more concerned with interpolating along the surface of a sphere instead of along a line. As part of this, we'll revisit the representations we covered in Chapter 5, discussing the pros and cons of each representation for handling the task of interpolation.

6.3.1 General Discussion

Our interpolation problem for position was to find a space curve—a function given a time parameter that returns a position—that passes through our sample points and maintains our desired curvature at each sample point. The same is true of interpolating orientation, except that our curve doesn't pass through a series of positions, but a series of orientations.

We can think of this as wanting to interpolate from one coordinate frame to another. If we were simply interpolating two vectors $\mathbf{v}_1$ and $\mathbf{v}_2$, we could find the rotation between them via the axis–angle representation $(\theta, \hat{\mathbf{r}})$, and then interpolate by rotating $\mathbf{v}_1$ as

$$\mathbf{v}(t) = R(t\theta, \hat{\mathbf{r}})\mathbf{v}_1$$

In other words, we linearly interpolate the angle from 0 to θ and continually apply the newly generated rotation to $\mathbf{v}_1$ to get our interpolated orientations. But for a coordinate frame, we need to interpolate three vectors simultaneously. We could use the same process for all three basis vectors, but it's not guaranteed that they will remain orthogonal. What we would need to do is find the overall rotation in axis–angle form from one coordinate frame to another, and then apply the process described. This is not a simple thing to do, and as it turns out, there are better ways.

However, for Euler angles and axis–angle formats, we can use this to interpolate simple cases of rotation around a single axis. For instance, if we're interpolating from $(90, 0, 0)$ to $(180, 0, 0)$, we can linearly interpolate the first angle from 90 degrees to 180 degrees. Or, with an axis–angle format, if the rotation is from the reference orientation to another orientation, again we only need to interpolate the angle. Using this method also allows for interpolations over angles greater than 360 degrees. Suppose we want to rotate twice around the z-axis and represent this as only two values. We could interpolate between the two x-y-z Euler angles $(0, 0, 0)$ and $(0, 0, 4\pi)$. As we interpolate from 0 to 1, our object will rotate twice. More sample orientations are needed to do this with matrices and quaternions.

Source Code
Demo
Euler

But extending this to more complex cases does not work. Suppose we are using Euler angles, with a starting orientation of $(0, 90, 0)$ and an ending orientation of $(90, 45, 90)$. If we linearly interpolate the angles to find a value halfway between them, we get $(45, 67.5, 45)$. But this is wrong. One possible value that is correct is $(90, 22.5, 90)$. The consequence of interpolating linearly from one sequence of Euler angles to another is that the object tends to sidle along, rotating around mostly one axis and then switching to rotations around mostly another axis, instead of rotating around a single axis, directly from one orientation to another.

We can mitigate this problem by defining Hermite or higher-order splines to better control the interpolation, and some 3D modeling packages provide output to do just that. However, you may not want to dedicate the space for the intermediary key frames or the processing power to perform the spline interpolation, and it's still an approximation. For more complex cases, the only two formats that are practical are matrices and quaternions, and as we'll see, this is where quaternions truly shine.

There are generally two approaches used when interpolating matrices and quaternions in games: linear interpolation and spherical linear interpolation. Both methods are usually applied piecewise between each orientation sample pair, and even though this will generate discontinuities at the sample points, the artifacts are rarely noticeable. While we will mention some ways of computing cubic curves, they generally are just too expensive for the small gain in visual quality.

6.3.2   Linear Interpolation

Source Code
Demo
LerpSlerp

By using the scalar multiplication and addition operations, we can linearly interpolate rotation matrices and quaternions just as we did vectors. Let's look at a matrix example first. Consider two orientations: one represented as the identity matrix and the other by a rotation of 90 degrees around the z-axis. Using linear interpolation to find the orientation halfway between the start and end orientations, we get

$$\frac{1}{2}\begin{bmatrix} 1 & 0 & 0 \\ 0 & 1 & 0 \\ 0 & 0 & 1 \end{bmatrix} + \frac{1}{2}\begin{bmatrix} 0 & 1 & 0 \\ -1 & 0 & 0 \\ 0 & 0 & 1 \end{bmatrix} = \begin{bmatrix} \frac{1}{2} & \frac{1}{2} & 0 \\ -\frac{1}{2} & \frac{1}{2} & 0 \\ 0 & 0 & 1 \end{bmatrix}$$

The result is not a well-formed rotation matrix. The basis vectors are indeed perpendicular, but they are not unit length. In order to restore this, we need to perform Gram–Schmidt orthogonalization, which is a rather expensive operation to perform every time we want to perform an interpolation.

With quaternions we run into some problems similar to those encountered with matrices. Suppose we perform the same interpolation, from the identity quaternion to a rotation of 90 degrees around z. This second quaternion is $(\sqrt{2}/2, 0, 0, \sqrt{2}/2)$. The resulting interpolated quaternion when $t = 1/2$ is

$$\mathbf{r} = \frac{1}{2}(1, 0, 0, 0) + \frac{1}{2}\left(\frac{\sqrt{2}}{2}, 0, 0, \frac{\sqrt{2}}{2}\right)$$
$$= \left(\frac{2 + \sqrt{2}}{4}, 0, 0, \frac{\sqrt{2}}{4}\right)$$

The length of $\mathbf{r}$ is 0.9239—clearly, not 1. With matrices, we had to reorthogonalize after performing linear interpolation; with quaternions we will have to renormalize. Fortunately, this is a cheaper operation than orthogonalization, so when interpolating orientation, quaternions are our preferred format.

In both cases, this happens because linear interpolation has the effect of cutting across the arc of rotation. If we compare a vector in one orientation with its equivalent in the other, we can get some sense of this. In the ideal case, as we rotate from one vector to another, the tips of the interpolated vectors trace an arc across the surface of a sphere (Figure 6.18). But as we can see in Figure 6.19, the linear interpolation is following a line segment between the two tips of the vectors, which causes the interpolated vectors to shrink to a length of 1/2 at the halfway point, and then back up to 1.

Another problem with linear interpolation is that it doesn't move at a constant rate of rotation. Let's divide our interpolation at the t values $0, 1/4, 1/2, 3/4,$ and 1. In the ideal case, we'll travel one-quarter of the arc length to get from orientation to orientation.

However, when we use linear interpolation, the t value doesn't interpolate along the arc, but along the chord that passes between the start and end orientations. When we divide

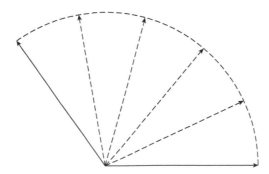

Figure 6.18. Ideal orientation interpolation, showing intermediate vectors tracing a path along the arc.

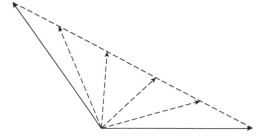

Figure 6.19. Linear orientation interpolation, showing intermediate vectors tracing a path along the line.

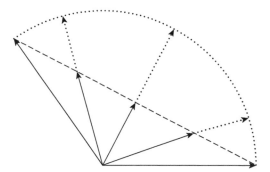

Figure 6.20. Effect of linear orientation interpolation on arc length when interpolating over 1/4 intervals.

the chord into four equal parts, the corresponding arcs on the surface of the sphere are no longer equal in length (Figure 6.20). Those closest to the center of interpolation are longer. The effect is that instead of moving at a constant rate of rotation throughout the interpolation, we will move at a slower rate at the endpoints and faster in the middle. This is particularly noticeable for large angles, as the figure shows. What we really want is a constant change in rotation angle as we apply a constant change in t.

One way to solve both of these issues is to insert one or two additional sample orientations and use quadratic or cubic interpolation. However, these are still only approximations to the spherical curve, and they involve storing additional orientation key frames.

Even if you are willing to deal with nonconstant rotation speed, and eat the cost of orthogonalization, linear interpolation does create other problems. Suppose we use linear interpolation to find the orientation midway between these two matrices:

$$\frac{1}{2}\begin{bmatrix} 0 & 0 & 1 \\ 0 & 1 & 0 \\ -1 & 0 & 0 \end{bmatrix} + \frac{1}{2}\begin{bmatrix} 0 & 0 & -1 \\ 0 & 1 & 0 \\ 1 & 0 & 0 \end{bmatrix} = \begin{bmatrix} 0 & 0 & 0 \\ 0 & 1 & 0 \\ 0 & 0 & 0 \end{bmatrix} \qquad (6.14)$$

This is clearly not a rotation matrix, and no amount of orthogonalization will help us. The problem is that our two rotations (a rotation of $\pi/2$ around y and a rotation of $-\pi/2$ around y, respectively) produce opposing orientations—they're 180 degrees apart. As we interpolate between the pairs of transformed $\mathbf{i}$ and $\mathbf{k}$ basis vectors, we end up passing through the origin.

Quaternions are no less susceptible to this. Suppose we have a rotation of π radians counterclockwise around the y-axis, and a rotation of π radians clockwise around y. Interpolating the equivalent quaternions gives us

$$\mathbf{r} = \frac{1}{2}(0,0,1,0) + \frac{1}{2}(0,0,-1,0)$$
$$= (0,0,0,0)$$

Again, no amount of normalization will turn this into a unit quaternion. The problem here is that we are trying to interpolate between two quaternions that are negatives of each other. They represent two rotations in the opposite direction that rotate to the *same* orientation. Rotating a vector 180 degrees counterclockwise around y will end up in the same place as rotating the same vector 180 degrees clockwise (or -180 degrees counterclockwise) around y. Even if we considered this an interpolation that runs entirely around the sphere, it is not clear which path to take—there are infinitely many.

This problem with negated quaternions shows up in other ways. Let's look at our first example again, interpolating from the identity quaternion to a rotation of $\pi/2$ around z. Recall that our result with $t = 1/2$ was $(2 + \sqrt{2}/4, 0, 0, \sqrt{2}/4)$. This time we'll negate the second quaternion, giving us a rotation of $-3\pi/2$ around z. We get the result

$$\mathbf{r} = \frac{1}{2}(1,0,0,0) + \frac{1}{2}\left(-\frac{\sqrt{2}}{2}, 0, 0, -\frac{\sqrt{2}}{2}\right)$$
$$= \left(\frac{2 - \sqrt{2}}{4}, 0, 0, -\frac{\sqrt{2}}{4}\right)$$

This new result is not the negation of the original result, nor is it the inverse. What is happening is that instead of interpolating along the shortest arc along the sphere, we're interpolating all the way around the other way, via the longest arc. This will happen when the dot product between the two quaternions is negative, so the angle between them is greater than 90 degrees.

This may be the desired result, but usually it's not. What we can do to counteract it is to negate the first quaternion and reinterpolate. In our example, we end up with

$$\mathbf{r} = \frac{1}{2}(-1,0,0,0) + \frac{1}{2}\left(-\frac{\sqrt{2}}{2}, 0, 0, -\frac{\sqrt{2}}{2}\right)$$
$$= \left(-\frac{2 + \sqrt{2}}{4}, 0, 0, -\frac{\sqrt{2}}{4}\right)$$

This gives us the negation of our original result, but this isn't a problem as it will rotate to the same orientation.

This also takes care of the case of interpolating from a quaternion to its negative, so, for example, interpolating from $(0, 0, 1, 0)$ to $(0, 0, -1, 0)$ is

$$\mathbf{r} = -\frac{1}{2}(0, 0, 1, 0) + \frac{1}{2}(0, 0, -1, 0)$$
$$= (0, 0, -1, 0)$$

Negating the first one ends up interpolating to and from the same quaternion, which is a waste of processing power, but won't give us invalid results. Note that we will have to do this even if we are using spherical linear interpolation, which we will address next. All in all, it is better to avoid such cases by culling them out of our data beforehand.

6.3.3   Spherical Linear Interpolation

To better solve the nonconstant rotation speed and normalization issues, we need an interpolation method known as *spherical linear interpolation* (usually abbreviated as *slerp*[2]). Slerp is similar to linear interpolation except that instead of interpolating along a line, we're interpolating along an arc on the surface of a sphere. Figure 6.21 shows the desired result. When using spherical interpolation at quarter intervals of t, we travel one-quarter of the arc length to get from orientation to orientation. We can also think of slerp as interpolating along the angle, or in this case, dividing the angle between the orientations into quarter intervals.

One interesting aspect of orientations is that operations appropriate for positions move up one step in complexity when applied to orientations. For example, to concatenate positions we add, whereas to concatenate orientations we multiply. Subtraction becomes division,

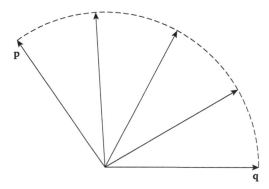

Figure 6.21. Effect of spherical linear interpolation when interpolating at quarter intervals. Interpolates equally along arc and angle.

---

[2] As Shoemake [136] says, because it's fun.

and scalar multiplication becomes exponentiation. Using this knowledge, we can take our linear interpolation function for two rotations P and Q,

$$\text{lerp}\,(P, Q, t) = P + (P - Q)t$$

and convert it to the slerp function,

$$\text{slerp}\,(P, Q, t) = P(P^{-1}Q)^t$$

For matrices, the question is how to take a matrix $\mathbf{R}$ to a power t. We can use a method provided by Eberly [36] as follows. Since we know that $\mathbf{R}$ is a rotation matrix, we can pull out the axis $\mathbf{v}$ and angle θ of rotation for the matrix as we've described, multiply θ by t to get a percentage of the rotation, and convert back to a matrix to get $\mathbf{R}^t$. This is an extraordinarily expensive operation. However, if we want to use matrices, it does give us the result we want of interpolating smoothly along arc length from one orientation to another.

For quaternions, we can derive slerp in another way. Figure 6.22 shows the situation. We have two quaternions $\mathbf{p}$ and $\mathbf{q}$, and an interpolated quaternion $\mathbf{r}$. The angle between $\mathbf{p}$ and $\mathbf{q}$ is θ, calculated as $\theta = \arccos{(\mathbf{p} \cdot \mathbf{q})}$. Since slerp interpolates the angle, the angle between $\mathbf{p}$ and $\mathbf{r}$ will be a fraction of θ as determined by t, or $t\theta$. Similarly, the angle between $\mathbf{r}$ and $\mathbf{q}$ will be $(1 - t)\theta$.

The general interpolation of $\mathbf{p}$ and $\mathbf{q}$ can be represented as

$$\mathbf{r} = a(t)\mathbf{p} + b(t)\mathbf{q} \tag{6.15}$$

The goal is to find two interpolating functions $a(t)$ and $b(t)$ so that they meet the criteria for slerp.

We determine these as follows. If we take the dot product of $\mathbf{p}$ with Equation 6.15 we get

$$\mathbf{p} \cdot \mathbf{r} = a(t)\mathbf{p} \cdot \mathbf{p} + b(t)\mathbf{p} \cdot \mathbf{q}$$
$$\cos{(t\theta)} = a(t) + b(t)\cos{\theta}$$

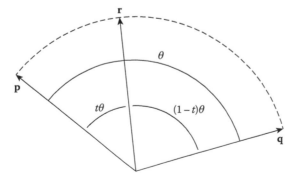

Figure 6.22. Construction for quaternion slerp. Angle θ is divided by interpolant t into subangles $t\theta$ and $(1 - t)\theta$.

Similarly, if we take the dot product of $\mathbf{q}$ with Equation 6.15 we get

$$\cos{(1-t)\theta} = a(t)\cos\theta + b(t)$$

We have two equations and two unknowns. Solving for $a(t)$ and $b(t)$ gives us

$$a(t) = \frac{\cos{(t\theta)} - \cos{(1-t)\theta}\cos\theta}{(1+\cos^2\theta)}$$

$$b(t) = \frac{\cos{(1-t)\theta} - \cos{(t\theta)}\cos\theta}{(1+\cos^2\theta)}$$

Using trigonometric identities, these simplify to

$$a(t) = \frac{\sin{(1-t)\theta}}{\sin\theta}$$

$$b(t) = \frac{\sin{(t\theta)}}{\sin\theta}$$

Our final slerp equation is

$$\mathrm{slerp}(\mathbf{p},\mathbf{q},t) = \frac{\sin{((1-t)\theta)}\mathbf{p} + \sin{(t\theta)}\mathbf{q}}{\sin\theta} \tag{6.16}$$

As we can see, this still is an expensive operation, consisting of three sines and a floating-point divide, not to mention the precalculation of the arccosine. But at 16 multiplications, 8 additions, 1 divide, and 4 transcendentals, it is much cheaper than the matrix method. It is clearly preferable to use quaternions versus matrices (or any other form) if you want to interpolate orientation.

One thing to notice is that as θ approaches 0 (i.e., as $\mathbf{p}$ and $\mathbf{q}$ become close to equal), $\sin\theta$ and thus the denominator of the slerp function approach 0. Testing for equality is not enough to catch this case, because of finite floating-point precision. Instead, we should test $\cos\theta$ before proceeding. If it's close to 1 (greater than $1-\epsilon$, say), then we use linear interpolation or *lerp* instead, since it's reasonably accurate for small angles and avoids the undesirable case of dividing by a very small number. It also has the nice benefit of helping our performance; lerp is much cheaper. In fact, it's generally best only to use slerp in the cases where it is obvious that rotation speed is changing.

Just as we do with linear interpolation, if we want to make sure that our path is taking the shortest route on the sphere and to avoid problems with opposing quaternions, we also need to test $\cos\theta$ to ensure that it is greater than 0 and negate the start quaternion if necessary. While slerp does maintain unit length for quaternions, it's still useful to normalize afterwards to handle any variation due to floating-point error.

6.3.3.1 Cubic Methods

Just as with lerp, if we do piecewise slerp we will have discontinuities at the sample orientations, which may lead to visible changes in orientation rather than the smooth curve we want. And just as we had available when interpolating points, there are cubic methods for interpolating quaternions. One such method is *squad*, which uses the formula

$$\text{squad}(\mathbf{p}, \mathbf{a}, \mathbf{b}, \mathbf{q}, t) = \text{slerp}(\text{slerp}(\mathbf{p}, \mathbf{q}, t), \text{slerp}(\mathbf{a}, \mathbf{b}, t), 2(1 - t)t) \tag{6.17}$$

This is a modification of a technique of using linear interpolation to do Bézier curves, described by Boehm [18]. It performs a Bézier interpolation from $\mathbf{p}$ to $\mathbf{q}$, using $\mathbf{a}$ and $\mathbf{b}$ as additional control points (or control orientations, to be more precise).

We can use similar techniques for other curve types, such as B-splines and Catmull–Rom curves. However, these methods usually are not used in games. They are more expensive than slerp (which is expensive enough), and most of the time the data being interpolated have been generated by an animation package or exist as samples from motion capture. Both of these tend to smooth the data out and insert additional samples at places where orientation is changing sharply, so smoothing the curve isn't that necessary. For those who are interested, Shoemake [136, 137] covers some of these spline methods in more detail.

But beyond the simple cost of operations, blending more than two quaternions with slerp is order dependent—that is, we will most likely get a different result if we blend $\mathbf{p}$ with $\mathbf{q}$, then with $\mathbf{r}$, versus blending $\mathbf{p}$ with $\mathbf{r}$, then $\mathbf{q}$. This becomes an important issue when smoothly blending between animation sets (walking to running, for example). We'd be blending between two animation keys as well as between the two sets, giving us an interpolation between four different quaternions. No matter the order, given four orientations and four weights, we want to end up with the same result, and because of this need, linear interpolation is the only solution.

6.3.4 Performance Improvements

Source Code
Demo
LerpSlerp

As we've seen, using slerp for interpolation, even when using quaternions, can take quite a bit of time—something we don't usually have. A typical character can have 70+ bones, all of which are being interpolated once a frame. If we have a team of characters in a room, there can be up to 20 characters being rendered at one time. The less time we spend interpolating, the better.

The simplest speedup is to use lerp plus normalization all the time. It's very fast: ignoring the setup time (checking angles and adjusting quaternions) and normalization, only 12 basic floating-point operations are necessary on a serial processor, and on a vector processor this drops to 3. We do have the problems with inconsistent rotational speeds, but for animation data this is not an issue for two reasons. First of all, in animation data the angles between the keys are usually less than 90 degrees, so any error is not visually apparent. Secondly, as Blow [15] points out, it's not clear that slerp is the correct solution for animation data in any case. The assumption of slerp is that you want uniform angular speed between orientations, but it's highly likely that's not the case with key-framed data. And if slerp is as much of an approximation of the correct result as lerp, one might as well pick the more efficient method. So in most cases, lerp is a fine solution.

However, if we need to interpolate angles larger than 90 degrees or we are truly concerned with accurate orientations, then we need to try something else. One solution is to improve the speed of slerp. If we assume that we're dealing with a set of stored quaternions for key-framed animation, there are some things we can do here. First of all, we can precompute θ and $1/\sin\theta$ for each quaternion pair and store them with the rest of our animation data.

In fact, if we're willing to give up the space, we could prescale **p** and **q** by $1/\sin\theta$ and store those values instead. This would mean storing up to two copies for each quaternion: one as the starting orientation of an interpolation and one as the ending orientation. Finally, if t is changing at a constant rate, we can use forward differencing to reduce our operations further. Shoemake [137] states that this can be done in eight multiplies, six adds, and two table lookups for the two remaining sines.

Another possibility is to use an alternative slerp formula derived by Blow [15], namely,

$$\mathbf{q}' = \frac{\mathbf{q} - (\mathbf{p}\bullet\mathbf{q})\mathbf{p}}{\|\mathbf{q} - (\mathbf{p}\bullet\mathbf{q})\mathbf{p}\|}$$

$$\mathrm{slerp}(\mathbf{p}, \mathbf{q}', t) = \cos{(t\theta)}\mathbf{p} + \sin{(t\theta)}\mathbf{q}'$$

The idea here is that we create an orthonormal basis by orthogonalizing **q** against **p**, then sweep an angle $t\theta$ from **p** to **q**' to get our final result. Normalizing **q**' is much cheaper than dividing by $\sin\theta$, and $\sin{(t\theta)}$ and $\cos{(t\theta)}$ can be approximated in a single library call.

A third alternative is also proposed by Blow [14]. His idea here is that instead of trying to change our interpolation method to fix our variable rotation speeds, we adjust our t values to counteract the variations. So, in the section where an object would normally rotate faster with a constantly increasing t, we slow t down. Similarly, in the section where an object would rotate slower, we speed t up. Blow uses a cubic spline to perform this adjustment:

$$t' = 2kt^3 - 3kt^2 + (1+k)t$$

where

$$k = 0.5069269(1 - 0.7878088\cos\theta)^2$$

and $\cos\theta$ is the dot product between the two quaternions. This technique tends to diverge from the slerp result when $t > 0.5$, so Blow recommends detecting this case and swapping the two quaternions (i.e., interpolate from **q** to **p** instead of from **p** to **q**). In this way our interpolant always lies between 0 and 0.5.

The nice thing about this method is that it requires very few floating-point operations, doesn't involve any transcendental functions or floating-point divides, and fits in nicely with our existing lerp functions. It gives us slerp interpolation quality with close to lerp speed, which can considerably speed up our animation system.

Further possibilities are provided by Busser [20], who approximates $a(t)$ and $b(t)$ in Equation 6.15 by polynomial equations, and Thomason [145], who explores a variety of techniques. Whether these would be necessary would depend on your data, although in practice we've found Blow's second approach to be sufficient.

6.4 Sampling Curves

Source Code
Library
IvCurves
Filename
IvHermite

Given a parametric curve, it is only natural that we might want to determine a point on it, or *sample* it. We've already stated one reason when motivating interpolation in the introduction: we may have created a curve from a low-resolution set of points, and now want to resample at a higher resolution to match frame rate or simply to provide a better-quality animation. Another purpose is to sample the curve at various points, or tesselate it, so that it might be rendered. After all, artists will want to see, and thus more accurately control, the animation paths that they are creating. Finally, we may also want to sample curves for length calculations, as we'll see later.

Sampling piecewise linear splines is straightforward. For rendering we can just draw lines between the sample points. For animation, the function `EvaluatePiecewiseLinear` will do just fine in computing the 'tween points. A similar approach works well when slerping piecewise quaternion curves.

Things get more interesting when we use a cubic curve. For simplicity's sake, we'll only consider one curve segment Q and a parameter u within that segment—determining those are similar to our linear approach. The most direct method is to take the general function for our curve segment $Q(u) = \mathbf{a}u^3 + \mathbf{b}u^2 + \mathbf{c}u + D$ and evaluate it at our u values. Assuming that we're generating points in $\mathbb{R}^3$, this will take 11 multiplies and 9 adds per point (we save 3 multiplies by computing u^3 as $u \bullet u^2$).

An alternative which is slightly faster is to use Horner's rule [80], which expresses the same cubic curve as

$$Q(u) = ((\mathbf{a}u + \mathbf{b})u + \mathbf{c})u + D$$

This will take only nine multiplies and nine adds per point. In addition, it can actually improve our floating-point accuracy under certain circumstances, and on many systems the $\mathbf{a}u + \mathbf{b}$ can be performed as a series of multiply–add instructions, if not as a single-vector operation.

One issue with Horner's method, however, is that for deeply pipelined processors it can stall the pipeline waiting for one result before proceeding to the next step. For example, we must wait for $\mathbf{a}u + \mathbf{b}$ before calculating $(\mathbf{a}u + \mathbf{b})u + \mathbf{c}$. Estrin's scheme [42] can manage this better at the cost of a few more operations:

$$Q(u) = (\mathbf{a}u + \mathbf{b})u^2 + \mathbf{c}u + D$$

This can be staged as

$$\mathbf{t}_0 = \mathbf{a}u + \mathbf{b}$$
$$\mathbf{t}_1 = u^2$$
$$\mathbf{t}_2 = \mathbf{c}u + D$$
$$Q(u) = \mathbf{t}_0\mathbf{t}_1 + \mathbf{t}_2$$

As with Horner's method, many of these steps can be performed as a series of multiply–add operations.

6.4.1 Forward Differencing

Previously we assumed that there is no pattern to how we evaluate our curve. Suppose we know that we want to sample our curve at even intervals of u, say at a time step of every h. This gives us a list of $n + 1$ parameter values: $0, h, 2h, \ldots, nh$. In such a situation, we can use a technique called *forward differencing*.

For the time being, let's consider computing only the x values for our points. For a given value x_i, located at parameter u, we can compute the next value x_{i+1} at parameter $u + h$. Subtracting x_i from x_{i+1}, we get

$$x_{i+1} - x_i = x(u + h) - x(u) = \Delta x_1(u)$$

We'll label this difference between x_{i+1} and x_i as $\Delta x_1(u)$. For a cubic curve this equals

$$
\begin{aligned}
\Delta x_1(u) &= a(u+h)^3 + b(u+h)^2 + c(u+h) + d - (au^3 + bu^2 + cu + d) \\
&= a(u^3 + 3hu^2 + 3h^2u + h^3) + b(u^2 + 2hu + h^2) + c(u+h) + d \\
&\quad - au^3 - bu^2 - cu - d \\
&= au^3 + 3ahu^2 + 3ah^2u + ah^3 + bu^2 + 2bhu + bh^2 + cu + ch + d \\
&\quad - au^3 - bu^2 - cu - d \\
&= 3ahu^2 + 3ah^2u + ah^3 + 2bhu + bh^2 + ch \\
&= (3ah)u^2 + (3ah^2 + 2bh)u + (ah^3 + bh^2 + ch)
\end{aligned}
$$

Pseudocode to compute the set of values might look like the following:

```
u = 0;
x = d;
output(x);
dx1 = ah^3 + bh^2 + ch;
for ( i = 1; i <= n; i++ )
{
    u += h;
    x += dx1;
    output(x);
    dx1 = (3ah)u^2 + (3ah^2 + 2bh)u + (ah^3 + bh^2 + ch);
}
```

While we have removed the cubic equation, we have introduced evaluation of a quadratic equation $\Delta x_1(u)$. Fortunately, we can perform the same process to simplify this equation. Computing the difference between $\Delta x_1(u + h)$ and $\Delta x_1(u)$ as $\Delta x_2(u)$, we get

$$
\begin{aligned}
\Delta x_2(u) &= \Delta x_1(u + h) - \Delta x_1(u) \\
&= (3ah)(u+h)^2 + (3ah^2 + 2bh)(u+h) + (ah^3 + bh^2 + ch) \\
&\quad - [(3ah)u^2 + (3ah^2 + 2bh)u + (ah^3 + bh^2 + ch)] \\
&= 3ahu^2 + 6ah^2u + 3ah^3 + (3ah^2 + 2bh)u + 3ah^3 + 2bh^2 \\
&\quad + (ah^3 + bh^2 + ch) - [(3ah)u^2 + (3ah^2 + 2bh)u + (ah^3 + bh^2 + ch)] \\
&= 6ah^2u + (6ah^3 + 2bh^2)
\end{aligned}
$$

This changes our pseudocode to the following:

```
u = 0;
x = d;
```

```
output(x);
dx1 = ah^3 + bh^2 + ch;
dx2 = 6ah^3 + 2bh^2;
for ( i = 1; i <= n; i++)
{
    u += h;
    x += dx1;
    output(x);
    dx1 += dx2;
    dx2 = 6ah^2u + (6ah^3 + 2bh^2);
}
```

We can carry this one final step further to remove the linear equation for Δx_2. Computing the difference between $\Delta x_2(u+h)$ and $\Delta x_2(u)$ as $\Delta x_3(u)$, we get

$$
\begin{aligned}
\Delta x_3(u) &= \Delta x_2(u+h) - \Delta x_2(u) \\
&= 6ah^2(u+h) + (6ah^3 + 2bh^2) \\
&\quad - 6ah^2u + (6ah^3 + 2bh^2) \\
&= 6ah^2u + 6ah^3 + (6ah^3 + 2bh^2) \\
&\quad - 6ah^2u + (6ah^3 + 2bh^2) \\
&= 6ah^3
\end{aligned}
$$

Our final code for forward differencing becomes the following:

```
x = d;
output(x);
dx1 = ah^3 + bh^2 + ch;
dx2 = 6ah^3 + 2bh^2;
dx3 = 6ah^3;
for ( i = 1; i <=n; i++ )
{
    x += dx1;
    output(x);
    dx1 += dx2;
    dx2 += dx3;
}
```

We have simplified our evaluation of x from three multiplies and three adds, down to three adds. We'll have to perform similar calculations for y and z, with differing deltas and a, b, c, and d values for each coordinate, giving a total of nine adds for each point.

Note that forward differencing is only possible if the time steps between each point are equal. Because of this, in general we can't use it for animating along a curve, as time between frames may vary from frame to frame. In this case, the appropriate Horner's rule for our degree of polynomial is the most efficient solution. Another issue to be aware of is as the value h gets small, floating-point error may accumulate and cause the endpoint to be miscalculated. In this case, one of the other subdivision methods may be more appropriate.

6.4.2   Midpoint Subdivision

An alternative method for generating points along a curve is to recursively subdivide the curve until we have a set of subcurves, each of which can be approximated by a line segment. This subdivision is usually set to stop at a certain resolution that depends on our needs. This may end up with a more accurate and more efficient representation of the curve than forward differencing, since more curve segments will be generated in areas with high curvature (areas that we might cut across with forward differencing) and fewer in areas with lower curvature.

We can perform this subdivision by taking a curve $Q(u)$ and breaking it into two new curves $L(s)$ and $R(t)$, usually at the midpoint $Q(1/2)$. In this case, $L(s)$ is the subcurve of $Q(u)$ where $0 \leq u \leq 1/2$, and $R(t)$ is the subcurve where $1/2 \leq u \leq 1$. The parameters s and t are related to u by

$$s = 2u$$
$$t = 2u - 1$$

Each subcurve is then tested for relative "straightness"—if it can be approximated well by a line segment, we stop subdividing; otherwise, we keep going. The general algorithm looks like the following:

```
void
RenderCurve( Q )
{
    if ( Straight( Q ) )
        DrawLine( Q(0), Q(1) );
    else
    {
        MidpointSubdivide( Q, &L, &R );
        RenderCurve( L );
        RenderCurve( R );
    }
}
```

There are a few ways of testing how straight a curve is. The most accurate is to measure the length of the curve and compare it to the length of the line segment between the curve's two extreme points. If the two lengths are within a certain tolerance ϵ, then we can say the curve is relatively straight. This assumes that we have an efficient method for computing the arc length of a curve. We discuss some ways of calculating this in the next section.

Another method is to use the two endpoints and the midpoint (Figure 6.23a). If the distance between the midpoint and the line segment formed by the two endpoints is close to 0, then we can usually say that the curve is relatively close to a line segment. The one exception is when the curve crosses the line segment between the two endpoints (Figure 6.23b), which will result in a false positive when clearly the curve is not straight. To avoid the worst examples of this case, Parent [117] recommends performing forward differencing down to a certain level and only then adaptively subdividing.

The convex hull properties of the Bézier curve lead to a particularly efficient method for testing straightness, with no need to calculate a midpoint. If the interior control points are incident with the line segment formed by the two exterior control points, the area of the convex hull is 0, and the curve generated is itself a line segment. So for a cubic Bézier

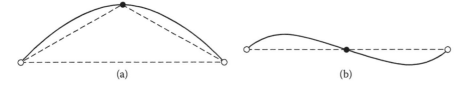

Figure 6.23. Midpoint test for curve straightness. (a) Total distance from endpoints to midpoint (black dot) is compared to distance between endpoints. (b) Example of midpoint test failure.

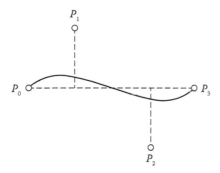

Figure 6.24. Test of straightness for Bézier curve. Measure distance of P_1 and P_2 to line segment P_0P_3.

curve, we can test the squared distance between the line segment formed by P_0 and P_3 and the two control points P_1 and P_2 (Figure 6.24). If both squared distances are less than some tolerance value, then we can say that the curve is relatively straight.

How we subdivide the curve if it fails the test depends on the type of curve. The simplest curves to subdivide are Bézier curves. To achieve this, we will generate new control points for each subcurve from our existing control points. So for a cubic curve, we will compute new control points L_1, L_2, L_3, and L_4 for curve L, and new control points R_1, R_2, R_3, and R_4 for curve R. These can be built by using a technique devised by de Casteljau. This method—known as *de Casteljau's method*—geometrically evaluates a Bézier curve at a given parameter u, and as a side effect creates the new control points needed to subdivide the curve at that point.

Figure 6.25 shows the construction for a cubic Bézier curve. L_0 and R_3 are already known: They are the original control points P_0 and P_3, respectively. Point L_1 lies on segment $\overline{P_0P_1}$ at position $(1-u)P_0 + uP_1$. Similarly, point H lies on segment $\overline{P_1P_2}$ at $(1-u)P_1 + uP_2$, and point R_2 at $(1-u)P_2 + uP_3$. We then linearly interpolate along the newly formed line segments $\overline{L_1H}$ and $\overline{HR_2}$ to form $L_2 = (1-u)L_1 + uH$ and $R_1 = (1-u)H + uR_2$. Finally, we split segment $\overline{L_2R_1}$ to find $Q(u) = L_3 = R_1 = (1-u)L_2 + uR_1$.

Using the midpoint to subdivide is particularly efficient in this case. It takes only six adds and six multiplies (to perform the division by 2).

```
L0 = P0;
R3 = P3;
L1 = (P0 + P1) * 0.5f;
```

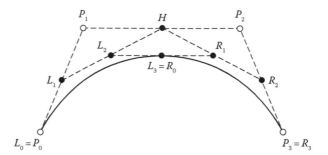

Figure 6.25. De Casteljau's method for subdividing Bézier curves.

```
 H = (P1 + P2) * 0.5f;
R2 = (P2 + P3) * 0.5f;
L2 = (L1 + H) * 0.5f;
R1 = (H + R2) * 0.5f;
L3 = R0 = (L2 + R1) * 0.5f;
```

Subdividing other types of curves, in particular B-splines, can be handled by using an extension of this method devised by Boehm [17]. More information on Boehm subdivision and knot insertion can be found in Bartels et al. [10].

6.4.3 Computing Arc Length

We can informally define s, the *arc length* between points $Q(u_1)$ and $Q(u_2)$ on a continuous curve Q, as the distance along the curve between those two points. At first glance, computing the length of a curve may not appear to be very related to sampling and tessellation. However, as mentioned above, some methods for subdividing a curve require knowing the arc lengths of subsections of that curve. Also, as we'll see, some arc length methods require sampling the curve to obtain their results.

The most accurate method of computing the length of a smooth curve $Q(u)$ from $Q(a)$ to $Q(b)$ is to directly compute the line integral:

$$ s = \int_a^b \left\| Q'(u) \right\| du $$

Unfortunately, for most cubic polynomial curves, it is not possible to find an analytic solution to this integration. For quadratic curves, there is a closed-form solution, but evaluating the resulting functions is more expensive than using a numerical method that gives similar accuracy. In any case, if we wish to vary our curve types, we would have to redo the calculation and so it is not always practical.

The usual approach is to use a numerical method to solve the integral. There are many methods, which Burden and Faires [19] cover in some detail. In this case, the most efficient for its accuracy is Gaussian quadrature, since it attempts to minimize the number of function evaluations, which can be expensive. It approximates a definite integral from -1 to 1 by a weighted sum of unevenly spaced function evaluations, or

$$\int_{-1}^{1} f(x)dx \approx \sum_{i=1}^{n} c_i f(x_i)$$

The actual c_i and x_i values depend on n and are carefully selected to give the best approximation to the integral. The first few values are

n	x_i	c_i
2	$\pm\sqrt{1/3}$	1
3	0	8/9
	$\pm\sqrt{3/5}$	5/9
4	± 0.3399810436	0.6521451549
	± 0.8611363116	0.3478548451
5	0.0000000000	0.5688888889
	± 0.5384693101	0.4786286705
	± 0.9061798459	0.2369268850

Burden and Faires [19] describe in detail how these are derived for arbitrary values of n.

The restriction that we have to integrate over $[-1, 1]$ is not a serious obstacle. For a general definite integral over $[a, b]$, we can remap to $[-1, 1]$ by

$$\int_{a}^{b} f(x)dx = \int_{-1}^{1} f\left(\frac{(b-a)t+b+a}{2}\right)\frac{b-a}{2}dt$$

Guenter and Parent [67] describe a method that uses Gaussian quadrature in combination with adaptive subdivision to get very efficient results when computing arc length. Similar to using adaptive subdivision for rendering, we cut the current curve segment in half. We use Gaussian quadrature to measure the length of each half, and compare their sum to the length of the entire curve, again computed using Gaussian quadrature. If the results are close enough, we stop and return the sum of lengths of the two halves. Otherwise, we recursively compute their lengths via subdivision.

There are other arc length methods that don't involve computing the integral in this manner. One is to subdivide the curve and use the sums of the lengths of the line segments created to approximate arc lengths at each of the subdivision points. We can create a sorted table of pairs (u_i, s_i), where u_i is the parameter for each subdivision, and s_i is the corresponding length at the point $Q(u_i)$. Since both u and len are monotonically increasing, we can sort by either parameter. An example of such a table can be seen in Table 6.1.

To find the length from the start of the curve for a given u, we search through the table to find the two neighboring entries with parameters u_k and u_{k+1} such that $u_k \leq u \leq u_{k+1}$.

Table 6.1. Mapping Parameter Value to Arc Length

u	s
0.0	0.0
0.1	0.2
0.15	0.3
0.29	0.7
0.35	0.9
0.56	1.1
0.72	1.6
0.89	1.8
1.00	1.9

Since the entries are sorted, this can be handled efficiently by a binary search. The length then can be approximated by linearly interpolating between the two entries:

$$s \approx \frac{u_{k+1} - u}{u_{k+1} - u_k} s_k + \frac{u - u_k}{u_{k+1} - u_k} s_{k+1}$$

A higher-order curve can be used to get a better approximation.

To find the length between two parameters a and b where $a \le b$, we compute the length for each and subtract one from the other, or

$$\text{length}(Q, a, b) = \text{length}(Q, b) - \text{length}(Q, a)$$

If we are using cubic Bézier curves, we can use a method described by Gravesen [61]. First of all, given a parameter u, we can subdivide the curve (using de Casteljau's method) to be the subcurve from $[0, u]$. The new control points for this new subcurve can be used to calculate bounds on the length. The length of the curve is bounded by the length of the chord $\overline{P_0 P_3}$ as the minimum, and the sum of the lengths of the line segments $\overline{P_0 P_1}$, $\overline{P_1 P_2}$, and $\overline{P_2 P_3}$ as the maximum. We can approximate the arc length by the average of the two, or

$$L_{min} = \|P_3 - P_0\|$$
$$L_{max} = \|P_1 - P_0\| + \|P_2 - P_1\| + \|P_3 - P_2\|$$
$$L \approx \frac{1}{2}(L_{min} + L_{max})$$

The error can be estimated by the square of the difference between the minimum and maximum:

$$\xi = (L_{max} - L_{min})^2$$

If the error is judged to be too large, then the curve can be subdivided and the length becomes the sum of the lengths of the two halves. Gravesen [61] states that for m subdivisions the error drops to 0 as 2^{-4m}.

A final alternative is presented by Vincent and Forsey [140]. Their method notes that for three neighboring curve points P_0, P_1, and P_2, the length of arc through them can be approximated by $D_2 + (D_2 - D_1)/3$, where

$$D_1 = \|\overline{P_0P_2}\|$$
$$D_2 = \|\overline{P_0P_1}\| + \|\overline{P_1P_2}\|$$

This assumes that P_1 is relatively equidistant from P_0 and P_2, and the arc has low curvature. To improve accuracy, estimates for a given segment (say, $\overline{P_1P_2}$) can be computed using the neighboring point on either side (i.e., in our case, one estimate is computed using P_0 and another using P_3). These are averaged to give the final result.

The general algorithm begins by sampling an odd number of points across the curve. For each interior segment, it determines if more samples need to be taken due to high curvature. If so, it computes those recursively until a low-enough curvature is reached. Otherwise, it computes the segment length and returns. For general curves, this technique is slightly more expensive than Gaussian quadrature, but it does handle certain pathological cases (such as cusps) better.

6.5 Controlling Speed along a Curve

6.5.1 Moving at Constant Speed

Source Code
Demo
SpeedControl

One common requirement for animation is that the object animated move at a constant speed along a curve. However, in most interesting cases, using a given curve directly will not achieve this. The problem is that in order to achieve variety in curvature, the first derivative must vary as well, and hence the distance we travel in a constant time will vary depending on where we start on the curve. For example, Figure 6.26 shows a curve subdivided at equal intervals of the parameter u. The lengths of the subcurves generated vary greatly from one to another.

Ideally, given a constant rate of travel r and time of travel t, we'll want to cover a distance of $s = rt$. So given a starting parameter u_1 on the curve, we want to find the parameter u_2 such that the arc length between $Q(u_1)$ and $Q(u_2)$ equals s.

We've already discussed a number of methods for computing the arc length of a curve. Regardless of the method we use, we'll assume we have some function $G(u)$ that returns the length s from $Q(0)$ to $Q(u)$. So, for the case where $u_1 = 0$, we can use the inverse function $G^{-1}(s)$ to determine the parameter u_2, given an input length s. This is known as a reparameterization by arc length. Unfortunately, in general the arc length function for

Figure 6.26. Parameter-based subdivision of curve, showing nonequal segment lengths.

a parameterized curve is impossible to invert in terms of a finite number of elementary functions, so numerical methods are used instead.

One way is to note that finding u_2 is equivalent to the problem of finding the solution u of the equation

$$s - \text{length}(u_1, u) = 0 \qquad (6.18)$$

A method that allows us to solve this is Newton–Raphson root finding. Burden and Faires [19] present a derivation for this using the Taylor series expansion.

Suppose we have a function $f(x)$ where we want to find p such that $f(p) = 0$. We begin with a guess for p, which we'll call $\bar{x}$, such that $f'(\bar{x}) \neq 0$ and $|p - \bar{x}|$ is relatively small. In other words, $\bar{x}$ may not quite be p but it's a pretty good guess. If we use $\bar{x}$ as a basis for the Taylor series polynomial, we get

$$f(x) = f(\bar{x}) + (x - \bar{x})f'(\bar{x}) + \frac{1}{2}(x - \bar{x})^2 f''(\xi(x))$$

We assume that $\xi(x)$ is bounded by x and $\bar{x}$, so we can ignore the remainder of the terms. If we substitute p for x, then $f(p) = 0$ and

$$0 = f(\bar{x}) + (p - \bar{x})f'(\bar{x}) + \frac{1}{2}(p - \bar{x})^2 f''(\xi(x))$$

Since $|p - \bar{x}|$ is relatively small, we assume that $(p - \bar{x})^2$ is small enough that we can ignore it, and so

$$0 \approx f(\bar{x}) + (p - \bar{x})f'(x)$$

Solving for p gives us

$$p \approx \bar{x} - \frac{f(\bar{x})}{f'(\bar{x})} \qquad (6.19)$$

This gives us our method. We make an initial guess $\bar{x}$ at the solution and use the result of Equation 6.19 to get a more accurate result p. If p still isn't close enough, then we feed it back into the equation as $\bar{x}$ to get a still more accurate result, and so on until we reach a solution of sufficient accuracy or after a given number of iterations is performed.

For our initial guess in solving Equation 6.18, Eberly [35] recommends taking the ratio of our traveled length to the total arc length of the curve and mapping it to our parameter space. Assuming our curve is normalized so that u is in $[0, 1]$, then pseudocode for our root-finding method will look like the following:

```
float FindParameterByDistance( float u1, float s )
{
    // ensure that we remain within valid parameter space
    if (s > ArcLength(u1,1.0f))
        return 1.0f;

    // get total length of curve
    float len = ArcLength(0.0f,1.0f);
```

```
    // make first guess
    float p = u1 + s/len;

    for (int i = 0; i < MAX_ITER; ++i)
    {
        // compute function value and test against zero
        float func = ArcLength(u1,p) - s;
        if ( fabsf(func) < EPSILON )
        {
            return p;
        }

         // perform Newton-Raphson iteration step
        p -= func/Length(Derivative(p));
    }

    // done iterating, return last guess
    return p;
}
```

The first test ensures that the distance we wish to travel is not greater than the remaining length of the curve. In this case, we assume that this is the last segment of a piecewise curve and just jump to the end. A more robust implementation should subtract the remaining length from the distance and restart at the beginning of the next segment.

Computing the derivative of the curve is simple, as this is easily derived from the definition of the curve, as we did for clamped and natural splines. However, there is a serious problem if `Length(Derivative(p))` is zero or near zero. This will lead to a division by zero and we will end up subtracting infinity or NaN from p, which will give us a garbage result.

The solution is to use an alternative root-finding technique known as bisection. It makes use of the mean value theorem, which states that if you have a function $f(x)$ that's continuous on $[a, b]$ and $f(a)f(b) < 0$ (i.e., $f(a)$ and $f(b)$ have opposite signs), then there is some value p between a and b where $f(p) = 0$. This is definitely true in our case. The length of the curve is monotonically increasing, so there will be only one zero. If it's not at the beginning, then $f(a) = length(a) - s = -s$, which is less than 0. If it's not at the end, then $f(b) = length(b) - s$, which is greater than 0. Our endpoints have differing signs, so we can use the bisection method.

This is solved by doing a binary search: we cut the problem interval in two and search further in the more fruitful half. The problem with bisection is that it converges considerably slower than Newton–Raphson, so sometimes we want to use Newton–Raphson and sometimes bisection. Therefore, our hybrid approach will look like (for brevity's sake we have only included the parts that are different) the following:

```
float FindParameterByDistance( float u1, float s )
{
    // set endpoints for bisection
    float a = u1;
    float b = 1.0f;
```

```
        // ensure that we remain within valid parameter space

        // get total length of curve

        // make first guess

        for (int i = 0; i < MAX_ITER; ++i)
        {
            // compute function value and test against zero

            // update endpoints for bisection
            if (func < 0.0f)
                a = p;
            else
                b = p;

            // compute speed
            speed = Length(Derivative(p));
            if (bisection)
                // do bisection step
                p = 0.5f*(a+b);
            else
                // perform Newton-Raphson iteration step
                p -= func/speed;
        }

        // done iterating, return last guess
        return p;
}
```

The only remaining question is how we determine to use bisection over Newton–Raphson. One obvious possibility is to check whether the speed is zero, as that got us into trouble in the first place. However, that's not enough. If the speed is nonzero but sufficiently small, func/speed could be sufficiently large to cause us to step outside the bisection interval or even the valid parameter space of the curve. So that gives us our test: if p - func/speed is less than a or greater than b, use bisection. We can write this as follows:

```
        if (p - func/speed < a || p - func/speed > b)
            // do bisection step
        else
            // perform Newton-Raphson iteration step
```

Multiplying by speed and rearranging terms gives us the following:

```
        if ((p-a)*speed < func || (p-b)*speed > func)
            // do bisection step
        else
            // perform Newton-Raphson iteration step
```

Press et al. [126] further recommend the following so as to be floating-point safe:

```
if (((p-a)*speed - func)*((p-b)*speed - func) > 0.0f)
    // do bisection step
else
        // perform Newton-Raphson iteration step
```

That should solve our problem.

A few other implementation notes are in order at this point. As we've seen, computing `ArcLength()` can be a nontrivial operation. Because of this, if we're going to be calling `FindParameterByDistance()` many times for a fixed curve, it is more efficient to precompute `ArcLength(0.0f, 1.0f)` and use this stored value instead of recomputing it each time. Also, the constants `MAX_ITER` and `EPSILON` will need to be tuned depending on the type of curve and the number of iterations we can feasibly calculate due to performance constraints. Reasonable starting values for this tuning process are 32 for `MAX_ITER` and 1.0e-06f for `EPSILON`.

As a final note, there is an alternative approach if we've used the table-driven method for computing arc length. Recall that we used Table 6.1 to compute s given a parameter u. In this case, we invert the process and search for the two neighboring entries with lengths s_j and s_{j+1} such that $s_j \leq s \leq s_{j+1}$.

Again, we can use linear interpolation to approximate the parameter u, which gives us length s as

$$u \approx \frac{s_{j+1} - s}{s_{j+1} - s_j} u_j + \frac{s - s_j}{s_{j+1} - s_j} u_{j+1}$$

To find the parameter b given a starting parameter a and a length s, we compute the length at a and add that to s. We then use the preceding process with the total length to find parameter b.

The obvious disadvantage of this scheme is that it takes additional memory for each curve. However, it is simple to implement, somewhat fast, and does avoid the Newton–Raphson iteration needed with other methods.

6.5.2 Moving at Variable Speed

In our original equation for computing the desired distance to travel, $s = rt$, we assumed that we were traveling at a constant rate of speed. However, it is often convenient to have an adjustable rate of speed over the length of the curve. We can represent this by a general distance–time function $s(t)$, which maps a time value t to the total distance traveled from t_0. As an example, Figure 6.27 shows $s(t) = rt$ as a distance–time graph.

Other than traveling at a constant rate, the most common distance–time function is known as *ease-in/ease-out*. Here, we start at a zero rate of speed, accelerate up to a constant nonzero rate of speed in the middle, and then decelerate down again to a stop. This feels natural, as it approximates the need to accelerate a physical camera, move it, and slow it down to a stop. Figure 6.28 shows the distance–time graph for one such function.

Parent [117] describes two methods for constructing ease-in/ease-out distance–time functions. One is to use sinusoidal pieces for the acceleration/deceleration areas of the function

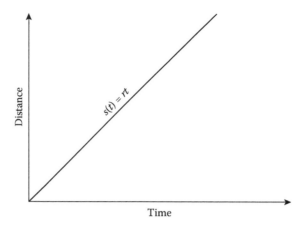

Figure 6.27. Example of distance–time graph: moving at constant speed.

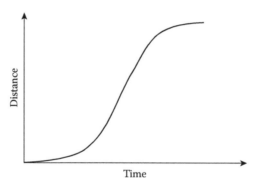

Figure 6.28. Example of distance–time graph: ease-in/ease-out function.

and a constant velocity in the middle. The pieces are carefully chosen to ensure C^1 continuity over the entire function. The second method involves setting a maximum velocity that we wish to attain in the center part of the function and assumes that we move with constant acceleration in the opening and closing ease-in/ease-out areas. This gives a velocity–time curve as in Figure 6.29. By integrating this, we get a distance–time curve. By assuming that we start at the beginning of the curve, this gives us a piecewise curve with parabolic acceleration and deceleration.

However, there is no reason to stop with an ease-in/ease-out distance–time function. We can define any curve we want, as long as the curve remains within the positive d and t axes for the valid time and distance intervals. One possibility is to let the user trace out a curve, but that can lead to invalid inputs and difficulty of control. Instead, animation packages such as those in 3D Studio Max and Maya allow artists to create these curves by setting keys with particular arrival and departure characteristics. Standard parlance includes such terms as fast-in, fast-out, slow-in, and slow-out. *In* and *out* in this case refer to the incoming and outgoing speed at the key point, respectively; *fast* means that the speed is greater than 1, and *slow* that it is less than 1. An example curve with both fast-in/fast-out and slow-in/slow-out can be seen in Figure 6.30. There also can be linear keys, which represent the linear rate

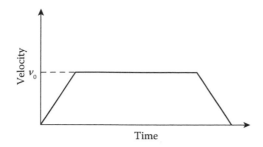

Figure 6.29. Example of velocity–time function: ease-in/ease-out with constant acceleration/deceleration.

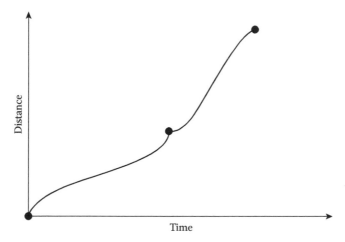

Figure 6.30. Example of distance–time graph: fast-out/fast-in followed by slow-out/slow-in.

seen in Figure 6.27, and step-keys, where distance remains constant for a certain period of time and then abruptly changes, as in Figure 6.31. Alternatively, the user may specify no speed characteristics and just expect the program to pick an appropriately smooth curve.

With all of these, the final distance–time curve can be easily generated with the techniques described in Section 6.2.3. More detail can be found in Van Verth [151].

6.6 Camera Control

Source Code
Demo
CameraControl

One common use for a parametric curve is as a path for controlling the motion of a virtual camera. We'll discuss camera models in more detail in Chapter 7, but for now assume that a camera is like any other object, with a position and orientation. In games controlling camera motion comes into play most often when setting up in-game cinematics, where we want to play a series of scripted events in engine while giving a game a cinematic feel via the clever use of camera control. For example, we might want to have a camera track around a pair of characters as they dance about a room. Or, we might want to simulate a crane shot zooming from a far point of view right down into a close-up. While either of these

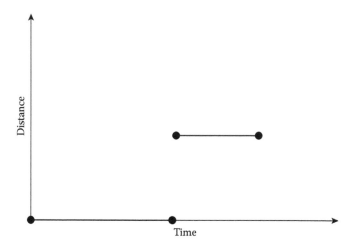

Figure 6.31. Example of distance–time graph: step-key transition.

could be done programmatically, it would be better to provide external control to the artist, who most likely will be setting up the shot. The artist sets the path for the camera—all the programmer needs to do is provide code to move the camera along the given path.

Determining the position of the camera isn't a problem. Given the start time t_s for the camera and the current time t_c, we compute the parameter $t = t_c - t_s$ and then use our time controls together with our curve description to determine the current position at $Q(t)$.

Computing orientation is another matter. The most basic option is to set a fixed orientation for the entire path. This might be appropriate if we are trying to create the effect of a panning shot, but is rather limiting and somewhat static. Another way would be to set orientations at each sample time as well as positions, and interpolate both. However, this can be quite time-consuming and may require more keys to get the effect we want.

A further possibility is to use the Frenet frame for the curve. This is an orthonormal frame with an origin of the current position on the curve, and a basis $\{\widehat{\mathbf{T}}, \widehat{\mathbf{N}}, \widehat{\mathbf{B}}\}$, where $\widehat{\mathbf{T}}$ (the tangent) points in the direction of the first derivative, $\widehat{\mathbf{N}}$ (the normal) points roughly in the direction of the second derivative, and $\widehat{\mathbf{B}}$ (the binormal) is the cross product of the first two. The vector $\widehat{\mathbf{T}}$ acts as our camera's forward (or view direction) vector, $\widehat{\mathbf{N}}$ acts as our side vector, and $\widehat{\mathbf{B}}$ acts as our up vector.

For any curve specified by the matrix form $Q(u) = \mathbf{UMG}$, we can easily compute the first derivative by using the form $\mathbf{Q}'(u) = \mathbf{U}'\mathbf{MG}$, where for a cubic curve

$$\mathbf{U}' = \begin{bmatrix} 3u^2 & 2u & 1 & 0 \end{bmatrix}$$

Similarly, we can compute the second derivative as $\mathbf{Q}''(u) = \mathbf{U}''\mathbf{MG}$, where

$$\mathbf{U}'' = \begin{bmatrix} 6u & 2 & 0 & 0 \end{bmatrix}$$

As mentioned, we set $\mathbf{T} = \mathbf{Q}'(u)$. We compute $\mathbf{B}$ as the cross product of the first and second derivatives:

$$\mathbf{B} = \mathbf{Q}'(u) \times \mathbf{Q}''(u)$$

Then, finally, $\mathbf{N}$ is the cross product of the other two:

$$\mathbf{N} = \mathbf{B} \times \mathbf{T}$$

Normalizing $\mathbf{T}$, $\mathbf{N}$, and $\mathbf{B}$ gives us our orthonormal basis.

Parent [117] describes a few flaws with using the Frenet frame directly. First of all, the second derivative may be $\mathbf{0}$, which means that $\widehat{\mathbf{B}}$ and hence $\widehat{\mathbf{N}}$ will be $\mathbf{0}$. One solution is to interpolate between two frames on either side of our current location. Since the second derivative is 0, or near 0, the first derivative won't be changing much, so we're really interpolating between two frames in $\mathbb{R}^2$. This consists of finding the angle between them and interpolating along that angle (Figure 6.32). The one flaw with this is that when finding these frames we're still using $\mathbf{Q}''$, which may be near 0 and hence lead to floating-point issues. In particular, if we are moving with linear motion, there will be no valid neighboring values for estimating $\mathbf{Q}''$.

Then, too, it assumes that the second derivative exists for all values of t, namely, that $Q(t)$ is C^2 continuous. Many of the curves we've discussed, in particular the piecewise curves, do not meet this criterion. In such cases, the camera will rather jarringly change orientation. For example, suppose we have two curve segments as seen in Figure 6.33, where the second

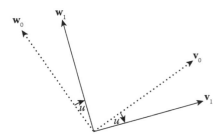

Figure 6.32. Interpolating between two path frames.

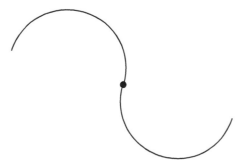

Figure 6.33. Frame interpolation issues. Discontinuity of second derivative at point.

derivative instantly changes to the opposite direction at the join between the segments. In the Frenet frame for the first segment, the **w** vector points out of the page. In the second segment, it points into the page. As the camera crosses the join, it will instantaneously flip upside down. This is probably not what the animator had in mind.

Finally, we may not want to use the second derivative at all. For example, if we have a path that heads up and then down, like a hill on a roller coaster, the direction of the second derivative points generally down along that section of path. This means that our view up vector will end up parallel to the ground for that section of curve—again, probably not the intention of the animator.

A further refinement of this technique is to use something called the *parallel transport frame* [70]. This is an extension of the interpolation technique shown in Figure 6.32. We begin at a position with a valid frame. At the next time step, we compute the derivative, which gives us our view direction vector as before. To compute the other two vectors, we rotate the previous frame by the angle between the current derivative and the previous derivative. If the vectors are parallel, we won't rotate at all, which solves the problem where the second derivative may be 0. This will generate a smooth transition in orientation across the entire path, but doesn't provide much control over expected behavior, other than setting the initial orientation.

An alternative solution is to adopt a technique from Chapter 7. Again, we use the first derivative as our forward vector, but instead generate the camera's up vector from this and the world up vector. The camera's side vector is the cross product of these two. This solves the problem, but does mean that if we have a fixed up vector we can't roll our camera through a banking turn—its up vector will remain relatively aligned with the given up vector.

A refinement of this is to allow user-specified up vectors at each sample position, which default to the world up vector. The program would interpolate between these up vectors just as it interpolates between the positions. Alternatively, the user could set a path $U(t)$ that is used to calculate the up vector: $\mathbf{v}_{up} = U(t) - Q(t)$. The danger here is that the user may specify two up vectors of opposing directions that end up interpolating to **0**, or an up vector that aligns with the view direction vector, which would lead to a cross product of **0**. If the user is allowed this kind of flexibility, recovery cases and some sort of error message will be needed.

We can take this one step further by separating our view direction from the Frenet frame and focusing on a point that we want to keep in the center of the frame. The method for computing the corresponding orientation can be found in Chapter 7. The choice of what we use as our so-called look-at point can depend on the camera effect desired. For example, we might pick a fixed point on the ground and then perform a fly-by. We could use the position of an object or the centroid of positions for a set of objects. We could set an additional path, and use the position along that path at our current time, to give the effect of a moving point of view without tying it to a particular object.

Another possibility is to look ahead along our current path a few steps in time, as if we were following an object a few seconds ahead of us. So, if we're at position $Q(t)$, we use as our look-at point the position $Q(t + \delta t)$. In this situation, we have to be sure to reparameterize the curve based on arc length, because otherwise the distance $\|Q(t) - Q(t + \delta t)\|$ may change depending on where we are on the curve, which may lead to odd changes in the view direction.

An issue with this technique is that it may make the camera seem clairvoyant, which can ruin the drama in some situations. Also, if our curve is particularly twisty, looking ahead may lead to sudden changes in direction. We can smooth this by averaging a set of points ahead of our position on the curve. How separated the points are makes a difference: too separated and our view direction may not change much; too close together and the smoothing effect will be nullified. It's usually best to make the amount of separation another setting available to the animator so that he or she can control the effect desired.

6.7 Chapter Summary

In this chapter we have touched on some of the issues involved with using parametric curves to aid in animation. We have discussed the most commonly used of the many possible curve types and how to subdivide these curves. Possible interfaces have been presented that allow animators and designers to create curves that can be used in the games they create. We have also covered some of the most common animation tasks beyond simple interpolation: controlling travel speed along curves and maintaining a logical camera orientation.

For rotations, fixed and Euler and axis–angle formats interpolate well only under simple circumstances. Matrices can be interpolated, but at significantly greater cost than quaternions. If you need to interpolate orientation, the clear choice is to use quaternions.

For further reading, Rogers and Adams [131] and Bartels et al. [10] present much of this material in greater detail, in particular focusing on B-splines. Parent [117] covers the use of splines in animation, as well as additional animation techniques. Burden and Faires [19] have a chapter on interpolation and explain some of the numerical methods used with curves, in particular integration techniques and the Newton–Raphson method.

We have not discussed parametric surfaces, but many of the same principles apply: surfaces are approximated or interpolated by a grid of points and are usually rendered using a subdivision method. Rogers [130] is an excellent resource for understanding how NURBS surfaces, the most commonly used parametric surfaces, are created and used.

7 Viewing and Projection

7.1 Introduction

In previous chapters we've discussed how to represent objects, basic transformations we can apply to these objects, and how we can use these transformations to move and manipulate our objects within our virtual world. With that background in place, we can begin to discuss the mathematics underlying the techniques we use to display our game objects on a monitor or other visual display medium.

It doesn't take much justification to understand why we might want to view the game world—after all, games are primarily a visual media. Other sensory outputs are of course possible, particularly sound and haptic (or touch) feedback. Both have become more sophisticated and in their own way provide another representation of the relative three-dimensional (3D) position and orientation of game objects. But in the current market, when we think of games, we first think of what we can see.

To achieve this, we'll be using a continuation of our transformation process known as the *graphics pipeline*. Figure 7.1 shows the situation. We already have a transformation that takes our model from its local space to world space. At each stage of the graphics pipeline, we continue to concatenate matrices to this matrix. Our goal is to build a single matrix to transform the points in our object from their local configuration to a two-dimensional (2D) representation suitable for displaying.

The first part of the display process involves setting up a virtual viewer or camera, which allows us to control which objects lie in our current view. As we'll see, this camera is just like any other object in the game; we can set the camera's position and orientation based on an affine transformation. Inverting this transformation is the first stage of our pipeline: it allows us to transform objects in the world frame into the point of view of the camera object.

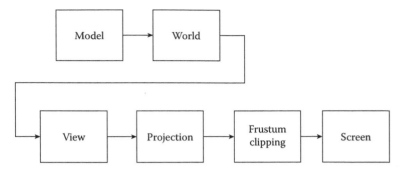

Figure 7.1. The graphics pipeline.

From there we will want to build and concatenate a matrix that transforms our objects in view into coordinates so they can be represented in an image. This flattening or *projection* takes many forms, and we'll discuss several of the most commonly used projections. In particular, we'll derive perspective projection, which most closely mimics our viewpoint of the real world.

At this point, it is usually convenient to cull out any objects that will not be visible on our screen, and possibly cut, or clip, others that intersect the screen boundaries. This will make our final rendering process much faster.

The final stage is to transform our projected coordinates and stretch and translate them to fit a specific portion of the screen, known as the viewport. This is known as the screen transformation.

In addition, we'll cover how to reverse this process so we can take a mouse click on our 2D screen and use it to select objects in our 3D world. This process, known as *picking*, can be useful when building an interface with 3D elements. For example, selecting units in a 3D real-time strategy game is done via picking.

As with other chapters, we'll be discussing how to implement these transformations in production code. Because our primary platform is OpenGL, for the most part we'll be focusing on its traditional transformation pipeline. However we will also cover the cases where it may differ from other graphics APIs, particularly Direct3D.

One final note before we begin: There is no standard representation for this process. In other books you may find these stages broken up in different ways, depending on the rendering system the authors are trying to present. However, the ultimate goal is the same: take an object in the world and transform it from a viewer's perspective onto a 2D medium.

7.2 View Frame and View Transformation

7.2.1 Defining a Virtual Camera

In order to render objects in the world, we need to represent the notion of a viewer. This could be the main character's viewpoint in a first-person shooter, or an over-the-shoulder view in a third-person adventure game, or a zoomed-out wide shot in a strategy game. We may want to control properties of our viewer to simulate a virtual camera; for example, we may want to create an in-game scripted sequence where we pan across a screen or follow a set path through a space. We encapsulate these properties into a single entity, commonly called the *camera*.

For now, we'll consider only the most basic properties of the camera needed for rendering. We are trying to answer two questions [12]: Where am I? Where am I looking? We can think of this as someone taking an actual camera, placing it on a tripod, and aiming it at an object of interest.

The answer to the first question is the camera's position, E, which is variously called the *eyepoint*, the *view position*, or the *view-space origin*. As we mentioned, this could be the main character's eye position, a location over his shoulder, or a spot pulled back from the action. While this position can be placed relative to another object's location, it is usually cleaner and easier to manage if we represent it in the world frame.

A partial answer to the second question is a vector called the *view direction vector*, or $\mathbf{v}_{dir}$, which points along the facing direction for the camera. This could be a vector from the camera position to an object or point of interest, a vector indicating the direction the main character is facing, or a fixed direction if we're trying to simulate a top-down view for a strategy game. For the purposes of setting up the camera, this is also specified in the world frame.

Having a single view direction vector is not enough to specify our orientation, since there are an infinite number of rotations around that vector. To constrain our possibilities down to one, we specify a second vector orthogonal to the first, called the *view up vector*, or $\mathbf{v}_{up}$. This indicates the direction out of the top of the camera. From these two we can take the cross product to get the *view side vector*, or $\mathbf{v}_{side}$, which usually points out toward the camera's right. Normalizing these three vectors and adding the view position gives us an orthonormal basis and an origin, or an affine frame. This is the camera's local frame, also known as the *view frame* (Figure 7.2).

The three view vectors specify where the view orientation is relative to the world frame. However, we also need to define where these vectors are from the perspective of the camera. The standard order used by most viewing systems is to make the camera's y-axis represent the view up vector in the camera's local space, and the camera's x-axis represent the corresponding view side vector. This aligns our camera's local coordinates so that x values vary left and right along the plane of the screen and y values vary up and down, which is very intuitive.

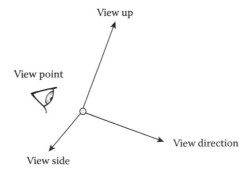

Figure 7.2. View frame relative to the world frame.

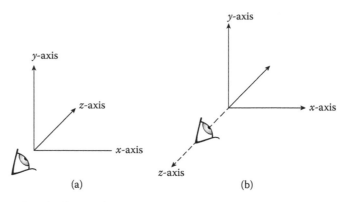

Figure 7.3. (a) Standard view frame axes. (b) OpenGL-style view frame axes.

The remaining question is what to do with z and the view direction. In most systems, the z-axis is treated as the camera-relative view direction vector (Figure 7.3a). This has a nice intuitive feel: as objects in front of the viewer move farther away, their z values relative to the camera will increase. The value of z can act as a measure of the distance between the object and the camera, which we can use for hidden object removal. Note, however, that this is a left-handed system, as $(\hat{\mathbf{v}}_{side} \times \hat{\mathbf{v}}_{up}) \cdot \hat{\mathbf{v}}_{dir} < 0$.

A slightly different approach was traditionally suggested for use with OpenGL. The OpenGL-style view frame is a right-handed system where the camera-relative view direction is aligned with the negative z-axis (Figure 7.3b). So in this case, the farther away the object is, its $-z$ coordinate gets larger relative to the camera. This is not as convenient for distance calculations, but it does allow us to remain in a right-handed coordinate system. This avoids having to worry about reflections when transforming from the world frame to the view frame, as we'll see below. Note that despite the name, we can still use left-handed systems with OpenGL, just as we can use right-handed systems with Direct3D and other APIs—we just have to be careful how we construct our projection matrices, as we'll see.

7.2.2 Constructing the View-to-World Transformation

Now that we have a way of representing and setting camera position and orientation, what do we do with it? The first step in the rendering process is to move all of the objects in our world so that they are no longer relative to the world frame, but are relative to the camera's view. Essentially, we want to transform the objects from the world frame to the view frame. This gives us a sense of what we can see from our camera position. In the view frame, those objects along the line of the view direction vector (i.e., the $-z$-axis in the case of the OpenGL-style frame) are in front of the camera and so will most likely be visible in our scene. Those on the other side of the plane formed by the view position, the view side vector, and the view up vector are behind the camera, and therefore not visible. In order to achieve this situation, we need to create a transformation from world space to view space, known as the world-to-view transformation or, more simply, the view transformation. We can represent this transformation as $\mathbf{M}_{world \to view}$.

However, rather than building this transformation directly, we usually find it easier to build $\mathbf{M}_{world \to view}^{-1}$, or $\mathbf{M}_{view \to world}$, first, and then invert to get our final world-to-view frame transformation. In order to build this, we'll make use of the principles we introduced in Chapter 4. If we look again at Figure 7.2, we note that we have an affine frame—the view frame—represented in terms of the world frame.

We can use this information to define the transformation from the view frame to the world frame as a 4×4 affine matrix. The origin E of the view frame is translated to the view position, so the translation vector $\mathbf{y}$ is equal to $E - O$. We'll abbreviate this as $\mathbf{v}_{pos}$. Similarly, the view vectors represent how the standard basis vectors in view space are transformed into world space and become columns in the upper left 3×3 matrix $\mathbf{A}$. To build $\mathbf{A}$, however, we need to define which standard basis vector in the view frame maps to a particular view vector in the world frame.

Recall that in the standard case, the camera's local x-axis represents $\hat{\mathbf{v}}_{side}$, the y-axis represents $\hat{\mathbf{v}}_{up}$, and the z-axis represents $\hat{\mathbf{v}}_{dir}$. This mapping indicates which columns the view vectors should be placed in, and the view position translation vector takes its familiar place in the rightmost column. The corresponding transformation matrix is

$$\mathbf{C}_s = \left[\begin{array}{cccc} \hat{\mathbf{v}}_{side} & \hat{\mathbf{v}}_{up} & \hat{\mathbf{v}}_{dir} & \mathbf{v}_{pos} \\ 0 & 0 & 0 & 1 \end{array} \right] \qquad (7.1)$$

Note that in this case we are mapping from a left-handed view frame to the right-handed world frame, so the upper 3×3 is not a pure rotation but a rotation concatenated with a reflection.

For the OpenGL-style frame, the only change is that we want to look down the $-z$-axis. This is the same as the z-axis mapping to the negative view direction vector. So, the corresponding matrix is

$$\mathbf{C}_{ogl} = \left[\begin{array}{cccc} \hat{\mathbf{v}}_{side} & \hat{\mathbf{v}}_{up} & -\hat{\mathbf{v}}_{dir} & \mathbf{v}_{pos} \\ 0 & 0 & 0 & 1 \end{array} \right] \qquad (7.2)$$

In this case, since we are mapping from a right-handed frame to a right-handed frame, no reflection is necessary, and the upper 3×3 matrix is a pure rotation. Not having a reflection can actually be a benefit, particularly with some culling methods.

7.2.3 Controlling the Camera

It's not enough that we have a transformation for our camera that encapsulates position and orientation. More often we'll want to move it around the world. Positioning our camera is a simple enough matter of translating the view position, but controlling view orientation is another problem. One way is to specify the view vectors directly and build the matrix as described. This assumes, of course, that we already have a set of orthogonal vectors we want to use for our viewing system.

The more usual case is that we only know the view direction. For example, suppose we want to continually focus on a particular object in the world (known as the look-at object). We can construct the view direction by subtracting the view position from the object's position. But whether we have a given view direction or we generate it from the look-at object, we still need two other orthogonal vectors to properly construct an orthogonal basis. We can calculate them by using one additional piece of information: the world up vector.

Source Code
Demo
LookAt

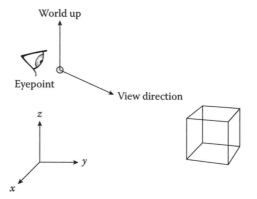

Figure 7.4. Look-at representation.

This is a fixed vector representing the up direction in the world frame. In our case, we'll use the z-axis basis vector $\mathbf{k}$ (Figure 7.4), although in general, any vector that we care to call "up" will do. For example, suppose we had a mission on a boat at sea and wanted to give the impression that the boat was rolling from side to side, without affecting the simulation. One method is to change the world up vector over time, oscillating between two keeled-over orientations, and use that to calculate your camera orientation.

For now, however, we'll use $\mathbf{k}$ as our world up vector. Our goal is to compute orthonormal vectors in the world frame corresponding to our view vectors, such that one of them is our view direction vector $\hat{\mathbf{v}}_{dir}$, and our view up vector $\hat{\mathbf{v}}_{up}$ matches the world up vector as closely as possible. Recall that we can use Gram–Schmidt orthogonalization to create orthogonal vectors from a set of nonorthogonal vectors, so

$$\mathbf{v}_{up} = \mathbf{k} - (\mathbf{k} \bullet \hat{\mathbf{v}}_{dir})\hat{\mathbf{v}}_{dir}$$

Normalizing gives us $\hat{\mathbf{v}}_{up}$. We can take the cross product to get the view side vector:

$$\hat{\mathbf{v}}_{side} = \hat{\mathbf{v}}_{dir} \times \hat{\mathbf{v}}_{up}$$

We don't need to normalize in this case because the two vector arguments are orthonormal. The resulting vectors can be placed as columns in the transformation matrix as before.

One problem may arise if we are not careful: What if $\hat{\mathbf{v}}_{dir}$ and $\mathbf{k}$ are parallel? If they are equal, we end up with

$$\begin{aligned}
\mathbf{v}_{up} &= \mathbf{k} - (\mathbf{k} \bullet \hat{\mathbf{v}}_{dir})\hat{\mathbf{v}}_{dir} \\
&= \mathbf{k} - 1 \cdot \hat{\mathbf{v}}_{dir} \\
&= \mathbf{0}
\end{aligned}$$

If they point in opposite directions we get

$$\begin{aligned}
\mathbf{v}_{up} &= \mathbf{k} - (\mathbf{k} \bullet \hat{\mathbf{v}}_{dir})\hat{\mathbf{v}}_{dir} \\
&= \mathbf{k} - (-1) \cdot \hat{\mathbf{v}}_{dir} \\
&= \mathbf{0}
\end{aligned}$$

Clearly, neither case will lead to an orthonormal basis.

The recovery procedure is to pick an alternative vector that we know is not parallel, such as **i** or **j**. This will lead to what seems like an instantaneous rotation around the z-axis. To understand this, raise your head upward until you are looking at the ceiling. If you keep going, you'll end up looking at the wall behind you, but upside down. To maintain the view looking right-side up, you'd have to rotate your head 180 degrees around your view direction.[1] This is not a very pleasing result, so avoid aligning the view direction with the world up vector whenever possible.

Source Code
Demo
Rotation

There is a third possibility for controlling camera orientation. Suppose we want to treat our camera just like a normal object and specify a rotation matrix and translation vector. To do this we'll need to specify a starting orientation Ω for our camera and then apply our rotation matrix to find our camera's final orientation, after which we can apply our translation. Which orientation is chosen is somewhat arbitrary, but some are more intuitive and convenient than others. In our case, we'll say that in our default orientation the camera has an initial view direction along the world x-axis, an initial view up along the world z-axis, and an initial view side along the $-y$-axis. This aligns the view up vector with the world up vector, and using the x-axis as the view direction fits the convention we set for objects' local space in Chapter 4.

Substituting these values into the view-to-world matrix for the standard left-handed view frame (Equation 7.1) gives

$$\Omega_s = \begin{bmatrix} 0 & 0 & 1 & 0 \\ -1 & 0 & 0 & 0 \\ 0 & 1 & 0 & 0 \\ 0 & 0 & 0 & 1 \end{bmatrix}$$

The equivalent matrix for the right-handed OpenGL-style view frame (using Equation 7.2) is

$$\Omega_{ogl} = \begin{bmatrix} 0 & 0 & -1 & 0 \\ -1 & 0 & 0 & 0 \\ 0 & 1 & 0 & 0 \\ 0 & 0 & 0 & 1 \end{bmatrix}$$

Whichever system we are using, after this we apply our rotation to orient our frame in the direction we wish and, finally, the translation for the view position. If the three column vectors in our rotation matrix are **u**, **v**, and **w**, then the final OpenGL-style transformation matrix is

$$\mathbf{M}_{view \to world} = \mathbf{T}\mathbf{R}\Omega_{ogl}$$

$$= \begin{bmatrix} \mathbf{i} & \mathbf{j} & \mathbf{k} & \mathbf{v}_{pos} \\ 0 & 0 & 0 & 1 \end{bmatrix} \begin{bmatrix} \mathbf{u} & \mathbf{v} & \mathbf{w} & \mathbf{0} \\ 0 & 0 & 0 & 1 \end{bmatrix} \begin{bmatrix} -\mathbf{j} & \mathbf{k} & -\mathbf{i} & \mathbf{0} \\ 0 & 0 & 0 & 1 \end{bmatrix}$$

$$= \begin{bmatrix} -\mathbf{v} & \mathbf{w} & -\mathbf{u} & \mathbf{v}_{pos} \\ 0 & 0 & 0 & 1 \end{bmatrix}$$

[1] Don't try this at home.

7.2.4 Constructing the World-to-View Transformation

Using the techniques in the previous two sections, now we can create a transformation that takes us from view space to world space. To create the reverse operator, we need only to invert the transformation. Since we know that it is an affine transformation, we can invert it as

$$\mathbf{M}_{world \to view} = \begin{bmatrix} \mathbf{R}^{-1} & -(\mathbf{R}^{-1}\mathbf{v}_{pos}) \\ \mathbf{0}^T & 1 \end{bmatrix}$$

where $\mathbf{R}$ is the upper 3×3 block of our view-to-world transformation. And since $\mathbf{R}$ is the product of either a reflection and rotation matrix (in the standard case) or two rotations (in the OpenGL-style case), it is an orthogonal matrix, so we can compute its inverse by taking the transpose:

$$\mathbf{M}_{world \to view} = \begin{bmatrix} \mathbf{R}^T & -(\mathbf{R}^T\mathbf{v}_{pos}) \\ \mathbf{0}^T & 1 \end{bmatrix}$$

Source Code
Demo
LookAt

In practice, this transformation is usually calculated directly, rather than taking the inverse of an existing transformation. One possible implementation for an OpenGL-style matrix is as follows:

```
void LookAt( const IvVector3& eye,
             const IvVector3& lookAt,
             const IvVector3& up )
{
  // compute view vectors
  IvVector3 viewDir = lookAt - eye;
  IvVector3 viewSide;
  IvVector3 viewUp;
  viewDir.Normalize();
  viewUp = up - up.Dot(viewDir)*viewDir;
  viewUp.Normalize();
  viewSide = viewDir.Cross(viewUp);

  // now set up matrices
  // build transposed rotation matrix
  IvMatrix33 rotate;
  rotate.SetRows( viewSide, viewUp, -viewDir );

  // transform translation

  IvVector3 eyeInv = -(rotate*eye);

  // build 4x4 matrix
  IvMatrix44 matrix;
  matrix.Rotation(rotate);
  matrix(0,3) = eyeInv.x;
  matrix(1,3) = eyeInv.y;
  matrix(2,3) = eyeInv.z;

  // set view to world transformation
  SetViewMatrix( matrix.mV );
}
```

Note that we use the method `IvMatrix33::SetRows()` to set the transformed basis vectors since we're setting up the inverse matrix, namely, the transpose. There is also no recovery code if the view direction and world up vectors are collinear—it is assumed that any external routine will ensure this does not happen. The renderer method `SetViewMatrix()` stores the calculated view transformation and is discussed in more detail in Section 7.7.

7.3 Projective Transformation

7.3.1 Definition

Now that we have a method for controlling our view position and orientation, and for transforming our objects into the view frame, we can look at the second stage of the graphics pipeline: taking our 3D space and transforming it into a form suitable for display on a 2D medium. This process of transforming from $\mathbb{R}^3$ to $\mathbb{R}^2$ is called *projection*.

We've already seen one example of projection: using the dot product to project one vector onto another. In our current case, we want to project the points that make up the vertices of an object onto a plane, called the *projection plane* or the *view plane*. We do this by following a *line of projection* through each point and determining where it hits the plane. These lines could be perpendicular to the plane, but as we'll see, they don't have to be.

To understand how this works, we'll look at a very old form of optical projection known as the *camera obscura* (Latin for "dark room"). Suppose one enters a darkened room on a sunny day, and there is a small hole allowing a fraction of sunlight to enter the room. This light will be projected onto the opposite wall of the room, displaying an image of the world outside, albeit upside down and flipped left to right (Figure 7.5). This is the same principle that allows a pinhole camera to work; the hole is acting like the focal point of a lens. In this case, all the lines of projection pass through a single *center of projection*. We can determine where a point will project to on the plane by constructing a line through both the original point and the center of projection and calculating where it will intersect the plane of projection. The virtual film in this case is a rectangle on the view plane, known as the *view window*. This will eventually get mapped to our display.

This sort of projection is known as *perspective projection*. Note that this relates to our perceived view in the real world. As an object moves farther away, its corresponding projection will shrink on the projection plane. Similarly, lines that are parallel in view space will appear to converge as their extreme points move farther away from the view position.

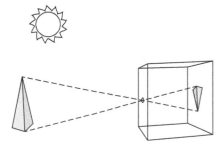

Figure 7.5. Camera obscura.

This gives us a result consistent with our expected view in the real world. If we stand on some railroad tracks and look down a straight section, the rails will converge in the distance, and the ties will appear to shrink in size and become closer together. In most cases, since we are rendering real-world scenes—or at least, scenes that we want to be perceived as real world—this will be the projection we will use.

There is, of course, one minor problem: the projected image is upside down and backwards. One possibility is just to flip the image when we display it on our medium. This is what happens with a camera: the image is captured on film upside down, but we can just rotate the negative or print to view it properly. This is not usually done in graphics. Instead, the projection plane is moved to the other side of the center of projection, which is now treated as our view position (Figure 7.6). As we'll see, the mathematics for projection in this case are quite simple, and the objects located in the forward direction of our view will end up being projected right-side up. The objects behind the view will end up projecting upside down, but we don't want to render them anyway, and as we'll see, there are ways of handling this situation.

An alternate type of projection is *parallel projection*, which can be thought of as a perspective projection where the center of projection is infinitely distant. In this case, the lines of projection do not converge; they always remain parallel (Figure 7.7)—hence the name. The placement of the view position and view plane is irrelevant in this case, but we place them in the same relative location to maintain continuity with perspective projection.

Parallel projection produces a very odd view if used for a scene: objects remain the same size no matter how distant they are, and parallel lines remain parallel. Parallel projections are usually used for computer-assisted design (CAD) programs, where maintaining parallel lines is important. They are also useful for rendering 2D elements like interfaces; no matter how far from the eye a model is placed, it always will be the same size, presumably the size we expect.

A parallel projection where the lines of projection are perpendicular to the view plane is called an *orthographic projection*. By contrast, if they are not perpendicular to the view plane, this is known as an *oblique projection* (Figure 7.8). Two common oblique projections

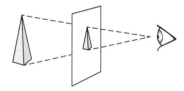

Figure 7.6. Perspective projection.

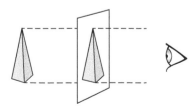

Figure 7.7. Orthographic parallel projection.

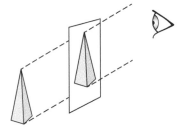

Figure 7.8. Oblique parallel projection.

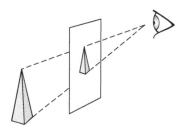

Figure 7.9. Oblique perspective projection.

are the *cavalier projection*, where the projection angle is 45 degrees, and the *cabinet projection*, where the projection angle is $\cot^{-1}(1/2)$. When using cavalier projections, projected lines have the same length as the original lines, so there is no perceived foreshortening. This is useful when printing blueprints, for example, as any line can be measured to find the exact length of material needed to build the object. With cabinet projections, lines perpendicular to the projection plane foreshorten to half their length (hence the $\cot^{-1}(1/2)$), which gives a more realistic look without sacrificing the need for parallel lines.

We can also have oblique perspective projections where the line from the center of the view window to the center of projection is not perpendicular to the view plane. For example, suppose we need to render a mirror. To do so, we'll render the space using a plane reflection transformation and clip it to the boundary of the mirror. The plane of the mirror is our projection plane, but it may be at an angle to our view direction (Figure 7.9). For now, we'll concentrate on constructing projective transformations perpendicular to the projection plane and examine these special cases later.

As a side note, oblique projections can occur in the real world. The classic pictures we see of tall buildings, shot from the ground but with parallel sides, are done with a "view camera." This device has an accordion-pleated hood that allows the photographer to bend and tilt the lens up while keeping the film parallel to the side of the building. Ansel Adams also used such a camera to capture some of his famous landscape photographs.

7.3.2 Normalized Device Coordinates

Before we begin projecting, our objects have passed through the view stage of the pipeline and so are in view frame coordinates. We will be projecting from this space in $\mathbb{R}^3$ to the view plane, which is in $\mathbb{R}^2$. In order to accomplish this, it will be helpful to define a frame for the

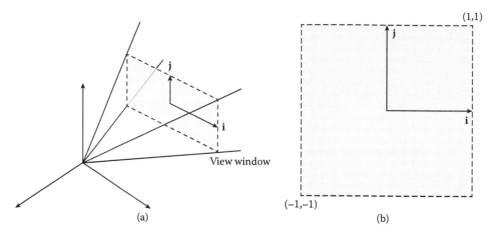

Figure 7.10. (a) NDC frame in view window and (b) view window after NDC transformation.

space of the view plane. We'll use as our origin the center of the view window, and create basis vectors that align with the sides of the view window, with magnitudes of half the width and height of the window, respectively (Figure 7.10a). For the purposes of this discussion, we'll be using the OpenGL-style viewing matrix, where we look down the $-z$-axis. Within this frame, our view window is transformed into a square two units wide and centered at the origin, bounded by the $x = 1$, $x = -1$, $y = 1$, and $y = -1$ lines (Figure 7.10b).

Using this as our frame provides a certain amount of flexibility when mapping to devices of varying size. Rather than transform directly to our screen area, which could be of variable width and height, we use this normalized form as an intermediate step to simplify our calculations and then do the screen conversion as our final step. Because of this, coordinates in this frame are known as *normalized device coordinates*.

To take advantage of the normalized device coordinate frame, or *NDC space*, we'll want to create our projection so that it always gives us the -1 to 1 behavior, regardless of the exact view configuration. This helps us to compartmentalize the process of projection (just as the view matrix did for viewing). When we're done projecting, we'll stretch and translate our NDC values to match the width and height of our display.

To simplify this mapping to the NDC frame, we will begin by using a view window in the view frame with a height of two units. This means that for the case of a centered view window, xy coordinates on the view plane will be equal to the projected coordinates in the NDC frame. In this way we can consider the projection as related to the view plane in view coordinates and not worry about a subsequent transformation.

7.3.3   View Frustum

The question remains: How do we determine what will lie within our view window? We could, naively, project all of the objects in the world to the view plane and then, when converting them to pixels, ignore those pixels that lie outside of the view window. However, for a large number of objects this would be very inefficient. It would be better to constrain our space to a convex volume, specified by a set of six planes. Anything inside these planes

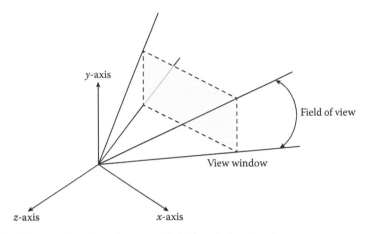

Figure 7.11. Perspective view frustum (right-handed system).

will be rendered; everything outside them will be ignored. This volume is known as the *view frustum*, or *view volume*.

To constrain what we render in the view frame *xy* directions, we specify four planes aligned with the edges of the view window. For perspective projection each plane is specified by the view position and two adjacent vertices of the view window (Figure 7.11), producing a semi-infinite pyramid. The angle between the upper plane and the lower plane is called the vertical *field of view*.

There is a relationship between field of view, view window size, and view plane distance: given two, we can easily find the third. For example, we can fix the view window size, adjust the field of view, and then compute the distance to the view plane. As the field of view gets larger, the distance to the view plane needs to get smaller to maintain the view window size. Similarly, a small field of view will lead to a longer view plane distance. Alternatively, we can set the distance to the view plane to a fixed value and use the field of view to determine the size of our view window. The larger the field of view, the larger the window and the more objects are visible in our scene. This gives us a primitive method for creating telephoto (narrow field of view) or wide-angle (wide field of view) lenses. We will discuss the relationship among these three quantities in more detail when we cover perspective projection.

In our case, the view window size is fixed, so when adjusting our field of view, we will move the view plane relative to the center of projection. This continues to match our camera analogy: the film size is fixed and the lens moves in and out to create a telephoto or wide-angle effect.

Usually the field of view chosen needs to match the display medium, as the user perceives it, as much as possible. For a standard widescreen monitor placed about 3 ft away, the monitor only covers about a 40- to 45-degree field of view from the perspective of the user, so we would expect that we would use a field of view of that size in the game. However, this constrains the amount we can see in the game to a narrow area, which feels unnatural because we're used to a 180-degree field of view in the real world. The usual compromise is to set the field of view to the range of 60–90 degrees. The distortion is not that perceptible,

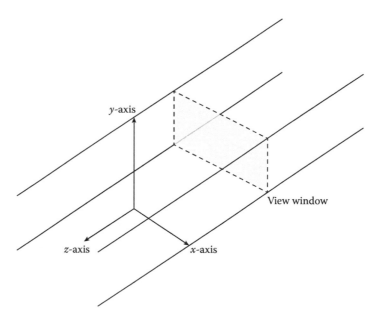

Figure 7.12. Parallel view frustum (right-handed system).

and it allows the user to see more of the game world. If the display were stretched to cover more of your personal field of view, as in a dual-monitor system or some virtual reality systems, a larger field of view would be appropriate. And of course, if the desired effect is of a telephoto or wide-angle lens, a narrower or wider field of view, respectively, is appropriate.

For parallel projection, the xy culling planes are parallel to the direction of projection, so opposite planes are parallel and we end up with a parallelopiped that is open at two ends (Figure 7.12). There is no concept of field of view in this case.

In both cases, to complete a closed view frustum we also define two planes that constrain objects in the view frame z direction: the near and far planes (Figure 7.13). With perspective projection it may not be obvious why we need a near plane, since the xy planes converge at the center of projection, closing the viewing region at that end. However, as we will see when we start talking about the perspective transformation, rendering objects at the view frame origin (which in our case is the same as the center of projection) can lead to a possible division by zero. This would adversely affect our rendering process. We could also, like some viewing systems, use the view plane as the near plane, but not doing so allows us a little more flexibility.

In some sense, the far plane is optional. Since we don't have an infinite number of objects or an infinite amount of game space, we could forego using the far plane and just render everything within the five other planes. However, the far plane is useful for culling objects and area from our rendering process, so having a far plane is good for efficiency's sake. It is also extremely important in the hidden surface removal method of z-buffering; the distance between the near and far planes is a factor in determining the precision we can expect in our z values. We'll discuss this in more detail in Chapter 10.

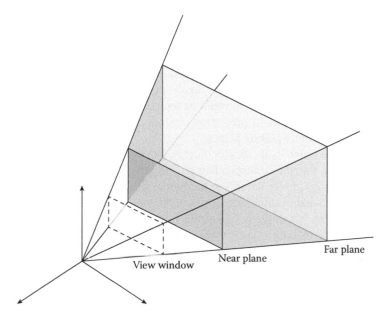

Figure 7.13. View frustum with near plane and far plane.

7.3.4 Homogeneous Coordinates

There is one more topic we need to cover before we can start discussing projection. Previously we stated that a point in $\mathbb{R}^3$ can be represented by $(x, y, z, 1)$ without explaining much about what that might mean. This representation is part of a more general representation for points known as homogeneous coordinates, which prove useful to us when handling perspective projections. In general, homogeneous coordinates work as follows: if we have a standard representation in n-dimensional space, then we can represent the same point in a $(n + 1)$–dimensional space by scaling the original coordinates by a single value and then adding the scalar to the end as our final coordinate. Since we can choose from an infinite number of scalars, a single point in $\mathbb{R}^n$ will be represented by an infinite number of points in the $(n + 1)$–dimensional space: a line. This $(n + 1)$–dimensional space is called a *real projective space* or $\mathbb{R}P^n$. In computer graphics parlance, the real projective space $\mathbb{R}P^3$ is also often called *homogeneous space*.

Suppose we start with a point (x, y, z) in $\mathbb{R}^3$, and we want to map it to a point (x', y', z', w) in homogeneous space. We pick a scalar for our fourth element w, and scale the other elements by it, to get (xw, yw, zw, w). As we might expect, our standard value for w will be 1, so (x, y, z) maps to $(x, y, z, 1)$. To map back to 3D space, divide the first three coordinates by w, so (x', y', z', w) goes to $(x'/w, y'/w, z'/w)$. Since our standard value for w is just 1, we could just drop the w: $(x', y', z', 1) \rightarrow (x', y', z')$. However, in the cases that we'll be concerned with next, we need to perform the division by w.

What happens when $w = 0$? In this case, a point in $\mathbb{R}P^3$ doesn't represent a point in $\mathbb{R}^3$, but a vector. We can think of this as a "point at infinity." While we will try to avoid cases where $w = 0$, they do creep in, so checking for this before performing the homogeneous division is often wise.

7.3.5 Perspective Projection

Source Code
Demo
Perspective

Since this is the most common projective transform we'll encounter, we'll begin by constructing the mathematics necessary for the perspective projection. To simplify things, let's take a 2D view of the situation on the yz plane and ignore the near and far planes for now (Figure 7.14). We have the y-axis pointing up, as in the view frame, and the projection direction along the negative z-axis as it would be with our OpenGL-style matrix. The point on the left represents our center of projection, and the vertical line our view plane. The diagonal lines represent our y culling planes.

Suppose we have a point P_v in view coordinates that lies on one of the view frustum planes, and we want to find the corresponding point P_s that lies on the view plane. Finding the y coordinate of P_s is simple: We follow the line of projection along the plane until we hit the top of the view window. Since the height of the view window is 2 and is centered on 0, the y coordinate of P_s is half the height of the view window, or 1. The z coordinate will be negative since we're looking along the negative z-axis and will have a magnitude equal to the distance d from the view position to the projection plane. So, the z coordinate will be $-d$.

But how do we compute d? As we see, the cross section of the y view frustum planes are represented as lines from the center of projection through the extents of the view window $(1, d)$ and $(-1, d)$. The angle between these lines is our field of view θ_{fov}. We'll simplify things by considering only the area that lies above the negative z-axis; this bisects our field of view to an angle of $\theta_{fov}/2$. If we look at the triangle bounded by the negative z-axis, the cross section of the upper view frustum plane, and the cross section of the projection plane, we can use trigonometry to compute d. Since we know the distance between the negative z-axis and the extreme point P_s is 1, we can say that

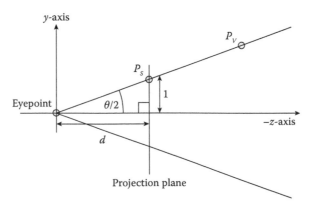

Figure 7.14. Perspective projection construction.

$$\frac{1}{d} = \tan\left(\frac{\theta_{fov}}{2}\right)$$

Rewriting this in terms of d, we get

$$d = \frac{1}{\tan\left(\frac{\theta_{fov}}{2}\right)}$$

$$= \cot\left(\frac{\theta_{fov}}{2}\right)$$

So for this fixed-view window size, as long as we know the angle of field of view, we can compute the distance d, and vice versa.

This gives the coordinates for any point that lies on the upper y view frustum plane; in this 2D cross section they all project down to a single point $(1, -d)$. Similarly, points that lie on the lower y frustum plane will project to $(-1, -d)$. But suppose we have a general point (y_v, z_v) in view space. We know that its projection will lie on the view plane as well, so its z_{ndc} coordinate will be $-d$. But how do we find y_{ndc}?

We can compute this by using similar triangles (Figure 7.15). If we have a point (y_v, z_v), the lengths of the sides of the corresponding right triangle in our diagram are y_v and $-z_v$ (since we're looking down the $-z$-axis, any visible z_v is negative, so we need to negate it to get a positive value). The lengths of the sides of the right triangle for the projected point are y_{ndc} and d. By similar triangles (both have the same angles), we get

$$\frac{y_{ndc}}{d} = \frac{y_v}{-z_v}$$

Solving for y_{ndc}, we get

$$y_{ndc} = \frac{dy_v}{-z_v}$$

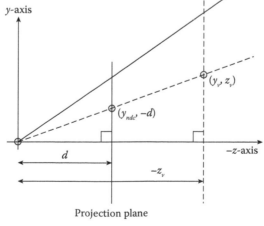

Figure 7.15. Perspective projection of similar triangles.

This gives us the coordinate in the y direction. If our view region was square, then we could use the same formula for the x direction. Most, however, are rectangular to match the relative dimensions of a computer monitor or other viewing device. We must correct for this by the aspect ratio of the view region. The aspect ratio a is defined as

$$a = \frac{w_v}{h_v}$$

where w_v and h_v are the width and height of the view rectangle, respectively. We're going to assume that the NDC view window height remains at 2 and correct the NDC view width by the aspect ratio. This gives us a formula for similar triangles of

$$\frac{a x_{ndc}}{d} = \frac{x_v}{-z_v}$$

Solving for x_{ndc}:

$$x_{ndc} = \frac{d x_v}{-a z_v}$$

So, our final projection transformation equations are

$$x_{ndc} = \frac{d x_v}{-a z_v}$$
$$y_{ndc} = \frac{d y_v}{-z_v}$$

The first thing to notice is that we are dividing by a z coordinate, so we will not be able to represent the entire transformation by a matrix operation, since it is neither linear nor affine. However, it does have some affine elements—scaling by d and d/a, for example— which can be performed by a transformation matrix. This is where the conversion from homogeneous space comes in. Recall that to transform from $\mathbb{R}P^3$ to $\mathbb{R}^3$ we need to divide the other coordinates by the w value. If we can set up our matrix to map $-z_v$ to our w value, we can take advantage of the homogeneous divide to handle the nonlinear part of our transformation. We can write the situation before the homogeneous divide as a series of linear equations,

$$x' = \frac{d}{a} x$$
$$y' = dy$$
$$z' = dz$$
$$w' = -z$$

and treat this as a four-dimensional (4D) linear transformation. Looking at our basis vectors, $\mathbf{e}_0$ will map to $(d/a, 0, 0, 0)$, $\mathbf{e}_1$ to $(0, d, 0, 0)$, $\mathbf{e}_2$ to $(0, 0, d, -1)$, and $\mathbf{e}_3$ to $(0, 0, 0, 0)$, since w is not used in any of the equations.

Based on this, our homogeneous perspective matrix is

$$\begin{bmatrix} \frac{d}{a} & 0 & 0 & 0 \\ 0 & d & 0 & 0 \\ 0 & 0 & d & 0 \\ 0 & 0 & -1 & 0 \end{bmatrix}$$

As expected, our transformed w value no longer will be 1. Also note that the rightmost column of this matrix is all zeros, which means that this matrix has no inverse. This is to be expected, since we are losing one dimension of information. Individual points in view space that lie along the same line of projection will project to a single point in NDC space. Given only the points in NDC space, it would be impossible to reconstruct their original positions in view space.

Let's see how this matrix works in practice. If we multiply it by a generic point in view space, we get

$$\begin{bmatrix} \frac{d}{a} & 0 & 0 & 0 \\ 0 & d & 0 & 0 \\ 0 & 0 & d & 0 \\ 0 & 0 & -1 & 0 \end{bmatrix} \begin{bmatrix} x_v \\ y_v \\ z_v \\ 1 \end{bmatrix} = \begin{bmatrix} \frac{dx_v}{a} \\ dy_v \\ dz_v \\ -z_v \end{bmatrix}$$

Dividing out the w (also called the reciprocal divide), we get

$$x_{ndc} = \frac{dx_v}{-az_v}$$
$$y_{ndc} = \frac{dy_v}{-z_v}$$
$$z_{ndc} = -d$$

which is what we expect.

So far, we have dealt with projecting x and y and completely ignored z. In the preceding derivation all z values map to $-d$, the negative of the distance to the projection plane. While losing a dimension makes sense conceptually (we are projecting from a 3D space down to a 2D plane, after all), for practical reasons it is better to keep some measure of our z values around for z-buffering and other depth comparisons (discussed in more detail in Chapter 10). Just as we're mapping our x and y values within the view window to an interval of $[-1, 1]$, we'll do the same for our z values within the near plane and far plane positions. We'll specify the near and far values n and f relative to the view position, so points lying on the near plane have a z_v value of $-n$, which maps to a z_{ndc} value of -1. Those points lying on the far plane have a z_v value of $-f$ and will map to 1 (Figure 7.16).

We'll derive our equation for z_{ndc} in a slightly different way than our xy coordinates. There are two parts to mapping the interval $[-n, -f]$ to $[-1, 1]$. The first is scaling the interval to a width of 2, and the second is translating it to $[-1, 1]$. Ordinarily, this would be a straightforward linear process; however, we also have to contend with the final w divide.

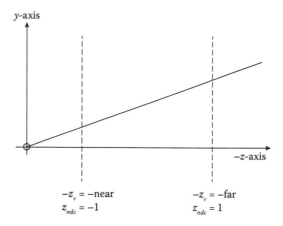

Figure 7.16. Perspective projection: z values.

Instead, we'll create a perspective matrix with unknowns for the scaling and translation factors and use the fact that we know the final values for $-n$ and $-f$ to solve for the unknowns. Our starting perspective matrix, then, is

$$
\begin{bmatrix}
\frac{d}{a} & 0 & 0 & 0 \\
0 & d & 0 & 0 \\
0 & 0 & A & B \\
0 & 0 & -1 & 0
\end{bmatrix}
$$

where A and B are our unknown scale and translation factors, respectively. If we multiply this by a point $(0, 0, -n)$ on our near plane, we get

$$
\begin{bmatrix}
\frac{d}{a} & 0 & 0 & 0 \\
0 & d & 0 & 0 \\
0 & 0 & A & B \\
0 & 0 & -1 & 0
\end{bmatrix}
\begin{bmatrix}
0 \\
0 \\
-n \\
1
\end{bmatrix}
=
\begin{bmatrix}
0 \\
0 \\
-An + B \\
n
\end{bmatrix}
$$

Dividing out the w gives

$$
z_{ndc} = -A + \frac{B}{n}
$$

We know that any point on the near plane maps to a normalized device coordinate of -1, so we can substitute -1 for z_{ndc} and solve for B, which gives us

$$
B = (A - 1)n \tag{7.3}
$$

Now we'll substitute Equation 7.3 into our original matrix and multiply by a point $(0, 0, -f)$ on the far plane:

$$\begin{bmatrix} \frac{d}{a} & 0 & 0 & 0 \\ 0 & d & 0 & 0 \\ 0 & 0 & A & (A-1)n \\ 0 & 0 & -1 & 0 \end{bmatrix} \begin{bmatrix} 0 \\ 0 \\ -f \\ 1 \end{bmatrix} = \begin{bmatrix} 0 \\ 0 \\ -Af + (A-1)n \\ f \end{bmatrix}$$

This gives us a z_{ndc} of

$$z_{ndc} = -A + (A-1)\frac{n}{f}$$

$$= -A + A\left(\frac{n}{f}\right) - \frac{n}{f}$$

$$= A\left(\frac{n}{f} - 1\right) - \frac{n}{f}$$

Setting z_{ndc} to 1 and solving for A, we get

$$A\left(\frac{n}{f} - 1\right) - \frac{n}{f} = 1$$

$$A\left(\frac{n}{f} - 1\right) = \frac{n}{f} + 1$$

$$A = \frac{(n/f) + 1}{(n/f) - 1}$$

$$= \frac{n+f}{n-f}$$

If we substitute this into Equation 7.3, we get

$$B = \frac{2nf}{n-f}$$

So, our final perspective matrix is

$$\mathbf{M}_{persp} = \begin{bmatrix} \frac{d}{a} & 0 & 0 & 0 \\ 0 & d & 0 & 0 \\ 0 & 0 & \frac{n+f}{n-f} & \frac{2nf}{n-f} \\ 0 & 0 & -1 & 0 \end{bmatrix}$$

It is important to be aware that this matrix will not work for all viewing systems. For one thing, the standard left-handed view frame looks down the positive z-axis, so this affects both our xy and z transformations. For example, in this case we have mapped $[-n, -f]$ to $[-1, 1]$. With the standard system we would want to begin by mapping $[n, f]$ to the NDC

z range. In addition, this range is not always set to $[-1, 1]$. Direct3D, for one, has a mapping to $[0, 1]$ in the z direction.

Using the standard left-handed view frame and a $[0, 1]$ mapping in z gives us a perspective transformation matrix of

$$
\mathbf{M}_{pD3D} = \begin{bmatrix} \frac{d}{a} & 0 & 0 & 0 \\ 0 & d & 0 & 0 \\ 0 & 0 & \frac{f}{f-n} & -\frac{nf}{f-n} \\ 0 & 0 & 1 & 0 \end{bmatrix}
$$

This matrix can be derived using the same principles described above.

When setting up a perspective matrix, it is good to be aware of the issues involved in rasterizing z values. In particular, to maintain z precision, keep the near and far planes as close together as possible. More details on managing perspective z precision can be found in Chapter 10.

7.3.6   Oblique Perspective

The matrix we constructed in the previous section is an example of a basic perspective matrix, where the direction of projection through the center of the view window is perpendicular to the view plane. A more general example of perspective, based on the deprecated OpenGL glFrustum() call, takes six parameters: the near and far z distances, as before, and four values that define our view window on the near z plane: the x interval $[l, r]$ (left, right) and the y interval $[b, t]$ (bottom, top). Figure 7.17a shows how this looks in $\mathbb{R}^3$, and Figure 7.17b shows the cross section on the yz plane. As we can see, these values need not be centered around the z-axis, so we can use them to generate an oblique projection.

To derive this matrix, once again we begin by considering similar triangles in the y direction. Remember that given a point $(y_v, -z_v)$, we project to a point on the view plane $(dy_v/-z_v, -d)$, where d is the distance to the projection. However, since we're using our near plane as our projection plane, this is just $(ny_v/-z_v, -n)$. The projection remains the same, we're just moving the window of projected points that lie within our view frustum.

With our previous derivation, we could stop at this point because our view window on the projection plane was already in the interval $[-1, 1]$. However, our new view window lies in the interval $[b, t]$. We'll have to adjust our values to properly end up in NDC space. The first step is to translate the center of the window, located at $(t + b)/2$, to the origin. Applying this translation to the current projected y coordinate gives us

$$
y' = y - \frac{(t+b)}{2}
$$

We now need to scale to change our interval from a magnitude of $(t - b)$ to a magnitude of 2 by using a scale factor $2/(t - b)$:

$$
y_{ndc} = \frac{2y}{t-b} - \frac{2(t+b)}{2(t-b)}
\tag{7.4}
$$

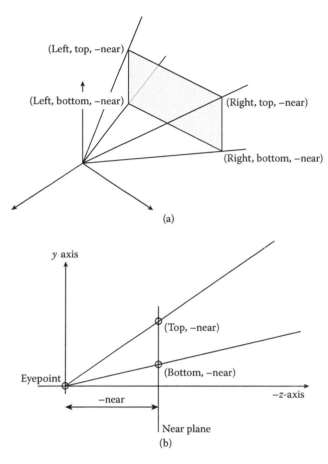

(a)

(b)

Figure 7.17. (a) View window for general perspective, 3D view. (b) View window for general perspective, cross section.

If we substitute $ny_v/-z_v$ for y and simplify, we get

$$
\begin{aligned}
y_{ndc} &= \frac{2n(y_v/-z_v)}{t-b} - \frac{2(t+b)}{2(t-b)} \\
&= \frac{2n(y_v/-z_v)}{t-b} - \frac{(t+b)(-z_v/-z_v)}{t-b} \\
&= \frac{1}{-z_v}\left(\frac{2n}{t-b}y_v + \frac{t+b}{t-b}z_v\right)
\end{aligned}
$$

A similar process gives us the following for the x direction:

$$
x_{ndc} = \frac{1}{-z_v}\left(\frac{2n}{r-l}x_v + \frac{r+l}{r-l}z_v\right)
$$

We can use the same A and B from our original perspective matrix, so our final projection matrix is

$$\mathbf{M}_{oblpersp} = \begin{bmatrix} \frac{2n}{r-l} & 0 & \frac{r+l}{r-l} & 0 \\ 0 & \frac{2n}{t-b} & \frac{t+b}{t-b} & 0 \\ 0 & 0 & \frac{n+f}{n-f} & \frac{2nf}{n-f} \\ 0 & 0 & -1 & 0 \end{bmatrix}$$

A casual inspection of this matrix gives some sense of what's going on here. We have a scale in the x, y, and z directions, which provides the mapping to the interval $[-1, 1]$. In addition, we have a translation in the z direction to align our interval properly. However, in the x and y directions, we are performing a z-shear to align the interval, which provides us with the oblique projection.

The equivalent left-handed Direct3D matrix is

$$\mathbf{M}_{opD3D} = \begin{bmatrix} \frac{2n}{r-l} & 0 & -\frac{r+l}{r-l} & 0 \\ 0 & \frac{2n}{t-b} & -\frac{t+b}{t-b} & 0 \\ 0 & 0 & \frac{f}{f-n} & -\frac{nf}{f-n} \\ 0 & 0 & 1 & 0 \end{bmatrix}$$

As unusual as it might appear, there are a number of applications of oblique perspective projection in real-time graphics. First of all, it can be used in mirrors: we treat the mirror as our view window, the mirror plane as our view plane, and the viewer's location as our view position. If we apply a plane reflection to all of our objects, flipping them around the mirror plane, and then render with the appropriate visual effects, we will end up with a result in the view window that emulates a mirror.

Another application is stereo. By using a single view plane and view window, but separate view positions for each eye that are offset from the standard center of projection, we get slightly different projections of the world. By using either a red-blue system to color each view differently, or some sort of goggle system that displays the left and right views in each eye appropriately, we can provide a good approximation of stereo vision. We have included an example of this in the sample code.

Finally, this can be used for a system called *fishtank VR*. Normally we think of VR as a helmet attached to someone's head with a display for each eye. However, by attaching a tracking device to a viewer's head we can use a single display and create an illusion that we are looking through a window into a world on the other side. This is much the same principle as the mirror: the display is our view window and the tracked location of the eye is our view position. Add stereo and this gives a very pleasing effect.

7.3.7 Orthographic Parallel Projection

Source Code
Demo ■
Orthographic

After considering perspective projection in two forms, orthographic projection is much easier. Examine Figure 7.18, which shows a side view of our projection space as before,

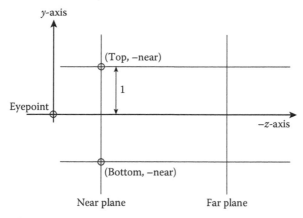

Figure 7.18. Orthographic projection construction.

with the lines of projection passing through the view plane and the near and far planes shown as vertical lines. This time the lines of projection are parallel to each other (hence this is a parallel projection) and parallel to the z-axis (hence an orthographic projection).

We can use this to help us generate a matrix similar to the result from the deprecated OpenGL glOrtho() call. Like the generalized perspective matrix, this also uses six parameters: the near and far z distances, and four values l, r, b, and t that define our view window on the near z plane. As before, the near plane is our projection plane, so a point (y_v, z_v) projects to a point $(y_v, -n)$. Note that since this is a parallel projection, there is no division by z or scale by d; we just use the y value directly. As with our general perspective matrix, we now need to consider only values between t and b and scale and translate them to the interval $[-1, 1]$. Substituting y_v into our range transformation equation (7.4), we get

$$y_{ndc} = \frac{2y_v}{t-b} - \frac{t+b}{t-b}$$

A similar process gives us the equation for x_{ndc}. We can do the same for z_{ndc}, but since our viewable z values are negative and our values for n and f are positive, we need to negate our z value and then perform the range transformation. The result of all three equations is

$$\mathbf{M}_{ortho} = \begin{bmatrix} \frac{2}{r-l} & 0 & 0 & -\frac{r+l}{r-l} \\ 0 & \frac{2}{t-b} & 0 & -\frac{t+b}{t-b} \\ 0 & 0 & -\frac{2}{f-n} & -\frac{f+n}{f-n} \\ 0 & 0 & 0 & 1 \end{bmatrix}$$

There are a few things we can notice about this matrix. First of all, multiplying by this matrix gives us a w value of 1, so we don't need to perform the homogeneous division. This means that our z values will remain linear; that is, they will not compress as they approach the far plane. This gives us better z resolution at far distances than the perspective matrices. It also means that this is a linear transformation matrix and possibly invertible.

Secondly, in the x and y directions, what was previously a z-shear in the oblique perspective matrix has become a translation. Before, we had to use shear, because for a given point the displacement was dependent on the distance from the view position. Because the lines of projection are now parallel, all points displace equally, so only a translation is necessary.

The left-handed Direct3D equivalent matrix is

$$\mathbf{M}_{orthoD3D} = \begin{bmatrix} \frac{2}{r-l} & 0 & 0 & -\frac{r+l}{r-l} \\ 0 & \frac{2}{t-b} & 0 & -\frac{t+b}{t-b} \\ 0 & 0 & \frac{1}{f-n} & -\frac{n}{f-n} \\ 0 & 0 & 0 & 1 \end{bmatrix}$$

7.3.8   Oblique Parallel Projection

Source Code
Demo
Oblique

While most of the time we'll want to use orthographic projection, we may from time to time need an oblique parallel projection. For example, suppose for part of our interface we wish to render our world as a set of schematics or display particular objects with a 2D CAD/CAM feel. This set of projections will achieve our goal. We will give our projection a slight oblique angle ($\cot^{-1}(1/2)$, which is about 63.4 degrees), which gives a 3D look without perspective. More extreme angles in x and y tend to look strangely flat.

Figure 7.19 is another example of our familiar cross section, this time showing the lines of projection for our oblique projection. As we can see, we move one unit in the y direction for every two units we move in the z direction. Using the formula of $\tan(\theta) = opposite/adjacent$, we get

$$\tan(\theta) = \frac{2}{1}$$
$$\cot(\theta) = \frac{1}{2}$$
$$\theta = \cot^{-1}\frac{1}{2}$$

which confirms the expected value for our oblique angle.

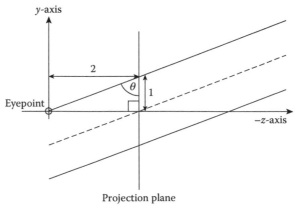

Figure 7.19. Example of oblique parallel projection.

As before, we'll consider the yz case first and extrapolate to x. Moving one unit in y and two units in $-z$ gives us the vector $(1, -2)$, so the formula for the line of projection for a given point P is

$$L(t) = P + t(1, -2)$$

We're only interested in where this line crosses the near plane, or where

$$P_z - 2t = -n$$

Solving for t, we get

$$t = \frac{1}{2}(n + P_z)$$

Plugging this into the formula for the y coordinate of $L(t)$, we get

$$y' = P_y + \frac{1}{2}(n + P_z)$$

Finally, we can plug this into our range transformation equation (6.4) as before to get

$$y_{ndc} = 2\frac{\left[y_v + \frac{1}{2}(n + z_v)\right]}{t - b} - \frac{t + b}{t - b}$$

$$= \frac{2y_v}{t - b} - \frac{t + b}{t - b} + \frac{z_v + n}{t - b}$$

Once again, we examine our transformation equation more carefully. This is the same as the orthographic transformation we had before, with an additional z-shear, as we'd expect for an oblique projection. In this case, the shear plane is the near plane rather than the xy plane, so we add an additional factor of $\frac{n}{t - b}$ to take this into account.

A similar process can be used for x. Since the oblique projection has a z-shear, z is not affected and so,

$$\mathbf{M}_{obl} = \begin{bmatrix} \frac{2}{r-l} & 0 & \frac{1}{r-l} & -\frac{r+l-n}{r-l} \\ 0 & \frac{2}{t-b} & \frac{1}{t-b} & -\frac{t+b-n}{t-b} \\ 0 & 0 & -\frac{2}{f-n} & -\frac{n+f}{f-n} \\ 0 & 0 & 0 & 1 \end{bmatrix}$$

The left-handed Direct3D equivalent matrix is

$$\mathbf{M}_{oblD3D} = \begin{bmatrix} \frac{2}{r-l} & 0 & -\frac{1}{r-l} & -\frac{r+l-n}{r-l} \\ 0 & \frac{2}{t-b} & -\frac{1}{t-b} & -\frac{t+b-n}{t-b} \\ 0 & 0 & \frac{1}{f-n} & -\frac{n}{f-n} \\ 0 & 0 & 0 & 1 \end{bmatrix}$$

7.4 Culling and Clipping

7.4.1 Why Cull or Clip?

We will now take a detour from discussing the transformation aspect of our pipeline to discuss a process that often happens at this point in many renderers. In order to improve rendering, both for speed and appearance's sake, it is necessary to cull and clip objects. Culling is the process of removing objects from consideration for some process, whether it be rendering, simulation, or collision detection. In this case, that means we want to ignore any models or whole pieces of geometry that lie outside of the view frustum, since they will never end up being projected to the view window. In Figure 7.20, the lighter objects lie outside of the view frustum and so will be culled for rendering.

Clipping is the process of cutting geometry to match a boundary, whether it be a polygon or, in our case, a plane. Vertices that lie outside the boundary will be removed and new ones generated for each edge that crosses the boundary. For example, in Figure 7.21 we see a box being clipped by a plane, showing the extra vertices created where each edge intersects the plane. We'll use this for any models that cross the view frustum, cutting the geometry

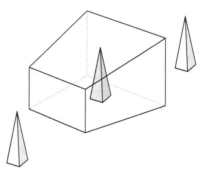

Figure 7.20. View frustum culling.

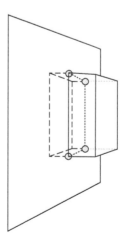

Figure 7.21. View frustum clipping.

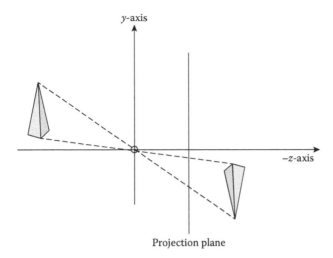

y-axis

−z-axis

Projection plane

Figure 7.22. Projection of objects behind the eye.

so that it fits within the frustum. We can think of this as slicing a piece of geometry off for every frustum plane.

Why should we want to use either of these for rendering? For one thing, it is more efficient to remove any data that will not ultimately end up on the screen. While copying the transformed object to the frame buffer (a process called *rasterization*) is almost always done in hardware and thus is fast, it is not free. Anywhere we can avoid unnecessary work is good.

But even if we had infinite rasterization power, we would still want to cull and clip when performing perspective projection. Figure 7.22 shows one example why. Recall that we finessed the problem of the camera obscura inverting images by moving the view plane in front of the center of projection. However, we still have the same problem if an object is behind the view position; it will end up projected upside down. The solution is to cull objects that lie behind the view position.

Figure 7.23a shows another example. Suppose we have a polygon edge $\overline{PQ}$ that crosses the $z = 0$ plane. Endpoint P projects to a point P' on the view plane, and Q to Q'. With the correct projection, the intermediate points of the line segment should start at the middle of the view, move up, and wrap around to reemerge at the bottom of the view. In practice, however, the rasterizing hardware has only the two projected vertices as input. It will take the vertices and render the shortest line segment between them (Figure 7.23b). If we clip the line segment to only the section that is viewable and then project the endpoints (Figure 7.23c), we end with only a portion of the line segment, but at least it is from the correct projection.

There is also the problem of vertices that lie on the $z = 0$ plane. When transformed to homogeneous space by the perspective matrix, a point $(x, y, 0, 1)$ will become $(x', y', z', 0)$. The resulting transformation into NDC space will be a division by 0, which is not valid.

To avoid all of these issues, at the very least we need to set a near plane that lies in front of the eye so that the view position itself does not lie within the view frustum. We first cull any objects that lie on the same side of the near plane as the view position. We then clip any

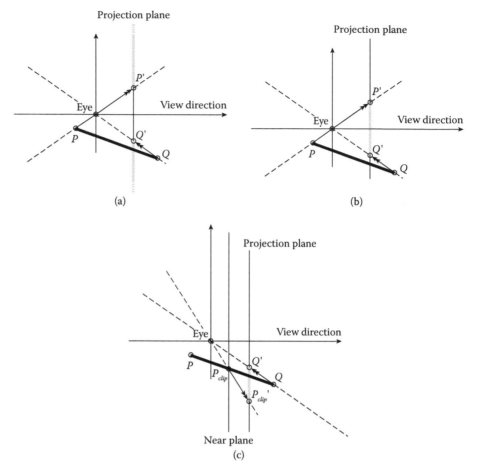

Figure 7.23. (a) Projection of line segment crossing behind view point. (b) Incorrect line segment rendering based on projected endpoints. (c) Line segment rendering when clipped to near plane.

objects that cross the near plane. This avoids both the potential of dividing by 0 (although it is sometimes prudent to check for it anyway, at least in a debug build) and trying to render any line segments passing through infinity.

While clipping to a near plane is a bare minimum, clipping to the top, bottom, left, and right planes is useful as well. While the windowing hardware will usually ignore any pixels that lie outside of a window's visible region (this is commonly known as *scissoring*), it is faster if we can avoid unnecessary rasterization. Also, if we want to set a viewport that covers a subrectangle of a window, not clipping to the border of the viewport may lead to spurious geometry being drawn (although most hardware allows for adjustable scissoring regions; in particular, OpenGL and D3D provide interfaces to set this).

Finally, some hardware has a limited range for screen-space positions, for example, 0 to 4,095. The viewable area might lie in the center of this range, say from a minimum point of (1,728, 1,808) to a maximum point of (2,688, 2,288). The area outside of the viewable area is known as the *guard band*—anything rendered to this will be ignored, since it won't be displayed. In some cases we can avoid clipping in x and y, since we can just render objects whose screen-space projection lies within the guard band and know that they will be handled automatically by the hardware. This can improve performance considerably, since clipping can be quite expensive. However, it's not entirely free. Values that lie outside the maximum range for the guard band will wrap around. So, a vertex that would normally project to coordinates that should lie off the screen, say (6,096, 6,096), will wrap to (2,000, 2,000)—right in the middle of the viewable area. Unfortunately, the only way to solve this problem is what we were trying to avoid in the first place: clipping in the x and y directions. However, now our clip window encompasses the much larger guard band area, so using the guard band can still reduce the amount of clipping that we have to do overall.

7.4.2 Culling

A naive method of culling a model against the view frustum is to test each of its vertices against each of the frustum planes in turn. We designate the plane normal for each plane as pointing toward the inside half-space. If for one plane $ax + by + cz + d < 0$ for every vertex $P = (x, y, z)$, then the model lies outside of the frustum and we can ignore it. Conversely, if for all the frustum planes and all the vertices $ax + by + cz + d > 0$, then we know the model lies entirely inside the frustum and we don't need to worry about clipping it.

While this will work, for models with large numbers of vertices this becomes expensive, probably outweighing any savings we might gain by not rendering the objects. Instead, culling is usually done by approximating the object with a convex bounding volume, such as a sphere, that contains all of the vertices for the object. Rather than test each vertex against the planes, we test only the bounding object. Since it is a convex object and all the vertices are contained within it, we know that if the bounding object lies outside of the view frustum, all of the model's vertices must lie outside as well. More information on computing bounding objects and testing them against planes can be found in Chapter 12.

Bounding objects are usually placed in the world frame to aid with collision detection, so culling is often done in the world frame as well. One approach generates the clip planes from the view parameters. We can find each x or y clipping plane in the view frame by using the view position and two corners of the view window to generate the plane. The two z planes (in OpenGL-style projections) are $z = -near$ and $z = -far$, respectively. Transforming them to the world frame is a simple case of using the technique for transforming plane normals, as described in Chapter 4. This can work in many cases, but does involve knowing the view parameters. A better and more general approach generates the clip planes directly from the transformation matrix, which we discuss in Section 7.4.5.

While view frustum culling can remove a large number of objects from consideration, it's not the only culling method. Another is backface culling, which allows us to determine which polygons are pointing away from the camera (acting as the "back faces" of objects, hence the name) so we can ignore them. There also are a large number of culling methods that break up the scene in order to cull objects that aren't visible. This can help with interior levels, so you don't render rooms that may be within the view frustum but not visible because they're blocked by a wall. Such methods are out of the purview of this book but are described in detail in many of the references cited in the following sections.

7.4.3 General Plane Clipping

To clip polygons, we first need to know how to clip a polygon edge (i.e., a line segment) to a plane. As we'll see, the problem of clipping a polygon to a plane degenerates to handling this case. Suppose we have a line segment $\overline{PQ}$, with endpoints P and Q, that crosses a plane. We'll say that P is inside our clip space and Q is outside. Our clipped line segment will be $\overline{PR}$, where R is the intersection of the line segment and the plane (Figure 7.24).

To find R, we take the line equation $P + t(Q - P)$, plug it into our plane equation $ax + by + cz + d = 0$, and solve for t. To simplify the equations, we'll define $\mathbf{v} = Q - P$. Substituting the parameterized line coordinates for x, y, and z, we get

$$
\begin{aligned}
0 &= a(P_x + tv_x) + b(P_y + tv_y) + c(P_z + tv_z) + d \\
&= aP_x + tav_x + bP_y + tbv_y + cP_z + tcv_z + d \\
&= aP_x + bP_y + cP_z + d + t(av_x + bv_y + cv_z) \\
t &= \frac{-aP_x - bP_y - cP_z - d}{av_x + bv_y + cv_z}
\end{aligned}
$$

And now, substituting in $Q - P$ for $\mathbf{v}$:

$$
t = \frac{(aP_x + bP_y + cP_z + d)}{(aP_x + bP_y + cP_z + d) - (aQ_x + bQ_y + cQ_z + d)}
$$

We can use Blinn's notation [12], slightly modified, to simplify this to

$$
t = \frac{BCP}{BCP - BCQ}
$$

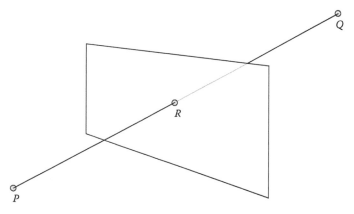

Figure 7.24. Clipping edge to plane.

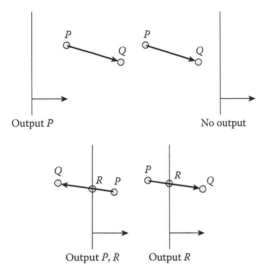

Figure 7.25. Four possible cases of clipping an edge against a plane.

where *BCP* is the result from the plane equation (the boundary coordinate) when we test *P* against the plane, and *BCQ* is the result when we test *Q* against the plane. The resulting clip point *R* is

$$R = P + \frac{BCP}{BCP - BCQ}(Q - P)$$

To clip a polygon to a plane, we need to clip each edge in turn. A standard method for doing this is to use the Sutherland–Hodgeman algorithm [143]. We first test each edge against the plane. Depending on what the result is, we output particular vertices for the clipped polygon. There are four possible cases for an edge from *P* to *Q* (Figure 7.25). If both are inside, then we output *P*. The vertex *Q* will be output when we consider it as the start of the next edge. If both are outside, we output nothing. If *P* is inside and *Q* is outside, then we compute *R*, the clip point, and output *P* and *R*. If *P* is outside and *Q* is inside, then we compute *R* and output just *R*—as before, *Q* will be output as the start of the next edge. The sequence of vertices generated as output will be the vertices of our clipped polygon.

We now have enough information to build a class for clipping vertices, which we'll call `IvClipper`. We can define this as

```
class IvClipper
{
public:
    IvClipper()
    {
        mFirstVertex = true;
    }
    ~IvClipper();

    void ClipVertex( const IvVector3& end )
```

```
        inline void StartClip() { mFirstVertex = true; }
        inline void SetPlane( const IvPlane& plane ) { mPlane = plane; }

private:
    IvPlane    mPlane;    // current clipping plane
    IvVector3 mStart;    // current edge start vertex
    float      mBCStart; // current edge start boundary condition
    bool    mStartInside; // whether current start vertex is inside
    bool    mFirstVertex; // whether expected vertex is start vertex
};
```

Note that `IvClipper::ClipVertex()` takes only one argument: the end vertex of the edge. If we send the vertex pair for each edge down to the clipper, we'll end up duplicating computations. For example, if we clip P_0 and P_1, and then P_1 and P_2, we have to determine whether P_1 is inside or outside twice. Rather than do that, we'll feed each vertex in order to the clipper. By storing the previous vertex (`mStart`) and its plane test information (`mBCStart`) in our `IvClipper` class, we need to calculate data only for the current vertex. Of course, we'll need to prime the pipeline by sending in the first vertex, not treating it as part of an edge, and just storing its boundary information.

Using this, clipping an edge based on the current vertex might look like the following code:

```
void IvClipper::ClipVertex( const IvVector3& end )
{
  float BCend = mPlane.Test(end);
  bool endInside = ( BCend >= 0 );
  if (!mFirstVertex)
 ,{
    // if one of the points is inside
    if ( mStartInside || endInside )
    {
      // if the start is inside, just output it
      if (mStartInside)
        Output( mStart );
      // if one of them is outside, output clip point
      if ( !(mStartInside && endInside) )
      {
        if (endInside)
        {
          float t = BCend/(BCend - mBCStart);
          Output( end - t*(end - mStart) );
        }
        else
        {
          float t = mBCStart/(mBCStart - BCend);
          Output( mStart + t*(end - mStart) );
        }
      }
    }
  }
```

```
    mStart = end;
    mBCStart = BCend;
    mStartInside = endInside;
    mFirstVertex = false;
}
```

Note that we generate t in the same direction for both clipping cases—from inside to outside. Polygons will often share edges. If we were to clip the same edge for two neighboring polygons in different directions, we may end up with two slightly different points due to floating-point error. This will lead to visible cracks in our geometry, which is not desirable. Interpolating from inside to outside for both cases avoids this situation.

To clip against the view frustum, or any other convex volume, we need to clip against each frustum plane. The output from clipping against one plane becomes the input for clipping against the next, creating a clipping pipeline. In practice, we don't store the entire clipped polygon, but pass each output vertex down as we generate it. The current output vertex and the previous one are treated as the edge to be clipped by the next plane. The Output() call above becomes a ClipVertex() for the next stage.

Note that we have only generated new positions at the clip boundary. There are other parameters that we can associate with an edge vertex, such as colors, normals, and texture coordinates (we'll discuss exactly what these are in Chapters 8–10). These will have to be clipped against the boundary as well. We use the same t value when clipping these parameters, so the clip part of our previous algorithm might become as follows:

```
// if one of them is outside, output clip vertex
if ( !(mStartInside && endInside) )
{
    ...
    clipPosition = startPosition + t*(endPosition - startPosition);
    clipColor = startColor + t*(endColor - startColor);
    clipTexture = startTexture + t*(endTexture - startTexture);
    // Output new clip vertex
}
```

This is only one example of a clipping algorithm. In most cases, it won't be necessary to write any code to do clipping. The hardware will handle any clipping that needs to be done for rendering. However, for those who have the need or interest, other examples of clipping algorithms are the Liang-Barsky [96], Cohen–Sutherland (found in Hughes et al. [82] as well as other graphics texts), and Cyrus–Beck [28] methods. Blinn [12] describes an algorithm for lines that combines many of the features from the previously mentioned techniques; with minor modifications it can be made to work with polygons.

7.4.4 Homogeneous Clipping

In the presentation above, we clip against a general plane. When projecting, however, Blinn and Newell [11] noted that we can simplify our clipping by taking advantage of some properties of our projected points prior to the division by w. Recall that after the division by w, the visible points will have normalized device coordinates lying in the interval $[-1, 1]$, or

$$-1 \leq x/w \leq 1$$
$$-1 \leq y/w \leq 1$$
$$-1 \leq z/w \leq 1$$

Multiplying these equations by w provides the intervals prior to the w division:

$$-w \leq x \leq w$$
$$-w \leq y \leq w$$
$$-w \leq z \leq w$$

In other words, the visible points are bounded by the six planes:

$$w = x$$
$$w = -x$$
$$w = y$$
$$w = -y$$
$$w = z$$
$$w = -z$$

Instead of clipping our points against general planes in the world frame or view frame, we can clip our points against these simplified planes in $\mathbb{R}P^3$ space. For example, the plane test for $w = x$ is $w - x$. The full set of plane tests for a point P are

$$BCP_{-x} = w + x$$
$$BCP_x = w - x$$
$$BCP_{-y} = w + y$$
$$BCP_y = w - y$$
$$BCP_{-z} = w + z$$
$$BCP_z = w - z$$

The previous clipping algorithm can be used, with these plane tests replacing the `IvPlane::Test()` call. While these tests are cheaper to compute in software, their great advantage is that since they don't vary with the projection, they can be built directly into hardware, making the clipping process very fast.

There is one potential wrinkle to homogeneous clipping, however. Figure 7.26 shows the visible region for the x coordinate in homogeneous space. However, our plane tests will clip to the upper triangle region of that hourglass shape—any points that lie in the lower region will be inadvertently removed. With the projections that we have defined, this will happen only if we use a negative value for the w value of our points. And since we've chosen 1 as the standard w value for points, this shouldn't happen. However, if you do have points that for some reason have negative w values, Blinn [12] recommends the following procedure: transform, clip, and render your points normally; then multiply your projection matrix by -1; and then transform, clip, and render again.

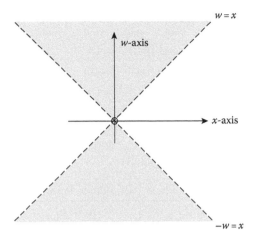

Figure 7.26. Homogeneous clip regions for NDC interval $[-1, 1]$.

7.4.5 Extracting Clip Planes

With our knowledge of homogeneous clipping, we can now discuss how to extract clip planes directly from a perspective transformation matrix [64]. Given a perspective transformation matrix $\mathbf{P}$ and a point in view space represented as a vector $\mathbf{v} = (x_v, y_v, z_v, 1)$, we can multiply them to get a point in homogeneous space (x, y, z, w) or

$$
\begin{bmatrix} x \\ y \\ z \\ w \end{bmatrix} = \begin{bmatrix} \mathbf{p}_0^T \\ \mathbf{p}_1^T \\ \mathbf{p}_2^T \\ \mathbf{p}_3^T \end{bmatrix} \begin{bmatrix} x_v \\ y_v \\ z_v \\ 1 \end{bmatrix} = \begin{bmatrix} \mathbf{p}_0^T \bullet \mathbf{v} \\ \mathbf{p}_1^T \bullet \mathbf{v} \\ \mathbf{p}_2^T \bullet \mathbf{v} \\ \mathbf{p}_3^T \bullet \mathbf{v} \end{bmatrix}
$$

As we stated when we clip homogeneous space, we are clipping within the region bounded by six planes. Let's take the first one, $w = x$. This is equivalent to the plane equation $0 = w - x$. From our matrix multiplication above, we can substitute for x and w, and so

$$
\begin{aligned}
0 &= w - x \\
&= \mathbf{p}_3^T \bullet \mathbf{v} - \mathbf{p}_0^T \bullet \mathbf{v}
\end{aligned}
$$

From the additivity property of the dot product we can simplify this to

$$
0 = (\mathbf{p}_3^T - \mathbf{p}_0^T) \bullet \mathbf{v}
$$

Expanding this out, we get a plane equation in terms of $\mathbf{v}$:

$$
0 = (p_{3,0} - p_{0,0})x_v + (p_{3,1} - p_{0,1})y_v + (p_{3,2} - p_{0,2})z_v + (p_{3,3} - p_{0,3})
$$

where our a, b, c, d values for the plane are

$$a = (p_{3,0} - p_{0,0})$$
$$b = (p_{3,1} - p_{0,1})$$
$$c = (p_{3,2} - p_{0,2})$$
$$d = (p_{3,3} - p_{0,3})$$

So to get our plane constants we only need to subtract the top row from the bottom row. The other clip planes can be computed similarly:

Homogeneous Plane	Clip Plane Parameters
$w = x$	$\mathbf{p}_3^T - \mathbf{p}_0^T$
$w = -x$	$\mathbf{p}_3^T + \mathbf{p}_0^T$
$w = y$	$\mathbf{p}_3^T - \mathbf{p}_1^T$
$w = -y$	$\mathbf{p}_3^T + \mathbf{p}_1^T$
$w = z$	$\mathbf{p}_3^T - \mathbf{p}_2^T$
$w = -z$	$\mathbf{p}_3^T + \mathbf{p}_2^T$

Note that the parameters above are for the OpenGL NDC z range of $[-1, 1]$. For the Direct3D NDC z range of $[0, 1]$, the last homogeneous plane is $w = 0$, so the corresponding clip parameters are just $\mathbf{p}_3^T$.

Given a perspective transformation matrix this will give us our clip planes in view space. However, this same approach will work if the perspective matrix is concatentated with other transformations, or even with orthographic projections. So if we want the clip planes in world space for a general projection, we need only concatenate our world-to-view matrix with our projection matrix, run this process, and our computed clip plane parameters will be appropriate for world space.

7.5 Screen Transformation

Now that we've covered viewing, projection, and clipping, our final step in transforming our object in preparation for rendering is to map its geometric data from the NDC frame to the screen or device frame. This could represent a mapping to the full display, a window within the display, or an offscreen pixel buffer.

Remember that our coordinates in the NDC frame range from a lower left corner of $(-1, -1)$ to an upper right corner of $(1, 1)$. Real device space coordinates usually range from an upper left corner $(0, 0)$ to a lower right corner (w_s, h_s), where w_s (screen width) and h_s (screen height) are usually not the same. In addition, in screen space the y-axis is commonly flipped so that y values increase as we move down the screen. Some windowing systems allow you to use the standard y direction, but we'll assume the default (Figure 7.27).

What we'll need to do is map our NDC area to our screen area (Figure 7.28). This consists of scaling it to the same size as the screen, flipping our y direction, and then translating it so that the upper left corner becomes the origin.

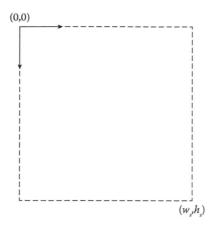

Figure 7.27. View window in standard screen-space frame.

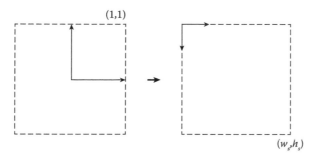

Figure 7.28. Mapping NDC space to screen space.

Let's begin by considering only the y direction, because it has the special case of the axis flip. The first step is scaling it. The NDC window is two units high, whereas the screen-space window is h_s high, so we divide by 2 to scale the NDC window to unit height, and then multiply by h_s to scale to screen height:

$$y' = \frac{h_s}{2} y_{ndc}$$

Since we're still centered around the origin, we can do the axis flip by just negating:

$$y'' = -\frac{h_s}{2} y_{ndc}$$

Finally, we need to translate downwards (which is now the positive y direction) to map the top of the screen to the origin. Since we're already centered on the origin, we need to translate only half the screen height, so

$$y_s = -\frac{h_s}{2} y_{ndc} + \frac{h_s}{2}$$

Another way of thinking of the translation is that we want to map the extreme point $-h_s/2$ to 0, so we need to add $h_s/2$.

A similar process, without the axis flip, gives us our x transformation:

$$x_s = \frac{w_s}{2}x_{ndc} + \frac{w_s}{2}$$

This assumes that we want to cover the entire screen with our view window. In some cases, for example, in a split-screen console game, we want to cover only a portion of the screen. Again, we'll have a width and height of our screen-space area, w_s and h_s, but now we'll have a different upper left corner position for our area: (s_x, s_y). The first part of the process is the same; we scale the NDC window to our screen-space window and flip the y-axis. Now, however, we want to map $(-w_s/2, -h_s/2)$ to (s_x, s_y), instead of $(0, 0)$. The final translation will be $(w_s/2 + s_x, h_s/2 + s_y)$. This gives us our generalized screen transformation in xy as

$$x_s = \frac{w_s}{2}x_{ndc} + \frac{w_s}{2} + s_x \tag{7.5}$$

$$y_s = -\frac{h_s}{2}y_{ndc} + \frac{h_s}{2} + s_y \tag{7.6}$$

Our z coordinate is a special case. As mentioned, we'll want to use z for depth testing, which means that we'd really prefer it to range from 0 to d_s, where d_s is usually 1. This mapping from $[-1, 1]$ to $[0, d_s]$ is

$$z_s = \frac{d_s}{2}z_{ndc} + \frac{d_s}{2} \tag{7.7}$$

We can, of course, express this as a matrix:

$$\mathbf{M}_{ndc \to screen} = \begin{bmatrix} \frac{w_s}{2} & 0 & 0 & \frac{w_s}{2} + s_x \\ 0 & -\frac{h_s}{2} & 0 & \frac{h_s}{2} + s_y \\ 0 & 0 & \frac{d_s}{2} & \frac{d_s}{2} \\ 0 & 0 & 0 & 1 \end{bmatrix}$$

7.5.1 Pixel Aspect Ratio

Recall that in our projection matrices, we represented the shape of our view window by setting an aspect ratio a. Most of the time it is expected that the value of a chosen in the projection will match the aspect ratio w_s/h_s of the final screen transformation. Otherwise, the resulting image will be distorted. For example, if we use a square aspect ratio ($a = 1.0$) for the projection and a standard aspect ratio of 4:3 for the screen transformation, the image will appear compressed in the y direction. If your image does not quite look right, it is good practice to ensure that these two values are the same.

An exception to this practice arises when your final display has a different aspect ratio than the offscreen buffers that you're using for rendering. For example, NTSC televisions have 448 scan lines, with 640 analog pixels per scan line, so it was common practice to render to a 640 × 448 area and then send that to the NTSC converter to be displayed. Using the offscreen buffer size would give an aspect ratio of 10:7. But the physical CRT television screen has a 4:3 aspect ratio, so the resulting image would be distorted, producing stretching in the y direction. The solution was to set $a = 4/3$ despite the aspect ratio of the offscreen buffer. The image in the offscreen buffer was compressed in the y direction, but is then proportionally stretched in the y direction when the image is displayed on the television, thereby producing the correct result.

7.6 Picking

Source Code
Demo
Picking

Now that we understand the mathematics necessary for transforming an object from world coordinates to screen coordinates, we can consider the opposite case. In our game we may have enemy objects that we'll want to target. The interface we have chosen involves tracking them with our mouse and then clicking on the screen. The problem is: How do we take our click location and use that to detect which object we've selected (if any)? We need a method that takes our 2D screen coordinates and turns them into a form that we can use to detect object intersection in 3D game space. Effectively we are running our pipeline backwards, from the screen transformation to the projection to the viewing transformation (clipping is ignored as we're already within the boundary of our view window).

For the purposes of discussion, we'll assume that we are using the basic OpenGL perspective matrix. Similar derivations can be created using other projections. Figure 7.29 is yet another cross section showing our problem. Once again, we have our view frustum, with our top and bottom clipping planes, our projection plane, and our near and far planes. Point P_s indicates our click location on the projection plane. If we draw a ray (known as a pick ray) from the view position through P_s, we pass through every point that lies underneath our click location. So to determine which object we have clicked on, we need only generate this point on the projection plane, create the specific ray, and then test each object for intersection with the ray. The closest object to the eye will be the object we're seeking.

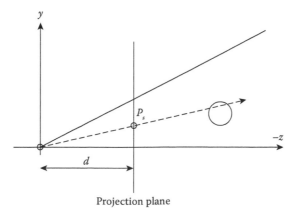

Figure 7.29. Pick ray.

To generate our point on the projection plane, we'll have to find a method for going backwards from screen space into view space. To do this we'll have to find a means to "invert" our projection. Matrix inversion seems like the solution, but it is not the way to go. The standard projection matrix has zeros in the rightmost column, so it's not invertible. But even using the z-depth projection matrix doesn't help us, because (1) the reciprocal divide makes the process nonlinear, and in any case, (2) our click point doesn't have a z value to plug into the inversion.

Instead, we begin by transforming our screen-space point (x_s, y_s) to an NDC space point (x_{ndc}, y_{ndc}). Since our transform from NDC to screen space is affine, this is easy enough: we need only invert our previous Equations 7.5 and 7.6. That gives us

$$x_{ndc} = \frac{2(x_s - s_x)}{w_s} - 1$$

$$y_{ndc} = -\frac{2(y_s - s_y)}{h_s} + 1$$

Now the tricky part. We need to transform our point in the NDC frame to the view frame. We'll begin by computing our z_v value. Looking at Figure 7.29 again, this is straightforward enough. We'll assume that our point lies on the projection plane so the z value is just the z location of the plane or $-d$. This leaves our x and y coordinates to be transformed. Again, since our view region covers a rectangle defined by the range $[-a, a]$ (recall that a is our aspect ratio) in the x direction and the range $[-1, 1]$ in the y direction, we only need to scale to get the final point. The view window in the NDC frame ranges from $[-1, 1]$ in y, so no scale is needed in the y direction and we scale by a in the x direction. Our final equations of screen space to view space are

$$x_v = \frac{2a}{w_s}(x_s - s_x) - 1$$

$$y_v = -\frac{2}{h_s}(y_s - s_y) + 1$$

$$z_v = -d$$

Since this is a system of linear equations, we can express this as a 3×3 matrix:

$$\begin{bmatrix} x_v \\ y_v \\ z_v \end{bmatrix} = \begin{bmatrix} \frac{2a}{w_s} & 0 & -\frac{2a}{w_s}s_x - 1 \\ 0 & -\frac{2}{h_s} & \frac{2}{h_s}s_y + 1 \\ 0 & 0 & -d \end{bmatrix} \begin{bmatrix} x_s \\ y_s \\ 1 \end{bmatrix}$$

From here we have a choice. We can try to detect intersection with an object in the view frame, we can detect in the world frame, or we can detect in the object's local frame. The first involves transforming every object into the view frame and then testing against our pick ray. The second involves transforming our pick ray into the world frame and testing against the world coordinates of each object. For simulation and culling purposes, often we're already pregenerating our world location and bounding information. So, if we're only concerned with testing for intersection against bounding information, it can be more efficient to go with testing in world space. However, usually we test in local space so we can check for intersection within the frame of the stored model vertices. Transforming these vertices into the world frame or the view frame every time we did picking could be prohibitively expensive.

In order to test in the model's local space, we'll have to transform our view-space point by the inverse of the viewing transformation. Unlike the perspective transformation, however, this inverse is much easier to compute. Recall that since the view transformation is an affine matrix, we can invert it to get the view-to-world matrix $\mathbf{M}_{view \to world}$. So, multiplying $\mathbf{M}_{view \to world}$ by our click point in the view frame gives us our point in world coordinates:

$$P_w = \mathbf{M}_{view \to world} \, P_v$$

We can transform this and our view position E from world coordinates into model coordinates by multiplying by the inverse of the model-to-world matrix:

$$P_l = \mathbf{M}_{world \to model} \, P_w$$
$$E_l = \mathbf{M}_{world \to model} \, E$$

Then, the formula for our pick ray in model space is

$$R(t) = E_l + t(P_l - E_l)$$

We can now use this ray in combination with our objects to find the particular one the user has clicked on. Chapter 12 discusses how to determine intersection between a ray and an object and other intersection problems.

7.7 Management of Viewing Transformations

Source Code
Library
IvGraphics
Filename
IvRenderer

Up to this point we have presented a set of transformations and corresponding matrices without giving some sense of how they would fit into a game engine. While the thrust of this book is not about writing renderers, we can still provide a general sense of how some renderers and application programming interfaces (APIs) manage these matrices, and how to set transformations for a standard API.

The view, projection, and screen transformations change only if the camera is moved. As this happens rarely, these matrices are usually computed once, stored, and then concatenated with the new world transformation every time a new object instance is rendered. How this is handled depends on the API used. The most direct approach is to concatenate the newly set world transform matrix with the others, creating a single transformation all the way from model space to prehomogeneous divide screen space:

$$M_{model \to screen} = M_{ndc \to screen} \bullet M_{projection} \bullet M_{world \to view} \bullet M_{model \to world}$$

Multiplying by this single matrix and then performing three homogeneous divisions per vertex generates the screen coordinates for the object. This is extremely efficient, but ignores any clipping we might need to do. In this case, we can concatenate up to homogeneous space, also known as clip space:

$$M_{model \to clip} = M_{projection} \bullet M_{world \to view} \bullet M_{model \to world}$$

Then we transform our vertices by this matrix, clip against the view frustum, perform the homogeneous divide, and either calculate the screen coordinates using Equations 7.5 through 7.7 or multiply by the NDC to the screen matrix, as before.

With more complex renderers, we end up separating the transformations further. For example, in Chapter 9 we will need to handle some calculations in world space or view space. For that purpose we'll need to store concatenated matrices that only include the transformations up to that point.

This leaves the NDC-to-screen-space transformation. Usually the graphics API will not require a matrix but will perform this operation directly. In the xy directions the user is only expected to provide the dimensions and position of the screen window area, also known as the viewport. In OpenGL this is set by using the call `glViewport()`. For the z direction, OpenGL provides a function `glDepthRange()`, which maps $[-1, 1]$ to $[near, far]$, where the defaults for $near$ and far are 0 and 1, respectively. Similar methods are available for other APIs.

In our case, we have decided not to overly complicate things and are providing simple convenience routines in the `IvRenderer` class:

```
SetWorldMatrix()
SetViewMatrix()
SetProjectionMatrix()
SetViewport()
```

The first three routines update an internal copy of the given matrix in `IvRenderer`, and then concatenate all three together into a fourth internal matrix that contains the full transformation matrix from model to homogeneous space. Setting the world matrix also updates a fifth matrix called the normal matrix, which contains the inverse transpose of the world matrix, suitable for transforming normals into world space. All of these matrices are then available to be loaded into our rendering system when needed (we'll discuss how in Chapter 8). The last routine passes the viewport parameters directly to our API.

7.8 Chapter Summary

Manipulating objects in the world frame is only useful if we have appropriate techniques for presenting those data. In this chapter we have discussed the viewing, projection, and screen transformations necessary for rendering objects on a screen or image. While we have focused on OpenGL as our rendering API, the same principles apply to Direct3D or any other rendering system. We transform the world to the perspective of a virtual viewer, project it to a view plane, and then scale and translate the result to fit our final display. We also covered how to reverse those transformations to allow one to select an object in view or world space by clicking on the screen. In the following chapters we will discuss how to use the data generated by these transformations to actually set pixels on the screen.

For those who are interested in reading further, most graphics textbooks—such as Akenine-Möller et al. [1] and Hughes et al. [82]—describe the graphics pipeline in great detail. In addition, one of Blinn's collections [12] is almost entirely dedicated to this subject. Various culling techniques are discussed in Akenine-Möller et al. [1]. Finally, the OpenGL graphics system specification [135] discusses the particular implementation of the graphics pipeline used in OpenGL.

8 Geometry and Programmable Shading

8.1 Introduction

Having discussed in detail in the preceding chapters how to represent, transform, view, and animate geometry, the next three chapters form a sequence that describes the second half of the rendering pipeline. The second half of the rendering pipeline is specifically focused on visual matters: the representation, computation, and usage of color.

This chapter will discuss how we connect the points we have been transforming and projecting to form solid surfaces, as well as the extra information we use to represent the unique appearance of each surface. All visual representations of geometry require the computation of colors; this chapter will discuss the data structures used to store colors and perform basic color computations.

Having shown how to build these renderable surface objects and described the methods of storing and computing colors, we will then lay out the foundations of the rest of the rendering section: the programmable shading and rasterization pipeline. Note that this chapter, unlike the others in the rendering section, is by comparison devoid of pure mathematics. This chapter serves to lay out the fundamental pipeline within which the mathematical work is done: the rendering pipeline. The stages of the framework described in this chapter will be detailed in the later chapters (and to some degree in the previous viewing chapter), where the fascinating mathematical issues that arise within them can be explored. By its nature, this chapter focuses on the framework itself, the rendering pipeline, and its two most interesting components, the programmable vertex and fragment shader units.

We will also introduce some of the simpler methods of using this programmable pipeline to render colored geometry by introducing the basics of a common high-level shading language, OpenGL's GLSL. Common inputs and outputs to and from the shading pipeline will be discussed, concluding in a detailed introduction to the most complex and powerful of programmable shader source values—image-based texturing. However, this chapter includes only the most basic of programmable shaders, seeking mainly to introduce the rendering pipeline itself.

In Chapter 9, we will simultaneously explain the mathematics of real-time light simulation for rendering and demonstrate how to use the programmable shading pipeline to implement dynamic coloring of surfaces. In this chapter we will mix geometric intuitions, the basics of light-related physics, and simulated lighting equations and common approximations thereof with a discussion of more advanced uses of programmable shading.

As the concluding chapter in this sequence, Chapter 10 covers details of the final step in the overall rendering pipeline—rasterization, or the method of determining how to draw the colored surfaces as pixels on the display device. This will complete the discussion of the rendering pipeline.

In each section in these chapters we will relate the basic programming concepts, data structures, and functions that affect the creation, rendering, and coloring of geometry. As we move from geometry representation through shading, lighting, and rasterization, implementation information will become increasingly frequent, as the implementation of the final stages of the rendering pipeline is very much system dependent. While we will select a particular rendering application programming interface (API) (the book's basic Iv engine) and shading language (OpenGL's GLSL), the basic rendering concepts discussed will apply to most rendering systems.

As a note, we use the phrase *implementation* to refer to the underlying software or *driver* that maps our application calls to a given standard rendering API such as OpenGL or Direct3D into commands for a particular piece of graphics hardware (a graphics processing unit, or GPU, a term coined to recognize the CPU-like rising complexity and performance of modern graphics hardware). OpenGL and Direct3D implementations for a particular piece of graphics hardware are generally supplied with the device by the hardware vendor. A low-level hardware driver is not something that users of these APIs will have to write or even use directly. In fact, the main purpose of OpenGL and other such APIs is to provide a standard interface on top of these widely varying hardware/software three-dimensional (3D) systems. To avoid doubling the amount of implementation-related text in these chapters, most of the code examples in this and the following rendering chapters will describe the book's Iv rendering APIs, supplied as full source code on the book's accompanying CD-ROM. Interested readers may look at the implementations of the referenced Iv functions to see how each operation can be written in OpenGL or Direct3D.

8.2 Color Representation

8.2.1 RGB Color Model

To represent color, we will use the additive *RGB* (red, green, blue) color model that is almost universal in real-time 3D systems. Approximating the physiology of the human visual system (which is tuned to perceive color based on three primitives that are close to these red, green, and blue colors), the RGB system is used in all common display devices used by real-time 3D graphics systems. Color cathode ray tubes (or CRTs, such as traditional televisions and computer monitors), flat-panel liquid crystal displays (LCDs), plasma displays, and video projector systems are for the most part based upon the additive

RGB system. While some colors cannot be accurately displayed using the RGB model, it does support a very wide range of colors, as proven by the remarkable color range and accuracy of modern television and computer displays. For a detailed discussion of color vision and the basis of the RGB color model, see Malacara [97].

The RGB color model involves mixing different amounts of three predefined *primary* colors of light. These carefully defined primary colors are each named by the colors that most closely match them: red, green, and blue. By mixing independently controlled levels of these three colors of light, a wide range of *brightnesses*, *tones*, and *shades* may be created. In the next few sections we will define much more specifically how we build and represent colors using this method.

8.2.2 Colors as "Points"

The levels of each of the three primary colors are independent. In a sense, this is similar to a subset of points in $\mathbb{R}^3$, but with a "frame" consisting of the red, green, and blue axes (or components), and an origin representing black. While these can be thought of as a frame for our particular display device's color space, they are not a frame in any true sense for color in general. The behavior of colors does not always map directly into the concept of a real affine space. However, many of the concepts of real vector and affine spaces are useful in describing color representation and operations.

Our colors will be represented by points in 3-space, with the following frame vectors:

$$(1, 0, 0) \rightarrow red$$
$$(0, 1, 0) \rightarrow green$$
$$(0, 0, 1) \rightarrow blue$$

Often, as a form of shorthand, we will refer to the red component of a color C as C_r and to the green and blue components as C_g and C_b, respectively.

8.2.3 Color-Range Limitation

The theoretical RGB color space is semi-infinite in all three axes. There is an absolute zero value for each component, bounding the negative directions, but the positive directions are (theoretically) unbounded. Throughout much of the discussions of coloring, lighting, and shading, we will implicitly assume (or actually declare in the shading language) that the colors are nonnegative real values, potentially represented in the shading system as floating-point numbers.

However, the reality of physical display devices imposes severe limitations on the final output color space. When limited to the colors that can be represented by a specific display device, the RGB color space is not infinite in any direction. Real display devices, such as CRTs (standard tube monitors), LCD panel displays, and video projectors all have limits of both brightness and darkness in each color component; these are basic physical limitations of the technologies that these displays use to emit light. For details on the functionality and limitations of display device hardware, Hearn and Baker [73] detail many popular display devices.

Displays have minimum and maximum brightnesses in each of their three color axes, defining the range of colors that they can display. This range is generally known as a display device's *gamut*. The minimum of all color components combine to the device's darkest color, which represents black, and the maximum of all color components combine to the device's brightest color, which represents white. While it might be possible to create extrema that are not pure black and pure white, these are unlikely to be useful in a general display device.

Every display device is likely to have different exact values for its extrema, so it is convenient to use a standard color space for all devices as sort of normalized device colors. This color space is built such that

$$(0, 0, 0) \rightarrow \textit{darkest black}$$

$$(1, 1, 1) \rightarrow \textit{brightest white}$$

In the rest of this chapter and the following chapter we will work in these normalized color coordinates. This space defines an RGB color cube, with black at the origin, white at $(1, 1, 1)$, gray levels down the main diagonal between them (v, v, v), and the other six corners representing pure, maximal red $(1, 0, 0)$, green $(0, 1, 0)$, blue $(0, 0, 1)$, cyan $(0, 1, 1)$, magenta $(1, 0, 1)$, and yellow $(1, 1, 0)$.

When discussing colors, we often refer to their luminance. More specifically, when using this term with a normalized device color, we mean a mapping from the color to a specific gray value's v—this is known as relative luminance. The larger our component values, in general the closer to a luminance of 1 (which again represents pure white) we will get. We will describe this mapping below, and define luminance more formally in Chapter 9.

The following sections will describe some of the point and vector operations (and point- and vector-like operations) we will apply to colors, as well as discuss how these abstract color points map onto their final destinations, namely, hardware display devices.

8.2.4 Operations on Colors

Adding RGB colors is done using vector addition; the colors are added componentwise. Adding two colors has the same effect as combining the light from two colored light sources, for example, adding red $(R = (1, 0, 0))$ and green $(G = (0, 1, 0))$ gives yellow:

$$R + G = (1, 0, 0) + (0, 1, 0) = (1, 1, 0)$$

The operation of adding colors will be used through our lighting computations to represent the addition of light from multiple light sources and to add the multiple forms of light that each source can apply to a surface.

Scalar multiplication of RGB colors (sC) is computed in the same way as with vectors, multiplying the scalar times each component, and is ubiquitous in lighting and other color computations. It has the result of increasing $(s > 1.0)$ or decreasing $(s < 1.0)$ the luminance of the color by the amount of the scalar factor. Scalar multiplication is most frequently used to represent light attenuation due to various physical and geometric lighting properties.

One important vector operation that is used somewhat rarely with colors is vector length. While it might seem that vector length would be an excellent (if expensive) way to compute the luminance of a color, the nature of human color perception does not match the Euclidean norm of the linear RGB color space. Luminance is a "norm" that is affected by human physiology. The human eye is most sensitive to green, less to red, and least sensitive to blue. As a result, the equal weighting given to all components by the Euclidean norm means that blue contributes to the Euclidean norm far more than it contributes to luminance.

Although there are numerous methods used to compute the luminance of RGB colors as displayed on a screen, a common method (assuming nonnegative color components) is

$$luminance(C) = 0.2126C_r + 0.7152C_g + 0.0722C_b$$

or basically, the dot product of the color with a luminance reference color. The three color-space transformation coefficients used to scale the color components are basically constant for modern, standard monitors and HDTV screens. However, they may not apply to older NTSC television screens, which use a different set of luminance conversions. Discussion of these may be found in Poynton [124]. Note that luminance is *not* equivalent to perceived brightness. The luminance as we've computed it is linear with respect to the source linear RGB values. Brightness as perceived by the human visual system is nonlinear and subject to the overall brightness of the viewing environment, as well as the viewer's adaptation to it. See Cornsweet [26] for a related discussion of the physiology of human visual perception.

An operation that is rarely applied to geometric vectors but is used very frequently with colors is componentwise multiplication. Componentwise multiplication takes two colors as operands and produces another color as its result. We will represent the operation of componentwise multiplication of colors as "∘", or in shorthand by placing the colors next to one another (as we would multiply scalars), and the operation is defined as follows:

$$C \circ D = CD = (C_r D_r, C_g D_g, C_b D_b)$$

This operation is often used to represent the *filtering* of one color of light through an object of another color, such as white light passing through a stained glass window. In such a situation, one operand is assumed to be the light color, while the other operand is assumed to be the amount of light of each component that is passed by the filter. Another use of componentwise color multiplication is to represent the reflection of light from a surface—one color represents the incoming light and the other represents the amount of each component that the given surface reflects (the surface's reflectivity). We will use this frequently in Chapter 9 when computing lighting. For example, a color C and a filter (or surface) $F = (1, 0, 0)$ result in

$$C \circ F = (C_r, 0, 0)$$

or the equivalent of a pure red filter; only the red component of the light was passed, while all other light was blocked.

8.2.5   Alpha Values

Frequently, RGB colors are augmented with a fourth component, called *alpha*. Such colors are often written as *RGBA* colors. Unlike the other three components, the alpha component does not represent a specific color basis, but rather defines how the combined color interacts with other colors. The most frequent use of the alpha component is an opacity value, which defines how much of the surface's color is controlled by the surface itself and how much is controlled by the colors of objects that are behind the given surface. When alpha is at its maximum (we will define this as 1.0), then the color of the surface is independent of any objects behind it. The red, green, and blue components of the surface color may be used directly, for example, in representing a solid concrete wall. At its minimum (0.0), the RGB color of the surface is ignored and the object is invisible, as with a pane of clear glass, for instance. At an intermediate alpha value, such as 0.5, the colors of the two objects are blended together; in the case of alpha equaling 0.5, the resulting color will be the componentwise average of the colors of the surface and the object behind the surface.

We will discuss the uses of the alpha value when we cover mixing colors or *color blending* in Chapter 10. In a few cases, rendering APIs handle alpha a little differently from other color components (mention will be made of these situations as needed). In addition, it is often convenient to multiply the RGB values of a color by the alpha value, and store those new values along with the original alpha. This is known as *premultiplied alpha*, and makes it easier to perform more advanced blending operations. This will be covered in more detail when discussing blending.

8.2.6   Remapping Colors into the Unit Cube

Although devices cannot display colors outside of the range defined by their $(0, 0, 0) \dots$ $(1, 1, 1)$ cube, colors outside of this cube are often seen during intermediate color computations such as lighting. In fact, the very nature of lighting can lead to final colors with components outside of the $(1, 1, 1)$ limit. During lighting computations, these are generally allowed, but prior to assigning final colors to the screen, all colors must be within the normalized cube. This requires either the hardware, the device driver software, or the application to somehow remap or limit the values of colors so that they fall within the unit cube.

The simplest and easiest method is to clamp the color on a per-component basis:

$$\text{safe}(C) = (clamp(C_r), clamp(C_g), clamp(C_b))$$

where

$$clamp(x) = \max(\min(x, 1.0), 0.0)$$

However, it should be noted that such an operation can cause significant perceptual changes to the color. For example, the color $(1.0, 1.0, 10.0)$ is predominantly blue, but its clamped version is pure white $(1.0, 1.0, 1.0)$. In general, clamping a color can lead to the color becoming less *saturated,* or less colorful. While this might seem unsatisfactory, it actually can be beneficial in some forms of simulated lighting, as it tends to make overly bright objects appear to "wash out," an effect that can perceptually appear rather natural under the right circumstances.

Another, more computationally expensive method is to rescale all three color components of any color with a component greater than 1.0 such that the maximal component is 1.0. This may be written as

$$\text{safe}(C) = \frac{(\max(C_r, 0), \max(C_g, 0), \max(C_b, 0))}{\max(C_r, C_g, C_b, 1)}$$

Note the appearance of 1 in the max function in the denominator to ensure that colors already in the unit cube will not change—it will never increase the color components.

While this method does tend to avoid changing the overall saturation of the color, it can produce some unexpected results. The most common issue is that extremely bright colors that are scaled back into range can actually end up appearing darker than colors that did not require scaling. For example, comparing the two colors $C = (1, 1, 0)$ and $D = (10, 5, 0)$, we find that after scaling, $D = (1, 0.5, 0)$, which is significantly darker than C. As a result, this is almost never used in practice.

Scaling works best when it is applied equally (or at least coherently) to all colors in a scene, not to each color individually. There are numerous methods for this, but one such method involves finding the maximum color component of any object in the scene, and scaling all colors equally such that this maximum maps to 1.0. This is somewhat similar to a camera's auto exposure system. By scaling the entire scene by a single scalar, color ratios between objects in the scene are preserved. Figure 8.1 shows two different color-range limitation methods for the same source image. In Figure 8.1a, we clamp the values that are too large to display. Note that this results in a loss of image detail in the brightest sections of the image, which become pure white. In Figure 8.1b, we rescale all of the colors in the image based on the maximum value method described above. The details in the brightest areas of the screen are retained. However, even this method is not perfect. The rescaling of the colors does sacrifice some detail in the darker shadows of the image.

(a)   (b)

Figure 8.1. Color-range limitation methods: (a) image colors clamped and (b) image colors rescaled.

A more advanced method generally known as *tone mapping* uses nonlinear functions (often based on luminance) to remap an image. The simplest is global tone mapping, which applies the same function to all colors in an image. Local tone mapping, on the other hand, remaps regions of an image differently; a very bright section of the scene may be darkened to fit the range (e.g., a bright, cloud-streaked sky), while the shadowed sections of the image actually may be scaled to be brighter so that details are not lost in the shadows. The scaling may be different for different sections of the image, but the remapping is done in a regionally coherent method so that the relative brightnesses of related objects are reasonable. Regionally coherent means that we take the brightness of the region surrounding any point on the screen and try to keep the relative bright–dark relationships. A common trick in a daytime image of buildings and sky would be to darken the sky to fit in range and brighten the buildings to be less in shadow. While we are applying different scalings to different parts of the image (darkening to the sky and brightening to the buildings), the relative brightnesses within the buildings' region of the image are kept intact, and the relative brightnesses within the sky's regions of the image are kept intact. Thus, the sky and the buildings each look like what we'd expect, but the overall image fits within the limited brightness range.

These techniques are often used in high dynamic range (HDR) rendering, in which wide orders of magnitude exist in the computed lighting, but are then mapped down to the unit cube in a manner that forms a vibrant image. Figure 8.2 shows the same image for Figure 8.1,

Figure 8.2. A tone-mapped image.

but tone-mapped to retain details in both the shadows and highlights. The shadowed and highlighted areas are processed independently to avoid losing detail in either.

HDR rendering is now a standard feature in 3D games and other applications as GPU feature sets and performance have improved. Many examples of HDR rendering may be found at the developers' web sites of the major GPU vendors [3, 112].

8.2.7 Color Storage Formats

A wide range of color storage formats are used by modern rendering systems, both floating point and fixed point (as well as one or two hybrid formats). Common RGBA color formats include:

- Single-precision floating-point components (128 bits for RGBA color).

- Half-precision floating-point components (64 bits for RGBA color).

- 16-bit unsigned integer components (64 bits for RGBA color).

- 8-bit unsigned integer components (32 bits for RGBA color).

- Shared exponent extended-range formats. In the most common of these formats, red, green, and blue represent 0-dot-8 fixed-point mantissas, while a final 8-bit shared exponent is used to scale all three components. This is not as flexible as a floating-point value per color component (since all components share a single exponent), but it can represent a huge dynamic range of colors using only 32 bits for an RGB color.

In general, the floating-point formats are used as would be expected (in fact, on modern systems, the single-precision floating-point colors are now IEEE 754 compliant, making them useful for noncolor computations as well). However, the integer formats have a special mapping in most graphics systems. An integer value of 0 maps to 0, but the maximal value maps to 1.0. Thus, the integer formats are slightly different than those seen in any fixed-point format.

While a wide range of color formats are available to applications, a small subset of them cover most use cases. Internal to the programmable rendering pipeline, floating-point values are the most popular intermediate result format. As mentioned in Chapter 1, these can be of high, medium, or low precision (or even fixed point), depending on the platform and how they are specified by the programmer.

However, floating-point values are *not* the most popular format for shading output, the values that are stored in the frame buffer or other image buffer. Perhaps the most popular format for final color storage is unsigned 8-bit values per component, leading to 3 bytes per RGB color, a system known as *24-bit color* or, in some cases, by the misnomer *true color.* With an alpha value, the format becomes 32 bits per pixel, which aligns well on modern 32- and 64-bit CPU architectures. Another format that is sometimes used, particularly on extremely low-memory systems (such as mobile devices), has 5 bits each for red and blue and 6 bits for green, or 16 bits per pixel. This system, which sometimes goes by the name *high color*, is interesting in that it includes different amounts of precision for green than for red or blue. As we've discussed, the human eye is most sensitive to green, so the additional bit in the 16-bit format is assigned to it. This does have the downside that it is not possible to exactly represent grayscale values—the closest representable values tend to be tinted slightly green or magenta.

The historical reasons for using these lower-precision formats are storage space requirements, computational expense, and the fact that display devices often have the ability to display only 5–8 bits of precision per component. Even 32 bits per pixel requires one-quarter the amount of storage that is needed for floating-point RGBA values. Using full floating-point numbers for output colors (the colors that are drawn to the output LCD or CRT screen) is actually overkill, due to the limitations of current display device color resolution. For example, a good quality LCD display has a dynamic range (the ratio of luminance between the brightest and darkest levels that can be displayed by the devices) of around 1,000:1. This ratio means that current display devices cannot deliver anywhere near the eye's full range of perceived brightness or darkness. There are display technologies that can represent more than 24-bit color, but these are still the exception, rather than the rule. As these display devices become more common, device-level color representations will require more bits per component in order to avoid wasting the added precision available from these new displays.

Research has shown that the human visual system (depending on lighting conditions, etc.) can perceive between 1 million and 7 million colors, which leads to the (erroneous) theory that 24-bit color display systems, with their $2^{24} \approx 16.7$ million colors, are more than sufficient. While it is true that the number of different color "names" in a 24-bit system (where a color is named by its 24-bit RGB triple) is a greater number than the human visual system can discern, this does not take into account the fact that the colors being generated on current display devices do not map directly to the 1 million to 7 million colors that can be discerned by the human visual system. In addition, in some color ranges, different 24-bit color names appear the same to the human visual system (the colors are closer to one another than the human eye's *just noticeable difference*, or JND). In other words, 24-bit color wastes precision in some ranges, while lacking sufficient precision in others. Current 24-bit "true color" display systems are not sufficient to cover the entire range of human vision, either in range or in precision. As mentioned above, once the so-called deep color displays that support 10, 12, or 16 bits per color component become more ubiquitous, this will be less of an issue.

8.2.8 Nonlinear Color

In our discussion so far, we have assumed that color values act linearly; that is, the output value increases as a linear function of the given input color value. Physically if one were to simulate light that would produce an accurate result. However, when dealing with both display technology and human perception, that is not the case. Let's look at display technology first.

When CRT displays were more prevalent, it was noted that an increase in voltage did not produce an equivalently linear increase in phosphor output; rather, it behaved with a roughly exponential scale or

$$O = I^\gamma$$

Different monitor technologies could have a different γ value, or simply be calibrated differently—hence each monitor would be described as having a particular *gamma*.

To correct for the nonlinear response, the input color values need to have a corresponding inverse correction applied. The end result is linear inputs producing a linear response:

$$O = (I^{1/\gamma})^{\gamma}$$

This is called *gamma correction*.

With the rapid growth of web technology in the late 1990s, it became clear that the large preponderance of different color spaces and gamma values was having a negative effect on both good color reproduction and network bandwidth. To reconstruct a color correctly, one would have to bake the original monitor's parameters into the image, which would both increase the size of the image and take time converting to the destination monitor's parameters.

In 1996, Microsoft and Hewlett-Packard proposed a new color-space standard. The idea was that all new monitors would use the same color parameters, with a gamma of approximately 2.2. In this way, there would be no need to convert from one color space to another—the same color values would produce the same result on all monitors. Hence, there would be no need to attach the original color parameters, and no time taken for conversion. This new standard is called sRGB, and while there are other RGB color spaces in play (Adobe RGB, for example), sRGB is currently the de facto standard for artists working with real-time systems.

What this means for games is that any color values you are given are probably in sRGB. Since this is a nonlinear format, linearly interpolating or blending colors as is will not produce the correct result—one way or another, you will have to convert to linear color, blend, then convert back to sRGB.

To convert from sRGB to the linear RGB equivalent, we use the formula

$$C_{linear} = \begin{cases} \frac{C_{srgb}}{12.92} & C_{srgb} \leq 0.04045 \\ \left(\frac{C_{srgb}+0.055}{1.055}\right)^{2.4} & C_{srgb} > 0.04045 \end{cases}$$

and to convert back from linear to sRGB, we use

$$C_{srgb} = \begin{cases} 12.92\, C_{linear} & C_{linear} \leq 0.0031308 \\ 1.055\, C_{linear}^{1/2.4} - 0.055 & C_{linear} > 0.0031308 \end{cases}$$

Note that this is not a pure exponential. Both curves have a small linear piece to prevent the slope of the sRGB curve from becoming infinite near 0.

A more efficient solution might seem to be to convert all color inputs to linear, store them internally as linear, and then when we are done rendering, convert the resulting image back to sRGB. The problem is—particularly when using 8-bit color components—that we need more bits to represent the same values in linear as we do in sRGB. Consider 1, the smallest possible positive 8-bit value. This represents a color component value of 1/255, or 0.00392157. Converting this from sRGB to linear, we end up with 0.000303527. Rounding to the closest 8-bit value, we end up with 0. As it happens, the first seven color values will end up being rounded to 0, which is a significant loss of information. We would need 13 bits per color channel in order to represent these same distinct values in a linear format as we do in sRGB. And at the high end, an 8-bit value of 254 represents 254/255, or 0.996078. Converting from sRGB to linear, and then back to 8-bit values, gives 253. We actually need fewer bits to represent the higher bits in linear—and we've just added more to represent the low end, wasting many unnecessary bits.

Another question might be: Since CRTs are rarely used these days and modern LCD displays have closer to linear response, why not change the standard? One unhelpful answer might be that changing an established standard is not that easy. But there is another reason: the sRGB color space also roughly follows the human eye's response curve to luminance. Our eyes can distinguish better between darker tones than we can brighter ones. So sRGB colors naturally dedicate more bits to colors at the dark end of the space where the just noticable difference between colors is smaller, and fewer to the colors at the brighter end where the JND is larger. Were we to use linear 24-bit colors, we would notice distinct jumps in the dark values known as banding. Using sRGB, particularly with 24-bit color, has compressed our colors to match our visual response.

Fortunately, in most modern hardware there is an easier and more efficient solution than having to explicitly convert from and to sRGB. In both OpenGL and Direct3D it is possible to specify sRGB format for both input and output images. The system will then, upon reading a color from an image, efficiently convert it to linear RGB, and upon writing, convert it back. Our temporary variables should be in floating point or 16-bit fixed point, which is enough to handle the extra precision needed without having to add those bits to our storage formats. This way our programs can continue to assume linear colors without worrying about the headaches and inefficiencies of conversion.

8.3 Points and Vertices

So far, we have discussed points as our sole geometry representation. As we begin to abstract to the higher level of a surface, points will become insufficient for representing the attributes of an object or, for that matter, the object itself. The first step in the move toward a way of defining an object's surface is to associate additional data with each point. Combined together (often into a single data structure), each point and its additional information form what is often called a *vertex*. In a sense, a vertex is a heavy point: a point with additional information that defines some properties of the surface around it.

8.3.1 Per-Vertex Attributes

Within a vertex, the most basic value is the position of the vertex, generally a 3D point that we will refer to as P_V in later sections.

Other than vertex position, perhaps the most basic of the standard vertex attributes are colors. Common additions to a vertex data structure, vertex colors are used in many different ways when drawing geometry. Much of the remainder of this chapter will discuss the various ways that per-vertex colors can be assigned to geometry, as well as the different ways that these vertex colors are used to draw geometry to the screen. When referring to a color attribute we will use C_V (and will sometimes specifically refer to the vertex alpha as A_V, even though it is technically a component of the overall color).

Another data element that can add useful information to a vertex is a vertex normal. This is a unit-length 3-vector that defines the orientation of the surface in an infinitely small neighborhood of the vertex. If we assume that the surface passing through the vertex is locally planar (at least in an infinitely small neighborhood of the vertex), the surface normal is the normal vector to this plane (recall the discussion of plane normal vectors from Chapter 2). In most cases, this vector is defined in the same space as the vertices, generally model (aka object) space. As will be seen later, the normal vector is a pivotal component in lighting computations. We will generally refer to the normal as $\hat{\mathbf{n}}_V$.

A vertex attribute that we will use frequently later in this chapter is a texture coordinate. This will be discussed in detail in the sections in this chapter on texturing and in parts of the following two chapters; basically, a set of texture coordinates is a real-valued 2-vector (most frequently, although they also may be scalars or 3-vectors) per vertex that defines the position of the vertex within a smooth parameterization of the overall surface. These are used to map two-dimensional (2D) images onto the surface in a shading process known as texturing. A vertex may have more than one set of texture coordinates, representing the mapping of the vertex in several different parameterizations.

Finally, owing to the general and extensible nature of programmable shading, an object's vertices may have other sets of per-vertex attributes. Most common are additional values similar to the ones listed above: per-vertex color values, per-vertex directional vectors of some sort, or per-vertex texture coordinates. However, other programmable shaders could require a wealth of different vertex attributes; most shading systems support scalar vertex attributes as well as generic 2D, 3D, and 4D vectors. The meaning of these vectors is dependent upon the shading program itself.

8.3.2 An Object's Vertices

For any geometric object, its set of vertices can be represented as an array of structures. Each array element contains the value for each of the vertex attributes supported by the object. Note that for a given object, all of the vertices in the array have the same type of structure. If one vertex has a particular attribute, they all will contain that attribute (likely with a different value). An example of the vertex structure for an object with position values, a color, and one set of texture coordinates is shown below.

```
struct IvTCPVertex
{
    IvVector2 texturecoord;
    IvColor color;
    IvVector3 position;
};
```

A smaller, simpler vertex with just position and normal might be as follows:

```
struct IvNPVertex
{
    IvVector3 normal;
    IvVector3 position;
};
```

Along with the C or C++ representation of a vertex, an application must be able to communicate to the rendering API how the vertices are laid out. Each rendering API uses its own system, but two different methods are common; the simpler (but less flexible) method is for the API to expose some fixed set of supported vertex formats explicitly and use an enumerated type label to represent each of these formats. All of an application's geometry must be formatted to fit within the fixed set of supported vertex formats in this case. The more general system is for an API to allow the application to specify the type (float, etc.), usage (position, color, etc.), dimension (1D, 2D, etc.), and stride (bytes between the attribute for one vertex and the next) of each active attribute. This system is far more flexible, but can greatly increase the complexity of the rendering API implementation. The latter is common in modern graphics APIs, such as Direct3D and OpenGL. The former method is used in `Iv` for the purposes of simplicity and ease of cross-platform support. `Iv` uses the following enumeration to define the vertex formats it supports:

```
enum IvVertexFormat
{
    kCPFormat,       // color, position
    kNPFormat,       // normal, position
    kTCPFormat,      // texture coord, color, position
    kCNPFormat,      // color, normal, position
    kTNPFormat       // texture coord, normal, position
};
```

This enumeration is used in various places in the `Iv` rendering engine to declare the format of a given vertex or array of vertices to the system.

Some rendering APIs allow for the vertex attributes to be non interleaved; that is, the application keeps independent packed arrays of each vertex attribute. This so-called structure of arrays format has fallen out of favor, as the interleaved formats provide better cache coherence—in an interleaved format, accessing one attribute in a vertex is likely to load the entire vertex into cache. We will assume an interleaved vertex format for the remainder of the rendering discussions.

8.3.2.1 Vertex Buffers

Programmable shaders and graphics rendering pipelines implemented entirely in dedicated hardware have made it increasingly important for as much rendering-related data as possible to be available to the GPU in device local memory, rather than system memory. Modern graphics APIs all include the concept of a *vertex buffer* or *vertex buffer object*, an opaque handle that represents source vertex data resident in GPU memory.

In order to use vertex buffers to render an object, an application must make calls to the rendering API to allocate enough storage for the object's array of vertices in GPU memory. Then, some method is used to transfer the vertex array from system memory to GPU memory. Having transferred the data, the application can then use the opaque handle to render the geometry at peak performance. Note that once vertex array data are in GPU memory, it is usually computationally expensive to modify them. Thus, vertex buffers are most frequently used for data that the CPU does not need to modify on a per-frame basis. Over time, as programmable shaders have become more and more powerful, there have been fewer and fewer (if any) per-vertex operations that need to be done on the CPU, thus making it more easily possible to put all vertex data in static vertex buffers.

A common vertex buffer creation sequence in many APIs is to create the vertex buffer, passing in the vertex format and number of vertices, and any associated vertex data. After creation, the given vertex array data are transferred to GPU-accessible memory, and the vertex buffer can be used repeatedly without any further need to copy data to the GPU. In Iv, the sequence is as follows:

```
IvResourceManager& manager;
// ...

    // Create vertex data with 1024 vertices
    // Each vertex has a color and position
    IvCPVertex verts[1024];
    // Loop over all 1024 vertices in verts and
    // fill in the data...
    // ...
    // Create the vertex buffer and upload the vertex data
    IvVertexBuffer* buffer
        = manager.CreateVertexBuffer(kCPFormat, 1024, verts, kImmutableUsage);
```

Once this process is complete, the vertex buffer is filled with data and ready to be used to render.

The last parameter to CreateVertexBuffer is a hint that will be passed to the graphics driver so it can efficiently allocate and manage the vertex buffer memory. In this case, we have indicated with kImmutableUsage that we never want to change our vertex data. However, at times we need to be able to update the data contained inside, usually once a frame. Most graphics APIs also allow for a client to request to map or lock a vertex buffer, which returns a pointer to system memory that can be loaded with new data. Finally, the buffer is unmapped or unlocked, which releases access to the system memory pointer and (if needed) transfers the vertex data to GPU-accessible memory. It's also common to allow for uploading to subsets of the full buffer, but in our case we are going to simplify our implementation and only lock the entire buffer. The corresponding sequence is as follows:

```
IvResourceManager& manager;
// ...

    // Create vertex data with 1024 vertices
    // Each vertex has a color and position
    IvVertexBuffer* buffer
        = manager.CreateVertexBuffer(kCPFormat, 1024, NULL, kDynamicUsage);
    // Lock the vertex buffer and cast to the correct
    // vertex format
    IvCPVertex* verts = (IvCPVertex*) buffer->BeginLoadData();
    // Loop over all 1024 vertices in verts and
    // fill in the data...
    // ...
    // Unlock the buffer, so it can be used
    buffer->EndLoadData();
```

Note that in the CreateVertexBuffer function we have specified NULL for our original buffer data, and a new usage parameter. In the first case, we are going to use the pointer that `BeginLoadData` gives us, so we don't need to specify an initial data pointer here—the driver will automatically create a vertex buffer of the correct size. The parameter `kDynamicUsage` indicates that we are going to be changing the vertex buffer around once a frame, and allows us to use the `BeginLoadData–EndLoadData` interface (which `kImmutableUsage` does not). A third alternative is `kDefaultUsage`, which splits the difference between the other two: it allows us to use the locking–unlocking interface, but tells the driver we will rarely be updating the texture. There are other usage patterns available in both OpenGL and Direct3D, but these three are the most commonly used.

8.4   Surface Representation

In this section we will discuss another important concept used to represent and render objects in real-time 3D graphics: the concept of a surface and the most common representation of surfaces in interactive 3D systems, sets of triangles. These concepts will allow us to build realistic-looking objects from the sets of vertices that we have discussed thus far.

In Chapter 2 we introduced the concept of a triangle, a subset of a plane defined by the convex combination of three noncollinear points. In this chapter we will build upon this foundation and make frequent use of triangles, the normal vector to a triangle, and barycentric coordinates. A quick review of the sections of Chapter 2 covering these topics is recommended.

While most of the remainder of this chapter focuses only on the assignment of colors to objects for the purposes of rendering, the object and surface representations we will discuss are useful for far more than just rendering. Collision detection, picking, and even artificial intelligence all make use of these representations.

8.4.1   Vertices and Surface Ambiguity

Unstructured collections of vertices (sometimes called *point clouds*) generally cannot represent a surface unambiguously. For example, draw a set of 10 or so dots representing points on a piece of paper. There are numerous ways one could connect these 2D points into a closed curve (a 1D surface) or even into several smaller curves. This is true even if the vertices include normal vectors, as these normal vectors only define the orientation of the surface in an infinitely small neighborhood of the vertex. Without additional structure, either implicit or explicit, a finite set of points rarely defines an unambiguous surface.

A cloud of points that is infinitely dense on the desired surface can represent that surface. Obviously, such a directly stored collection of unstructured points would be far too large to render in real time (or even store) on a computer. We need a method of representing an infinitely dense surface of points that requires only a finite amount of representational data.

There are numerous methods of representing surfaces, depending on the intended use. Our requirements are that we can make direct use of the conveniently defined vertices that our geometry pipeline generates, and that the representation we use is efficient to render. As it turns out, we have already been introduced to such a representation in one of the earliest sections of the book: planar triangles.

8.4.2   Triangles

The most common method used to represent 3D surfaces in real-time graphics systems is simple, scalable, requires little additional information beyond the existing vertices, and allows for direct rendering algorithms; it is the approximation of surfaces with simple shapes, or *tessellation*. The shape almost always used is a triangle. Tessellation refers not only to the process that generates a set of triangles from a surface but also to the triangles and vertices that result.

Triangles, each represented and defined by only three points (vertices) on the surface, are connected point to point and edge to edge to create a piecewise flat (or faceted) approximation of the surface. By varying the number and density of the vertices (and thus the triangles) used to represent a surface, an application may make any desired trade-off between compactness/rendering speed and accuracy of representation. Representing a surface with more and more vertices and triangles will result in smaller triangles and a smoother surface, but will add rendering expense and storage overhead owing to the increased amount of data representing the surface.

One concept that we will use frequently with triangles is that of barycentric coordinates. From the discussion in Chapter 2, we know that any point in a triangle may be represented by an element of $\mathbb{R}^2$ (s, t) such that $0.0 \leq s, t \leq 1.0$. These coordinates uniquely define each point on a nondegenerate triangle (i.e., a triangle with nonzero area). We will often use barycentric coordinates as the domain when mapping functions defined across triangles, such as color.

8.4.3   Connecting Vertices into Triangles

To create a surface representation from the set of vertices on the surface, we will simply "connect the dots." That is, we will generate additional information for rendering that joins sets of three vertices by spanning them with a triangle. As an example, Figure 8.3a depicts a fan-shaped arrangement of six triangles (defining a hexagon) that meet in a single point. The vertex array for this geometry is an array of seven vertices: six around the edge and

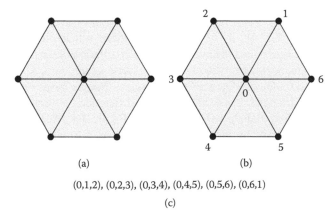

(a)　　　　　　　　(b)

(0,1,2), (0,2,3), (0,3,4), (0,4,5), (0,5,6), (0,6,1)

(c)

Figure 8.3. A hexagonal configuration of triangles: (a) configuration, (b) seven shared vertices, and (c) index list for shared vertices.

one in the center. Figure 8.3b shows these seven vertices, numbered with their array indices in the vertex array. However, this array alone does not define any information about the triangles in the object.

One solution allowed by most APIs is to pass in a list of vertices and treat every three vertices as representing a triangle—this is called a triangle list. However, in our example above that would mean duplicating vertices, so a triangle list with T triangles would use a vertex array with $3T$ vertices. This is generally suboptimal for memory usage, bus traffic, and processing time (this last because we have to transform each vertex more than once).

Indexed geometry, or an indexed triangle list, is a better solution. It defines an object with two arrays: the vertex array we have already discussed, and a second array of integral values for the triangle connectivities, called the index (or element) array. The *index array* is an array of integers that represent indices (offsets) into the vertex array; there are three times as many indices in the index array as there are triangles in the object. Each set of three adjacent indices represents a triangle. The indices are used to look up vertices in the vertex array; the three vertices are joined into a triangle. Figure 8.3c shows the index list for the hexagon example.

Note the several benefits of indexed geometry. First, vertices can be reused in as many triangles as desired simply by using the same index value several times in the index array. This is shown clearly by the hexagon example. One of the vertices (the central vertex) appears in every single triangle! If we had to duplicate a vertex each time it was used in a triangle, the memory requirements would be much higher, since even small vertex structures take more space than an index value. Index values are generally 16- or 32-bit unsigned integers. A 16-bit index value can represent a surface made up of up to 65,536 vertices, more than enough for the objects in many applications, while a 32-bit index array can represent a surface with more than 4 billion vertices (essentially unlimited).

Most rendering APIs support a wide range of nonindexed and indexed geometry. Triangle lists, such as the ones we've just introduced, are simple to understand but are not as optimal as other representations. The most popular of these more optimal representations are *triangle strips*, or *tristrips*. In a triangle strip, the first three vertices represent a triangle, just as they do in a triangle list. However, in a triangle strip, each additional vertex (the fourth, fifth, etc.) generates another triangle out of itself and the two indices that preceded it (e.g., 0-1-2, 1-2-3, 2-3-4, ...). This forms a ladderlike strip of triangles (note that each triangle is assumed to have the reverse orientation of the previous triangle—counterclockwise, then clockwise, then counterclockwise again, etc.). Then, too, whereas triangle lists require $3T$ indices to generate T triangles, triangle strips require only $T + 2$ indices to generate T triangles. An example of the difference between the size of index arrays for triangle lists and triangle strips is shown in Figure 8.4. Much research has gone into generating optimal strips by maximizing the number of triangles while minimizing the number of strips, since there is a two-vertex overhead to generate the first triangle in a strip. The longer the strip, the lower the average number of indices required per strip. Most consumer 3D hardware that is available today renders triangle strips at peak performance, because each new triangle reuses two previous vertices, requiring only one new vertex (and in the case of indexed primitives, one new index) per triangle. This minimizes transform work on the GPU, as well as potential traffic over the bus that connects the CPU to the GPU.

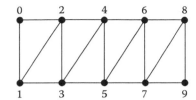

Index array for triangle list:
0,1,2, 1,3,2, 2,3,4, 3,5,4, 4,5,6, 5,7,6, 6,7,8, 7,9,8
(24 indices)

Index array for triangle strip:
0,1,2,3,4,5,6,7,8,9
(10 indices)

Figure 8.4. The same object as a triangle list and a triangle strip.

8.4.3.1 Index Buffers

Most GPUs can link vertices and indices into triangles without any CPU intervention. Thus, it is useful to be able to place index arrays into GPU-accessible memory. These objects are called *index buffers*, and they are directly analogous to the vertex buffers discussed previously. The only difference is that the format of an index buffer is far more limited; in Iv, only 32-bit indices are supported and are assumed. Iv code to create and fill an index buffer is shown below.

```
IvResourceManager& manager,
// ...
    // Create an index buffer with 999 indices
    // With triangle lists, this would be 333 triangles
    unsigned int indices[999];
    // Loop over all 999 indices and fill in the data...
    // ...
    IvIndexBuffer* buffer = manager.CreateIndexBuffer
    (999, indices, kImmutableUsage);
```

We can, similar to vertex buffers, also use `BeginLoadData` and `EndLoadData` for index buffers with dynamic and default usage.

8.4.4 Drawing Geometry

Source Code
Demo
BasicDrawing

The final step toward rendering geometry from an application point of view is to pass the required information into the rendering API to initiate the draw operation. Submitting geometry to the rendering API generally takes the form of a draw call. APIs differ on which subset of the geometry information is passed to the draw call and which is set as the current state beforehand, but the basic pieces of information that define the inputs to the draw call include at least the array of vertices, array of indices, type of primitive (list, strip, etc.), and rendering state defining the appearance of the object. Some APIs may also require the application to specify the location of each component (normal, position, etc.) within the vertex structure. The `Iv` rendering engine sets up the geometry and connectivity, and renders in a single call, as follows:

```
IvRenderer& renderer;
IvVertexBuffer* vertexBuffer;
IvIndexBuffer* indexBuffer;
// ...
```

```
    renderer.Draw(kTriangleListPrim, vertexBuffer, indexBuffer);
```

Note the enumerated type used to specify the primitive. In this case, we are drawing an indexed triangle list (`kTriangleListPrim`), but we could have specified a triangle strip (`kTriangleStripPrim`) or other primitive as listed in `IvPrimType`, assuming that the index data were valid for that type of primitive (each primitive type uses its index list a little differently, as discussed previously).

Once the geometry is submitted for rendering, the work really begins for the implementation and 3D hardware itself. The implementation passes the object geometry through the rendering pipeline and finally (if the geometry is visible) onto the screen. The following sections will detail the most common structure of the rendering pipeline in modern graphics APIs.

8.5   Rendering Pipeline

The basic rendering pipeline is shown in Figure 8.5. The flow is quite simple and will be the basis for much of the discussion in this chapter. Some of the items in the diagram will not yet be familiar. In the remainder of this chapter we will fill in these details. The flows are as follows:

1. **Primitive processing.** The pipeline starts with the triangle indices, which determine on a triangle-by-triangle basis which vertices in the array are required to define each triangle.

2. **Per-vertex operations.** All required vertices (which contain surface positions in model space along with the additional vertex attributes) are processed as follows:

   a. The positions are transformed into homogeneous space using the model view and projection matrices.

   b. Additional per-vertex items such as lit vertex colors are computed based on the positions, normals, and so forth.

3. **Triangle assembly.** The transformed vertices are grouped into triples representing the triangles to be rendered.

4. **Triangle clipping.** Each homogeneous-space triangle is clipped or culled as required to fall within the view rectangle.

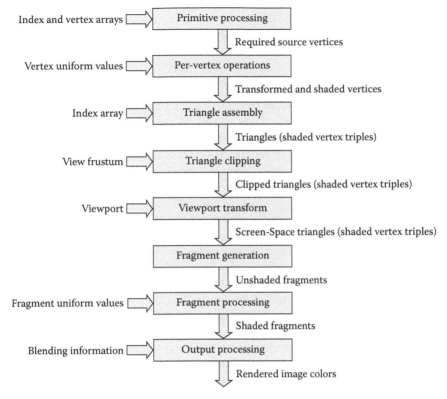

Figure 8.5. Details of the basic rendering pipeline.

5. **Viewport transform.** The resulting clipped triangles are transformed into screen space.

6. **Fragment generation.** Triangles are "sampled," generating pixel-aligned samples, called fragments.

7. **Fragment processing.** The final color and other properties of the surface are computed for each fragment.

8. **Output processing.** The final fragments are combined with those from other objects that are a part of the scene to generate the final rendered image.

The rendering section of this book covers all of these steps in various levels of detail. In this chapter we have already discussed the basics of indexed triangle primitives (primitive processing and triangle assembly). In Chapter 7 we discussed projection of vertices (per-vertex operations), clipping and culling (triangle clipping), and transformation into screen space (viewport transform). In this chapter we will provide an overview of other per-vertex operations and fragment processing. In Chapter 9, we will provide details on how light–surface interaction can be simulated in per-vertex operations and fragment processing.

Finally, the details of how fragments are generated and processed (fragment generation and processing), as well as how they are output to the device (output processing), are discussed in Chapter 10.

8.5.1   Fixed-Function versus Programmable Pipelines

The above pipeline has been common to rendering systems and APIs for over a decade. Initially, the major rendering APIs such as OpenGL 1.x (and OpenGL ES 1.x) and Direct3D's DX3 through DX7 implemented each stage with basically fixed functionality, modified only by a limited number of settings and switches. As features multiplied in commercial 3D systems, the switches and settings became more and more complex and often began to interact in confusing ways.

As a result, starting with APIs like OpenGL 2.0 and Direct3D's DX8, graphics systems have added flexibility. The APIs included new interfaces that allowed several of the most important fixed-function stages to be replaced with application-provided *shader* code. The major stages that were replaced with programmability were the per-vertex operations and fragment processing. Later, support for shaders in the triangle assembly stage was added as well. With the advent of Direct3D's DX10, the OpenGL Core interface, and the mobile 3D API OpenGL ES 2.0 (along with other APIs of that generation), the fixed-function per-vertex and fragment stages are eschewed entirely; *only* shaders are supported.

The two major high-level shading languages used for interactive 3D graphics are Microsoft's HLSL (High-Level Shading Language) and OpenGL's GLSL (GL Shading Language) [132]. While both of these languages have significant differences, they are remarkably similar. They all have the basic feel of C or C++, and thus switching between them is generally quite easy. Since OpenGL's GLSL is widely available, has been supported for some time by both OpenGL and OpenGL ES (the latter with some limitations, known as GLSL ES), and is quite clean, we will use it exclusively for in-text shading language examples. However, HLSL is capable of the same operations in relatively similar ways.

The remainder of this book will deal exclusively with shader-based pipelines. For the examples we will use, shaders are more illustrative and simpler. As we shall see in the lighting chapter (Chapter 9), high-level shading languages make it possible to directly translate shading and lighting equations into shader code. This is the additional value of shaders; while they make complex effects possible, they also make simple shading equations quite efficient by avoiding all of the conditionals and flag checking required by a fixed-function pipeline's settings.

8.6   Shaders

8.6.1   Using Shaders to Move from Vertex to Triangle to Fragment

The core shader types are vertex shaders (VSs) and fragment shaders (FSs), also known in some APIs as pixel shaders. They are, at their core function, very similar. In fact, on modern processors they are executed on the same hardware units—this is called unified shader hardware. They each take input values that represent a single entity, and output values that define additional properties of that entity. In the case of a vertex shader, the entity in question is a vertex, or source surface position and additional attributes as discussed previously in this chapter. In the case of a fragment shader, the entity is a fragment or sample representing an infinitesimally small region of the surface being rendered. When we discuss rasterization in Chapter 10, we will see that there is actually a much more precise definition of a fragment, but for now, the basic concept is that it is a sample somewhere on the surface of the object, generally at a point in the interior of one of the triangles, not coincident with any single vertex defining the surface.

The "one in, one out" nature of both types of shader is an inherent limitation that is simplifying yet at times frustrating. A vertex shader has access to the attributes of the current vertex only. It has no knowledge of surface continuity and cannot access other vertex array elements. Similarly, the fragment shader receives and can write to only the properties of the current fragment and cannot change the screen-space position of that fragment. It cannot access neighboring fragments or the source vertices of the triangle that contains the fragment. The sole deviation from this standard is that in many shading systems, the fragment shader can generate one or zero fragments. In other words, the fragment shader can choose to "kill" the current fragment, leaving a hole in the surface. This is useful for creating intratriangle cutouts to the surface.

Looking at the pipeline depicted in Figure 8.5 in reverse (bottom to top), from a single-shaded fragment backwards gives an understanding of the overall pipeline as a function. Starting from the end, the final, shaded fragment was computed in the fragment shader based on input values that are interpolated to the fragment's position within the triangle that contains it. This containing triangle is based upon three transformed and processed vertices that were each individually output from the vertex shader. These vertices were provided, along with the triangle connectivity, as a part of the geometry object being drawn. Thus, the entire shading pipeline is, in a sense, one long function.

8.6.2 Shader Input and Output Values

Both vertex and fragment shaders receive their inputs in roughly the same types, the most common being floating-point scalars (`float` in GLSL), vectors (`vec2`, `vec3`, and `vec4` in GLSL), matrices (`mat2`, `mat3`, `mat4`, etc., in GLSL), and arrays of each of these types of values. Colors are an extremely common type passed in to both forms of shaders and are generally represented in the shaders as floating-point 4-vectors, just as discussed in the introductory material in this chapter (although they are usually accessed in the shader as `v.r`, `v.g`, etc., instead of `v.x`, `v.y`, etc.). Signed and unsigned integers and associated vectors and arrays are often supported as well.

One additional type of input to a shader is a texture sampler, which represents image-based lookup within the shader. This is an extremely powerful shader input and will garner its own section later in this chapter and in the chapters to come. Most modern graphics systems and APIs allow samplers as inputs to both vertex and fragment shaders. However, for the purposes of this book, we will discuss them as inputs to fragment shaders, where they are more commonly used.

8.6.3 Shader Operations and Language Constructs

The set of shader operations in modern shading languages is generally the same in both vertex and fragment shaders. The operations and functions are too broad to list here, but include the most common infix operations (addition, subtraction, multiplication, division, negation) for scalar, vector, and matrix types and the sensible mixing thereof. A wide range of standard mathematical functions are also available, such as dot and cross products, vector normalization, trigonometric functions, and so forth.

Functions, procedures, conditionals, and loops are also provided in the high-level shading languages. However, since shaders are in essence SIMD (single-instruction multiple-data) systems, looping and branching can be expensive, especially on older hardware. However, the overall shading languages are exceedingly powerful. A full list of functionality for GLSL can be found on the OpenGL reference card [66].

8.6.4   Other Shader Types

In recent years, other shader types have been introduced that expand the functionality of the basic rendering pipeline, and most of which fall between the vertex and fragment shader stages.

The *geometry shader* executes during the triangle assembly stage. It takes vertices as input, and can modify them into more complex primitives. For example, you could send a 2D rectangle as two vectors: a 2-vector to the center of the rectangle, and a 2-vector representing the horizontal and vertical half-extents. In geometry shader this would be turned into a quad, with two triangles and four vertices. In this way you could upload less vertex data to the GPU, and thereby save transfer time.

Tessellation shaders also execute during the triangle assembly stage. They fall between the vertex shader and the geometry shader, and provide a more efficient way to create a large number of primitives from a given input. There are three stages: a user-defined tesselation control shader that controls the level of tesselation, a fixed-function tessellation primitive generator that performs the tesselation, and a user-defined tesselation evaluation shader that takes the output of the tesselation and produces the final vertex values. This can be used to generate the level of detail for objects, or to tessellate (or triangulate) curves and curved surfaces.

Finally, the *compute shader* lies outside of the standard graphics pipeline, and allows the programmer to treat the GPU as a general processor and run highly parallel computations.

At the time of printing, geometry and tessellation shaders are not generally supported in OpenGL ES. And for the purposes of simplicity, we will focus only on vertex and fragment shaders in this text. However, once these are understood, the reader is advised to investigate the other shader types.

8.7   Vertex Shaders

8.7.1   Vertex Shader Inputs

Vertex and fragment shaders do have slightly different sources of input, owing to their different locations in the rendering pipeline. Vertex shaders receive three basic sources of input: per-vertex attributes, per-object uniforms, and global constants. The first two can be thought of as properties of the geometry object being rendered, while the lattermost are properties and limits of the rendering hardware.

The per-vertex attributes are the elements of the object's vertex structure described above and will likely differ from vertex to vertex. In past versions of OpenGL, some per-vertex attributes corresponding to some of the fixed-function pipeline were standard and accessed via built-in variables in the vertex shader. These have been deprecated in the current OpenGL Core Profile, were never supported in OpenGL ES 2.0 or greater, and are also not available in DirectX's HLSL. For this reason, the Iv library has its own system for managing standard variables such as position, normal vector, color, and texture coordinates.

A mapping needs to be defined between the register locations of the vertex inputs in the shader and the position of the attributes in the vertex layout. OpenGL and DirectX have different mechanisms for managing this; since Iv has a fixed set of vertex layouts, it handles this internally. For GLSL the user only needs to declare the input variables in the shader with the correct keywords (IV_POSITION, IV_COLOR, IV_NORMAL, IV_TEXCOORD0). These match similar keywords in HLSL.

The per-object uniforms can be thought of as global variables and are the same value (or uniform) across the entire object being drawn. As with attributes, some uniforms were standard in OpenGL but are no longer; common examples include the model view and projection matrices. So again, Iv has defined a standard set of uniforms, which are explicitly set by the application.

The constants are provided by the rendering API and represent hardware limits that may be of use to shaders attempting to deal with running on different platforms. Constants are just that—constant over all rendered objects.

8.7.2 Vertex Shader Outputs

One required vertex shader output value is the homogeneous (postprojection transform) vertex position. It must be written by all vertex shaders. The projected positions are required in order to generate screen-space triangles from which fragment samples can be generated.

Vertex shaders provide their other output values by writing to specific variables (in older versions of GLSL these were called *varyings*). Standard (or built-in) vertex output variables differ by API and shading language. User-defined or custom vertex shader outputs may be declared as well, although platforms may differ in the limited number of custom output parameters that can be used.

8.7.3 Basic Vertex Shaders

The simplest vertex shader simply transforms the incoming model-space vertex by the model view and projection matrix, and places the result in the required output register, as follows:

```
// GLSL
uniform mat4 IvModelViewProjectionMatrix;
layout(location = IV_POSITION) in vec3 position;
void main()
{
    gl_Position = IvModelViewProjectionMatrix * position;
}
```

The layout keyword in GLSL indicates the location for a particular vertex attribute within our vertex data. As mentioned, for GLSL Iv provides an internal mapping to the correct attribute location, similar to that found in Direct3D. So this shader uses a vertex input (position) placed at the Iv IV_POSITION attribute location, a standard Iv uniform (IvModelViewProjectionMatrix), and a standard OpenGL vertex shader output (gl_Position). It transforms a floating-point 4-vector (vec4) by a floating-point 4×4 matrix (mat4) and assigns the result to a 4-vector. However, this simple vertex shader provides no additional information about the surface—no normals, colors, or additional attributes. In general, we will use more complex vertex shaders.

8.8 Fragment Shaders

8.8.1 Fragment Shader Inputs

Unlike vertex shaders, which are invoked on application-supplied vertices, fragment shaders are invoked on dynamically generated fragments. Thus, there is no concept of per-fragment attributes being passed into the fragment shader by the application.

Instead, shader-custom output values written by the previous shader in the pipeline (usually a vertex shader) are simply interpolated and provided as inputs to the linked fragment shader. They must be declared in the fragment shader using the same name and type as they were declared in the previous shader, so they can be linked together. Some of the built-in output values written by a shader are provided in a similarly direct manner. However, others are provided in a somewhat different manner as is appropriate to the primitive and value. For example, in GLSL, the vertex shader built-in output position (which is specified in homogeneous coordinates) and the fragment shader's built-in fragment coordinate (which is in a window-relative coordinate) are in different spaces.

Fragment shaders support constants and uniforms. A set of fragment shader–relevant constants may be provided by the implementation. In addition, fragment shaders can access uniform values in the same way they are accessed in vertex shaders. Fragment shaders also support an extremely powerful type of uniform value: texture image samplers (as mentioned above, most implementations support texture samplers in vertex shaders as well). These types of uniforms are so useful that we will dedicate entire sections to them in several of the rendering chapters.

8.8.2 Fragment Shader Outputs

The basic goal of the fragment shader is to compute the color of the current fragment. The entire pipeline, in essence, comes down to this single output value per fragment. The fragment shader cannot change the other values of the fragment, such as the position of the fragment, which remains locked in screen space. However, some shading systems do allow for a fragment to cancel itself, causing that fragment to go no further in the rendering pipeline. This is useful for cutout effects, but can have performance consequences on some architectures.

Each shading language defines a variable into which the final color must be written; in GLSL, this variable has its default specification at output location 0. An extremely basic shader that takes an application-set per-object color and applies it to the entire surface is shown below.

```
// GLSL
uniform vec4 objectColor;
out vec4 fragColor;
void main()
```

```
{
    fragColor = objectColor;
}
```

The fragment shader above is compatible with the simple vertex shader above—the two could be linked and used together.

Note that in the latest shading systems, a shader may output more than one color or value per fragment. This functionality is known as *multiple render targets* (MRTs) and will not be discussed in this text, as it does not directly affect the basic pipeline or mathematics of the system. However, the technique is extremely powerful and allows for many high-end rendering effects to be done efficiently. One example is deferred lighting, which we discuss briefly in Chapter 9. For more details and examples of the use of MRTs, see Gray [62].

8.8.3 Linking Vertex and Fragment Shaders

As described above, the triangle assembly stage takes sets of three processed vertices and generates triangles in screen space. Fragments on the surface of these triangles are generated, and the fragment shader is invoked upon each of these fragments. The connection between vertices and fragments is basically unbounded. Three vertices generate a triangle, but that triangle may generate many fragments (as will be discussed in Chapter 10). Or, the triangle may generate no fragments at all (e.g., if the triangle is outside of the view rectangle).

In defining the values and types in its output parameters, the vertex shader also provides one-half of the interface between itself and the fragment shader. In many cases, vertex and fragment shaders are written independently. As long as the input values required by a fragment shader are all supplied by a given vertex shader, those two shaders may be "linked" at runtime and used together. This ability to reuse a vertex or fragment shader with more than one of the other type of shader cuts down on the number of shaders that need to be written, avoiding a combinatorial explosion.

Real applications like large-scale 3D games often spend a lot of development time having to manage the many different shaders and shading paths that exist in a complex rendering engine. Some applications use very large shaders that include all of the possible cases, branching between the various cases using conditionals in the shader code. This can lead to large, complex shaders with a lot of conditionals (known as ubershaders) whose results will differ only at the per-object level, a potentially wasteful option. Other applications generate shader source code in the application itself, as needed, compiling their shaders at runtime. This can be problematic as well, as the shader compilation takes significant CPU cycles and can stall the application visibly. Finally, some applications use a hybrid approach, generating the required shaders offline and keeping them in a lookup table, loading the required shader based on the object being rendered.

8.8.4 Compiling, Linking, and Using Shaders

Source Code
Demo
BasicShaders

Programmable shaders are analogous to many other computer programs. They are written in a high-level language (GLSL, in our case), built from multiple source files or sections (e.g., a vertex shader and a fragment shader), compiled into the GPU's microcode, and linked (the vertex shader together with the fragment shader). The resulting program then can be used.

This implies several stages. The first stage, compilation, can be done at runtime in the application, or may be done as an offline process. The availability of runtime compilation is at this point universally expected. OpenGL and OpenGL ES 3.0 drivers include a GLSL compiler. Direct3D ships a runtime compiler as an independent library. And while OpenGL ES 2.0 doesn't require one, all vendors provide one. Hence, we will assume the availability of a runtime compiler in our Iv code examples. In either case, the source vertex and fragment shaders must be compiled into compiled shader objects. If there are syntax errors in the source files, the compilation will fail.

A pair of compiled shaders (a vertex shader and a fragment shader) must then be linked into an overall shader or program. Most platforms support performing this step at runtime. Linking can fail if the vertex shader does not declare all of the input parameters that the fragment shader requires.

For details of how OpenGL and Direct3D implement shader compilation and linking, see the source code for Iv. Depending on the rendering API, some or all of these steps may be grouped into fewer function calls. In order to compile and link source shaders into a program in Iv, the steps are shown below. Iv supports loading and compiling shaders from text file or from string. The latter case is useful for simple shaders, as they can be simply compiled into the application itself as a static string, per the following code:

```
// Shader compilation code
IvShaderProgram* LoadProgram(IvResourceManager& manager)
{
    IvVertexShader* vertexShader
        = manager.CreateVertexShaderFromFile("vert.txt");
    IvFragmentShader* fragmentShader
        = manager.CreateFragmentShaderFromFile("frag.txt");

    IvShaderProgram* program
        = manager.CreateShaderProgram(vertexShader, fragmentShader);

    return program;
}
```

The resulting program object then must be set as the current shading program before an object can be rendered using it. In Iv, the code to set the current shading program is as follows. Other APIs use similar function calls, as follows:

```
IvRenderer& renderer;
IvShaderProgram* program;
// ...

    // Shader apply code
    renderer.SetShaderProgram( program );
```

8.8.5   Setting Uniform Values

As mentioned previously, uniform shader parameters form the most immediate application-to-shader communication. These values provide the global variables required inside of a shader and can be set on a per-object basis. Since they cannot be set during the course of a draw call, there is no way to change uniforms at a finer grain than the per-object level. Only vertex and fragment shader input variables will differ at that fine-grained level.

The most flexible way to set a uniform value for a shader is to query the uniform value by name from the application. Rendering APIs that support high-level shading languages commonly support some method of mapping string names for uniforms into the uniforms themselves. The exact method differs from API to API—in particular, Direct3D 11 only allows this kind of access via an auxiliary library.

However, querying by string can be expensive and should not be done every time an application needs to access a uniform in a shader. As a result, the rendering APIs can, given a string name and a shading program object, return a handle or pointer to an object that represents the uniform. While the initial lookup still requires a string match, the returned handle allows the uniform to be changed later without a string lookup each time. In Iv, the query function is as follows:

```
IvShaderProgram* program;
// ...

    IvUniform* uniform = program->GetUniform("myShaderUniformName");
```

The handle variable uniform now represents that uniform *in that shader* from this point onward. Note that uniforms are in the scope of a given shading program. Thus, if you need to set a uniform in multiple shading programs, you will need to query the handles and set the values independently for each shading program, even if the uniform has the same name in all of the programs. Although the application will generally know the type of the uniform already (since the application developer likely wrote the shader code), rendering APIs make it possible to retrieve the type (float; integer; Boolean; 2-, 3-, and 4- vectors of each; and float matrices) and array count (one or more of each type) for a uniform. Finally, the rendering API will include functions to set (and perhaps get) the values of each uniform. Iv code that demonstrates querying the type and count of a uniform as well as setting the value is as follows. The code below queries a handle for a uniform that is known to be a two-element array of 4D vectors, perhaps representing a pair of basis vectors.

```
IvUniform* uniform;
// ...

    IvUniformType uniformType = uniform->GetType();
    unsigned int uniformCount = uniform->GetCount();

    // We're expecting an array of two float vector-4's
    if ((uniformType == kFloat4Uniform) &&
        (uniformCount == 2))
    {
        // Set the vectors to the Z and X axes
        uniform->SetValue(IvVector4(0, 0, 1, 0), 0);
        uniform->SetValue(IvVector4(1, 0, 0, 0), 1);
    }
```

A more efficient alternative to querying by name is to use a similar structure to vertex and index buffers called a *uniform buffer* (in OpenGL) or *constant buffer* (in Direct3D).

These are simply other buffers of values that can be transferred to the GPU and shared among vertex and fragment shaders. If we have a common set of uniforms used by many shaders (for example, transformation matrices), we can place them in a uniform buffer, and have each shader access that uniform buffer. Then when the values change, we only need to update the uniform buffer once, rather than having to update each uniform independently for each shading program.

The downside of uniform buffers is that the shader programs need a little more knowledge about the layout of the uniform data, and hence are less flexible than simply querying by name. You also need to update the entire buffer even if you're changing only one variable. For this reason, shader programmers usually create multiple uniform buffers, each containing variables that require different rates of update (for example, per frame vs. per object). In our case, we would like the flexibility and are working with only one or two shader programs at a time, so we will continue querying by name—but for more complex and performance-critical shaders, uniform buffers are essential.

Regardless of access method, these uniform interfaces make it possible to pass a wide range of data items down from the application code to a shader. We will use uniforms extensively in Chapter 9 as we discuss lighting. Uniforms will form the basis of how we pass information regarding the number, type, and configuration of lights and surfaces to the shaders that will actually compute the lit colors.

8.9   Basic Coloring Methods

The following sections describe a range of simple methods to assign colors to surface geometry. Note that the cases described below are designed to best explain how to pass the desired colors to the fragment shader and are overly simplified. These basic methods can be (and will be in later sections and chapters) used to pass other noncolor values into the fragment shader for more complex shading. However, this initial discussion will focus simply on passing different forms of color values to the fragment shader, which will in turn simply write the color value being discussed directly as its output.

The simplest and generally highest-performing methods of coloring geometry are to use constant colors. Constant colors involve passing through colors that were assigned to the geometry prior to rendering. These colors may have been generated by having an artist assign colors to every surface during content creation time. Alternatively, an offline process may have been used to generate static colors for all geometry. With these static colors assigned, there is relatively little that must be done to select the correct color for a given fragment. Constant colors mean that for a given piece of geometry, the color at a fixed point on the surface will never change. No environmental information like dynamic lighting will be factored into the final color.

The following examples will show simple cases of constant color. These will serve as building blocks for later dynamic coloring methods, such as lighting.

8.9.1   Per-Object Colors

Source Code
Demo
UniformColors

The simplest form of useful coloring is to assign a single color per object. Constant coloring of an entire object is of very limited use, since the entire object will appear to be flat, with no color variation. At best, only the filled outline of the object will be visible against the backdrop. As a result, except in some special cases, per-object color is rarely used as the final shading function for an object.

Per-object color requires no special work in the vertex shader (other than basic projection). The vertex–fragment shader pair below implements per-object colors. The application need only specify the desired color by setting the color into the named uniform `objectColor`. The `objectColor` uniform must be declared in the fragment shader, and the application must set its value for the current object prior to rendering the object; it is not a built-in uniform.

```
// GLSL
uniform mat4 IvModelViewProjectionMatrix;
layout(location = IV_POSITION) in vec3 position;

void main()
{
    gl_Position = IvModelViewProjectionMatrix * vec4(position, 1.0);
}


// GLSL
uniform vec4 objectColor;
out vec4 fragColor;

void main() // fragment shader
{
    fragColor = objectColor;
}
```

8.9.2 Per-Triangle Colors

Another primitive-level coloring method is per-triangle coloring, which simply assigns a color to each triangle. This is also known as *faceted*, or *flat*, shading, because the resulting geometry appears planar on a per-triangle basis. Technically, this requires adding a color attribute for each triangle. However, explicit per-triangle attributes are not supported in most current rendering systems. As a result, in order to support per-triangle colors, rendering APIs tend to allow for a mode in which the color value computed for one of a triangle's vertices is used as the input value for the entire triangle, with no interpolation.

There are two common ways of specifying flat shading in programmable shading APIs. The original method was to use a shader-external render-state setting to place the rendering pipeline in flat-shaded mode. This was used in older versions of OpenGL and Direct3D, but is no longer available as a core feature in the latest versions. The current method of specifying per-triangle constant colors is built into the shading language itself, whereby an input value is declared in the shader with a `flat` or `nointerpolation` modifier. Such values will not be interpolated before being passed down to the fragment shader.

8.9.3 Per-Vertex Colors

Many of the surfaces approximated by tessellated objects are smooth, meaning that the goal of coloring these surfaces is to emphasize the smoothness of the original surface, not the artifacts of its approximation with flat triangles. This fact makes flat shading a very poor

Source Code

VertexColors

choice for many tessellated objects. A shading method that can generate the appearance of a smooth surface is needed. Per-vertex coloring, along with a method called *Gouraud shading* (after its inventor, Henri Gouraud), does this. Gouraud shading is based on the existence of some form of per-vertex colors, assigning a color to any point on a triangle by linearly interpolating the three vertex colors over the surface of the triangle. As with the other shading methods we have discussed, Gouraud shading is independent of the source of these per-vertex colors; the vertex colors may be assigned explicitly by the application, or generated on the fly via per-vertex lighting or other vertex shader. This linear interpolation is both simple and smooth and can be expressed as a mapping of barycentric coordinates (s, t) as follows:

$$Color(O, T, (s, t)) = sC_{V1} + tC_{V2} + (1 - s - t)C_{V3}$$

Examining the terms of the equation, it can be seen that Gouraud shading is simply an affine transformation from barycentric coordinates (as homogeneous points) in the triangle to RGB color space.

An important feature of per-vertex smooth colors is that color discontinuities can be avoided at triangle edges, making the piecewise-flat tessellated surface appear smooth. Internal to each triangle, the colors are interpolated smoothly. At triangle edges, color discontinuities can be avoided by ensuring that the two vertices defining a shared edge in one triangle have the same color as the matching pair of vertices in the other triangle. It can be easily shown that at a shared edge between two triangles, the color of the third vertex in each triangle (the vertices that are not an endpoint of the shared edge) does not factor into the color along that shared edge. As a result, there will be no color discontinuities across triangle boundaries, as long as the shared vertices between any pair of triangles are the same in both triangles. In fact, with fully shared, indexed geometry, this happens automatically (since colocated vertices are shared via indexing). Figure 8.6 allows a comparison of geometry drawn with per-face colors and with per-vertex colors.

Per-vertex colors are generated in the vertex shader, through either computation or direct use of per-vertex attributes, or a combination of both. In the fragment shader, the vertex

(a)   (b)

Figure 8.6. (a) Flat (per-face) and (b) Gouraud (per-vertex) shading.

color input value (which has been interpolated to the correct value for the fragment using Gouraud interpolation) is used directly.

```
// GLSL
uniform mat4 IvModelViewProjectionMatrix;
layout(location = IV_COLOR) in vec4 inColor;
layout(location = IV_POSITION) in vec3 position;
out vec4 color;

void main() // vertex shader
{
    gl_Position = IvModelViewProjectionMatrix*vec4(position,1.0);
    color = inColor;
}

// GLSL
in vec4 color;
out vec4 fragColor;

void main() // fragment shader
{
    fragColor = color;
}
```

8.9.4 Sharp Edges and Vertex Colors

Source Code
Demo
SharpEdges

Many objects that we render will contain a mixture of smooth surfaces and sharp edges. One need only look at the outlines of a modern automobile to see this mixture of sloping surfaces (a rounded fender) and hard creases (the sharp edge of a wheel well). Such an object cannot be drawn using per-triangle colors, as per-triangle colors will correctly represent the sharp edges, but will not be able to represent the smooth sections. In these kinds of objects, *some* sharp geometric edges in the tessellation really do represent the original surface accurately, while other sharp edges are designed to be interpolated across to approximate a smooth section of surface.

In addition, the edge between two triangles may mark the boundary between two different colors on the surface of the object, such as an object with stripes painted upon it. In this context, a sharp edge is not necessarily a geometric property. It is nothing more than an edge that is shared by two adjacent triangles where the triangle colors on either side of the edge are different. This produces a visible, sharp line between the two triangles where the color changes.

In these situations, we must use per-vertex interpolated colors. However, interpolating smoothly across all triangle boundaries is not the desired behavior with a smooth/sharp object. The vertices along a sharp edge need to have different colors in the two triangles abutting the edge. In general, when Gouraud shading is used, these situations require coincident vertices to be duplicated, so that the two coincident copies of the vertex can have different colors. Figure 8.7 provides an example of a cube drawn with entirely shared vertices and with duplicated vertices to allow per-vertex, per-face colors. Note that the cube is not flat-shaded in either case—there are still color gradients across each face. The example with duplicated vertices and sharp shading edges looks more like a cube.

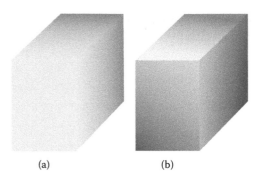

(a)   (b)

Figure 8.7. Sharp vertex discontinuities: (a) shared vertices lead to smooth-shaded edges, and (b) duplicated vertices allow the creation of sharp-shaded edges.

8.9.5 Limitations of Basic Shading Methods

Real-world surfaces often have detail at many scales. The shading/coloring methods described so far require that the fragment shader compute a final color based solely on sources assigned at tessellation-level features, either per triangle or per vertex. While this works well for surfaces whose colors change at geometric boundaries, many surfaces do not fit this restriction very well, making flat shading and Gouraud shading ineffective at best. While programmable shaders can be used to compute very complex coloring functions that change at a much higher frequency than per-vertex or per-triangle methods, doing so based only on these gross-scale inputs can be difficult and inefficient.

For example, imagine a flat sheet of paper with text written on it. The flat, rectangular sheet of paper itself can be represented by as few as two triangles. However, in order to use Gouraud shading (or even more complex fragment shading based on Gouraud-interpolated sources) to represent the text, the piece of paper would have to be subdivided into triangles at the edges of every character written on it. None of these boundaries represents geometric features, but rather are needed only to allow the color to change from white (the paper's color) to black (the color of the ink). Each character could easily require hundreds of vertices to represent the fine stroke details. This could lead to a simple, flat piece of paper requiring tens of thousands of vertices. Clearly, we require a shading method that is capable of representing detail at a finer scale than the level of tessellation.

8.10 Texture Mapping

8.10.1 Introduction

One method of adding detail to a rendered image without increasing geometric complexity is called *texture mapping*, or more specifically *image-based texture mapping*. The physical analogy for texture mapping is to imagine wrapping a flat, paper photograph onto the surface of a geometric object. While the overall shape of the object remains unchanged, the overall surface detail is increased greatly by the image that has been wrapped around it. From some distance away, it can be difficult to even distinguish what pieces of visual detail are the shape of the object and which are simply features of the image applied to the surface.

A real-world physical analogy to this is theatrical set construction. Often, details in the set will be painted on planar pieces of canvas, stretched over a wooden frame (so-called flats), rather than built out of actual, 3D wood, brick, or the like. With the right lighting and positioning, these painted flats can appear as very convincing replicas of their real, 3D counterparts. This is the exact idea behind texturing—using a 2D, detailed image placed upon a simple 3D geometry to create the illusion of a complex, detailed, fully 3D object.

An example of a good use of texturing is a rendering of a stucco wall; such a wall appears flat from any significant distance, but a closer look shows that it consists of many small bumps and sharp cracks. While each of these bumps could be modeled with geometry, this is likely to be expensive and unlikely to be necessary when the object is viewed from a distance. In a 3D computer graphics scene, such a stucco wall will be most frequently represented by a flat plane of triangles, covered with a detailed image of the bumpy features of lit stucco.

The fact that texture mapping can reduce the problem of generating and rendering complex 3D objects into the problem of generating and rendering simpler 3D objects covered with 2D paintings or photographs has made texture mapping very popular in real-time 3D. This, in turn, has led to the method being implemented in display hardware, making it even less expensive computationally. The following sections will introduce and detail some of the concepts behind texture mapping, some mathematical bases underlying them, and basics of how texture mapping can be used in 3D applications.

8.10.2 Shading via Image Lookup

The real power of texturing lies in the fact that it uses a dense plane of samples (an image) as its means of generating color. In a sense, texturing can be thought of as a powerful, general function that maps 2-vectors (the texture coordinates) into a vector-valued output (most frequently an RGBA color). To the shader it is basically irrelevant how the function is computed. Rather than directly interpolating colors that are stored in the vertices, the interpolated per-vertex texture coordinate values serve only to describe how an image is mapped to the triangle. While the mapping from the surface into the space of the image is linear, the lookup of the image value is not. By adding this level of indirection between the per-vertex values and the final colors, texturing can create the appearance of a very complex shading function that is actually no more than a lookup into a table of samples.

The process of texturing involves defining three basic mappings:

1. To map all points on a surface (smoothly in most neighborhoods) into a 2D (or in some cases, 1D or 3D) domain

2. To map points in this (possibly unbounded) domain into a unit square (or unit interval, cube, etc.)

3. To map points in this unit square to color values

The first stage will be done using a modification of the method we used for colors with Gouraud shading, an affine mapping. The second stage will involve methods such as min, max, and modulus. The final stage is the most unique to texturing and involves mapping points in the unit square into an image. We will begin our discussion with a definition of texture images.

8.10.3   Texture Images

The most common form of texture images (or *textures*, as they are generally known) are 2D, rectangular arrays of color values. Every texture has a width (the number of color samples in the horizontal direction) and a height (the number of samples in the vertical direction). Textures are similar to almost any other digital image, including the screen, which is also a 2D array of colors. Just as the screen has pixels (for picture elements), textures have *texels* (texture elements). While most graphics systems allow 1D textures (linear arrays of texels) and 3D textures (cubes or rectangular parallelepipeds of texels, also known as volume textures), by far the most commonly used are 2D, image-based textures. Textures can also be stored in arrays to support such features as cube mapping (six textures representing the faces of a cube surrounding an object). Cube mapping will come up again when we cover environment maps in Chapter 9. However, our discussion of texturing will focus entirely on single 2D textures.

We can refer to the position of a given texel via a 2D value (x, y) in texel units. (Note that these coordinates are (column, row), the reverse of how we generally refer to matrix elements in our row major matrix organization.) Figure 8.8 shows an example of a common mapping of texel coordinates into a texture. Note that while the left-to-right increasing mapping of x is universal in graphics systems, the mapping of y is not; top to bottom is used in Direct3D, and bottom to top is used in OpenGL.

As with most other features, while there are minor differences between the rendering APIs regarding how to specify texture images, all of the APIs require the same basic information:

- The per-texel color storage format of the incoming texture data

- The width and height of the image in texels

- An array of width × height color values for the image data

Put together, these define the image data and their basic interpretation in the same way that an array of vertices, the vertex format information, and the vertex count define vertex geometry to the rendering pipeline. As with vertex arrays, the array of texel data can be quite sizable. In fact, texture image data are one of the single-largest consumers of memory-related resources.

Rendering APIs generally include the notion of an opaque handle to a device-resident copy of a texture. For peak performance on most systems, texture image data need to reside in GPU device memory. Thus, in a process analogous to vertex buffer objects, rendering APIs include the ability to transfer a texture's image data to the device memory once. The opaque handle then can be used to reference the texture in later drawing calls, using the already-resident copy of the texture image data in GPU memory. In Iv, we use an object to wrap all of this state: IvTexture, which represents the texture image itself and the texture sampler state. Like most other resources (e.g., vertex and

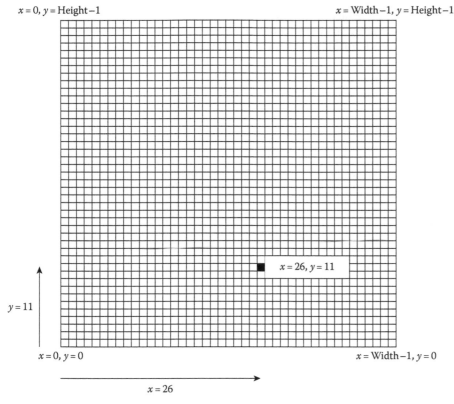

$x = 0, y = \text{Height}-1$ $x = \text{Width}-1, y = \text{Height}-1$

$x = 26, y = 11$

$y = 11$

$x = 0, y = 0$ $x = \text{Width}-1, y = 0$

$x = 26$

Figure 8.8. Texel-space coordinates in an image.

index buffers), `IvTexture` objects are created via the `IvResourceManager` object, as follows:

```
IvResourceManager* manager;
// image data
void* data;
// ...

{
    const unsigned int width = 256;
    const unsigned int height = 512;
    IvTexture* texture = manager->CreateTexture(kRGBA32TexFmt,
                                                width, height,
                                                data, kImmutableUsage);

    // ...
```

The preceding code creates an immutable texture object with a 32-bit-per-texel RGBA texture image that has a width of 256 texels and a height of 512 texels, and takes a pointer to the image data.

If we create with default or dynamic usage, we can similarly lock or map the texture analogous to our vertex and index buffers in order to fill the texture with texel data. The corresponding code to fill an RGBA texture with bright red texels is as follows:

```
IvTexture* texture;
// ...

{
    IvTexture* texture = manager->CreateTexture(kRGBA32TexFmt,
                                                width, height,
                                                NULL, kDynamicUsage);
    const unsigned int width = texture->GetWidth();
    const unsigned int height = texture->GetHeight();

    IvTexColorRGBA* texels = texture->BeginLoadData();

    for (int y = 0; y < height; y++) {
        for (int x = 0; x < width; x++) {
            IvTexColorRGBA& texel = texels[x + y * width];
            texel.r = 255;
            texel.g = 0;
            texel.b = 0;
            texel.a = 255;
        }
    }

// ...
    texture->EndLoadData();
```

8.10.4 Texture Samplers

Textures appear in the shading language in the form of a texture sampler object. Texture samplers are passed to a fragment shader as a uniform value (which is a handle that represents the sampler). The same sampler can be used multiple times in the same shader, passing different texture coordinates to each lookup. So, a shader can sample a texture at multiple locations when computing a single fragment. This is an extremely powerful technique that is used in many advanced shaders. From within a shader, a texture sampler is a sort of function object that can be evaluated as needed, each time with unique inputs.

8.10.4.1 Texture Samplers in Application Code

At the application C or C++ level, there is considerably more to a texture sampler. A texture sampler at the API level includes at least the following information:

- The texture image data

- Settings that control how the texture coordinates are mapped into the image

- Settings that control how the resulting image sample is to be post processed before returning it to the shader

All of these settings are passed into the rendering API by the application prior to using the texture sampler in a shader. As with other shader uniforms, we must include application C or C++ code to link a value to the named uniform; in this case, the uniform value represents a texture image handle. We will cover each of these steps in the following sections.

The book's rendering API uses the `IvTexture` object to represent texture samplers and all of their related rendering state. The code examples in the following section below all describe the `IvTexture` interfaces.

8.11 Texture Coordinates

While textures can be indexed by 2D vectors of nonnegative integers on a per-texel basis (texel coordinates), textures are normally addressed in a more general, texel-independent manner. The texels in a texture are most often addressed via width- and height-independent U and V values. These 2D real-valued coordinates are mapped in the same way as texel coordinates, except for the fact that U and V are normalized, covering the entire texture with the 0-to-1 interval. Figure 8.9 depicts the common mapping of UV coordinates into a texture. These normalized UV coordinates have the advantage that they are completely independent of the height and width of the texture, meaning that the texture resolution can change without having to change the mapping values. Almost all texturing systems use

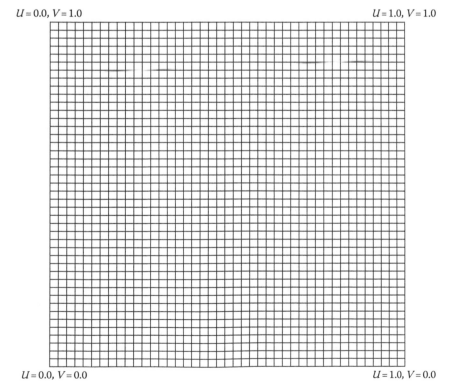

$U = 0.0, V = 1.0$ $U = 1.0, V = 1.0$

$U = 0.0, V = 0.0$ $U = 1.0, V = 0.0$

Figure 8.9. Mapping U and V coordinates into an image.

these normalized *UV* coordinates at the application and shading language level, and as a result, they are often referred to by the generic term of *texture coordinates*, or *texture UVs*.

8.11.1 Mapping Texture Coordinates onto Objects

The texture coordinates defined at the three vertices of a triangle define an affine mapping from barycentric coordinates to *UV* space. Given the barycentric coordinates of a point in a triangle, the texture coordinates may be computed as

$$
\begin{bmatrix} u \\ v \end{bmatrix} = \begin{bmatrix} (u_{V1} - u_{V3}) & (u_{V2} - u_{V3}) & u_{V3} \\ (v_{V1} - v_{V3}) & (v_{V2} - v_{V3}) & v_{V3} \end{bmatrix} \begin{bmatrix} s \\ t \\ 1 \end{bmatrix}
$$

Although there is a wide range of methods used to map textures onto triangles (i.e., to assign texture coordinates to the vertices), a common goal is to avoid distorting the texture. In order to discuss texture distortion, we need to define the *U* and *V* basis vectors in *UV* space. If we think of the *U* and *V* vectors as 2-vectors rather than the point-like texture coordinates themselves, then we compute the basis vectors as

$$
\mathbf{e}_u = (1, 0) - (0, 0)
$$
$$
\mathbf{e}_v = (0, 1) - (0, 0)
$$

The $\mathbf{e}_u$ vector defines the mapping of the horizontal dimension of the texture (and its length defines the size of the mapped texture in that dimension), while the $\mathbf{e}_v$ vector does the same for the vertical dimension of the texture.

If we want to avoid distorting a texture when mapping it to a surface, we must ensure that the affine mapping of a texture onto a triangle involves rigid transforms only. In other words, we must ensure that these texture-space basis vectors map to vectors in object space that are perpendicular and of equal length. We define *ObjectSpace*() as the mapping of a vector in texture space to the surface of the geometry object. In order to avoid distorting the texture on the surface, *ObjectSpace*() should obey the following guidelines:

$$
ObjectSpace(\mathbf{e}_u) \bullet ObjectSpace(\mathbf{e}_v) = 0
$$
$$
|ObjectSpace(\mathbf{e}_u)| = |ObjectSpace(\mathbf{e}_v)|
$$

In terms of an affine transformation, the first constraint ensures that the texture is not sheared on the triangle (i.e., perpendicular lines in the texture image will map to perpendicular lines in the plane of the triangle), while the second constraint ensures that the texture is scaled in a uniform manner (i.e., squares in the texture will map to squares, not rectangles, in the plane of the triangle). Figure 8.10 shows examples of texture-to-triangle mappings that do not satisfy these constraints.

Note that these constraints are by *no* means a requirement—many cases of texturing will stray from them, through either artistic desire or the simple mathematical inability to satisfy them in a given situation. However, the degree that these constraints do hold true for the texture coordinates on a triangle gives some measure of how closely the texturing across the triangle will reflect the original planar form of the texture image.

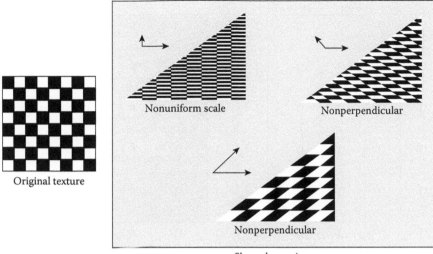

Figure 8.10. Examples of skewed texture coordinates.

8.11.2 Generating Texture Coordinates

Texture coordinates are often generated for an object by some form of projection of the object-space vertex positions in $\mathbb{R}^3$ into the per-vertex texture coordinates in $\mathbb{R}^2$. All texture coordinate generation—in fact, all 2D texturing—is a type of projection. For example, imagine the cartographic problem of drawing a flat map of the earth. This problem is directly analogous to mapping a 2D texture onto a spherical object. The process cannot be done without distortion of the texture image. Any 2D texturing of a sphere is an exercise in matching a projection of the sphere (or unwrapping it) onto a rectangular image (or several images) and the creation of 2D images that take this mapping into account. For example, a common, simple mapping of a texture onto a sphere is to use U and V as longitude and latitude, respectively, in the texture image. This leads to discontinuities at the poles, where more and more texels are mapped over smaller and smaller surface areas as we approach the poles.

The artist must take this into account when creating the texture image. Except for purely planar mappings (such as the wall of a building), most texturing work done by an artist is an artistic cycle between generating texture coordinates upon the object and painting textures that are distorted correctly to map in the desired way to those coordinates.

8.11.3 Texture Coordinate Discontinuities

As was the case with per-vertex colors, there are situations that require shared, collocated vertices to be duplicated in order to allow the vertices to have different texture coordinates. These situations are less common than in the case of per-vertex colors, due to the indirection that texturing allows. Pieces of geometry with smoothly mapped texture coordinates can still allow color discontinuities on a per-sample level by painting the color discontinuities

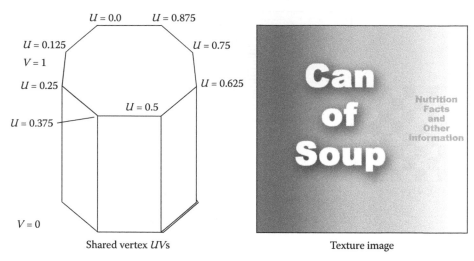

Figure 8.11. Texturing a can with completely shared vertices.

into the texture. Normally, the reason for duplicating collocated vertices in order to split the texture coordinates has to do with topology.

For example, imagine applying a texture as the label for a model of a tin can. For simplicity, we shall ignore the top and bottom of the can and simply wrap the texture as one would a physical label. The issue occurs at the texture's seam. Figure 8.11 shows a tin can modeled as an eight-sided cylinder containing 16 shared vertices—8 on the top and 8 on the bottom. The mapping in the vertical direction of the can (and the label) is simple, as shown in the figure. The bottom eight vertices set $V = 0.0$ and the top eight vertices set $V = 1.0$. So far, there is no problem. However, problems arise in the assignment of U. Figure 8.12 shows an obvious mapping of U to both the top and bottom vertices—U starts at 0.0 and increases linearly around the can until the eighth vertex, where it is 0.875, or $1.0 - 0.125$.

The problem is between the eighth vertex and the first vertex. The first vertex was originally assigned a U value of 0.0, but at the end of our circuit around the can, we would also like to assign it a texture coordinate of 1.0, which is not possible for a single vertex. If we leave the can as is, most of it will look perfectly correct, as we see in the front view of Figure 8.12. However, looking at the back view in Figure 8.12, we can see that the face between the eighth and first vertex will contain a squashed version of almost *the entire texture, in reverse!* Clearly, this is not what we want (unless we can always hide the seam). The answer is to duplicate the first vertex, assigning the copy associated with the first face $U = 0.0$ and the copy associated with the eighth face $U = 1.0$. This is shown in Figure 8.13 and looks correct from all angles.

8.11.4 Mapping Outside the Unit Square

So far, our discussion has been limited to texture coordinates within the unit square, $0.0 \leq u$ and $v \leq 1.0$. However, there are interesting options available if we allow texture coordinates to fall outside of this range. In order for this to work, we need to define how texture

Front side
(appears to be correctly mapped)

Back side
(incorrect, due to shared
vertices along the label "seam")

Figure 8.12. Shared vertices can cause texture coordinate problems.

Front side
(correct: unchanged from
previous mapping)

Back side
(correct, due to doubled
vertices along the label "seam")

Figure 8.13. Duplicated vertices used to solve texturing issues.

coordinates map to texels in the texture when the coordinates are less than 0.0 or greater than 1.0. These operations are per sample, not per vertex, as we shall discuss.

The most common method of mapping unbounded texture coordinates into the texture is known as *texture wrapping*, *texture repeating*, or *texture tiling*. The wrapping of a component u of a texture coordinate is defined as

$$wrap(u) = u - \lfloor u \rfloor$$

The result of this mapping is that multiple "copies" of the texture cover the surface, like tiles on a bathroom floor. Wrapping must be computed using the per-sample, not per-vertex, method. Figure 8.14 shows a square whose vertex texture coordinates are all outside of the unit square, with a texture applied via per-sample wrapping. Clearly, this is a very different result than if we had simply applied the wrapping function to each of the vertices, which can be seen in Figure 8.15. In most cases, per-vertex wrapping produces incorrect results.

Wrapping is often used to create the effect of a tile floor, paneled walls, and many other effects where obvious repetition of a texture is required. However, in other cases wrapping is used to create a more subtle effect, where the edges of each copy of the texture are not quite as obvious. In order to make the edges of the wrapping less apparent, texture images must be created in such a way that the matching edges of the texture image are equal.

Wrapping creates a toroidal mapping of the texture, as tiling matches the bottom edge of the texture with the top edge of the neighboring copy (and vice versa), and the left edge of the texture with the right edge of the neighboring copy (and vice versa). This is equivalent to rolling the texture into a tube (matching the top and bottom edges), and then bringing together the ends of the tube, matching the seams. Figure 8.16 shows this toroidal matching of texture edges. In order to avoid the sharp discontinuities at the texture repetition boundaries, the texture must be painted or captured in such a way so it has toroidal topology;

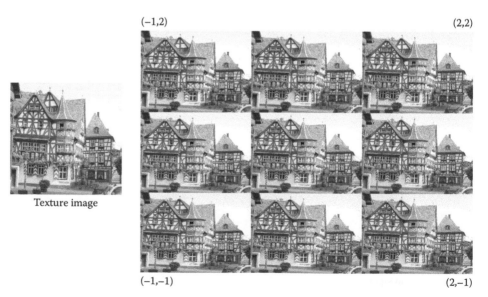

Texture image

(−1,2)   (2,2)

(−1,−1)   (2,−1)

Figure 8.14. An example of texture wrapping.

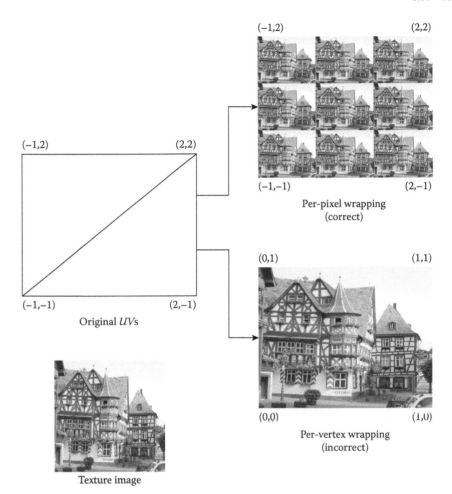

Figure 8.15. Computing texture wrapping.

that is, the neighborhood of its top edge is equal to the neighborhood of its bottom edge, and the neighborhood of its left edge must match the neighborhood of its right edge. Also, the neighborhood of the four corners must be all equal, as they come together in a point in the mapping. This can be a tricky process for complex textures, and various algorithms have been built to try to create toroidal textures automatically. However, the most common method is still to have an experienced artist create the texture by hand to be toroidal.

The other common method used to map unbounded texture coordinates is called *texture clamping*, and is defined as

$$clamp(u) = \max(\min(u, 1.0), 0.0)$$

Clamping has the effect of simply stretching the border texels (left, right, top, and bottom edge texels) out across the entire section of the triangle that falls outside of the unit square. An example of the same square we've discussed, but with texture clamping instead of

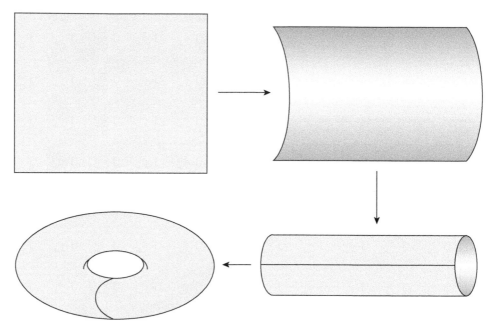

Figure 8.16. Toroidal matching of texture edges when wrapping.

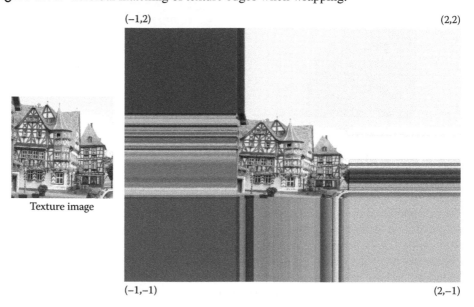

Figure 8.17. An example of texture clamping.

wrapping, is shown in Figure 8.17. Note that clamping the vertex texture coordinates is very different from texture clamping. An example of the difference between these two operations is shown in Figure 8.18. Texture clamping must be computed per sample and has no effect on any sample that would be in the unit square. Per-vertex coordinate clamping, on the

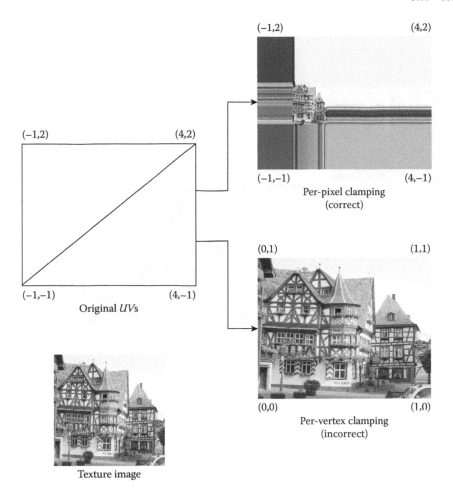

Figure 8.18. Computing texture clamping.

other hand, affects the entire mapping to the triangle, as seen in the lower-right corner of Figure 8.18.

Clamping is useful when the texture image consists of a section of detail on a solid-colored background. Rather than wasting large expanses of texels and placing a small copy of the detailed section in the center of the texture, the detail can be spread over the entire texture, but leaving the edges of the texture as the background color.

On many systems clamping and wrapping can be set independently for the two dimensions of the texture. For example, say we wanted to create the effect of a road: black asphalt with a thin set of lines down the center of the road. Figure 8.19 shows how this effect can be created with a very small texture by clamping the U dimension of the texture (to allow the lines to stay in the middle of the road with black expanses on either side) and wrapping in the V dimension (to allow the road to repeat off into the distance).

Most rendering APIs (including the book's Iv interfaces) support both clamping and wrapping independently in U and V. In Iv, the functions to control texture coordinate

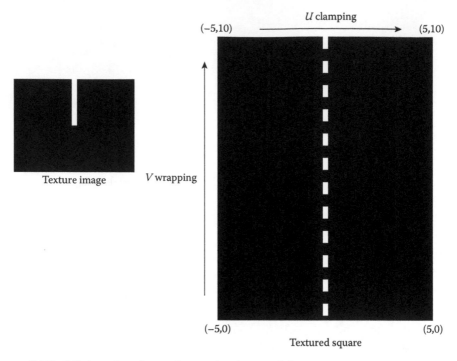

Figure 8.19. Mixing clamping and wrapping in a useful manner.

addressing are `SetAddressingU` and `SetAddressingV`. The road example above would be set up as follows using these interfaces:

```
IvTexture* texture;
// ...

{
    texture->SetAddressingU(kClampTexAddr);
    texture->SetAddressingV(kWrapTexAddr);

    // ...
```

8.11.5 Texture Samplers in Shader Code

Using a texture sampler in shader code is quite simple. As mentioned in Section 7.10.4, a fragment shader simply uses a declared texture sampler as an argument to a lookup function. The following shader code declares a texture sampler and uses it along with a set of texture coordinates to determine the fragment color:

```
// GLSL
layout(location = IV_POSITION) in vec4 position;
layout(location = IV_TEXCOORD0) in vec2 texCoord0;
out vec2 texCoords;
void main() // vertex shader
```

```
{
    // Grab the first set of texture coordinates
    // and pass them on
    texCoords = texCoord0;
    gl_Position = IvModelViewProjectionMatrix * position;
}

// GLSL - fragment shader
uniform sampler2D texture;
int vec2 texCoords;
out vec4 fragColor;

void main()
{
    // Sample the texture represented by "texture"
    // at the location "texCoords"
    fragColor = texture(texture, texCoords);
}
```

This is a simple example: the value passed in for the texture coordinate could be computed by other means, either in the vertex shader (and then interpolated automatically as an input value to the fragment shader) or in the fragment shader. However, applications should take care to remember that the vertex and fragment shaders are invoked at different frequencies. When possible, it is generally better to put computations that *can* be done in the vertex shader in the vertex shader. If a computation can be done in either the vertex or fragment shader with no difference in visual outcome, it may increase performance to have the shader units compute these values only at each vertex.

8.12   Steps of Texturing

Unlike basic, per-vertex (Gouraud) shading, texturing adds several levels of indirection between the values defined at the vertices (the *UV* values) and the final sample colors. This is at once the very power of the method and its most confusing aspect. This indirection means that the colors applied to a triangle by texturing can approximate an extremely complex function, far more complex and detailed than the planar function implied by Gouraud shading. However, it also means that there are far more stages in the method whereupon things can go awry. This section aims to pull together all of the previous texturing discussion into a simple, step-by-step pipeline. Understanding this basic pipeline is key to developing and debugging texturing use in any application.

8.12.1   Other Forms of Texture Coordinates

Real-valued, normalized texture coordinates would *seem* to add a continuity that does not actually exist across the domain of an image, which is a discrete set of color values. For example, in C or C++ one does not access an array with a floating-point value—the index must first be rounded to an integer value. For the purposes of the initial discussion of texturing, we will leave the details of how real-valued texture coordinates map to texture colors somewhat vague. This is actually a rather broad topic and will be discussed in detail in Chapter 10. Initially, it is easiest to think of the texture coordinate as referring to the color of the closest texel. For example, given our assumption, a texture coordinate of $(0.5, 0.5)$ in a texture with width and height equal to 128 texels would map to texel $(64, 64)$. This is referred to as *nearest-neighbor* texture mapping. While this is the simplest method of mapping real-valued texture coordinates into a texture, it is not necessarily the most commonly used in modern applications. We shall discuss more powerful and complex techniques in Chapter 10, but nearest-neighbor mapping is sufficient for the purposes of the initial discussion of texturing.

While normalized texture coordinates are the coordinates that most graphics systems use at the application and shading language level, they are not very useful at all when actually rendering with textures at the lowest level, where we are much more concerned with the texels themselves. We will use them very rarely in the following low-level rendering discussions. We notate normalized texture coordinates simply as (u, v).

The next form of coordinates is often referred to as *texel coordinates*. Like texture coordinates, texel coordinates are represented as real-valued numbers. However, unlike texture coordinates, texel coordinates are dependent upon the width ($w_{texture}$) and height ($h_{texture}$) of the texture image being used. We will notate texel coordinates as (u_{texel}, v_{texel}). The mapping from (u, v) to (u_{texel}, v_{texel}) is

$$(u_{texel}, v_{texel}) = (u \cdot w_{texture}, v \cdot h_{texture})$$

Figure 8.20 shows the coordinates for some texels. Note the edges of each texel lie on integer boundaries in texel coordinate space, and their centers are at the half coordinates.

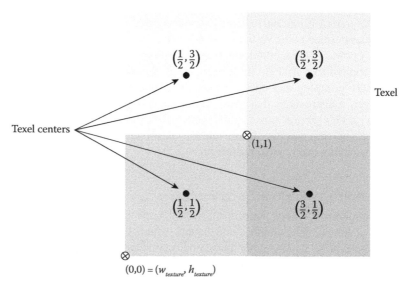

Figure 8.20. Texel coordinates and texel centers.

8.12.2 From Texture Coordinates to a Texture Sample Color

Texturing is a function that maps per-vertex 2-vectors (the texture coordinates), a texture image, and a group of settings into a per-sample color. The top-level stages are as follows:

1. Map the barycentric s and t values into u and v values using the affine mapping defined by the three triangle-vertex texture coordinates: (u_1, v_1), (u_2, v_2), and (u_3, v_3):

$$\begin{bmatrix} u \\ v \end{bmatrix} = \begin{bmatrix} (u_1 - u_3) & (u_2 - u_3) & u_3 \\ (v_1 - v_3) & (v_2 - v_3) & v_3 \end{bmatrix} \begin{bmatrix} s \\ t \\ 1 \end{bmatrix}$$

2. Using the texture coordinate mapping mode (either clamping or wrapping), map the U and V values into the unit square:

$$u_{unit}, v_{unit} = wrap(u), wrap(v)$$

or

$$u_{unit}, v_{unit} = clamp(u), clamp(v)$$

3. Using the width and height of the texture image in texels, map the U and V values into integral texel coordinates via simple scaling:

$$u_{int}, v_{int} = \lfloor u_{unit} \times width \rfloor, \lfloor v_{unit} \times height \rfloor$$

4. Using the texture image, map the texel coordinates into colors using image lookup:

$$C_T = Image(u_{int}, v_{int})$$

These steps compose to create the mapping from a point on a given triangle to a color value. The following inputs must be configured, regardless of the specific graphics system:

- The texture coordinate being sampled (from interpolated vertex attributes, interpolated from a computation in the vertex shader, or computed in the fragment shader)

- The texture image to be applied

- The coordinate mapping mode

8.13 Limitations of Static Shading

The shaders shown in this chapter are about as simple as shaders can possibly be. They project geometry to the screen and directly apply previously assigned vertex colors and textures to a surface. All of the methods described thus far assign colors that do not change for any given sample point at runtime. In other words, no matter what occurs in the scene, a fixed point on a given surface will always return the same color.

Real-world scenes are dynamic, with colors that change in reaction to changes in lighting, position, and even to the surfaces themselves. Any shading method that relies entirely on values that are fixed over both time and scene conditions will be unable to create truly convincing, dynamic worlds. Methods that can represent real-world lighting and the dynamic nature of moving objects are needed.

Programmable shading is tailor-made for these kinds of applications. A very popular method of achieving these goals is to use a simple, fast approximation of real-world lighting written into vertex and fragment shaders. Chapter 9 will discuss in detail many aspects of how lighting can be approximated in real-time 3D systems. The chapter will detail more and more complex shaders, adding increasing realism to the rendered scene. The shaders presented will use dynamic inputs, per-vertex and per-pixel math, and textures to simulate the dynamic and complex nature of real-world lighting. Shaders provide an excellent medium for explaining the mathematics of lighting, since in many cases, the mathematical formulas can be directly reflected in shader code. Finally, we will discuss the benefits and issues of computing lighting in the vertex or fragment shaders.

8.14 Chapter Summary

In this chapter we have discussed the basics of procedural shading and the most common inputs to the procedural shading pipeline. These techniques and concepts lay the foundation for the next two chapters, which will discuss popular shading techniques for assigning colors to geometry (dynamic lighting), as well as a detailed discussion of the low-level mathematical issues in computing these colors for display (rasterization). While we have already discussed the basics of the extremely popular shading method known as texturing, this chapter is not the last time we shall mention it. Both of the following two chapters will discuss the ways that texturing affects other stages in the rendering pipeline.

For further reading, popular graphics texts such as Hughes et al. [82] detail other aspects of shading, including methods used for high-end offline rendering, which are exactly the kinds of methods that are now starting to be implemented as pixel and vertex shaders in real-time hardware. Shader books such as Engel [40] and Pharr [120] also discuss and provide examples of specific programmable shaders that implement high-end shading methods and can serve as springboards for further experimentation.

# Lighting

9.1 Introduction

Much of the way we perceive the world visually is based on the way objects in the world react to the light around them. This is especially true when the lighting around us is changing or the lights or objects are moving. Given these facts, it is not surprising that one of the most common uses of programmable shading is to simulate the appearance of real-world lighting.

The coloring methods we have discussed so far have used colors that are statically assigned at content creation time (by the artist) or at the start of the application. These colors do not change on a frame-to-frame basis. At best, these colors represent a snapshot of the scene lighting at a given moment for a given configuration of objects. Even if we only intend to model scenes where the lights and objects remain static, these static colors cannot represent the view-dependent nature of lighting with respect to shiny or glossy surfaces.

Clearly, we need a dynamic method of rendering lighting in real time. At the highest level, this requires two basic items: a mathematical model for computing the colors generated by lighting and a high-performance method of implementing this model. We have already introduced the latter requirement; programmable shading pipelines were designed specifically with geometric and color computations (such as lighting) in mind. In this chapter we will greatly expand upon the basic shaders, data sources, and shader syntax that were introduced in Chapter 8. However, we must first address the other requirement—the mathematical model we will use to represent lighting.

The following sections will discuss the details of a basic set of methods for approximating lighting for real-time rendering, as well as examples of how these methods can be implemented as shaders. At the end of the chapter we will introduce several more advanced lighting techniques that take advantage of the unique abilities of programmable shaders.

We will refer to fixed-function lighting pipelines in many places in this chapter. Fixed-function lighting pipelines were the methods used in rendering application programming interfaces (APIs) to represent lighting calculations prior to the availability of programmable shaders. They are called fixed-function pipelines because the only options available to users of these pipelines were to change the values of predefined colors and settings. No other modifications to the lighting pipeline (and thus the lighting equation or representation) were available. In comparison, shaders make it possible to implement the exact lighting methods desired by the particular application (though many applications continue to use the simple but efficient lighting models found in the fixed-function pipeline).

9.2   Basics of Light Approximation

The physical properties of light are incredibly complex. Even relatively simple scenes never could be rendered realistically without—for lack of a better term—cheating. In a sense, all of computer graphics is little more than cheating—finding the cheapest-to-compute approximation for a given situation that will still result in a realistic image. Even non-real-time, photorealistic renderings are only approximations of reality, trading off accuracy for ease and speed of computation.

Real-time renderings are even more superficial approximations. Light in the real world reflects, scatters, refracts, and otherwise bounces around the environment. Historically, real-time three-dimensional (3D) lighting often modeled only direct lighting, the light that comes along an unobstructed path from light source to surface. Worse yet, many legacy real-time lighting systems did not support automatic shadowing. Shadowing involves computing light-blocking effects from objects located between the object being lit and the light source. These were ignored or loosely approximated in the name of efficiency. However, despite these limitations, even basic lighting can have a *tremendous* impact on the overall impression of a rendered 3D scene.

Lighting in real-time 3D generally involves data from at least three different sources: light emitter properties (the way the light sources emit light), the surface configuration (vertex position, normal vector), and the surface material (how the surface reacts to light). We will discuss each of these sources in terms of how they affect the lighting of an object and will then discuss how these values are passed to the shaders we will be constructing. All of the shader concepts from Chapter 8 (vertex and fragment shading, attributes, uniforms and input/output variables, etc.) will be pivotal in our creation of a lighting system.

For the purposes of introducing a real-time lighting equation, we will start by discussing an approximate lighting model. Initially, we will speak in terms of lighting a sample, or a generic point in space that may represent a fragment in a triangle or a vertex in a tessellation. We will attempt to avoid the concepts of fragments and vertices during this initial discussion, preferring to refer to a general point on a surface, along with a local surface normal and a surface material. (As will be detailed later, a surface material contains all of the information needed to determine how an object's surface reacts to lighting.) Once we have introduced the concepts, however, we will discuss how fragment and vertex shaders can be used to implement this model, along with the trade-offs of implementing it in one shading unit or another. As already mentioned, this simple lighting model does not accurately represent the real world—there are many simplifications required for real-time lighting performance. However, while simple, we will be applying physically based principles to our system,

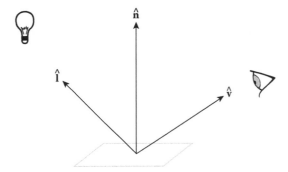

Figure 9.1. The basic geometry of lighting.

introducing these concepts at a very basic level. Later in the chapter we'll discuss some ways to improve this model to make it appear more realistic.

The geometry of our simple lighting model can be seen in Figure 9.1. We consider a portion of a surface, with normal $\hat{n}$. We illuminate this with one or more lights, and indicate the direction of a ray of light by the *light direction vector* $\hat{l}$. Note that $\hat{l}$ points toward the light—this is to simply some of our calculation. Finally, we indicate the direction toward the viewer with the *view direction vector* $\hat{v}$. We will see how all of these can be used to create a reasonable lighting approximation in later sections.

Our discussion of light sources will treat light from a light source as a collection of rays, or in some cases simply as vectors. These rays represent infinitely narrow "shafts" of light, or an amount of light energy per second in a given direction. This representation of light will make it much simpler to approximate light–surface interaction. Our light rays will often have RGB (red, green, blue) colors or scalars associated with them that represent the intensity (and in the case of RGB values, the color) of the light in a given direction. While this value is often described in rendering literature as brightness, this term is descriptive rather than physically based—as we will see, the correct term is *radiance*.

9.3 Measuring Light

In order to understand the mathematics of lighting, it is helpful to know more about how light is actually measured. The simplest way to appreciate how we measure light is in terms of an idealized lightbulb and an idealized surface being lit by that bulb. To explain the radiance of a lit surface, we need to measure the following:

- The amount of light generated by the bulb

- The amount of light generated by the bulb in a particular set of directions

- The amount of light reaching the surface from the bulb

- The amount of light reaching or leaving the surface in a particular set of directions

Each of these is measured and quantified differently.

First, we need a way of measuring the amount of light being generated by the light-bulb. The number most people think of with respect to lightbulbs is electrical wattage. For example, we think of a 100-watt lightbulb as being much brighter than a 25-watt lightbulb, and this is generally true when comparing bulbs of the same kind. The wattage in this case is a measure of the electrical power consumed by the bulb in order to create light. It is *not* a direct measure of the amount of light actually *generated* by the bulb. In other words, two lightbulbs may consume the same wattage (say, 100 watts) but produce different amounts of light—one type of bulb simply may be more efficient at converting electricity to light. For example, a 40-watt compact fluorescent bulb usually appears far brighter than a 40-watt incandescent bulb.

So rather than using electrical power to measure light output, we use a different form of power: light energy per unit time. This quantity is called *radiant flux*. Because radiant flux is power, the unit of radiant flux is also the *watt*. But to be clear, this is not the watts of electrical power that the bulb consumes, but the watts of light produced. Radiant flux is generally represented in equations as Φ.

Light energy can be emitted from a lightbulb in all directions, but it can be useful to measure power in a given direction or set of directions. To represent this, we use the concept of a *solid angle*. This can be thought of as a cone emitting from a central point, or a localized bundle of directions. Solid angles are measured in *steradians*, the value of which for a given solid angle is simply the cross-sectional area of the cone or bundle as it passes through a unit sphere (see Figure 9.2). A unit sphere has a total surface area of 4π, so there are 4π steradians in a sphere. Our measure of light in a given direction, then, is the density of the radiant flux per solid angle, or *radiant intensity*. This is measured in watts/steradian. Radiant intensity is generally represented as I.

We are, of course, interested in the light interacting with a surface. One measure is the radiant flux per area of the surface, or *radiant flux density*. If we're considering light arriving at a surface, we call the radiant flux density *irradiance*; if it's leaving a surface, we call it *radiant exitance*. All are measured in watts per meter squared. Irradiance is

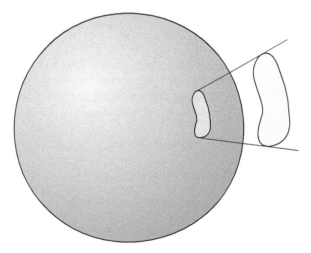

Figure 9.2. Geometry of a solid angle.

an important quantity because it measures not only the light power (in watts), but also the area over which this power is distributed (in square meters). Given a fixed amount of radiant flux, increasing the surface area over which it is distributed will decrease the irradiance proportionally. We will see this property again later, when we discuss the irradiance from a point light source. We represent irradiance in our equations with the term E.

Finally, we want to measure the contribution to radiant flux density from a given set of directions (e.g., from a light), and the resulting flux density leaving in a given set of directions (e.g., toward the viewer). The quantity used to measure this is *radiance*, which is defined as radiant flux density per unit solid angle, and measured in watts per steradian per projected area. Radiance thus takes into account how the reflected light is received and spread directionally. We represent radiance in our equations as L.

For irradiance and radiance, we have assumed (implicitly) that the surface in question is perpendicular to the light direction. However, the irradiance incident upon a surface is proportional to the radiant flux incident upon the surface, divided by the surface area over which it is distributed. If we define an infinitesimally narrow ray of light with direction $\hat{\mathbf{l}}$ to have radiant flux Φ and cross-sectional area δa (Figure 9.3), then the irradiance E incident upon a surface whose normal $\hat{\mathbf{n}} = \hat{\mathbf{l}}$ is

$$E \propto \frac{\Phi}{\delta a}$$

However, if $\hat{\mathbf{n}} \neq \hat{\mathbf{l}}$ (i.e., the surface is not perpendicular to the ray of light), then the configuration is as shown in Figure 9.4. The surface area intersected by the (now oblique) ray of

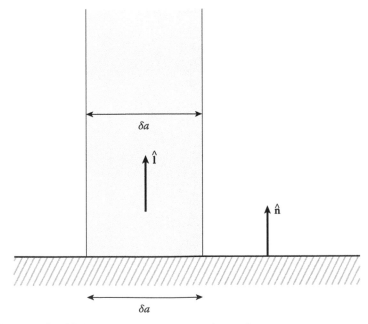

Figure 9.3. A shaft of light striking a perpendicular surface.

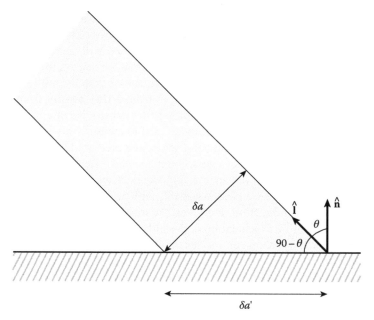

Figure 9.4. The same shaft of light at a glancing angle.

light is represented by $\delta a'$. From basic trigonometry and Figure 9.4, we can see that

$$\delta a' = \frac{\delta a}{\sin\left(\frac{\pi}{2} - \theta\right)}$$
$$= \frac{\delta a}{\cos\theta}$$
$$= \frac{\delta a}{\hat{\mathbf{l}} \cdot \hat{\mathbf{n}}}$$

And, we can compute the irradiance E' as follows:

$$E' \propto \frac{\Phi}{\delta a'}$$
$$\propto \Phi\left(\frac{\hat{\mathbf{l}} \cdot \hat{\mathbf{n}}}{\delta a}\right)$$
$$\propto \left(\frac{\Phi}{\delta a}\right)(\hat{\mathbf{l}} \cdot \hat{\mathbf{n}})$$
$$\propto \Phi(\hat{\mathbf{l}} \cdot \hat{\mathbf{n}})$$

So when computing irradiance and radiance, we'll have to keep this projected area in mind. Note that if we evaluate for the original special case $\hat{\mathbf{n}} = \hat{\mathbf{l}}$, the result is $E' = E$, as expected. Also, for radiance we can apply this projection to either area or solid angle—it is often convenient to do the latter.

Of the terms we've presented, *irradiance* and *radiance* are the most important to us. We'll discuss this in more detail when covering light reflection off of a surface later in this chapter, but we'll need to keep these quantities in mind when creating our light source models.

The preceding quantities are radiometric; that is, they are based on physical properties. For more detailed information, see Cohen and Wallace [24] or Pharr and Humphreys [121]. The field of photometry studies the measurement of analogous quantities that include a physiological weighting based on the human eye's response to different wavelengths of light. The photometric equivalent of radiant flux is luminous flux, and is measured in lumens. This quantity is generally listed on boxes of commercially available lightbulbs, near the wattage rating. The equivalent of irradiance is illuminance and the equivalent of radiance is *luminance*.

The unit of luminance is the *nit*, and this value is the closest of those we have discussed to representing brightness. However, brightness is a perceived value and is not linear with respect to luminance, due to the response curve of the human visual system. For details of the relationship between brightness and luminance, see Cornsweet [26].

9.4 Types of Light Sources

The next few sections will discuss some common types of light sources that appear in real-time 3D systems. Each section will open with a general discussion of a given light source, followed by coverage in mathematical terms, and close with the specifics of implementation in shader code (along with a description of the accompanying C code to feed the required data to the shader). Initially, we will look at one light source at a time, but will later discuss how to implement multiple simultaneous light sources.

For each type of light source, we will be computing two important quantities: the unit vector $\hat{\mathbf{l}}$ and the value L. As we mentioned above, the vector $\hat{\mathbf{l}}$ is the light direction vector—it points from the current surface sample point P_V *toward* the source of the light. This can also be seen in lighting equations as $\hat{\mathbf{L}}$ or $\hat{\omega}$; we will be using $\hat{\mathbf{l}}$ to distinguish it from the radiance L.

The value L represents the radiance from the light source across all wavelengths at the given surface location P_V. In our case, we will not be tracking a full spectrum of values, just the standard RGB colors. This is a sparse sampling of the full visible spectrum, but for our purposes it will be sufficient. One simplification we will start with is assuming that our light values are bounded and normalized to lie within $[0, 1]$, where a value of 1 represents our maximum representable value. We'll discuss how to handle more realistic semi-infinite values later. Finally, any color values we use are physical quantities, and so are assumed to be linear, not sRGB.

As Pharr and Humphreys [121] point out, strictly speaking, *radiance* is not the correct term for the output of the infinitely small (or *punctual*) light sources that we will be covering—rather, we should be using irradiance. However, much of the physically based lighting literature uses radiance in its equations, so we will abuse notation slightly for our discussion and note the places where this may affect our equations. However, it is good to be aware that there are assumptions in the literature that lights have a nonzero volume, which can have practical implications in more complex lighting models for both rendering and our equations. For the simplified light sources we are using, we will state that L then represents the radiance reflected from a white matte surface that is perpendicular to the light source. Or, to put it another way, L is the color produced if you were to shine the light straight at a white wall.

The values $\hat{\mathbf{l}}$ and L do not take any information about the surface orientation or material itself into account, only the relative positions of the light source and the sample point with respect to each other. As we've seen above, we'll need to modify L based on the surface orientation (i.e., the surface normal); this will become more relevant when we discuss reflection models.

9.4.1 Point Lights

Source Code
Demo
PointLight

A point or positional light source (also known as a local light source to differentiate it from an infinite source) is similar to a bare lightbulb, hanging in space. It illuminates equally in all directions. A point light source is defined by its location, the point P_L. The light source direction produced is

$$\hat{\mathbf{l}} = \frac{P_L - P_V}{|P_L - P_V|}$$

This is the normalized vector that is the difference from the sample position to the light source position. It is not constant across all samples, but rather forms a vector field that points toward P_L from all points in space. This normalization operation is one factor that often makes point lights more computationally expensive than some other light types. While this is not a prohibitively expensive operation to compute *once per light*, we must compute the subtraction of two points and normalize the result to compute this light vector for *each lighting sample* for every frame. Figure 9.5 shows the basic geometry of a point light.

We specify the location of a point light in the same space as the vertices, using a 3-vector. It is more straightforward to represent this in world space, but for certain shader calculations it can be more convenient to represent it in view space. The position of the light

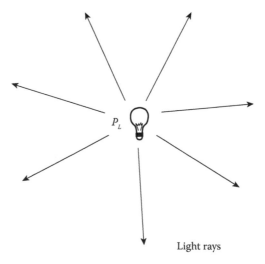

Figure 9.5. The basic geometry of a point light.

can be passed down as a uniform to the shader, but note that we cannot use that position directly as $\hat{\mathbf{l}}$. We must compute the value of $\hat{\mathbf{l}}$ per sample using the position of the current sample, which we will define to be the 3-vector `surfacePosition`. In a vertex shader, this would be the vertex position attribute, while in the fragment shader, it would be an interpolated input value representing the surface position. Because $\hat{\mathbf{l}}$ is a linear value, when using fragment shaders it can be computed in the vertex shader for each vertex in the triangle, and interpolated for each fragment. However, because vector length is not interpolated correctly during linear interpolation (see Chapter 6), we still need to normalize the result in the fragment shader.

We define a standard structure in GLSL code to hold the L and $\hat{\mathbf{l}}$ values:

```
struct lightSampleValues {
    vec3 L;
    vec3 dir;
};
```

And we define a function for each type of light that will return this structure:

```
// GLSL Code
uniform vec3 pointLightPosition;

// Later, in the code, we must compute L per sample...
// as described above, surfacePosition is passed in from a
// per-vertex attribute or a per-fragment input value
lightSampleValues computePointLightValues(in vec3 surfacePosition)
{
    lightSampleValues values;
    values.dir = normalize(pointLightPosition - surfacePosition).xyz;
    // we will add the computation of values.L later

    return values;
}
```

A point light has a nonconstant function defining L. This nonconstant intensity function approximates a basic physical property of light known as the *inverse-square law*: our idealized point light source radiates a constant amount of radiant flux Φ at all times. In addition, this light power is evenly distributed in all directions from the point source's location. Thus, any cone-shaped subset (i.e., a solid angle) of the light coming from the point source represents a constant fraction of this radiant flux (we will call this Φ_{cone}). An example of this conical subset of the sphere is shown in Figure 9.6.

Irradiance is measured as radiant flux per unit area. If we intersect the cone of light with a plane perpendicular to the cone, the intersection forms a disc (see Figure 9.6). This disc is the surface area illuminated by the cone of light. If we assume that this plane is at a distance d from the light center and the radius of the resulting disc is r, then the area of the disc is πr^2. The irradiance E_d is proportional to

$$E_d = \frac{power}{area} \propto \frac{\Phi_{cone}}{\pi r^2}$$

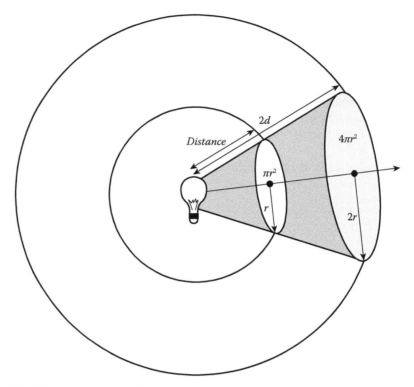

Figure 9.6. The inverse-square law.

However, at a distance of $2d$, then the radius of the disc is $2r$ (see Figure 9.6). The resulting radius is $\pi(2r)^2$, giving an irradiance E_{2d} proportional to

$$E_{2d} \approx \frac{\Phi_{cone}}{\pi(2r)^2} = \frac{\Phi_{cone}}{4\pi r^2} = \frac{E_d}{4}$$

Doubling the distance divides (or attenuates) the irradiance by a factor of 4, because the same amount of light energy is spread over four times the surface area. This is known as the inverse-square law (or more generally as *distance attenuation*), and it states that for a point source, the irradiance decreases with the square of the distance from the source. As an example of a practical application, the inverse-square law is the reason why a candle can illuminate a small room that is otherwise completely unlit, but will *not* illuminate an entire stadium. In both cases, the candle provides the same amount of radiant flux. However, the actual surface areas that must be illuminated in the two cases are vastly different due to distance.

For infinitely small lights like our point light radiance is proportional to irradiance, so the inverse-square law results in a basic L for a point light equal to

$$L = \frac{I}{d^2}$$

where I is the base radiant intensity and

$$d = |P_L - P_V|$$

which is the distance between the light position and the sample position.

While exact inverse-square law attenuation is technically correct for an infinitely small light source, it does not always work well artistically or perceptually. For example, a real lightbulb has an inverse-linear attenuation for objects relatively close to the light. The approximate rule for nonpoint lights is that the attenuation is roughly inverse-linear up to a distance away of five times the largest dimension of the light. Secondly, for distances less than 1 unit from the light, the attenuation starts to approach infinity. This can produce undesirable bright spots on objects. As a result, many lighting pipelines support a more general distance attenuation function for point lights: a general quadratic. Under such a system, the function L for a point light is

$$L = \frac{I}{k_c + k_l d + k_q d^2}$$

The distance attenuation constants k_c, k_l, and k_q are defined per light and determine the shape of that light's attenuation curve. Figure 9.7 is a visual example of constant, linear, and quadratic attenuation curves. The spheres in each row increase in distance linearly from left to right. One common set of values that provides good results is

$$L = \frac{I}{1 + d^2} \tag{9.1}$$

This function has a maximum value of 1, and is somewhat constant close to the light, while taking on the attributes of the inverse-quadratic function at distance.

One problem with both these approaches is that there is still some contribution from the light even at very far distances. To limit the number of lights that irradiate a certain object we would like to only consider those within a certain range r, and beyond that range have any light's contribution be equal to 0. Because of this, other attenuation functions are used. One common solution is from Dietrich [34]:

$$L = \max\left(1 - \frac{d^2}{r^2}, 0\right) I$$

This does have an issue in that there is a discontinuity at maximum range—one solution is to simply square the attenuation factor. This function does tend to flatten out the attenuation term, but has the advantage that it can be computed quite efficiently using two multiplies, a dot product, and a subtraction. However, in our case, we can use Equation 9.1, above. For our simple examples we are not concerned with lighting at distance, and we would like to keep closer to a physically based result.

Ideally, the distance value d should be computed in world coordinates (post–model transform); regardless, a consistent specification of the space used is important, as there may be scaling differences between model space, world space, and view space, which would change the scale of the attenuation. Most importantly, model-space scaling often differs per object, meaning the different objects whose model transforms have different

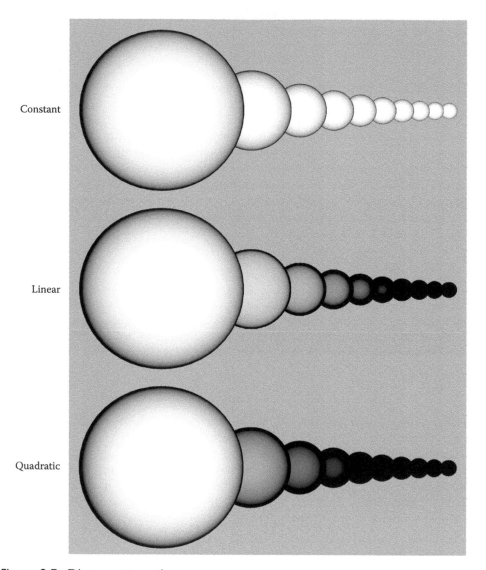

Figure 9.7. Distance attenuation.

scale would be affected differently by distance attenuation. This would not look correct. Distance attenuation must occur in a space that uses the same scale factor for all objects in a scene. One thing to be careful of when lighting in view space is when there is any scaling applied in the world-to-view transformation. For example, if we were to make the main character appear like he or she was shrinking by slowly scaling up the view transformation for a first-person camera, that would affect the computed distances between lights and surface positions, causing the attenuation to change. If a game uses this effect, the simplest solution is to compute attenuation in world space.

We can store the light's intensity in a single 3-vector uniform. Since the attenuation must be computed per sample and involves the length of the $P_L - P_V$ vector, we merge the L shader code into the previous $\hat{\mathbf{l}}$ shader code as follows:

```
// GLSL Code
uniform vec3 pointLightPosition;
uniform vec3 pointLightIntensity;

lightSampleValues computePointLightValues(in vec3 surfacePosition)
{
    lightSampleValues values;
    vec3 lightVec = pointLightPosition - surfacePosition;
    values.dir = normalize(lightVec);
    // Compute 1 + dist squared
    float distAtten = 1.0 + dot(lightVec);
    values.L = pointLightIntensity / distAtten;
    return values;
}
```

Some systems compute L in the vertex shader and then pass it the fragment shader to be interpolated (similar to what can be done for $\hat{\mathbf{l}}$). This is not strictly correct in most cases as the standard attenuation is inverse-quadratic and sharply changes near the light. However, for nearly linear attenuation functions or objects far from the light, the difference can be quite small, so if speed is a concern, it is a possibility.

9.4.2 Spotlights

A spotlight is like a point light source with the ability to limit its light to a cone-shaped region of the world. The behavior is similar to a theatrical spotlight with the ability to focus its light on a specific part of the scene.

In addition to the position P_L that defined a point light source, a spotlight can be defined by a direction vector $\hat{\mathbf{d}}$ and two scalar cone angles θ and ϕ. These additional values define the direction of the cone and the behavior of the light source as the sample point moves away from the central axis of the cone. The infinite cone of light generated by the spotlight has its apex at the light center P_L, an axis $\hat{\mathbf{d}}$ (pointing *toward the base* of the cone), and a half angle of ϕ. The angle θ is used to control how sharp the transition from the bright center spot to the edge of the cone is. If θ is 0, then you will get a soft spotlight. If θ is equal to ϕ, then you will get a very hard edge. For values in between, the spotlight will be at its brightest up to an angle of θ from the center (effectively acting like a point light), then will smoothly transition to no output at an angle of ϕ. Figure 9.8 illustrates this configuration.

The light vector is equivalent to that of a point light source:

$$\hat{\mathbf{l}} = \frac{P_L - P_V}{|P_L - P_V|}$$

For a spotlight, L is based on the point light function but adds an additional term to represent the focused, conical nature of the light emitted by a spotlight:

$$L = \frac{\text{spot}(\hat{\mathbf{l}}, \hat{\mathbf{d}}, \theta, \phi)}{1 + dist^2} I$$

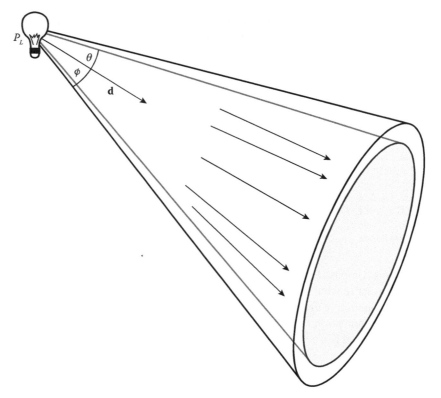

Figure 9.8. The basic geometry of a spotlight.

where

$$\text{spot}\,(\hat{\mathbf{l}}, \hat{\mathbf{d}}, \theta, \phi) = \begin{cases} 1 & \text{if}\,(-\hat{\mathbf{l}} \cdot \mathbf{d}) \geq \cos\theta \\ \text{smoothstep}(\cos\phi, \cos\theta, -\hat{\mathbf{l}} \cdot \hat{\mathbf{d}}) & \text{if}\,\cos\theta > (-\hat{\mathbf{l}} \cdot \hat{\mathbf{d}}) \geq \cos\phi \\ 0 & \text{otherwise} \end{cases}$$

where smoothstep(s, f, x) is a GLSL function that performs a cubic interpolation of $3t^2 - 2t^3$ with $t = (x - s)/(f - s)$, clamped to the interval $[0, 1]$. The end result is to blend smoothly from 1 down to 0 as $(-\hat{\mathbf{l}} \cdot \hat{\mathbf{d}})$ sweeps from $\cos\theta$ to $\cos\phi$.

As can be seen, the spot () function is 0 when the sample point is outside of the cone. It makes use of the fact that the light vector and the cone vector are normalized, causing $(-\hat{\mathbf{l}} \cdot \mathbf{d})$ to be equal to the cosine of the angle between the vectors. We must negate $\hat{\mathbf{l}}$ because it points toward the light, while the cone direction vector $\mathbf{d}$ points away from the light. Computing the cone term first can allow for performance improvements by skipping the rest of the light calculations if the sample point is outside of the cone. In fact, some graphics systems even check the bounding volume of an object against the light cone, avoiding any spotlight computation on a per-sample basis if the object is entirely outside of the light cone.

The multiplication of the spot () term with the distance attenuation term means that the spotlight will attenuate over distance within the cone. In this way, it acts exactly like a point light with an added conic focus. The fact that both of these expensive attenuation terms must be recomputed per sample makes the spotlight the most computationally expensive type of standard light in most systems. When possible, applications attempt to minimize the number of simultaneous spotlights (or even avoid their use altogether).

Spotlights with circular attenuation patterns are not universal. Another popular type of spotlight (see Warn [154]) models the so-called barn door spotlights that are used in theater, film, and television. However, because of these additional computational expenses, conical spotlights are by far the more common form in real-time graphics systems.

As described previously, $\hat{l}$ for a spotlight is computed as for a point light. In addition, the computation of L is similar, adding an additional term for the spotlight angle attenuation. The spotlight-specific attenuation requires two new uniform values per light, specifically:

- `spotLightDir`: A unit-length 3-vector representing the spotlight direction.

- `spotLightCosOuterInner`: The cosines of the half-angle of the spotlight's outer and inner cones.

These values and the previous formulas are then folded into the earlier shader code for a point light, giving the following computations:

```
// GLSL Code
uniform vec3 spotLightPosition;
uniform vec3 spotLightIntensity;
uniform vec3 spotLightDir; // unit-length
uniform vec2 spotLightCosOuterInner;

lightSampleValues computeSpotLightValues(in vec3 surfacePosition)
{
    lightSampleValues values;
    vec3 lightVec = spotLightPosition - surfacePosition;
    values.dir = normalize(lightVec);
    // Compute 1 + dist squared
    float distAtten = 1.0 + dot(lightVec);
    float spotAtten = dot(-spotLightDir, values.dir);
    spotAtten = smoothstep(spotLightOuterInner.x, spotLightOuterInner.y, spotAtten);
    values.L = spotLightIntensity * spotAtten / distAtten;

    return values;
}
```

9.4.3 Directional Lights

A directional light source (also known as an infinite or distant light source) is similar to the light of the sun as seen from the earth. Relative to the size of the earth, the sun seems almost infinitely far away, meaning that the rays of light reaching the earth from the sun are nearly parallel to one another, independent of position on the earth. Consider the source and the light it produces as a single vector. A directional light is defined by a *point at*

Source Code

DirectionalLight

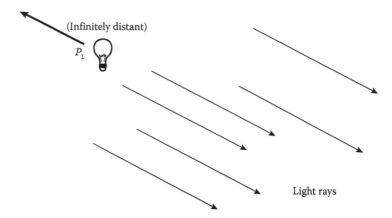

Figure 9.9. The basic geometry of a directional light.

infinity, P_L. The light source direction is produced by turning the point into a unit vector (by subtracting the position of the origin and normalizing the result):

$$\hat{\mathbf{l}} = \frac{P_L - O}{\|P_L - O\|}$$

Figure 9.9 shows the basic geometry of a directional light. Note that the light rays are the negative (reverse) of the light direction vector $\hat{\mathbf{l}}$, since $\hat{\mathbf{l}}$ points from the surface to the light source.

Dividing a value of I by a distance function to compute L for a directional light is not practical, as our distance is infinity. Consider the sun again. Because it is so far away and the size of the scene is small relative to that distance, the difference in attenuation for objects in our scene is practically negligible; that is, all contribution from the light will be nearly the same across the scene. So we will simply say that the value L for a directional light is constant for all sample positions.

Since both L and light vector $\hat{\mathbf{l}}$ are constant for a given light (and independent of the sample point P_V), directional lights are the least computationally expensive type of light source. Neither $\hat{\mathbf{l}}$ nor L needs to be recomputed for each sample, making them considerably cheaper than either the point light or spotlight. As a result, we can pass both of these values to the shader (fragment or vertex) as uniforms and use them directly.

```
// GLSL Code

uniform vec3 dirLightDirection;
uniform vec3 dirLightRadiance;

// Later, in the code, we can use these values directly...
lightSampleValues computeDirLightValues()
{
    lightSampleValues values;
    values.dir = dirLightDirection;
    values.L = dirLightRadiance;
    return values;
}
```

9.4.4 Ambient Lights

Ambient light is the term used in real-time lighting as an umbrella under which all forms of indirect lighting are grouped and approximated. Indirect lighting is light that is incident upon a surface not via a direct ray from light to surface, but rather via some other, more complex path. In the real world, light can be scattered by particles in the air, and light can reflect multiple times around a scene prior to reaching a given surface. Accounting for these multiple bounces and random scattering effects can be difficult to do in a real-time rendering system, so many systems approximate this by using a single constant value for ambient light. This is the least physical of all of our approximations and accounts for some of the most important (although subtle) visual differences between the real world and cheap renders. In particular, it makes the scene lack a certain level of variability that we expect in reality—corners look brighter than they should, and flat surfaces look a little too uniform. We'll discuss some methods for calculating more realistic indirect lighting later in the chapter.

The ambient light represents the radiance of the light from all sources that is to be scattered through the scene. Often a different ambient light is set per room or area, and the engine can interpolate between them as the player moves from space to space. Ambient light has no direction, so there is no associated $\hat{\mathbf{l}}$. Hence, we can't use our `lightSampleValues` structure and will have to separate our use of the ambient term as a special step in our calculations. We'll represent the light's contribution with a single uniform:

```
// GLSL Code
uniform vec3 ambientLightRadiance;
```

9.4.5 Other Types of Light Sources

The light sources above are only a few of the most basic that are seen in modern lighting pipelines, although they serve the purpose of introducing shader-based lighting quite well. There are many other forms of lights that are used in shader-based pipelines. We will discuss several of these at a high level and provide more detailed references in the advanced lighting sections at the end of the chapter.

9.5 Surface Materials and Light Interaction

Source Code
Demo
LightingComponents

Having discussed the various ways in which the light sources in our model generate light incident upon a surface, we must complete the model by discussing how this incoming light is converted (or reflected) into outgoing light as seen by the viewer or camera. This section will discuss a common real-time model of light–surface interaction.

In the presence of lighting, there is more to surface appearance than a single color. Surfaces respond differently to light, depending upon their composition, for example, unfinished wood, plastic, or metal. Gold-colored plastic, gold-stained wood, and actual gold all respond differently to light, even if they are all the same basic color. Most real-time 3D lighting models take these differences into account with the concept of a *material*.

How a material is defined depends on the needs of the system—in particular making a trade-off between simplicity and physical accuracy. In our case, we are going to use a very simple model, and discuss later some possibilities for improvement. Our material will have two major parts: any light the surface itself emits, and a representation of how the surface reflects any incoming light.

To see how we will use these components, let's consider the lighting process itself. The general equation that describes the outgoing radiance from a surface (called the *rendering equation*) is

$$L_o(\omega_o) = L_e(\omega_o) + \int_\Omega f(\omega_i, \omega_o) \circ L(\omega_i)(\hat{\mathbf{n}} \bullet \omega_i)\, d\omega_i \qquad (9.2)$$

This looks a bit like notation salad, but it is actually quite simple. We are trying to compute the outgoing radiance L_o from a surface in the direction ω_o. Part of that is any radiance that the surface emits itself in that direction, or $L_e(\omega_o)$. The remainder is radiance from light that strikes the surface. Consider a tiny hemisphere Ω over the surface point. For all directions ω_i in Ω, we take the incoming radiance $L_i(\omega_i)$, project it onto the surface (via $\hat{\mathbf{n}} \bullet \omega_i$), and then use the function $f(\omega_i, \omega_o)$, known as the *bidirectional reflection distribution function*, or BRDF, to compute the proportion reflected in direction ω_o (recall that the operator ∘ means componentwise multiplication). Adding this up gives us the total radiance in direction ω_o.

The emulation of a given surface's reflective properties is handled almost entirely by the BRDF. As mentioned, it gives the proportion of outgoing light in the direction ω_o to incoming light in direction ω_i. More formally, it is the ratio of the change in outgoing radiance to the change in incoming irradiance, or dL_o/dE_i, and hence has units of steradians^{-1}. In order for our equation to represent a physically based lighting model, the BRDF must meet a few properties. First, the result must always be greater than 0. Second, the reflection should be the same in both directions, or

$$f(\omega_i, \omega_o) = f(\omega_o, \omega_i)$$

This is called reciprocity. Finally, it must be energy preserving, so that the total energy reflected can't be greater than the total energy received. In our case, we can represent this as

$$\int_\Omega f(\omega_i, \omega_o)(\hat{\mathbf{n}} \bullet \omega_i)\, d\omega_i \leq 1$$

Note that this only says that the integral over a hemisphere of directions is less than or equal to 1—for a given direction the value of the BRDF could be greater than 1. Within these restrictions, BRDFs can be quite general in their reflective properties. However, we will be implementing a simplified version with two parts: a diffuse term, which represents highly scattered light, and a specular term, which represents narrowly reflected light. Even with this simplification, we will be meeting the conditions above to make it as physically based as possible.

The integral in Equation 9.2 is not practically solvable for a general scene. The incoming radiance includes any lights in the scene, but realistically includes light reflected off of other surfaces as well, and solving this for all the interreflections between surfaces can get quite complicated. In our case, we will simplify things considerably by using our ambient light to represent any interflections, and only consider incoming light from the idealized light sources we have discussed (ignoring any emissive surfaces in the scene other than the one under consideration). In this case, we end up with the following equation for light in the view direction $\hat{\mathbf{v}}$:

$$L(\hat{\mathbf{v}}) = L_e(\hat{\mathbf{v}}) + r_a(\hat{\mathbf{v}}) \circ L_a + \sum_{i}^{lights} \pi f(\hat{\mathbf{l}}_i, \hat{\mathbf{v}}) \circ L_i \max(0, \hat{\mathbf{n}} \cdot \hat{\mathbf{l}}_i) \qquad (9.3)$$

The derivation of the final term is due to Hoffman [79]. The summation indicates that we are adding over all lights, and L_i and $\hat{\mathbf{l}}_i$ are the incoming radiance and light direction vector for each light at the surface point. Clamping $\hat{\mathbf{n}} \cdot \hat{\mathbf{l}}_i$ to be positive acts as a self-shadowing term—the assumption here is that we are simulating opaque surfaces. As far as the ambient term, L_a represents ambient light, and the function r_a we will explain when we discuss the special case of the reflection of ambient light below.

Given this lighting equation, we now have the parts of our surface material. The $f(\hat{\mathbf{l}}, \hat{\mathbf{v}})$ term, or BRDF, broken into the diffuse and specular parts, represents how the surface reflects general light. The $L_e(\hat{\mathbf{v}})$, or emissive, term represents the light generated by the surface. And for our simple model, the $r_a(\hat{\mathbf{v}})$ term represents how the surface reflects our simplified indirect or ambient lighting. We will now discuss each in turn.

9.5.1 Diffuse

The diffuse reflection term of the BRDF treats the surface as a pure diffuse (or matte) surface, sometimes called a *Lambertian reflector*. Lambertian surfaces reflect light equally in all directions. These surfaces have the property that their outgoing radiance is independent of both the direction of any incoming radiance and the view direction. Hence, the corresponding BRDF will be constant.

When a surface receives light, a portion of the light energy, dependent on wavelength, will be reflected, and a portion will be absorbed. We represent this by assigning a diffuse color M_d to the surface. This will be modulated, or multiplied componentwise, with the irradiance on the surface to compute the total amount of radiant exitance. However, what we want is outgoing radiance. Since the surface reflects equally in all directions, we only need to divide by the projected solid angle of a hemisphere, or π. Our final BRDF is

$$f_d(\hat{\mathbf{v}}, \hat{\mathbf{l}}) = \frac{M_d}{\pi}$$

Substituting this into the nonemissive part of our lighting equation and simplifying, we end up with

$$L_o(\hat{\mathbf{v}}) = M_d \circ L_i \max(0, \hat{\mathbf{n}} \cdot \hat{\mathbf{l}}_i)$$

which is the base formula for Lambertian reflectance. Figure 9.10 provides a visual example of a sphere lit by a single light source that involves only diffuse lighting.

One thing to note is that the $1/\pi$ term in our BRDF has been canceled by the π term in our simple lighting equation. This is common when using physically weighted BRDFs with infinitely small light sources. Whenever you use such a BRDF with punctual lights, you must be sure that you haven't lost or added a π term somewhere. To keep it a little clearer, it can be helpful in formulas to continue to multiply radiance by π, and keep the π term in the denominator of the BRDF, but be sure to remove them for efficiency's sake when writing code.

Figure 9.10. Sphere lit by diffuse light.

The shader code to compute the diffuse component is as follows. We will store the diffuse color of an object's material in the 4-vector shader uniform value `materialDiffuseColor`. The diffuse material color is a 4-vector because it includes the alpha component of the surface as a whole. Note that adding the suffix `.rgb` to the end of a 4-vector creates a 3-vector out of the red, green, and blue components of the 4-vector. We separate out the BRDF calculation both to make it easier to replace if we wish to and because we'll be using it in the combined shader below. We also clamp the result of $\hat{\mathbf{n}} \cdot \hat{\mathbf{l}}_i$ to the interval $[0, 1]$ to meet our self-shading criteria, and to avoid any issues with floating-point precision.

We assume that the surface normal vector at the sample point, $\hat{\mathbf{n}}$, is passed into the function. This value may be either a per-vertex attribute in the vertex shader, an interpolated input value in the fragment shader, or perhaps even computed in either shader. The source of the normal is unimportant to this calculation.

```
// GLSL Code
uniform vec4 materialDiffuseColor;

// surfaceNormal is assumed to be unit-length
vec3 computeDiffuseBRDF(in vec3 surfaceNormal,
                        in vec3 lightDir,
                        in vec3 viewDir)
```

```
{
    return materialDiffuseColor.rgb;
}
vec3 computeDiffuseComponent(in lightSampleValues light,
                            in vec3 surfaceNormal,
                            in vec3 viewDir)
{
    return light.L * computeDiffuseBRDF(surfaceNormal, light.dir, viewDir)
                   * clamp(dot(surfaceNormal, light.dir), 0.0, 1.0);
}
```

9.5.2 Specular

A perfectly smooth mirror reflects all of the light from a given direction $\hat{\mathbf{l}}$ out along a single direction, the reflection direction $\hat{\mathbf{r}}$. While few surfaces approach completely mirrorlike behavior, most surfaces have at least some mirrorlike component to their lighting behavior. As a surface becomes rougher (at a microscopic scale), it no longer reflects all light from $\hat{\mathbf{l}}$ out along a single direction $\hat{\mathbf{r}}$, but rather in a distribution of directions centered about $\hat{\mathbf{r}}$. This tight (but smoothly attenuating) distribution around $\hat{\mathbf{r}}$ is often called a *specular highlight* and is often seen in the real world. A classic example is the bright white "highlight" reflections seen on smooth, rounded plastic objects. The specular component of real-time lighting is an entirely empirical approximation of this reflection distribution, specifically designed to generate these highlights.

Because specular reflection represents mirrorlike behavior, the intensity of the term is dependent on the relative directions of the light ($\hat{\mathbf{l}}$), the surface normal ($\hat{\mathbf{n}}$), *and* the viewer ($\hat{\mathbf{v}}$). Prior to discussing the specular term itself, we must introduce the concept of the light reflection vector $\hat{\mathbf{r}}$. Computing the reflection of a light vector $\hat{\mathbf{l}}$ about a plane normal $\hat{\mathbf{n}}$ involves negating the component of $\hat{\mathbf{l}}$ that is perpendicular to $\hat{\mathbf{n}}$. We do this by representing $\hat{\mathbf{l}}$ as the weighted sum of $\hat{\mathbf{n}}$ and a unit vector $\hat{\mathbf{p}}$ that is perpendicular to $\hat{\mathbf{n}}$ (but in the plane defined by $\hat{\mathbf{n}}$ and $\hat{\mathbf{l}}$) as follows and as depicted in Figure 9.11:

$$\hat{\mathbf{l}} = l_n\hat{\mathbf{n}} + l_p\hat{\mathbf{p}}$$

The reflection of $\hat{\mathbf{l}}$ about $\hat{\mathbf{n}}$ is then

$$\hat{\mathbf{r}} = l_n\hat{\mathbf{n}} - l_p\hat{\mathbf{p}}$$

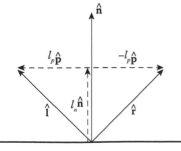

Figure 9.11. The relationship between the surface normal, light direction, and reflection vector.

We know that the component of $\hat{\mathbf{l}}$ in the direction of $\hat{\mathbf{n}}$ (l_n) is the projection of $\hat{\mathbf{l}}$ onto $\hat{\mathbf{n}}$, or

$$l_n = \hat{\mathbf{l}} \cdot \hat{\mathbf{n}}$$

Now we can compute $l_p \hat{\mathbf{p}}$ by substitution of our value for l_n:

$$\hat{\mathbf{l}} = l_n \hat{\mathbf{n}} + l_p \hat{\mathbf{p}}$$
$$\hat{\mathbf{l}} = (\hat{\mathbf{l}} \cdot \hat{\mathbf{n}})\hat{\mathbf{n}} + l_p \hat{\mathbf{p}}$$
$$l_p \hat{\mathbf{p}} = \hat{\mathbf{l}} - (\hat{\mathbf{l}} \cdot \hat{\mathbf{n}})\hat{\mathbf{n}}$$

So, the reflection vector $\hat{\mathbf{r}}$ equals

$$\hat{\mathbf{r}} = l_n \hat{\mathbf{n}} - l_p \hat{\mathbf{p}}$$
$$= (\hat{\mathbf{l}} \cdot \hat{\mathbf{n}})\hat{\mathbf{n}} - \omega_p \hat{\mathbf{p}}$$
$$= (\hat{\mathbf{l}} \cdot \hat{\mathbf{n}})\hat{\mathbf{n}} - (\hat{\mathbf{l}} - (\hat{\mathbf{l}} \cdot \hat{\mathbf{n}})\hat{\mathbf{n}})$$
$$= (\hat{\mathbf{l}} \cdot \hat{\mathbf{n}})\hat{\mathbf{n}} - \hat{\mathbf{l}} + (\hat{\mathbf{l}} \cdot \hat{\mathbf{n}})\hat{\mathbf{n}}$$
$$= 2(\hat{\mathbf{l}} \cdot \hat{\mathbf{n}})\hat{\mathbf{n}} - \hat{\mathbf{l}}$$

Computing the view vector involves having access to the camera location, so we can compute the normalized vector from the current sample location to the camera center. In an earlier section, view (or camera) space was mentioned as a common space in which we could compute our lighting. If we assume that the surface sample location is in view space, this simplifies the process, because the center of the camera is the origin of view space. Thus, the view vector is then the origin minus the surface sample location, that is, the zero vector minus the sample location. Thus, in view space, the view vector is simply the negative of the sample position treated as a vector and normalized. That said, if we already have the position in world space (e.g., if we are calculating distance for point lights), it may be more efficient to pass in the eye position in world space as a uniform and subtract the sample position in world space from that.

We wish to create a distribution that reaches its maximum when the view vector $\hat{\mathbf{v}}$ is equal to $\hat{\mathbf{r}}$, that is, when the viewer is looking directly at the reflection of the light vector. The standard distribution that uses the reflection vector in this way is known as the *Phong distribution*. It falls off toward 0 rapidly as the angle between the two vectors increases, with a "shininess" control that adjusts how rapidly it attenuates. The term is based on the following formula:

$$(\hat{\mathbf{r}} \cdot \hat{\mathbf{v}})^{m_\beta} = (\cos \theta)^{m_\beta}$$

where θ is the angle between $\hat{\mathbf{r}}$ and $\hat{\mathbf{v}}$. The shininess factor m_β controls the size of the highlight; a smaller value of m_β leads to a larger, more diffuse highlight, which makes the surface appear more dull and matte, whereas a larger value of m_β leads to a smaller, more intense highlight, which makes the surface appear shiny. This shininess factor is considered a property of the surface material and represents how smooth the surface appears.

One issue with the Phong distribution is that it does not work well on flat surfaces, for example, a ground surface with a light in the distance. The specular spot produced will be circular, when we'd expect an elliptical result. Jim Blinn's modification, known as the *Blinn–Phong distribution*, is much better at reproducing this effect. Rather than computing $\hat{\mathbf{r}}$ directly, this method uses what is known as a *halfway vector*. The halfway vector is the vector that is the normalized sum of $\hat{\mathbf{l}}$ and $\hat{\mathbf{v}}$:

$$\hat{\mathbf{h}} = \frac{\hat{\mathbf{l}} + \hat{\mathbf{v}}}{|\hat{\mathbf{l}} + \hat{\mathbf{v}}|}$$

The resulting vector bisects the angle between $\hat{\mathbf{l}}$ and $\hat{\mathbf{v}}$. This halfway vector is equivalent to the surface normal $\hat{\mathbf{n}}$ that would generate $\hat{\mathbf{r}}$ such that $\hat{\mathbf{r}} = \hat{\mathbf{v}}$. In other words, given fixed light and view directions, $\hat{\mathbf{h}}$ is the surface normal that would produce the maximum specular intensity. So, the highlight is brightest when $\hat{\mathbf{n}} = \hat{\mathbf{h}}$. Figure 9.12 is a visual representation of the configuration, including the surface orientation of maximum specular reflection. The resulting distribution is

$$(\hat{\mathbf{n}} \cdot \hat{\mathbf{h}})^{m_\alpha}$$

We have substituted m_α for m_β, to indicate that a slightly different exponent is needed to get similar results between the Phong and Blinn–Phong distributions.

Assuming the Blinn–Phong distribution, a potential BRDF for specular lighting is

$$f(\hat{\mathbf{l}}, \hat{\mathbf{v}}) = M_s(\hat{\mathbf{h}} \cdot \hat{\mathbf{v}})^{m_\alpha}$$

Like the diffuse term, the specular term includes a specular color defined on the material (M_s), which allows the highlights to be tinted a given color. Plastic and clear-coated surfaces (such as those covered with clear varnish), whatever their diffuse color, tend to have highlights that match the incoming light's color, while metallic surfaces tend to have highlights that match the diffuse color, modulated with the light's color. For a more detailed discussion of this and several other (more advanced) specular reflection methods, see Pharr and Humphreys [121] or Akenine-Möller et al. [1]. We will also have a general overview of these at the end of the chapter.

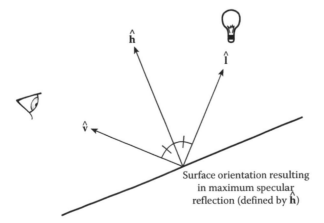

Figure 9.12. The specular halfway vector.

There is one problem with our equation above. Recall that we want a BRDF to be energy preserving—that the total weighting of all incoming irradiance must be no greater than 1. However, the Blinn–Phong distribution is not energy preserving—as [52] shows, its maximum reflectance is

$$\frac{8\pi (2^{-m_\alpha/2} + m_\alpha)}{(m_\alpha + 2)(m_\alpha + 4)}$$

We would need to divide by that factor in order to normalize our reflectance to be less than or equal to 1. A reasonable approximation to this normalization factor, presented by [1], is to multiply by $(m_\alpha + 8)/(8\pi)$. Substituting and canceling terms, this produces a BRDF of

$$f_s(\hat{\mathbf{l}}, \hat{\mathbf{v}}) = \frac{m_\alpha + 8}{8\pi} M_s (\hat{\mathbf{n}} \cdot \hat{\mathbf{h}})^{m_\alpha}$$

Our final specular reflection term is then

$$L_o = \frac{m_\alpha + 8}{8} M_s \circ L_i (\hat{\mathbf{n}} \cdot \hat{\mathbf{h}})^{m_\alpha} \max(0, \hat{\mathbf{n}} \cdot \hat{\mathbf{l}}_i)$$

Note that the standard Phong or Blinn–Phong lighting equations would not have the $\max(0, \hat{\mathbf{n}} \cdot \hat{\mathbf{l}}_i)$ term. In the traditional form, we would have to check whether $\hat{\mathbf{n}} \cdot \hat{\mathbf{l}}_i > 0$, and not generate specular lighting otherwise (simply clamping the specular term to be greater than 0 could allow objects whose normals point away from the light to generate highlights, which is not correct). However, treating in the fashion that we do both removes an unnecessary conditional and allows us to incorporate it as part of a general BRDF.

In our pipeline, we will store the specular color of an object's material in the first three elements of the 4-vector shader uniform value materialSpecularColorExp. Rather than perform any normalization in the shader, we will store the color as the preweighted value $\frac{m_\alpha + 8}{8} M_s$. The specular exponent material property will be stored in the a value. The end result can be seen in Figure 9.13.

Assuming we have a surfacePosition in camera space, the shader code to compute the specular component is as follows:

```
// GLSL Code
uniform vec4 materialSpecularColorExp;

vec3 computeSpecularBRDF(in vec3 surfaceNormal,
                         in vec3 lightDir,
                         in vec3 viewDir)
{
   vec3 halfVector = normalize(viewDir + light.dir);
   float nDotH = clamp(dot(surfaceNormal, halfVector), 0.0, 1.0);
   return materialSpecularColorExp.rgb
              * pow(nDotH, materialSpecularColorExp.a);
}

vec3 computeSpecularComponent(in lightSampleValues light,
                              in vec3 surfaceNormal,
                              in vec3 viewDir)
{
   return light.L * computeSpecularBRDF(surfaceNormal, viewDir, light)
              * clamp(dot(surfaceNormal, light.dir), 0.0, 1.0);
}
```

Figure 9.13. Sphere lit by specular light.

9.5.3   Emission

Emission, or emissive light, is the radiance produced by the surface itself, in the absence of any light sources. Put simply, it is the color and intensity with which the object "glows." Because this is purely a surface-based property, only surface materials (not lights) contain emissive colors. We will represent the emissive value of a material as a color M_e. As we mentioned, one approximation that is made in real-time systems is the (sometimes confusing) fact that this "emitted" light does not illuminate the surfaces of any other objects. In fact, another common (and perhaps more descriptive) term used for emission is *self-illumination*. The fact that emissive objects do not illuminate one another avoids the need for the graphics systems to take other objects into account when computing the light at a given point.

We will store the emissive color M_e in the 3-vector shader uniform value `materialEmissiveColor`.

9.5.4   Ambient

As mentioned in Section 9.4, we treat the basic ambient light source slightly differently from other light sources because it has no inherent direction. Hence, we introduced the ambient term $r_a(\hat{\mathbf{v}}) \circ L_a$ in Equation 9.3. Here L_a represents the radiance contribution from

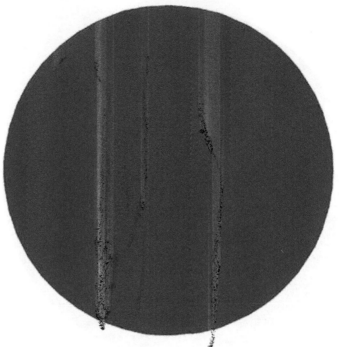

Figure 9.14. Sphere lit by ambient light.

our ambient light, and r_a is a reflection function dependent on only the view direction (again, because we have no corresponding $\hat{\mathbf{l}}$). In practice, $r_a(\hat{\mathbf{v}})$ is commonly set to be a constant, the ambient color M_a. Most often M_a is equal to the diffuse color M_d—another possibility recommended by [25] is to use a weighted blend of both diffuse and specular colors.

Since M_e, M_a, and L_a are all constant, it would be most efficient to store them in a single uniform. However, as we'll see later, for more advanced systems the ambient light value is not constant, so to be more general, we will keep them separate. We will store the ambient color of an object's material in the 3-vector shader uniform value `materialAmbientColor`, and the ambient light's value in `ambientLightRadiance`. Figure 9.14 provides a visual example of a sphere lit purely by ambient light.

The shader code to compute the ambient component is as follows:

```
// GLSL Code
uniform vec3 materialAmbientColor;
uniform vec3 ambientLightRadiance;

vec3 computeAmbientComponent()
{
    return ambientLightRadiance * materialAmbientColor;
}
```

9.5.5 Combined Lighting Equation

Having covered materials, lighting components, and light sources, we now have almost enough information to evaluate our full lighting model for a given light at a given point. The one piece remaining is to create a combined BRDF for diffuse and specular lighting. The obvious approach is to simply add their BRDFs together:

$$f(\hat{\mathbf{l}}, \hat{\mathbf{v}}) = \frac{M_d}{\pi} + \frac{m_\alpha + 8}{8\pi}(\hat{\mathbf{n}} \cdot \hat{\mathbf{h}})^{m_\alpha} M_s$$

The problem is that while the diffuse and specular terms are normalized, the total is not, and so again we're potentially adding energy into our system. The solution is to do a weighted sum:

$$f(\hat{\mathbf{l}}, \hat{\mathbf{v}}) = k_d \frac{M_d}{\pi} + k_s \frac{m_\alpha + 8}{8\pi}(\hat{\mathbf{n}} \cdot \hat{\mathbf{h}})^{m_\alpha} M_s$$

where $k_d + k_s \leq 1$. This gives us an energy-preserving BRDF. In practice, $k_d = 1 - k_s$, and both can be multiplied into `materialDiffuseColor` and `materialSpecularColor`, respectively. Our final lighting equation is then

$$C_V = M_e + M_a \circ L_a$$
$$+ \sum_i^{lights} \left[((1 - k_s)M_d) + \frac{m_\alpha + 8}{8}(\hat{\mathbf{n}} \cdot \hat{\mathbf{h}})^{m_\alpha}(k_s M_s) \right] \circ L_i \max(0, \hat{\mathbf{n}} \cdot \hat{\mathbf{l}}) \qquad (9.4)$$
$$A_V = M_{Alpha}$$

where the results are

1. C_V, the computed, lit RGB color of the sample

2. A_V, the alpha component of the RGBA color of the sample

The shader code to compute this for a single light, based upon the shader functions already defined previously, is as follows:

```
// GLSL Code

vec3 computeLitColor(in lightSampleValues light,
                     in vec4 surfaceNormal,
                     in vec3 viewDir)
{
    vec3 brdf = computeDiffuseBRDF(surfaceNormal, light.dir, viewDir)
              + computeSpecularBRDF(surfaceNormal, light.dir, viewDir);
    return light.L * brdf * clamp(dot(surfaceNormal, light.dir), 0.0, 1.0);
}

// ...

uniform vec3 materialEmissiveColor;
uniform vec4 materialDiffuseColor;
```

```
            vec4 finalColor;
            finalColor.rgb = materialEmissiveColor
                            + computeAmbientComponent()
                            + computeLitColor(light, normalize(normal),
                                                normalize(viewDir));
            finalColor.a = materialDiffuseColor.a;
```

For a visual example of all of these components combined, see the lit sphere in Figure 9.15.

Source Code
Demo
MultipleLights

Most interesting scenes will contain more than a single light source. In order to implement this equation in shader code, we need to compute L and $\hat{\mathbf{l}}$ per active light. The shader code for computing these values requires source data for each light. In addition, the type of data required differs by light type. The former issue can be solved by passing arrays of uniforms for each value required by a light type. The elements of the arrays represent the values for each light, indexed by a loop variable. For example, if we assume that all of our lights are directional, the code to compute the lighting for up to eight lights, again using some of the previous uniforms and routines, might be as follows:

```
// GLSL Code
uniform int dirLightCount;
uniform vec3 dirLightDirection[8];
uniform vec3 dirLightRadiance[8];
lightSampleValues computeDirLightValues(in int i)
{
    lightSampleValues values;
    values.dir = dirLightDirection[i];
    values.L = dirLightRadiance[i];
    return values;
}


{
    int i;
    vec4 finalColor;
    vec3 normalizedNormal = normalize(normal);
    vec3 normalizedViewDir = normalize(viewDir);
    finalColor.rgb = materialEmissiveColor
                    + computeAmbientComponent();
    finalColor.a = materialDiffuseColor.a;
    for (i = 0; i < dirLightCount; i++)
    {
        lightSampleValues light = computeDirLightValues(i);
        finalColor.rgb += computeLitColor(lightValues, normalizedNormal,
                                        normalizedViewDir);
    }
}
```

The code becomes even more complex when we must consider different types of light sources. One approach to this is to use independent arrays for each type of light and iterate over each array independently. The complexity of these approaches and the number of uniforms that must be sent to the shader can be prohibitive for some systems. As a result, it is common for rendering engines to either generate specific shaders for the lighting cases

Figure 9.15. Sphere lit by a combination of ambient, diffuse, and specular lighting.

they know they need, or generate custom shader source code in the engine itself, compiling these shaders at runtime as they are required. An alternative approach is to use deferred lighting, which we'll discuss below.

Clearly, many different values and components must come together to light even a single sample. This fact can make lighting complicated and difficult to use at first. A completely black rendered image or a flat-colored resulting object can be the result of many possible errors. However, an understanding of the lighting pipeline can make it much easier to determine which features to disable or change in order to debug lighting issues.

9.6 Lighting and Shading

Thus far, our lighting discussion has focused on computing color at a generic point on a surface, given a location, surface normal, view vector, and surface material. We have specifically avoided specifying whether these code snippets in our shader code examples are to be vertex or fragment shaders. Another aspect of lighting that is just as important as the basic lighting equation is the question of when and how to evaluate that equation to completely light a surface. Furthermore, if we do not choose to evaluate the full lighting equation at every sample point on the surface, how do we interpolate or reuse the explicitly lit sample points to compute reasonable colors for these other samples?

Ultimately, a triangle in view is drawn to the screen by coloring the screen pixels covered by that triangle (as will be discussed in more detail in Chapter 10). Any lighting system must be teamed with a shading method that can quickly compute colors for each and every pixel covered by the triangle. These shading methods determine when to invoke the shader to compute the lighting and when to simply reuse or interpolate already computed lighting results from other samples. In most cases, this is a performance versus visual accuracy trade-off, since it is normally more expensive computationally to evaluate the shader than it is to reuse or interpolate already computed lighting results.

The sheer number of pixels that must be drawn per frame requires that low- to mid-end graphics systems forego computing more expensive lighting equations for each pixel in favor of another method. For example, a sphere that covers 50 percent of a mid-sized $1{,}280 \times 1{,}024$ pixel screen will require the shading system to compute colors for over a half-million pixels, regardless of the tessellation. Next, we will discuss some of the more popular methods. Some of these methods will be familiar, as they are simply the shading methods discussed in Chapter 8, using results of the lighting equation as source colors.

9.6.1 Flat-Shaded Lighting

Historically, the simplest shading method applied to lighting was per-triangle, flat shading. This method involved evaluating the lighting equation once per triangle and using the resulting color as the constant triangle color. This color is assigned to every pixel covered by the triangle. In older, fixed-function systems, this was the highest-performance lighting–shading combination, owing to two facts: the more expensive lighting equation needed only to be evaluated once per triangle, and a single color could be used for all pixels in the triangle. Figure 9.16 shows an example of a sphere lit and shaded using per-triangle lighting and flat shading.

To evaluate the lighting equation for a triangle, we need a sample location and surface normal. The surface normal used is generally the triangle face normal (discussed in Chapter 2), as it accurately represents the plane of the triangle. However, the issue of sample position is more problematic. No single point can accurately represent the lighting across an entire triangle (except in special cases); for example, in the presence of a point light, different points on the triangle should be attenuated differently, according to their distance from the light. While the centroid of the triangle is a reasonable choice, the fact that it must be computed specifically for lighting makes it less desirable. For reasons of efficiency (and often to match with the graphics system), the most common sample point for flat shading is one of the triangle vertices, as the vertices already exist in the desired space. This can lead to artifacts, since a triangle's vertices are (by definition) at the edge of the area of the triangle. Flat-shaded lighting does not match quite as well with modern programmable shading pipelines, and the simplicity of the resulting lighting has meant that it is of somewhat limited interest in modern rendering systems.

9.6.2 Per-Vertex Lighting

Flat-shaded lighting suffers from the basic flaws and limitations of flat shading itself; the faceted appearance of the resulting geometry tends to highlight rather than hide the piecewise triangular approximation. In the presence of specular lighting, the tessellation is even more pronounced, causing entire triangles to be lit with bright highlights. With moving lights or geometry, this can cause gemstonelike flashing of the facets. Unless the goal is

Figure 9.16. Sphere lit and shaded by per-triangle lighting and flat shading.

an old-school gaming effect, for smooth surfaces such as the sphere in Figure 9.16, this faceting is often unacceptable.

The next logical step is to use per-vertex lighting with Gouraud interpolation of the resulting color values. The lighting equation is evaluated in the vertex shader, and the resulting color is passed as an interpolated input color to the simple fragment shader. The fragment shader can be extremely simple, doing nothing more than assigning the interpolated input color as the final fragment color.

Generating a single lit color that is shared by all colocated vertices leads to smooth lighting across surface boundaries. Even if colocated vertices are not shared (i.e., each triangle has its own copy of its three vertices), simply setting the normals to be the same in all copies of a vertex will cause all copies to be lit the same way. Figure 9.17 shows an example of a sphere lit and shaded using per-vertex lighting.

Per-vertex lighting only requires evaluating the lighting equation once per vertex. In the presence of well-optimized vertex sharing (where there are more triangles than vertices), per-vertex lighting can actually require *fewer* lighting equation evaluations than does true per-triangle flat shading. The interpolation method used to compute the per-fragment input values (Gouraud) is more expensive computationally than the trivial one used for flat shading, since it must interpolate between the three vertex colors on a per-fragment basis. However, modern shading hardware is heavily tuned for this form of fragment input interpolation, so the resulting performance of per-vertex lighting is generally close to peak.

Figure 9.17. Sphere lit and shaded by per-vertex lighting and Gouraud shading.

Gouraud-shaded lighting is a vertex-centric method—the surface positions and normals are used only at the vertices, with the triangles serving only as areas for interpolation. This shift to vertices as localized surface representations lends focus to the fact that we will need smooth surface normals at each vertex. The next section will discuss several methods for generating these vertex normals.

9.6.2.1 Generating Vertex Normals

In order to generate smooth lighting that represents a surface at each vertex, we need to generate a single normal that represents the surface at each vertex, not at each triangle. There are several common methods used to generate these per-vertex surface normals at content creation time or at load time, depending upon the source of the geometry data.

When possible, the best way to generate smooth normals during the creation of a tessellation is to use analytically computed normals based on the surface being approximated by triangles. For example, if the set of triangles represent a sphere centered at the origin, then for any vertex at location P_V, the surface normal is simply

$$\hat{\mathbf{n}} = \frac{P_V - O}{\|P_V - O\|}$$

This is the vertex position, treated as a vector (thus the subtraction of the origin) and normalized. Analytical normals can create very realistic impressions of the original surface, as the surface normals are pivotal to the overall lighting impression. Examples of surfaces for which analytical normals are available include implicit surfaces and parametric surface representations, which generally include analytically defined normal vectors at every point in their domain.

In the more common case, the mesh of triangles exists by itself, with no available method of computing exact surface normals for the surface being approximated. In this case, the normals must be generated from the triangles themselves. While this is unlikely to produce optimal results in all cases, simple methods can generate normals that tend to create the impression of a smooth surface and remove the appearance of faceting.

One of the most popular algorithms for generating normals from triangles takes the mean of all of the face normals for the triangles that use the given vertex. Figure 9.18 demonstrates a two-dimensional (2D) example of averaging triangle normal vectors. The algorithm may be pseudocoded as follows:

```
for each vertex V
{
    vector V.N = (0,0,0);
    for each triangle T that uses V
    {
      vector F = TriangleNormal(T);
      V.N += F;
    }
    V.N.Normalize();
}
```

Basically, the algorithm sums the normals of all of the faces that are incident upon the current vertex and then renormalizes the resulting summed vector. Since this algorithm is (in a sense) a mean-based algorithm, it can be affected by tessellation. Triangles are not weighted by area or other such factors, meaning that the face normal of each triangle incident upon the vertex has an equal "vote" in the makeup of the final vertex normal. While the method is far from perfect, any vertex normal generated from triangles will by its nature be an approximation. In most cases, the averaging algorithm generates convincing normals.

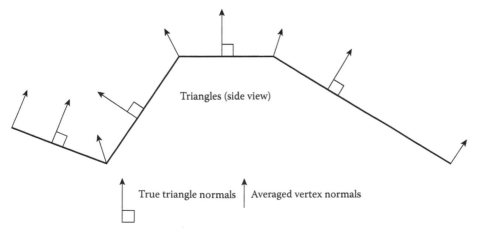

Figure 9.18. Averaging triangle normal vectors.

Note that in cases where there is no fast (i.e., constant-time) method of retrieving the set of triangles that use a given vertex (e.g., if only the OpenGL/Direct3D-style index lists are available), the algorithm may be turned "inside out" as follows:

```
for each vertex V
{
    V.N = (0,0,0);
}
for each triangle T
{
    // V1, V2, V3 are the vertices used by the triangle
    vector F = TriangleNormal(T);
    V1.N += F;
    V2.N += F;
    V3.N += F;
}
for each vertex V
{
    V.N.Normalize();
}
```

Basically, this version of the algorithm uses the vertex normals as accumulators, looping over the triangles, adding each triangle's face normal to the vertex normals of the three vertices in that triangle. Finally, having accumulated the input from all triangles, the algorithm goes back and normalizes each final vertex normal. Both algorithms will result in the same vertex normals, but each works well with different vertex/triangle data structure organizations.

9.6.2.2 Sharp Edges

As with Gouraud shading based on fixed colors, Gouraud-shaded lighting with vertices shared between triangles generates smooth triangle boundaries by default. In order to represent a sharp edge, vertices along a physical crease in the geometry must be duplicated so that the vertices can represent the surface normals on either side of the crease. By having different surface normals in copies of colocated vertices, the triangles on either side of an edge can be lit according to the correct local surface orientation. For example, at each vertex of a cube, there will be three vertices, each one with a normal of a different face orientation, as we see in Figure 9.19.

9.6.3 Per-Fragment Lighting

There are significant limitations to per-vertex lighting. Specifically, the fact that the lighting equation is evaluated only at the vertices can lead to artifacts. Even a cursory evaluation of the lighting equation shows that it is highly nonlinear. However, Gouraud shading interpolates linearly across polygons. Any nonlinearities in the lighting across the interior of the triangle will be lost completely. These artifacts are not as noticeable with diffuse and ambient lighting as they are with specular lighting, because diffuse and ambient lighting are closer to linear functions than is specular lighting (owing at least partially to the nonlinearity of the specular exponent term and to the rapid changes in the specular halfway vector $\hat{\mathbf{h}}$ with changes in viewer location).

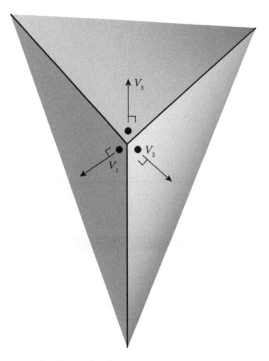

Figure 9.19. One corner of a faceted cube.

For example, let us examine the specular lighting term for the surface shown in Figure 9.20. We draw the 2D case, in which the triangle is represented by a line segment. In this situation, the vertex normals all point outward from the center of the triangle, meaning that the triangle is representing a somewhat curved (domed) surface. The point light source and the viewer are located at the same position in space, meaning that the view vector $\hat{v}$, the light vector $\hat{l}$, and the resulting halfway vector $\hat{h}$ will all be equal for all points in space. The light and viewer are directly above the center of the triangle. Because of this, the specular components computed at the two vertices will be quite dark (note the specular halfway vectors shown in Figure 9.20 are almost perpendicular to the normals at the vertices). Linearly interpolating between these two dark specular vertex colors will result in a polygon that is relatively dark.

However, if we look at the geometry that is being approximated by these normals (a domed surface as in Figure 9.20), we can see that in this configuration the interpolated normal at the center of the triangle would point straight up at the viewer and light. If we were to evaluate the lighting equation at a point near the center of the triangle in this case, we would find an extremely bright specular highlight there. The specular lighting across the surface of this triangle is highly nonlinear, and the maximum is internal to the triangle. Even more problematic is the case in which the surface is moving over time. In rendered images where the highlight happens to line up with a vertex, there will be a bright, linearly interpolated highlight at the vertex. However, as the surface moves so that the highlight falls between vertices, the highlight will disappear completely. This is a very fundamental

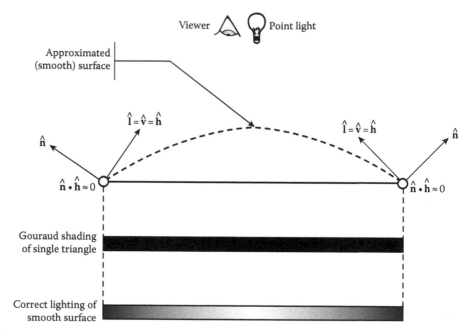

Figure 9.20. Per-vertex lighting can miss specular highlights.

problem with approximating a complex function with a piecewise linear representation. The accuracy of the result is dependent upon the number of linear segments used to approximate the function. In our case, this is equivalent to the density of the tessellation.

If we want to increase the accuracy of lighting on a general vertex-lit surface, we must subdivide the surface to increase the density of vertices (and thus lighting samples). However, this is an expensive process, and we may not know a priori which sections of the surface will require significant tessellation. Dependent upon the particular view at runtime, almost any tessellation may be either overly dense or too coarse. In order to create a more general, high-quality lighting method, we must find another way around this problem.

The solution is to evaluate the lighting equation once for each fragment covered by the triangle. This is called per-fragment lighting, or traditionally *Phong shading* (named after its inventor, Bui Tuong Phong [122]). The difference between per-vertex and per-fragment lighting may be seen in Figures 9.20 and 9.21. For each sample across the surface of a triangle, the vertex normals, positions, reflection, and view vectors are interpolated, and the interpolated values are used to evaluate the lighting equation. However, since triangles tend to cover more than 1–3 pixels, such a lighting method will result in far more lighting computations per triangle than do per-triangle or per-vertex methods.

Per-fragment lighting changes the balance of the work to be done in the vertex and fragment shaders. Instead of computing the lighting in the vertex shader, per-fragment lighting uses the vertex shader only to set up the source values (surface position, surface normal, view vector) and pass them down as input values to the fragment shader. As always, the fragment shader inputs are interpolated using Gouraud interpolation and passed to each invocation of the fragment shaders. These interpolated values now represent smoothly

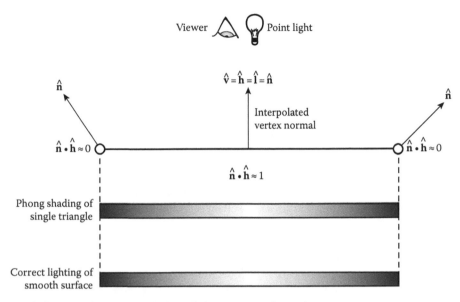

Figure 9.21. Per-fragment lighting of the same configuration.

interpolated position and normal vectors for the surface being represented. It is these values that are used as sources to the lighting computations, evaluated in the fragment shader.

There are several issues that make per-fragment lighting more computationally expensive than per-vertex lighting. The first of these is the actual normal vector interpolation, since as we saw in Chapter 6, basic barycentric interpolation of the three vertex normals will almost *never* result in a normalized vector. As a result, the interpolated normal vector will have to be renormalized *per fragment*, which is much more frequently than per vertex. The same is true of the view vector and the light vector.

Furthermore, the full lighting equation must be evaluated per sample once the interpolated normal is computed and renormalized. Not only is this operation expensive, but also it is not a fixed amount of computation. As we saw above, in a general engine, the complexity of the lighting equation is dependent on the number of lights and numerous graphics engine settings. This resulted in per-fragment shading being rather unpopular in game-centric consumer 3D hardware prior to the advent of pixel and vertex shaders. However, modern devices have become fast and flexible enough for per-fragment lighting to become the de facto standard for lighting quality.

9.7 Textures and Lighting

Of the methods we have discussed for coloring geometry, the two most powerful are texturing and dynamic lighting. However, they each have drawbacks when used by themselves. Texturing is normally a static method and looks flat and painted when used by itself in a dynamic scene. Lighting can generate very dynamic effects, but provides only gradual changes in detail. It is only natural that graphics systems would want to use the results of both techniques together on a single surface.

Source Code

Demo
TexturesAndLighting

(a)　　　　　　　　　　(b)　　　　　　　　　　(c)

Figure 9.22. Textures and lighting combined via modulation: (a) sphere with pure lighting, (b) sphere with pure texturing, and (c) same sphere with lighting and texturing combined.

9.7.1 Basic Modulation

The simplest methods of merging these two techniques involve simply viewing each of the two methods as generating a color per sample and merging them. With texturing, this is done directly via texture sampling; with lighting, it is done by evaluating the lighting equation. These two colors must be combined in a way that makes visual sense. The most common way of combining textures and lighting results is via multiplication, or modulation. In modulate lighting-texture combination, the texture color at the given sample C_T and the final lit or interpolated color are combined by per-component multiplication. This is similar to multiplying our BRDF by the incoming irradiance. If we set our material colors to white, our texture can act as a global material color. We'll discuss better ways to handle this in Section 9.7.3.

Assuming that our lighting values range between 0 and 1, the visual effect here is that the lit colors darken the texture (or vice versa). As a result, texture images designed to be used with modulate mode-texture combination are normally painted as if they were fully lit. The colors, representing the lighting in the scene, darken these fully lit textures to make them look more realistic in the given environment (Figure 9.22). The result of modulation can be very convincing, even though the lighting is rather simple and the textures are static paintings. In the presence of moving or otherwise animated lights, the result can be even more immersive, as the human perceptual system is very reliant upon lighting cues in the real world. This is, in a sense, using the texture as a factor in some or all of the surface material colors.

Assuming for the moment that the lit color is computed and stored in `litColor`, either by computation in the fragment shader or passed down as an input component, a simple textured, lit fragment shader would be as follows:

```
// GLSL - fragment shader
uniform sampler2D texture;
in vec2 texCoords;
out vec4 fragColor;

void main()
{
    // lit color is in vec3 litColor
```

```
    vec3 litColor;

    // ...

    // Sample the texture represented by "texture"
    // at the location "texCoords"
    fragColor.rgb = litColor * texture2D (texture, texCoords);
}
```

Until the advent of programmable shaders, modulation was the most popular and often the only method of combining lighting and textures.

9.7.2 Specular Lighting and Textures

As mentioned, if lit vertex colors are clamped to the range [0, 1], the full lighting equation (9.4) combined with the texture via multiplication can only *darken* the texture. While this looks correct for diffuse or matte objects, for shiny objects with bright specular highlights, it can look very dull. It is often useful to have the specular highlights "wash out" the texture. We cannot simply add the full set of lighting because the texture will almost always wash out and can *never* get darker. To be able to see the full range of effects with our limited light values requires an approximation where the diffuse colors darken the texture while the specular components of color add highlights. This is only possible if we split our general lighting calculation into separate diffuse and specular components.

Because the specular term is added after the texture is multiplied, this mode (sometimes called *modulate with late add*) causes the diffuse terms to attenuate the texture color, while the specular terms wash out the result. The differences between the separate and combined specular modes can be very striking, as Figure 9.23 makes clear.

The shader code to compute this involves computing the emissive, ambient, and diffuse lighting components into one color (which we'll call diffuseLighting) and

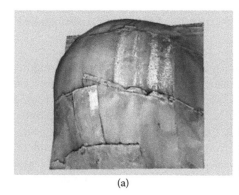
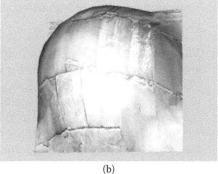

(a) (b)

Figure 9.23. Combining textures and lighting: (a) specular vertex color added to diffuse vertex color, then modulated with the texture and (b) diffuse vertex color modulated with the texture, then specular vertex color added.

the specular component into another (which we'll call `specularLighting`). Having computed these independently, we merge them as follows:

```
// GLSL - fragment shader
uniform sampler2D texture;
in vec2 texCoords;
out vec4 fragColor;

void main()
{
   vec3 diffuseLighting;
   vec3 specularLighting;

   // ...

   // Sample the texture represented by "texture"
   // at the location "texCoords"
   fragColor.rgb = diffuseLighting * texture2D (texture, texCoords)
                   + specularLighting;
}
```

In this case, the texture is providing a scaling factor for only the emissive, ambient, and diffuse material color. While not physically correct, the effect is simple to add to an existing lighting and texture shader and can make for a much more dynamic result.

9.7.3   Textures as Materials

The next step in using textures and lighting together involves using multiple textures on a single surface. As shown in the previous section, a texture can be used to modulate one or more material colors on a surface. In fact, textures also can be used to *replace* one or more surface material components. Common surface material colors to be replaced with textures are:

- **Material diffuse color.** Often called a *diffuse map,* this is extremely similar to basic modulation, as shown above. Frequently, the diffuse map is also applied as the ambient material color.

- **Material specular color.** This is frequently replaced with either an RGB texture (a specular map) or a single-channel grayscale texture, which is called a *gloss map*. The gloss map is a powerful technique: Wherever it is close to full brightness, the object appears glossy, because specular highlights can be seen at those points on the surface. Wherever it is close to black, the specular highlights do not appear. As a result, it can be used to mark shiny or worn sections of an object, independent of the color. Frequently, gloss maps are created in such a way that they are brightest on the edges and exposed areas, parts of a surface likely to be worn down by the elements and naturally polished.

- **Material specular exponent.** This can be replaced by a single-channel grayscale texture, which is called a *roughness map*. This allows the artists to model small changes across the surface in the size of the specular highlight for Phong lighting and for material roughness for more advanced lighting models.

- **Material emissive color.** Often called a *glow map*, this texture can be used to localize self-illumination of an object. These maps are frequently used to mark windows in nighttime views of vehicles or buildings, or taillights and running lights of vehicles.

Since multiple textures can be used in a single shader, any or all of these components can be easily replaced by individual textures. The uniform material color vectors simply become texture sampler uniforms. Many of these textures can reuse the same texture coordinates, as the mappings of each texture can be the same. Finally, optimizations are common. For example, it is common to use an RGBA texture in which the RGB components are used as the diffuse map, and the alpha component is used as a single-channel gloss map. The ease of painting these components into a single texture can assist the artists, and the reuse of an otherwise unused texture component can save graphics processing unit (GPU) resources. An example of a fragment shader using RGBA diffuse and gloss maps is shown below.

```
// GLSL - fragment shader
uniform sampler2D texture;
in vec2 texCoords;
out vec4 fragColor;

void main()
{
    vec3 diffuseLighting;
    vec3 specularLighting;

    // ...
    vec4 diffuseAndGlossMap = texture2D (texture, texCoords);

    // Sample the texture represented by "texture"
    // at the location "texCoords"
    fragColor.rgb = diffuseLighting * diffuseAndGlossMap.rgb
                  + specularLighting * diffuseAndGlossMap.a;
}
```

9.8 Advanced Lighting

Programmable shaders make an almost endless array of lighting effects possible. We will discuss a few of these methods and mention several others, citing references for additional information. Like the methods mentioned in the previous sections, many of these methods involve using textures as sources to the lighting equation.

9.8.1 Normal Mapping

So far, we have shown how one or more textures can be used to replace material colors or intensities in the lighting equation. However, even more advanced techniques are based on the fact that textures can be used to store more general values than mere colors. The most

Source Code
Demo
NormalMapping

popular of these techniques are *bump mapping* and *normal mapping*. As the names suggest, these methods simulate bumpy surfaces by storing the detailed "height offsets" (bump mapping) or normal vectors (normal mapping) for the surface in a texture. One of the basic limitations of dynamic lighting as discussed so far is that while we can evaluate lighting on a per-fragment basis, the source values describing the geometry of the surface are interpolated from per-vertex values—the position and normal. As a result, a rough or bumpy surface requires a very high density of vertices. We can simulate this effect at a finer level by adding bumpy, prelit colors to the diffuse map texture, but in the presence of moving or changing lighting conditions, the static nature of this trick is obvious and jarring. Bump mapping, the first of these techniques to be available, was actually present in some fixed-function rendering hardware in the late 1990s. However, bump mapping, since it represented a local height offset and generated surface normals implicitly by looking at the difference in height values between neighboring texels, was limited to surfaces that looked embossed. Very sharp changes in surface orientation were difficult with bump mapping. For a discussion of these limitations, see Theodore [144].

In order to add more detail to the lighting at this fine level, we must be able to generate a surface normal per fragment that contains real information (not just interpolated information) for each fragment. By storing the normal in a texture, we can generate normals that change very rapidly per fragment and respond correctly to changing lighting configurations. The normal vectors across the surface are stored in the RGB components of the texture (either as signed fixed-point values or as floating-point values). The exact space in which the normals are stored in the texture differs from method to method. Conceptually, the simplest space is object space, in which normals are computed in the model or object space of the geometry being mapped and then stored as (x, y, z) in the R, G, and B components of the texture, respectively. Object-space normal maps can be sampled as a regular texture into a `vec3` and then used in the same way as one would use normals passed into a fragment shader as an input value. An example of the effect that normal mapping can have on a simple object is shown in Figure 9.24.

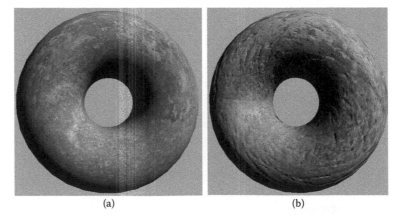

(a)   (b)

Figure 9.24. Normal mapping applied to simple objects: (a) low triangle—count object with diffuse map and per-vertex lighting and (b) same object with diffuse map and per-fragment normal mapping.

An example of an object-space normal map in use in a fragment shader is shown below. Note that since the sampled normal comes from the texture in object space, we must either do the transformation of the normal into world space in the fragment shader, or transform the light information into object space and light in object space. We pick the former (which is simpler to understand, but more expensive computationally) and transform the normal into world space using the `IvNormalMatrix` uniform, which `IvRenderer` will set up automatically based on our world matrix (see Chapter 7).

```
// GLSL Code
uniform sampler2D normalMap;
uniform mat3 IvNormalMatrix;
in vec2 texCoords;
out vec4 fragColor;
{
    lightSampleValues lightValues;
    // compute light values
    ....
    vec3 normal = texture2D(normalMap, texCoords);
    normal = normalize(IvNormalMatrix * normal);
    vec3 normalizedViewDir = normalize(viewDir);
    fragColor.rgb = materialEmissiveColor + computeAmbientComponent();
    fragColor.rgb += computeLitColor(lightValues, normal, normalizedViewDir);
    fragColor.a = materialDiffuseColor.a;
}
```

While object-space normal maps are conceptually easier to understand, the more common method now uses something called *tangent space*. If we store a normal and a single tangent vector (which is orthogonal to the normal) for each vertex, in the shader we can take a cross product of the two to produce a third orthogonal vector called the bitangent. This orthogonal basis plus the vertex position gives us a per-vertex coordinate frame that is the tangent space for that vertex. A texture can then store normals relative to this tangent space across the entire surface, and hence is known as a tangent space normal map. There are advantages and disadvantages to both object-space normal mapping and tangent space normal maps, which are discussed in various articles on normal mapping [45, 120].

9.8.1.1 Generating Normal Maps

Normal maps are rarely painted by an artist. The complexity of these maps and the esoteric spaces in which they reside mean that most normal maps are generated automatically from the geometry itself via commercial or open-source tools. Some real-time 3D engines and middleware provide tools that automatically convert a *very* high-polygon-count object (millions of triangles) into a low polygon-count geometry object and a high-resolution normal map. Put together, the low-resolution geometry object and the high-resolution normal map can be used to efficiently render what appears to be a very convincing facsimile of the original object.

9.8.2 Reflective Objects

While specular lighting can provide the basic impression of a shiny object, large expanses of a reflective surface are more convincing if they actually appear to reflect the other geometry in the scene. The best-known method for this is the (generally) non-real-time method of recursive ray tracing, which is not yet suitable for general interactive systems. However, we can once again use a mixture of lightinglike computations and textures to create very convincing reflections.

Environment mapping is a technique that uses a texture or set of textures that represents an "inside looking out" representation of the entire scene in all directions. It can be thought of as a spherical or cube-shaped set of images that represent a panorama of the scene. These images can be statically drawn offline, or on modern systems can even be rendered every frame to better represent the current scene. The environment map can be thought of as infinitely large or infinitely distant. Thus, any normalized direction vector maps to a single, fixed location in the environment map.

Environment maps are applied to a surface dynamically—they are not sampled via a priori texture coordinates; the mapping of the environment map will change as the scene and the view change. The most common method used is to compute the reflection of the view vector in a manner similar to that used for specular lighting earlier in this chapter. The reflected view vector represents the direction that the viewer sees in the reflective surface. By sampling the environment map in this view direction, we can apply what appears to be a convincing reflection to a surface with little more than a vector computation (per vertex or per fragment) and a texture lookup (per fragment). Figure 9.25 shows how this effect can be applied to a simple object. Note that in the two views of the object, the environment map moves like a reflection as the object rotates with respect to the viewer.

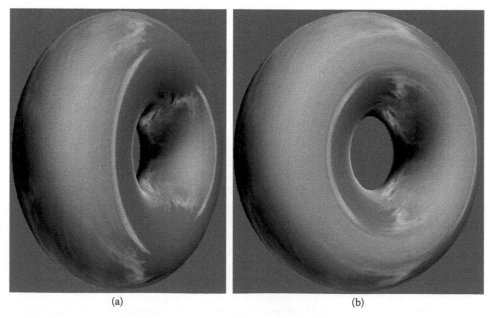

(a)   (b)

Figure 9.25. Environment mapping applied to simple objects.

9.8.3 Transparent Objects

One aspect of the rendering equation that we have neglected is that light can pass through objects as well as be reflected or absorbed. This is known as transmission, and can be incorporated into the rendering equation using an additional scattering function called the bidirectional transmissive distribution function, or BTDF. This includes both the partial absorption of light and the effect of refraction, where light bends as it passes from one surface type to another (for example, from air to glass).

In real-time systems, transmission is approximated in a variety of ways. The simplest is to ignore refractive effects and treat the object as a color filter on the background. We have included a simple approach to this by including an alpha value in our diffuse color. More complex approaches can take the thickness and other material properties of the object into account as well.

A common approximation for refraction is to use a *distortion map*. This indicates a warping of the background when a transparent or semitransparent object is rendered. It is commonly used to create heat distortion effects, or refraction through the surface of wavy glass or water. For more complex effects it's also possible to create a cubic environment map, simulating refraction from the center of the object rather than reflection.

Another approximation is for the specific case of subsurface scattering in semitranslucent surfaces, especially human skin. As one might expect, this is of particular interest to games because of the need to create more realistic-looking characters. One such technique is described in d'Eon and Luebke [31].

9.8.4 Area Lights

One thing all of the light sources we have discussed so far have in common is that a single vector can represent the direct lighting from each source at a particular sample on a surface. The lights described thus far either are infinitely distant or emit from a single point. Lights in the real world very often emit light not from a single point, but from a larger area. For example, the diffused fluorescent light fixtures that are ubiquitous in office buildings appear to emit light from a large, rectangular surface. There are two basic effects produced by these area light sources that are not represented by any of our lights above: a solid angle of incoming light upon the surface, and soft shadows.

One aspect of area light sources is that the direct lighting from them that is incident upon a single point on a surface comes from multiple directions. In fact, the light from an area light source on a surface point forms a complex, roughly cone-shaped volume whose apex is at the surface point being lit. Unless the area of the light source is large relative to its distance to the surface, the effect of this light coming from a range of directions can be very subtle. As the ratio of the area of the light source to the distance to the object (the projected size of the light source from the point of view of the surface point) goes down, the effect can rapidly converge to look like the single-vector cases we describe above (see [1]).

9.8.5 Physically Based Lighting

The lighting model that we have presented here is extremely simple, which is good for efficiency, but it is also extremely limited and cannot reproduce a broad range of materials. Because of this, researchers and developers have created more advanced lighting and reflection models, which more accurately represent real-world surfaces. On the lighting side we have area lights, mentioned above. Another common improvement in lighting is to replace the constant ambient light with a representation that also varies with incoming direction. One simple example of this is the hemispherical light. It has two colors representing ambient light from opposing directions, for example, the sky and the ground. Using the angle between the surface normal and a vector pointing toward the sky, we can blend between the two colors, giving a different ambient effect on the tops versus the bottoms of objects. Other examples are radiance or irradiance environment maps. These act like the reflective environment maps mentioned above, but store incoming radiance or irradiance in many directions at various points throughout a given space. Implementations of these include the radiosity normal mapping used by Valve in Half-Life 2 [108], and spherical harmonics irradiance maps [127]. Note that some implementations store radiance and some irradiance. We have assumed radiance for our ambient light—if irradiance is used, the leading π multiplier in the ambient term must be removed.

On the reflection side, there are far better BRDFs for representing materials. The most frequent change is to the specular term. If we look at our example, we can break it into two pieces: the reflection value (the specular color, in our case) and the distribution function $(\hat{\mathbf{n}} \cdot \hat{\mathbf{h}})^{\alpha}$. More advanced models add a third term—a microfaceting value, which models how tiny ridges in the surface scatter and self-shadow light that strikes the surface. The reflection value is often replaced with the Fresnel term, which controls how the outgoing radiance color changes with incoming radiance angles. And finally, the distribution function is made more complex, again being more flexible with incoming and outgoing angles. The most standard example of such a model is the Cook and Torrance reflection model [25].

There are also models that improve the diffuse term. The Lambertian model is highly idealized, and real surfaces rarely behave in such a uniform way. Oren and Nayar [113] created a new diffuse model that takes this into account, and recent games have used this to good effect.

The key point is to choose the lighting and reflection model that best suits your game—if your goal is realism, then one of the above systems is worth looking into. If not, then the simple model we have presented is a good compromise between realism, easy control, and efficiency. A good discussion of how a modern engine switched to use physically based lighting is [92].

9.8.6 High Dynamic Range Lighting

Throughout this chapter, we have assumed that our lighting values lie in the interval $[0, 1]$. However, in reality lighting values can be quite large, theoretically lying in the interval $[0, \infty)$. Our eyes adapt to this large range of values by adjusting the iris, closing for bright areas and opening for dark areas. A similar process is done in games, known as high dynamic range lighting. Rather than limiting our light values to lie in only $[0, 1]$, we allow a full range of values (to the limits of our floating-point representation). Note this affects the material values in textures (artists no longer paint them as fully lit, as they can now be both brightened and darkened) and allows us to get rid of our modulate with late add approximation for specular highlights.

This does mean we need to correct the resulting lighting to lie within $[0, 1]$ values for final display. In Chapter 8 we discussed a process called *tone mapping* that handles just this issue, dynamically mapping color values as we move between dark and light areas.

Most modern games support some form of HDR lighting with tone mapping, thereby giving much more realistic results.

9.8.7 Deferred Lighting and Shading

As mentioned, as we introduce multiple light sources the complexity of managing shaders grows. While we can have one so-called ubershader that handles a fixed number of light sources, we may have many cases where an object only really needs to be lit by one or two lights. And light types may vary between spotlights, point lights, and directional lights. Trying to handle all cases can create an explosion of shaders, while restricting them may be too limiting.

The other problem is that we can often be lighting objects that are in the back of the scene, and so are obscured by those closer to the viewer. We could end up lighting them, and then have those pixels written over by the lighting values for the objects in the front (see Chapter 10 for more information on this process, known as depth buffering). This is a waste of processing time.

One solution is to not light our objects as we render them into the scene, but rather to write out any lighting information per pixel into a temporary image, and then do a pass over the image for each light or a set of lights, using the stored information to update the image with lighting. This is known as *deferred lighting*. Deferred lighting has some advantages in that it does reduce our lighting calculations only to those areas that are visible, but as we have seen, there is quite a large amount of information needed for lighting, so the stored image (or *geometry buffer*) can get quite large. This makes this approach impractical for low-memory systems. That said, deferred lighting and its variants have been used to great effect in many commercial games.

9.8.8 Shadows

Shadows are an extremely important component of real-world lighting. However, while we think of them as a form of lighting, the challenge in rendering accurate shadows has little in common with the per-vertex and per-fragment direct lighting formulas we discussed earlier in this chapter. In a sense, shadowing is much less about lighting than it is about occlusion or intersection. Diffuse and specular lighting formulas are concerned mainly with determining how the light incident upon a surface is reflected toward the viewer. Shadowing, on the other hand, is far more concerned with determining whether light from a given light source reaches the surface in the first place. Unlike surface lighting calculations, shadowing is dependent upon all of the objects that *might* interpose between the light source and the surface being lit.

Since realistic light sources are area lights, they can be partially occluded and produce soft-edged shadows. This effect can be very significant, even if the area of the light source is quite small. Soft edged shadows occur at shadow boundaries, where the point in partial shadow is illuminated by part of the area light source but not all of it. The shadow becomes progressively darker as the given surface point is lit by less and less of the area light source. This soft shadow region (called the *penumbra*, as opposed to the fully shadowed region, called the *umbra*) is highly prized in non-real-time, photorealistic renderings for the realistic quality it lends to the results. Even when using point or directional lights, emulating soft shadows can greatly improve the realism of the scene.

A single surface shader is rarely, if ever, enough to implement shadowing. Shadowing is generally a multipass technique over the entire scene. The first pass involves determining which surfaces receive light from a given source, and the second pass involves applying this shadowing information to the shading of the objects in the scene. The many algorithms used to approximate real-time shadows differ in their approaches to both passes. With the advent of high-powered, programmable shading hardware, the push in shadowing methods over the past decade has focused on leveraging the rendering hardware as much as possible for both passes, avoiding expensive CPU-based computation. These algorithms have centered on the concept of using a first pass that involves rendering the scene from the point of view of the *light source*, as geometry visible from the point of view of the light is exactly the geometry that will be lit by that light source. Geometry that cannot be "seen" from a light's location is exactly the geometry that will be in shadow. Geometry that falls on the boundaries of these two cases is likely to be in the penumbra when rendering soft shadows.

Since the real core of shadowing methods lies in the structure of the two-pass algorithms rather than in the mathematics of lighting, the details of shadowing algorithms are beyond the scope of this book. A technique known as *ray tracing* (see Glassner [53]) uses ray–object intersection to track the way light bounces around a scene. Very convincing shadows (and reflections) could be computed using ray tracing, and the technique is very popular for non-real-time rendering. Owing to its computational complexity, this method is not generally used in real-time lighting (although modern shading languages and shading hardware are now capable of doing real-time ray tracing in some limited cases). Shadows are sometimes approximated using other tricks (see [3, 35, 111, 112]). Excellent references for real-time shadows, both sharp and soft edged, can be found in Fernando and Pharr [45, 120].

9.9   Chapter Summary

In this chapter we have discussed the basics of dynamic lighting, in terms of both geometric concepts and implementation using programmable shaders. Per-vertex and per-fragment lighting are very powerful additions to any 3D application, especially when mated with the use of multiple textures. Correct use of lighting can create compelling 3D environments at limited computational expense. As we have discussed, judicious use of lighting is important in order to maximize visual impact while minimizing additional computation.

For further information, there are numerous paths available to the interested reader. The growing wealth of shader resources includes websites [3, 112] and even book series [40]. Many of these new shaders are based on far more detailed and complex lighting models, such as those presented in computer graphics conference papers and journal articles like those of ACM SIGGRAPH or in books such as Akenine-Möller et al. [1] or Pharr and Humphreys [121].

⑩ Rasterization

10.1 Introduction

The final major stage in the rendering pipeline is called *rasterization*. Rasterization is the operation that takes screen-space geometry, a fragment shader, and the inputs to that shader and actually draws the geometry to the low-level two-dimensional (2D) display device. Once again, we will focus on drawing sets of triangles, as these are the most common primitive in three-dimensional (3D) graphics systems. In fact, for much of this chapter, we will focus on drawing an individual triangle. For almost all modern display devices, this low-level "drawing" operation involves assigning color values to each and every dot, or *pixel*, on the display device.

At the conceptual level, the entire topic of rasterization is simply an implementation detail. Rasterization is required because the display devices we use today are based on a dense rectangular grid of light-emitting elements, or pixels (a short version of the term *picture elements*), each of whose colors and intensities are individually adjustable in every frame. For historical reasons relating to the way that picture tube-based televisions work, these displays are called *raster displays*.

Raster displays require that the images displayed on them be discretized into a rectangular grid of color samples for each image. In order to achieve this, a computer graphics system must convert the projected, colored geometry representations into the required grid of colors. Moreover, in order to render real-time animation, the computer graphics system must do so many times per second. This process of generating a grid of color samples from a projected scene is called *rasterization*.

By its very nature, rasterization is time-consuming when compared to the other stages in the rendering pipeline. Whereas the other stages of the pipeline generally require per-object, per-triangle, or per-vertex computation, rasterization inherently requires computation of some sort for every pixel. At the time of this book's publication, displays 1,600 pixels wide by 1,200 pixels high—resulting in approximately *2 million* pixels on the screen—are quite popular. Add to this the fact that rasterization will in practice often require each pixel to be computed several times, and we come to the realization that the number of pixels that must be computed generally outpaces the number of triangles in a given frame by a factor of 10, 20, or more.

Historically, in purely software 3D pipelines, it is not uncommon to see as much as 80 to 90 percent of rendering time spent in rasterization. This level of computational demand has led to the fact that rasterization was the first stage of the graphics pipeline to be accelerated via purpose-built consumer hardware. In fact, most 3D computer games began to *require* some form of 3D hardware by the early 2000s. This chapter will not detail the methods and code required to write a software 3D rasterizer, since most game developers no longer have a need to write them. For the details on how to write a set of rasterizers, see Hecker's excellent series of articles on perspective texture mapping in *Game Developer Magazine* [76].

Despite the fact that few, if any, game developers will need to implement even a subset of the rasterization pipeline themselves in a modern game, the topic of rasterization is still extremely relevant, even today. The basic concepts of rasterization lead to discussions of some of the most interesting and subtle mathematical and geometric issues in the entire rendering pipeline. Furthermore, an understanding of these fundamental concepts can allow a game developer to better understand why and how rendering artifacts and performance bottlenecks occur, even when the rasterization implementation is in dedicated hardware. Many of these basic concepts and low-level details can have visually relevant results in almost any 3D game. This chapter will highlight some of the fundamental concepts of rasterization that are most pivotal to a deeper understanding of the process of using a rendering system, either graphics processing unit (GPU) or computer processing unit (CPU) based.

10.2   Displays and Framebuffers

Every piece of display device hardware, whether it be a computer monitor, television, or some other such device, requires a source of image data. For computer graphics systems, this source of image data is called a *framebuffer* (so called because it is a buffer of data that holds the image information for a frame, or a screen's worth of image). In basic terms, a framebuffer is a 2D digital image: a block of memory that contains numerical values that represent colors at each point on the screen. Each color value represents the color of the screen at a given point—a *pixel*. Each pixel has red, green, and blue components. Put together, this framebuffer represents the image that is to be drawn on the screen. The display hardware reads these colors from memory every time it needs to update the image on the screen, generally at least 30 times per second and often 60 or more times per second.

As we shall see, framebuffers often include more than just a single color per pixel. While it is the final per-pixel color that is actually used to set the color and intensity of light emitted by each point on the display, the other per-pixel values are used internally during the rasterization process. In a sense, these other values are analogous to per-vertex normals and per-triangle material colors; while they are never displayed directly, they have a significant effect on how the final color is computed.

10.3 Conceptual Rasterization Pipeline

The steps required to rasterize an entire frame are shown in Figure 10.1. The first step is to clear out any previous image from the framebuffer. This can in some cases be skipped; for example, if the scene geometry is known to cover the entire screen, then there is no need to clear the screen. The old image will be entirely overwritten by the new image in such a case. But for most applications, this step involves using the rendering application programming interface (API) to set all of the pixels in the framebuffer (in a single function call) to a fixed color.

The second step is to rasterize the geometry to the framebuffer. We will detail this stage in the rest of the chapter, as it is the most involved step of the three (by far).

The third step is to present the framebuffer image to the physical display. This stage is commonly known as *swapping* or *buffer swapping*, because historically it frequently involved (and in many cases still involves) switching between two buffers—drawing to one while the other is displayed, and then swapping the two buffers after each frame. This is to avoid flickering or other artifacts during rendering (specifically, to avoid having the user see a partially rendered frame). However, other techniques described later in the chapter will require additional work to be done during the presentation step. Therefore, we will refer to this step by the more general term *present*.

10.3.1 Rasterization Stages

There are several stages to even a simple rasterization pipeline. It should be noted that while these stages tend to exist in rasterization hardware implementations, hardware almost never

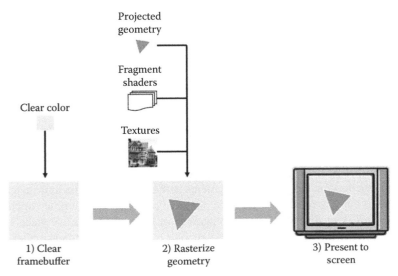

Figure 10.1. The steps to rasterizing a complete frame.

follows the order (or even the structure) of the conceptual stages in the list that follows. This simple pipeline rasterizes a single triangle as follows:

1. Determine the visible pixels covered by the triangle.

2. Compute a color for the visible triangle at each such pixel.

3. Determine a final color for each pixel and write to the framebuffer.

The first stage further decomposes into two separate steps:

1. Determining the pixels covered by a triangle

2. Determining which triangles are the ones visible at each pixel

The rest of this chapter will discuss each of these pipeline stages in detail.

10.4   Determining the Fragments: Pixels Covered by a Triangle

10.4.1   Fragments

In order to progress any further in the rasterization phase of rendering, we must break triangles (or more generally, geometry) in screen space into pieces that more directly match the pixels in the framebuffer. This involves determining the intersection of pixel rectangles or pixel center points with a triangle. In the color and lighting chapters, we used the term *fragment* to represent an infinitesimal piece of surface area around a given point on a polygonal surface. Fragment shaders were described as being evaluated on these tiny pieces of surface.

At the rasterization level, fragments have a much more explicit but related definition. They are the result of the aforementioned process of breaking down screen-space triangles to match pixels. These fragments can be thought of as pixel-sized pieces of a triangle in screen space. These can be visualized as a triangle diced into pieces by cutting along pixel boundaries. Many of these fragments (the interior of a triangle) will be square, the full size of the pixel square. We call these pixel-sized fragments *complete fragments*. However, along the edges of a triangle, these may be multisided polygons that fit inside of the pixel square and are thus smaller than a pixel. We call these smaller fragments *partial fragments*. In practice, these fragments may really be point samples of a triangle taken at the pixel center (similar to the concept we had of fragments in the lighting and shading chapters), but the basic idea is that fragments represent the pieces of a triangle that impinge upon a given pixel. We will think of pixels as being destinations or bins into which we place all of the fragments that cover the area of the pixel. As such, it is not a one-to-one mapping. A pixel may contain multiple fragments from different (or even the same) objects, or a pixel may not contain any fragments in the current view of the scene.

The remainder of this chapter will use this more specific definition of fragments. Figure 10.2 shows a triangle overlaid with pixel rectangle boundaries. Figure 10.3 shows

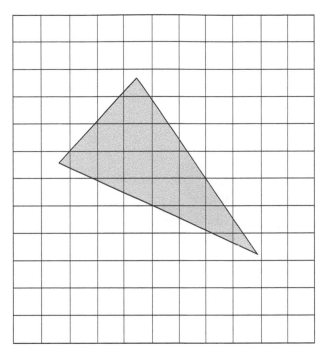

Figure 10.2. A screen-space triangle to be rasterized.

the same configuration broken into fragments, both complete and partial. The fragments are separated slightly in the figure to better demonstrate the shapes of the partial fragments.

10.4.2 Depth Complexity

The number of fragments in an entire scene can be much smaller or much greater than the number of pixels on the screen. If only a subset of the screen is covered by geometry, then there may be many pixels that contain no fragments from the scene. On the other hand, if a lot of triangles overlap one another in screen space, then many pixels on the screen may contain more than one fragment. The ratio of the number of fragments in the scene in a given frame to the number of pixels on the screen is called the *depth complexity* or *overdraw*, because this ratio represents how many full screens' worth of geometry comprises the scene. In general, scenes with a higher depth complexity are more expensive to rasterize. Note that this is an overall ratio for the whole view; a scene could have a depth complexity of two even if geometry only covers half of the screen. If, on average, the geometry on the half of the screen that is covered is four triangles deep, then the depth complexity would be two fragments per pixel amortized over the entire screen.

10.4.3 Converting Triangles to Fragments

Triangles are convex, no matter how they are projected by a projective transformation (in some cases, triangles may appear as a line or a point, but these are still convex objects). This is a very useful property, because it means that any triangle intersects a horizontal row of pixels (also called a *scan line*, for historical reasons having to do with CRT-based

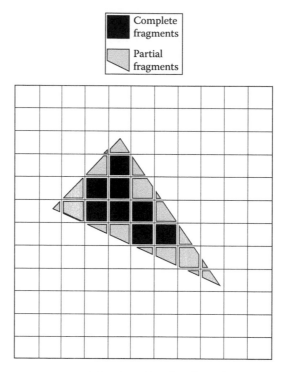

Figure 10.3. Fragments generated by the triangle. Complete fragments are dark gray; partial fragments are light gray.

television displays) in at most one contiguous segment. Thus, for any scan line that intersects a triangle, we can represent the intersection with only a minimum x value and a maximum x value, called a *span*. Thus, the representation of a triangle during rasterization consists of a set of spans, one per scan line, that the triangle intersects. Furthermore, the convexity of triangles also implies that the set of scan lines intersected by a triangle is contiguous in y; there is a minimum and a maximum y for a given triangle, which contains all of the nonempty spans. An example of the set of spans for a triangle is shown in Figure 10.4. The dark bands overlaid on the triangle represent the spans of adjacent fragments that will be used to draw the triangle.

The minimum y pixel coordinate for a triangle y_{min} is simply the minimum y value of the three triangle vertices. Similarly, the maximum y pixel coordinate y_{max} of the triangle is simply the maximum y value of the three vertices. Thus, a simple min/max computation among the three vertices defines the entire range of $(y_{max} - y_{min} + 1)$ spans that must be generated for a triangle.

The leftmost and rightmost fragments of each span may be partial fragments, since the edge of the triangle may not fall exactly on a pixel boundary. Also, the topmost and bottommost spans may contain partial fragments for the same reason. The remaining fragments for a triangle will be complete fragments.

Generating the spans themselves simply involves intersecting the horizontal scan line with the edges of the triangle. Owing to the convexity of the triangle, unless the scan line

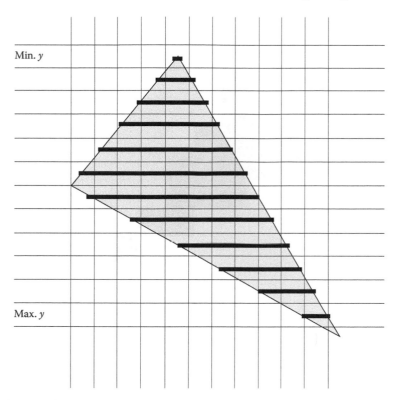

Figure 10.4. A triangle and its raster spans.

intersects a vertex, that scan line will intersect exactly two of the edges of the triangle: one to cross from outside the triangle into it, and one to leave again. These two intersection points will define the minimum and maximum x values of the span.

10.4.4 Handling Partial Fragments

Complete fragments always continue on to the next stage of the rasterization process. The fate of partial fragments, however, depends upon the particular rendering system. In more advanced systems, all partial fragments at a pixel are passed on as partial fragments, and the visibility and color of the final pixel may be influenced by all of them. However, simpler rasterization systems do not handle partial fragments, and must decide whenever a partial fragment is generated whether to drop the fragment or else promote it to a complete fragment. A common method for solving this is to keep partial fragments if and only if they contain the pixel's center point. This is sometimes called *point sampling* of geometry, as an entire fragment is generated or not generated based on a single-point sample within each pixel. Figure 10.5 shows the same triangle as in Figure 10.3, but with the partial fragments either dropped or promoted to complete fragments, based on whether the fragment contains the pixel's center point.

The behavior of such a graphics system when a triangle vertex or edge falls *exactly* on a pixel center is determined by a system-dependent *fill convention*, which ensures that if

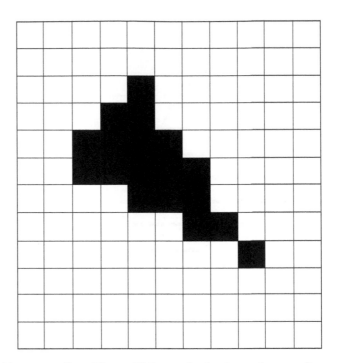

Figure 10.5. Fragments from Figure 10.3 rasterized using point sampling.

two triangles share a vertex or an edge, only one triangle will contribute a fragment to the pixel. This is very important, as without a well-defined fill convention, there may be holes (pixels where both triangles' partial fragments are dropped) or double-drawn pixels (where the partial fragments of both triangles are promoted to complete fragments) on the shared edges between triangles. Holes along a shared triangle edge allow the background color to show through what would otherwise be a continuous, opaque surface, making the surface appear to have cracks running through it. Double-drawn pixels along a shared edge result in more subtle artifacts, normally seen only when transparency or other forms of blending are used (see Section 10.8.1). For details on implementing point-sampled fill conventions, see Hecker's *Game Developer Magazine* article series [76].

10.5 Determining Visible Geometry

The overall goal in rendering geometry is to ensure that the final rendered images convincingly represent the given scene. At the highest level, this means that objects must appear to be correctly obscured by closer objects and must not be obscured by more distant objects. This process is known as *visible surface determination* (VSD), and there are numerous, very different ways of accomplishing it. The methods all involve comparing the depth of surfaces at one level of granularity or another and rendering them in such a way that the object of minimum depth (i.e., the closest object) at a given pixel is the one rendered to the screen.

Historically, numerous different methods have been used for VSD. Many of the early algorithms were based on clever sorting tricks, involving ordering the geometry back to front prior to rasterization. This was an expensive proposition normally computed per frame on the CPU. By far, the most common method in use today is a rasterization-based method: the depth buffer. Rasterizers were the first parts of the graphics pipeline to be accelerated with purpose-built hardware, meaning that a rasterizer-based visible surface determination system could achieve high performance. The depth buffer is also known as a *z-buffer*, which is actually a specific, special case of the more general depth buffering.

10.5.1 Depth Buffering

Source Code
Demo
DepthBuffering

Depth buffering is based on the concept that visibility should be output focused. In other words, since pixels are the final destination of our rendering pipeline, visibility should be computed on a per-pixel (or rather, per-fragment) basis. If the final color seen at each pixel is the color of the fragment with the minimum depth (of all fragments drawn to that pixel), the scene will appear to be drawn correctly. In other words, of all the fragments drawn to a pixel, the fragment with minimum depth should "win" the pixel and select that pixel's color. For the purposes of this discussion, we assume point-sampled geometry (i.e., there are no partial fragments).

Since common rasterization methods tend to render a triangle at a time, a given pixel may be redrawn several times over the course of a frame by fragments from different triangles. If we wish to avoid sorting the triangles by depth (and we do), then the fragment that should win a given pixel may not be the last one drawn to that pixel. We must have some method of storing the depth of the current nearest fragment at each pixel, along with the color of that fragment.

Having stored this information, we can compute a simple test each time a fragment is drawn to a pixel. If the new fragment's depth is closer than the currently stored depth value at that pixel, then the new fragment wins the pixel. The color of the new fragment is computed, and this new fragment color is written to the pixel. The fragment's depth value replaces the existing depth value for that pixel. If the new fragment has greater depth than the current fragment coloring the pixel, then the new fragment's color and depth are ignored, as the fragment represents a surface that is behind the closest known surface at the current pixel. In this case, we know that the new fragment will be obscured at that pixel, because we have already seen a fragment at that pixel that is closer than the newest fragment. Figure 10.6 represents the rendering of the fragments from two triangles to a small depth buffer. Note how the closer triangle's fragment always wins the pixel (the correct result), even if it is drawn first.

Because the method is per pixel and thus per fragment, the depth of each triangle is computed on a per-fragment granularity, and this value is used in the depth comparison. As a result of this finer subtriangle granularity, the depth buffer automatically handles triangle configurations that cannot be correctly displayed using per-triangle sorting. Geometry may be passed to the depth buffer in any order. The situation in which this random order can be problematic is when two fragments at a given pixel have equal depth. In this case, order will matter, depending on the exact comparison used to order depth (i.e., $<$ or $\leq$). However, such circumstances are problematic with almost any visible surface method.

There are several drawbacks to the depth buffer, although most of these are no longer significant on modern PCs or game consoles. One of the historical drawbacks of the depth buffering method is implied in the name of the method; it requires a *buffer* or array of depth

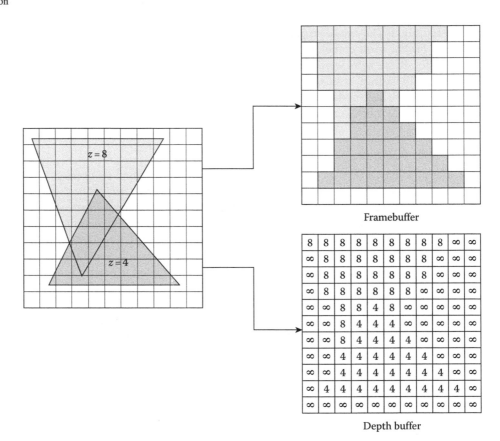

Figure 10.6. Two triangles rendered to a depth buffer.

values, one per pixel. This is a large block of memory, often requiring as much memory as the framebuffer itself. Also, just as the framebuffer must be cleared to the background color before each frame, the depth buffer must be cleared to the background depth, which is generally the maximum representable depth value. These issues can be significant on handheld and embedded 3D systems, where GPU memory is limited. Finally (and still relevant on PCs and consoles), the depth buffer requires the following work:

- Computation of a depth value for the fragment

- Lookup of the existing pixel depth in the depth buffer

- Comparison of these two values

- (For new "winner" fragments only) writing the new depth to the depth buffer

On many GPUs depth buffers are stored in a hierarchical, compressed data structure that allows for quick rejection tests using a large block of pixels. However, for a basic implementation, this is computed for *each fragment*. For most software rasterizers, this additional work per fragment can make depth buffering unsuitable for constant use. Fully software 3D systems have tended to use optimized geometry sorting wherever possible, reserving depth buffering for the few objects that truly require it. For example, early third-person-shooter game rendering engines put enormous work into specialized sorting of the environments, thereby avoiding any depth buffer testing for them. This left enough CPU cycles to render the animated characters, monsters, and small objects (which covered far fewer pixels than the scenery) using software depth buffering.

In addition, the depth buffer does not fix the potential performance problems of high-depth complexity scenes. We must still compute the depth of every fragment and compare it to the buffer. However, it can make overdraw less of an issue in some cases, since it is not necessary to compute or write the color of any fragment that fails the depth test. In fact, some applications will try to render their depth-buffered scenes in roughly near-to-far ordering (while still avoiding per-triangle, per-frame sorting on the CPU) so that the later geometry is likely to fail the depth buffer test and not require color computations.

Depth buffering is extremely popular in 3D applications that run on hardware-accelerated platforms, as it is easy to use and requires little application code or host CPU computation and produces quality images at high performance.

10.5.1.1 Computing Per-Fragment Depth Values

The first step in computing the visibility of a fragment using a depth buffer is to compute the depth value of the current fragment. As we shall see, z_{ndc} (which appeared to be a rather strange choice for z back in Chapter 7) will work quite well. However, the reason why z_{ndc} works well and the view-space value z_v does not is rather interesting.

In order to better understand the nature of how depth values change across a triangle in screen space, we must be able to map a point on the screen to the point in the triangle that projected to it. This is very similar to picking, and we will use several of the concepts we first discussed in Chapter 7. Owing to the nonlinear nature of perspective projection, we will find that our mapping from screen-space pixels to view-space points on a given triangle is somewhat complicated. We will follow this mapping through several smaller stages. For the discussion in this chapter, we'll be assuming that we are using the OpenGL-style matrices where we look down the $-z$ axis in view space.

A triangle in view space is simply a convex subset of a plane in view space. As a result, we can define the plane of a triangle in view space by a normal vector to the plane $\hat{\mathbf{n}} = (a, b, c)$ and a constant d, such that the points $P = (x_p, y_p, z_p)$ in the plane are those that satisfy

$$ax_p + by_p + cz_p + d = 0$$
$$(a, b, c) \bullet (x_p, y_p, z_p) + d = 0 \qquad (10.1)$$
$$\hat{\mathbf{n}} \bullet (x_p, y_p, z_p) + d = 0$$

Looking back at picking, a point in 2D normalized device coordinates (x_{ndc}, y_{ndc}) maps to the view-space ray $t\mathbf{r}$ such that

$$t\mathbf{r} = t(x_{ndc}, y_{ndc}, -d_{proj}), \quad t \geq 0$$

where d_{proj} is the projection distance (the distance from the view-space origin to the projection plane). Any point in view space that projects to the pixel at (x_{ndc}, y_{ndc}) must intersect this ray. Normally, we cannot invert the projection transformation, since a point on the screen maps to a ray in view space. However, by knowing the plane of the triangle, we can intersect the triangle with the view ray as follows. All points P in view space that fall in the plane of the triangle are given by Equation 10.1. In addition, we know that the point on the triangle that projects to (x_{ndc}, y_{ndc}) must be equal to $t\mathbf{r}$ for some t. Substituting the vector $t\mathbf{r}$ for the points (x_p, y_p, z_p) in Equation 10.1 and solving for t,

$$\hat{\mathbf{n}} \bullet (t\mathbf{r}) + d = 0$$
$$t(\hat{\mathbf{n}} \bullet \mathbf{r}) = -d$$
$$t = \frac{-d}{\hat{\mathbf{n}} \bullet \mathbf{r}}$$

From this value of t, we can compute the point along the projection ray $(x_v, y_v, z_v) = t\mathbf{r}$ that is the view-space point on the triangle that projects to (x_{ndc}, y_{ndc}). This amounts to finding

$$
\begin{aligned}
(x_v, y_v, z_v) &= t\mathbf{r} \\
&= t(x_{ndc}, y_{ndc}, -d_{proj}) \\
&= \frac{-d(x_{ndc}, y_{ndc}, -d_{proj})}{\hat{\mathbf{n}} \bullet \mathbf{r}} \\
&= \frac{-d(x_{ndc}, y_{ndc}, -d_{proj})}{\hat{\mathbf{n}} \bullet (x_{ndc}, y_{ndc}, -d_{proj})} \\
&= \frac{-d(x_{ndc}, y_{ndc}, -d_{proj})}{\hat{\mathbf{n}}_x x_{ndc} + \hat{\mathbf{n}}_y y_{ndc} - \hat{\mathbf{n}}_z d_{proj}}
\end{aligned}
\tag{10.2}
$$

However, we are only interested in z_v right now, since we are trying to compute a per-fragment value for depth buffering. The z_v component of Equation 10.2 is

$$z_v = \frac{d_{proj}\, d}{\hat{\mathbf{n}}_x x_{ndc} + \hat{\mathbf{n}}_y y_{ndc} - \hat{\mathbf{n}}_z d_{proj}} \tag{10.3}$$

As a quick check of a known result, note that in the special case of a triangle of constant depth $z_v = z_{const}$, we can substitute

$$\hat{\mathbf{n}} = (0, 0, 1)$$

and

$$d = -z_{const}$$

Substituted into Equation 10.3, this evaluates to the expected constant $z_v = z_{const}$:

$$
\begin{aligned}
z_v &= \frac{d_{proj}(-z_{const})}{0 \cdot x_{ndc} + 0 \cdot y_{ndc} - 1 \cdot d_{proj}} \\
&= \frac{-d_{proj}\, z_{const}}{-d_{proj}} \\
&= z_{const}
\end{aligned}
$$

As defined in Equation 10.3, z_v is an expensive value to compute per fragment (in the general, nonconstant depth case), because it is a fraction with a nonconstant denominator.

This would require a per-fragment division to compute z_v, which is more expensive than we would like. However, depth buffering requires only the ability to compare depth values against one another. If we are comparing z_v values, we know that they decrease with increasing depth (as the view direction is $-z$), giving a depth test of

$$z_v \geq DepthBuffer \rightarrow \text{New fragment is visible}$$
$$z_v < DepthBuffer \rightarrow \text{New fragment is not visible}$$

However, if we compute and store the reciprocal (the multiplicative inverse) of z_v, then a similar comparison still works in the same manner. If we use the reciprocal of all of the z_v values, we get

$$\frac{1}{z_v} \leq DepthBuffer \rightarrow \text{New fragment is visible}$$
$$\frac{1}{z_v} > DepthBuffer \rightarrow \text{New fragment is not visible}$$

If we reciprocate Equation 10.3, we can see that the per-fragment computation becomes simpler:

$$\frac{1}{z_v} = \frac{\hat{\mathbf{n}}_x x_{ndc} + \hat{\mathbf{n}}_y y_{ndc} - \hat{\mathbf{n}}_z d_{proj}}{d_{proj}\, d}$$
$$= \left(\frac{\hat{\mathbf{n}}_x}{d_{proj}\, d}\right) x_{ndc} + \left(\frac{\hat{\mathbf{n}}_y}{d_{proj}\, d}\right) y_{ndc} - \left(\frac{\hat{\mathbf{n}}_z d}{d_{proj}\, d}\right)$$

where all of the parenthesized terms are constant across a triangle. In fact, this forms an affine mapping of ND coordinates to $1/z_v$. Since we know that there is an affine mapping from pixel coordinates (x_s, y_s) to ND coordinates (x_{ndc}, y_{ndc}), we can compose these affine mappings into a single affine mapping from screen-space pixel coordinates to $1/z_v$. As a result, for a given projected triangle,

$$\frac{1}{z_v} = f x_s + g y_s + h \tag{10.4}$$

where f, g, and h are real values and are constant per triangle. We define the preceding mapping for a given triangle as

$$RecipZ(x_s, y_s) = f x_s + g y_s + h$$

An interesting property of $RecipZ(x_s, y_s)$ (or of any affine mapping, for that matter) can be seen from the derivation

$$RecipZ(x_s + 1, y_s) - RecipZ(x_s, y_s) = (f(x_s + 1) + g y_s + h) - (f x_s + g y_s + h)$$
$$= f(x_s + 1) - (f x_s)$$
$$= f$$

meaning that

$$RecipZ(x_s + 1, y_s) = RecipZ(x_s, y_s) + f$$

and similarly

$$RecipZ(x_s, y_s + 1) = RecipZ(x_s, y_s) + g$$

In other words, once we compute our *RecipZ* depth buffer value for any starting fragment, we can compute the depth buffer value of the next fragment in the span by simply adding f. Once we compute a base depth buffer value for a given span, as we step along the scan line, filling the span, all we need to do is add f to our current depth between each adjacent fragment (Figure 10.7). This makes the per-fragment computation of a depth value very fast indeed. And, once the base *RecipZ* of the first span is computed, we may add g to the previous span's base depth to compute the base depth of the next span. As we saw in Chapter 6, this technique is known as *forward differencing*, as we use the difference (or delta) between the value at a fragment and the value at the next fragment to step along, updating the current depth. This method will work for *any* value for which there is an affine mapping from screen space. We refer to such values as *affine in screen space*, or *screen affine*.

In fact, we can use the z_{ndc} value that we computed during projection as a replacement for *RecipZ*. In Chapter 7, on viewing and projection, we computed a z_{ndc} value that is equal to -1 at the near plane and 1 at the far plane and was of the form

$$z_{ndc} = \frac{a + bz_v}{z_v} = a\frac{1}{z_v} + b$$

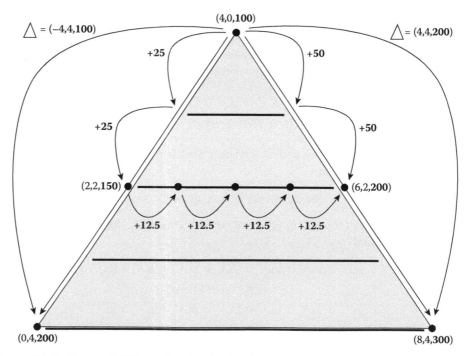

Figure 10.7. Forward differencing the depth value.

which is an affine mapping of *RecipZ*. As a result, we find that our existing value z_{ndc} is screen affine and is suitable for use as a depth buffer value. This is the special case of depth buffering we mentioned earlier, often called *z-buffering*, as it uses z_{ndc} directly.

10.5.1.2 Numerical Precision and z-Buffering

In practice, depth buffering in screen space has some numerical precision limitations that can lead to visual artifacts. As was mentioned earlier in the discussion of depth buffers, the order in which objects are drawn to a depth buffering system (at least in the case of opaque objects) is only an issue if the depth values of the two surfaces (two fragments) are equal at a given pixel. In theory, this is unlikely to happen unless the geometric objects in question are truly coplanar. However, because computer number representations do not have infinite precision (recall the discussion in Chapter 1), surfaces that are not coplanar can map to the same depth value. This can lead to objects being drawn in the wrong order.

If our depth values were mapped linearly into view space, then a 16-bit, fixed-point depth buffer would be able to correctly sort any objects whose surfaces differed in depth by about 1/60,000 of the difference between the near and far plane distances. This would seem to be more than enough for almost any application. For example, with a view distance of 1 km, this would be equal to about 1.5 cm of resolution. Moving to a higher-resolution depth buffer would make this value even smaller.

However, in the case of *z*-buffering, representable depth values are not evenly distributed in view space. In fact, the depth values stored to the buffer are, as we've seen, basically $1/z_v$, which is definitely not an even distribution of view space *z*. A graph of the depth buffer value over view space *z* is shown in Figure 10.8. This is a hyperbolic mapping of view space *z* into depth buffer values—notice how little the depth value changes with a change in *z* toward

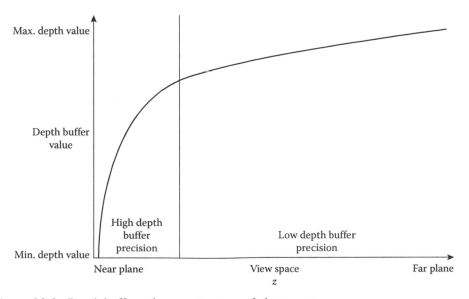

Figure 10.8. Depth buffer value as a function of view space *z*.

the far plane. Using a fixed-point value for this leads to very low precision in the distance, as large intervals of z map to the same fixed-point value of inverse z. In fact, a common estimate is that a z-buffer focuses 90 percent of its precision in the closest 10 percent of view space z. This means that the fragments of distant objects are often sorted incorrectly with respect to one another.

One method for handling precision issues that was popular in 3D hardware is known as the w-buffer. The w-buffer interpolates a screen-affine value for depth (often $1/w$) at a high precision, then computes the inverse of the interpolation at each pixel to produce a value that is linear in view space (i.e., $1/\frac{1}{w}$). It is this inverted value that is then stored in the depth buffer. By quantizing (dropping the extra precision used during interpolation) and storing a value that is linear in view space, the hyperbolic nature of the z-buffer can be avoided to some degree. However, as mentioned, w-buffers are no longer supported. They also have a problem in that the values stored are nonlinear in screen space per primitive, which doesn't work well with certain postprocessing algorithms.

Another solution uses floating-point depth buffers, which are available on most platforms. In combination with them, we flip the depth-buffered values such that the depth values map to 1.0 at the near plane and 0.0 at the far plane, and a comparison of $>$ or $\geq$ is used for the depth test [89]. By doing this, the natural precision characteristics of floating-point numbers end up counteracting *some* of the hyperbolic nature of z-buffer values. The increased dynamic range for floating-point values near 0 compensates for the loss of range in the z value at far distances, effectively acting much like the old w-buffer. That said, floating-point depth buffers can have other issues, overcorrecting and leaving the region of the scene closest to the camera with *too little* precision. This is particularly noticeable in rendered scenes because the geometry nearest the camera is the most obvious to the viewer.

Finally, the simplest way to avoid these issues is to maximize usage of the depth buffer by moving the near plane as far out as possible so that the accuracy close to the near plane is not wasted. All of these methods have scene- and application-dependent trade-offs.

10.5.2 Depth Buffering in Practice

Using depth buffering in most graphics systems requires additions to several points in rendering code:

- Creation of the depth buffer when the framebuffer is created

- Clearing the depth buffer each frame

- Enabling depth buffer testing and writing

The first step is to ensure that the rendering window or device is created with a depth buffer. This differs from API to API, with Iv automatically allocating a depth buffer in all cases. Having requested the creation of a depth buffer (and in most cases, it is just that—a *request* for a depth buffer, dependent upon hardware support), the buffer must be cleared at the start of each frame. The depth buffer is generally cleared using the same function as the framebuffer clear. Iv uses the `IvRenderer` function, `ClearBuffers`, but with a new argument, `kDepthClear`. While the depth buffer can be cleared independently of the framebuffer using

```
renderer->ClearBuffers(kDepthClear);
```

if you are clearing both buffers at the start of a frame, it can be faster on some systems to clear them both with a single call, which is done as follows in `Iv`:

```
renderer->ClearBuffers(kColorDepthClear);
```

To enable or disable depth testing we simply set the desired test mode using the `IvRenderer` function `SetDepthTest`. To disable testing, pass `kDisableDepth-Test`. To enable testing, pass one of the other test modes (e.g., `kLessDepthTest`). By default, depth testing is disabled, so the application should enable it explicitly prior to rendering. The most common depth testing modes are `kLessDepthTest` and `kLessEqualDepthTest`. The latter mode causes a new fragment to be used if its depth value is less than or equal to the current pixel depth.

The writing of depth values also can be enabled or disabled, independent of depth testing. As we shall see later in this chapter, it can be useful to enable depth testing while disabling depth buffer writing. A call to the `IvRenderer` function `SetDepthWrite` can enable or disable writing the *z*-buffer.

10.6 Computing Fragment Shader Inputs

The next stage in the rasterization pipeline is to compute the overall color (and possibly other shader output values) of a fragment by evaluating the currently active fragment shader for the current fragment. This in turn requires that the input values used by the shader be evaluated at the current fragment location. These inputs come in numerous forms, as discussed in the previous two chapters. Common sources include:

- Per-object uniform values set by the application

- Per-vertex attributes generated or passed through from the source vertices by the vertex shader

- Indirect per-fragment values, generally from textures

Note that as we saw in the lighting chapter (Chapter 9), numerous sources may exist for a given fragment. Each of them must be independently evaluated per fragment as a part of shader input source generation. Having computed the per-fragment source values, a final fragment color must be generated by running the fragment shader. Chapter 9 discussed various ways that per-fragment vertex color values, per-vertex lighting values, and texture colors can be combined in the fragment shader. The shader generates a final fragment color that is passed to the last stage of the rasterization pipeline, blending (which will be discussed later in this chapter).

The next few sections will discuss how shader source values are computed per fragment from the sources we have listed. While there are many possible methods that may be used, we will focus on methods that are fast to compute in screen space and are well suited to the scan line–centric nature of most rasterizer software and even some rasterizer hardware.

10.6.1 Uniform Values

As with all other stages in the pipeline, per-object values or colors are the easiest to rasterize. For each fragment, the constant uniform value may be passed down to the shader directly. No per-fragment evaluation or computation is required. As a result, uniform values can have minimal performance impact on the fragment shading process.

10.6.2 Per-Vertex Attributes

As we've discussed previously, per-vertex attributes are variables that are passed to the fragment shader from the last vertex processing stage (in our case, from the vertex shader). These values are defined only at the three vertices of each triangle, and thus must be interpolated to determine a value at each fragment center in the triangle. As we shall see, in the general case this can be an expensive operation to compute correctly for each of a triangle's fragments. However, we will first look at the special case of triangles of constant depth. The mapping in this case is not at all computationally expensive, making it a tempting approximation to use even when rendering triangles of nonconstant depth (especially in a software renderer).

10.6.2.1 Constant Depth Interpolation

To analyze the constant-depth case, we will determine the nature of the mapping of our constant-depth triangle from pixel space, through NDC space, into view space, through barycentric coordinates, and finally to the per-vertex source attributes. We start first with a special case of the mapping from pixel space to view space.

The overall projection equations derived in Chapter 7 (mapping from view space through NDC space to screen-space pixel coordinates) were all of the form

$$x_s = \frac{ax_v}{z_v} + b$$

$$y_s = \frac{cy_v}{z_v} + d$$

where both $a, c \neq 0$. If we assume that a triangle's vertices are all at the same depth (i.e., view space z_v is equal to a constant z_{const} for all points in the triangle), then the projection of a point in the triangle is

$$x_s = \frac{ax_v}{z_{const}} + b = \left(\frac{a}{z_{const}} \right) x_v + b = a'x_v + b$$

$$y_s = \frac{cy_v}{z_{const}} + d = \left(\frac{c}{z_{const}} \right) y_v + d = c'y_v + d$$

Note that $a, c \neq 0$ implies that $a', c' \neq 0$, so we can rewrite these such that

$$x_v = \frac{x_s - b}{a'}$$

$$y_v = \frac{y_s - d}{c'}$$

Thus, for triangles of constant depth z_{const},

- Projection forms an affine mapping from screen vertices to view-space vertices on the $z_v = z_{const}$ plane.

- Barycentric coordinates are an affine mapping of view-space vertices (as we saw in Chapter 2).

- Vertex attributes define an affine mapping from a barycentric coordinate to an attribute value (e.g., Gouraud shading, as seen in Chapter 8).

If we compose these affine mappings, we end up with an affine mapping from screen-space pixel coordinates to an attribute value. For example, we can write this affine mapping from pixel coordinates to color as

$$Color(x_s, y_s) = C_x x_s + C_y y_s + C_0$$

where C_x, C_y, and C_0 are all colors (each of which are possibly negative or greater than 1.0). For a derivation of the formula that maps the three screen-space pixel positions and corresponding trio of vertex colors to the three colors C_x, C_y, and C_0, see page 126 of Eberly [35]. From our earlier derivation of the properties of inverse z in screen space, we note that $Color(x_s, y_s)$ is screen affine *for triangles of constant z*:

$$
\begin{aligned}
Color(x_s + 1, y_s) - Color(x_s, y_s) &= (C_x(x_s + 1) + C_y y_s + C_0) - (C_x x_s + C_y y_s + C_0) \\
&= C_x(x_s + 1) - (C_x x_s) \\
&= C_x
\end{aligned}
$$

meaning that

$$Color(x_s + 1, y_s) = Color(x_s, y_s) + C_x$$

and similarly

$$Color(x_s, y_s + 1) = Color(x_s, y_s) + C_y$$

As with $1/z$, we can compute per-fragment values for per-vertex attributes for a constant-z triangle simply by computing forward differences of the color of a "base fragment" in the triangle.

10.6.2.2 Perspective-Correct Interpolation

When a triangle that does not have constant depth in camera space is projected using a perspective projection, the resulting mapping is not screen affine. From our discussion of depth buffer values, we can see that given a general (not necessarily constant-depth) triangle in view space, the mapping from NDC space to the view-space point on the triangle is of the form

$$x_v = \frac{-dx_{ndc}}{ax_{ndc} + by_{ndc} + c}$$

$$y_v = \frac{-dy_{ndc}}{ax_{ndc} + by_{ndc} + c}$$

$$z_v = \frac{d_{proj}d}{ax_{ndc} + by_{ndc} + c}$$

These are projective mappings, not affine mappings as we had in the constant-depth case. This means that the overall mapping from screen space to linearly interpolated per-vertex attributes is also projective. In order to correctly interpolate vertex attributes of a triangle in perspective, we must use this more complex projective mapping.

Most hardware rendering systems now interpolate all per-vertex attributes in a perspective-correct manner. However, this has not always been universal, and in the case of older software rendering systems running on lower-powered platforms, it was too expensive. If the per-vertex attributes being interpolated are colors from per-vertex lighting, such as in the case of Gouraud shading, it is possible to make an accuracy–speed trade-off. Keeping in mind that Gouraud shading is an approximation method in the first place, there is somewhat decreased justification for using the projective mapping on the basis of "correctness." Furthermore, Gouraud-shaded colors tend to interpolate so smoothly that it can be difficult to tell whether the interpolation is perspective correct or not. In fact, Heckbert and Moreton [75] mention that the New York Institute of Technology's offline renderer interpolated colors incorrectly in perspective for several years before anyone noticed! As a result, software graphics systems have often avoided the expensive, perspective-correct projective interpolation of Gouraud colors and have simply used the affine mapping and forward differencing.

That said, other per-vertex values, such as texture coordinates, are not as forgiving of issues in perspective-correct interpolation. The process of rasterizing a texture starts by interpolating the per-vertex texture coordinates to determine the correct value at each fragment. Actually, it is generally the texel coordinates (the texture coordinates multiplied by the texture image dimensions) that are interpolated in a rasterizer. This process is analogous to interpolating other per-vertex attributes. However, because texture coordinates are actually used somewhat differently than vertex colors in the fragment shader, we are not able to use the screen-affine approximation described previously. Texture coordinates require the correct perspective interpolation. The indirect nature of texture coordinates means that while the texture coordinates change smoothly and subtly over a triangle, the resulting texture color lookup does not.

The issue in the case of texture coordinates has to do with the properties of affine and projective transformations. Affine transformations map parallel lines to parallel lines, while projective transformations guarantee only to map straight lines to straight lines. Anyone who has ever looked down a long, straight road knows that the two lines that form the edges of the road appear to meet in the distance, even though they are parallel. Perspective, being a projective mapping, does not preserve parallel lines.

The classic example of the difference between affine and projective interpolations of texture coordinates is the checkerboard square, drawn in perspective. Figure 10.9 shows a checkered texture as an image, along with the image applied with wrapping to a square formed by two triangles (the two triangles are shown in outline, or *wire frame*). When the top is tilted away in perspective, note that if the texture is mapped using a projective mapping (Figure 10.10), the vertical lines converge into the distance as expected.

If the texture coordinates are interpolated using an affine mapping (Figure 10.11), we see two distinct visual artifacts. First, within each triangle, all of the parallel lines remain parallel, and the vertical lines do not converge the way we expect. Furthermore, note the obvious "kink" in the lines along the square's diagonal (the shared triangle edge). This might

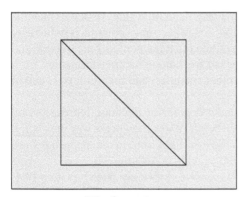

Wire-frame view   Textured view

Figure 10.9. Two textured triangles parallel to the view plane.

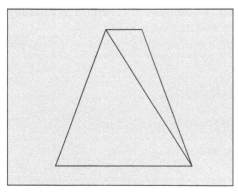

Wire-frame view   Textured view

Figure 10.10. Two textured triangles oblique to the view plane, drawn using a projective mapping.

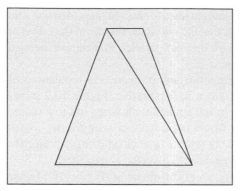 

Wire-frame view   Textured view

Figure 10.11. Two textured triangles oblique to the view plane, drawn using an affine mapping.

at first glance seem to be a bug in the interpolation code, but a little analysis shows that it is actually a basic property of an affine transformation. An affine transformation is defined by the three points of a triangle. As a result, having defined the three points of the triangle and their texture coordinates, there are no more degrees of freedom in the transformation. Each triangle defines its transform independent of the other triangles, and the result is a bend in what should be a set of lines across the square.

The projective transform, however, has additional degrees of freedom, represented by the depth values associated with each vertex. These depth values change the way the texture coordinate is interpolated across the triangle and allow straight lines in the mapped texture image to remain straight on-screen, even across the triangle boundaries.

Fortunately, the solution is relatively simple, if expensive. As we saw from Equation 10.4, $1/z_v$ can be computed using an affine mapping of screen-space positions. Since the texture coordinates themselves are affine mappings, we can compose them with the $1/z_v$ affine mapping, and find that u_{texel}/z_v and v_{texel}/z_v are affine mappings. Hence, these three quantities ($1/z_v$, u_{texel}/z_v, and v_{texel}/z_v) can be interpolated across the triangle using forward differencing. At each fragment, the final (u_{texel}, v_{texel}) values can be computed by inverting $1/z_v$ to get z_v, and then multiplying that by the interpolated u_{texel}/z_v and v_{texel}/z_v.

The downside of this projective mapping is that it requires the following operations per fragment for correct evaluation:

1. An affine forward difference operation to update $1/z_v$

2. An affine forward difference operation to update u_{texel}/z_v

3. An affine forward difference operation to update v_{texel}/z_v

4. A division to recover the perspective-correct z_v from $1/z_v$

5. A multiplication of u_{texel}/z_v by z_v to recover the perspective-correct u_{texel}

6. A multiplication of v_{texel}/z_v by z_v to recover the perspective-correct v_{texel}

Many PC games and some video game consoles in the 1990s used less expensive (and less correct) approximations of true perspective texturing. However, as mentioned, on modern hardware rasterization systems per-fragment perspective-correct texturing is simply assumed. Also, the fact that programmable fragment shaders can allow basically any per-vertex attribute to be used as a texture coordinate has further influenced hardware vendors in the move to interpolate all vertex attributes in correct perspective. In practice, many GPUs don't follow the procedure above for all vertex attributes. Rather, they compute a single set of barycentric coordinates using a perspective-correct interpolation, and then use those barycentric coordinates to map each attribute to the correct value.

10.6.3 Indirect Per-Fragment Values

Interpolation of per-vertex attributes is only one possible source of per-fragment values. Owing to the power of modern fragment shaders, texture coordinates and other values need not come directly from per-vertex attributes. A texture lookup may be evaluated from a set of coordinates generated *in the fragment shader itself* as the result of a computation involving other per-vertex attributes.

A texture coordinate generated in the fragment shader can even be the *result of an earlier texture lookup* in that same fragment shader. In this technique the texture image values in the first texture are not colors, but rather texture coordinates themselves. This is an extremely powerful technique called *indirect texturing*. The first texture lookup forms a table lookup, or indirection, that generates a new texture coordinate for the second texture lookup.

Indirect texturing is an example of a more general case of texturing in which evaluating a texture sample generates a "value" other than a color. Clearly, not all texture lookups are used as colors. However, for ease of understanding in the following discussion, we will assume that the texture image's values represent the most common case—colors.

10.7 Rasterizing Textures

The previous section described how to interpolate general per-vertex attributes for use in a fragment shader, and if these attributes were all we needed, we could simply evaluate or run the fragment shader and compute the fragment's color. However, if we have texture lookups, this is only the first step. Having computed or interpolated the texture coordinate for a given fragment, the texture coordinate must be mapped into the texture image itself to produce a color.

Some of the earliest shading languages required that the textures be addressed only by per-vertex attributes and, in some cases, actually computed the texture lookups before even invoking the fragment shader. However, as discussed above, modern shaders allow for texture coordinates to be computed in the fragment shader itself, perhaps even as the result of a texture lookup. Also, conditionals and varying loop iterations in a shader may cause texture lookups to be skipped for some fragments. As a result, we will consider the rasterization of textures to be a part of the fragment shader itself.

In fact, while the mathematical computations that are done inside of the fragment shader are interesting, the most (mathematically) complex part of an isolated fragment shader evaluation is the computation of the texture lookups. The texture lookups are, as we shall see, far more than merely grabbing and returning the closest texel to the fragment center.

The wide range of mappings of textures onto geometry and then geometry into fragments requires a much larger set of techniques to avoid glaring visual artifacts.

10.7.1 Texture Coordinate Review

We will be using a number of different forms of coordinates throughout our discussion of rasterizing textures. This includes the application-level, normalized, texel-independent *texture coordinates* (u, v), as well as the texture size-dependent texel coordinates (u_{texel}, v_{texel}), both of which are considered real values. We used these coordinates in our introduction to texturing.

A final form of texture coordinate is the *integer texel coordinate*, or *texel address*. These represent direct indexing into the texture image array. Unlike the other two forms of coordinates, these are (as the name implies) integral values. The mapping from texel coordinates to integer texel coordinates is not universal and is dependent upon the texture filtering mode, which will be discussed below.

10.7.2 Mapping a Coordinate to a Texel

When rasterizing textures, we will find that—due to the nature of perspective projection, the shape of geometric objects, and the way texture coordinates are generated—fragments will rarely correspond directly and exactly to texels in a one-to-one mapping. Any rasterizer that supports texturing needs to handle a wide range of texel-to-fragment mappings. In the initial discussions of texturing in Chapter 8, we noted that texel coordinates generally include precision (via either floating-point or fixed-point numbers) that is much more fine-grained than the per-texel values that would seem to be required. As we shall see, in several cases we will use this so-called subtexel precision to improve the quality of rendered images in a process known as *texture filtering*.

Texture filtering (in its numerous forms) performs the mapping from real-valued texel coordinates to final texture image values or colors through a mixture of texel coordinate mapping and combinations of the values of the resulting texel or texels. We will break down our discussion of texture filtering into two major cases: one in which a single texel maps to an area that is the size of multiple fragments (magnification), and one in which a number of texels map into an area covered by a single fragment (minification), as they are handled quite differently.

10.7.2.1 Magnifying a Texture

Our initial texturing discussion stated that one common method of mapping these subtexel precise coordinates to texture image colors was simply to select the texel containing the fragment center point and use its color directly. This method, called *nearest-neighbor texturing*, is very simple to compute. For any (u_{texel}, v_{texel}) texel coordinate, the integer texel coordinate (u_{int}, v_{int}) is the nearest integer texel center, computed via truncation:

$$(u_{int}, v_{int}) = (\lfloor u_{texel} \rfloor, \lfloor v_{texel} \rfloor)$$

Having computed this integer texel coordinate, we simply use a function *Image*(), which maps an integer texel coordinate to a texel value, to look up the value of the texel. The returned color is passed to the fragment shader for the current fragment. While this method

Figure 10.12. Nearest-neighbor magnification.

is easy and fast to compute, it has a significant drawback when the texture is mapped in such a way that a single texel covers more than 1 pixel. In such a case, the texture is said to be magnified, as a quadrilateral block of multiple fragments on the screen is entirely covered by a single texel in the texture, as can be seen in Figure 10.12.

With nearest-neighbor texturing, all (u_{texel}, v_{texel}) texel coordinates in the square

$$i_{int} \leq u_{texel} < i_{int} + 1$$
$$j_{int} \leq v_{texel} < j_{int} + 1$$

will map to the integer texel coordinates (i_{int}, j_{int}) and thus produce a constant fragment shader value. This is a square of height and width 1 in texel space, centered at the texel center. This results in obvious squares of constant color, which tends to draw attention to the fact that a low-resolution image has been mapped onto the surface. See Figure 10.12 for an example of a nearest-neighbor filtered texture used with a fragment shader that returns the texture as the final output color directly. In most cases, this blocky result is not the desired visual impression.

The problem lies with the fact that nearest-neighbor texturing represents the texture image as a piecewise constant function of (u, v). The resulting fragment shader attribute is constant across all fragments in a triangle until either u_{int} or v_{int} changes. Since the *floor* operation is discontinuous at integer values, this leads to sharp edges in the function represented by the texture over the surface of the triangle.

The common solution to the issue of discontinuous colors at texel boundaries is to treat the texture image values as specifying a different kind of function. Rather than creating a piecewise constant function from the discrete texture image values, we create a piecewise smooth color function. While there are many ways to create a smooth function from a set of discrete values, the most common method in rasterization hardware is linearly interpolating between the colors at each texel center in two dimensions. The method first computes the maximum texel center coordinate (u_{int}, v_{int}) that is less than (u_{texel}, v_{texel}), the texel coordinate (i.e., the floor of the texel coordinates minus a half-texel offset):

$$(u_{int}, v_{int}) = (\lfloor u_{texel} - 0.5 \rfloor, \lfloor v_{texel} - 0.5 \rfloor)$$

In other words, (u_{int}, v_{int}) defines the minimum (lower left in texture image space) corner of a square of four adjacent texel centers that "bound" the texel coordinate (Figure 10.13). Having found this square, we can also compute a *fractional texel coordinate* $0.0 \leq u_{frac}, v_{frac} < 1.0$ that defines the position of the texel coordinate within the 4-texel square.

$$(u_{frac}, v_{frac}) = (u_{texel} - u_{int} - 0.5, v_{texel} - v_{int} - 0.5)$$

We use *Image*() to look up the texel colors at the four corners of the square. For ease of notation, we define the following shorthand for the color of the texture at each of the four corners of the square (Figure 10.14):

$$C_{00} = Image(u_{int}, v_{int})$$
$$C_{10} = Image(u_{int} + 1, v_{int})$$
$$C_{01} = Image(u_{int}, v_{int} + 1)$$
$$C_{11} = Image(u_{int} + 1, v_{int} + 1)$$

Then, we define a smooth interpolation of the 4 texels surrounding the texel coordinate. We define the smooth mapping in two stages. First, we linearly interpolate

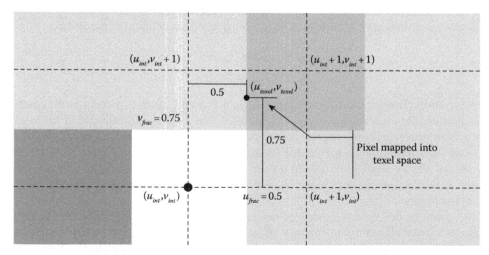

Figure 10.13. Finding the 4 texels that bound a pixel center and the fractional position of the pixel.

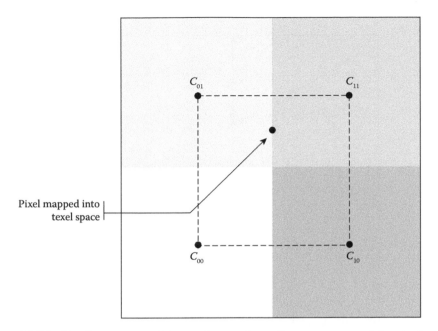

Figure 10.14. The four corners of the texel-space bounding square around the pixel center.

between the colors along the minimum-v edge of the square, based on the fractional u coordinate:

$$C_{MinV} = C_{00}(1 - u_{frac}) + C_{10}u_{frac}$$

and similarly along the maximum-v edge:

$$C_{MaxV} = C_{01}(1 - u_{frac}) + C_{11}u_{frac}$$

Finally, we linearly interpolate between these two values using the fractional v coordinate:

$$C_{Final} = C_{MinV}(1 - v_{frac}) + C_{MaxV}v_{frac}$$

See Figure 10.15 for a graphical representation of these two steps. Substituting these into a single, direct formula, we get

$$C_{Final} = C_{00}(1 - u_{frac})(1 - v_{frac}) + C_{10}u_{frac}(1 - v_{frac})$$
$$+ C_{01}(1 - u_{frac})v_{frac} + C_{11}u_{frac}v_{frac}$$

This is known as *bilinear texture filtering* because the interpolation involves linear interpolation in two dimensions to generate a smooth function from four neighboring texture image values. It is extremely popular in hardware 3D graphics systems. The fact that we interpolated along u first and then interpolated along v does not affect the result (other than by potential precision issues). A quick substitution shows that the results are the same either way. However, note that this is *not* an affine mapping. Affine mappings in 2D are uniquely

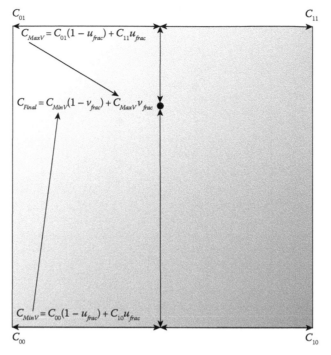

Figure 10.15. Bilinear filtering.

defined by three distinct points. The fourth source point of our bilinear texture mapping may not fit the mapping defined by the other three points.

Using bilinear filtering, the colors across the entire texture domain are continuous. An example of the visual difference between nearest-neighbor and bilinear filtering is shown in Figure 10.16. While bilinear filtering can greatly improve the image quality of magnified textures by reducing the visual blockiness, it will not add new detail to a texture. If a texture is magnified considerably (i.e., 1 texel maps to many pixels), the image will look blurry due to this lack of detail. The texture shown in Figure 10.16 is highly magnified, leading to obvious blockiness in the left image (a) and blurriness in the right image (b).

10.7.2.2 Texture Magnification in Practice

The `Iv` APIs use the `IvTexture` function `SetMagFiltering` to control texture magnification. `Iv` supports both bilinear filtering and nearest-neighbor selection. They are each set as follows:

```
IvTexture* texture;

// ...

{
    // Nearest-neighbor
    texture->SetMagFiltering(kNearestTexMagFilter);
```

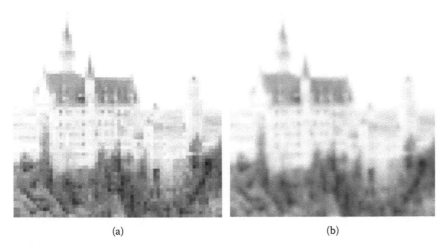

(a)   (b)

Figure 10.16. Extreme magnification of a texture using (a) nearest-neighbor filtering and (b) bilinear filtering.

```
// Bilinear interpolation
texture->SetMagFiltering(kBilerpTexMagFilter);

// ...
```

10.7.2.3 Minifying a Texture

Throughout the course of our discussions of rasterization so far, we have mainly referred to fragments by their centers—infinitesimal points located at the center of a square fragment (continuing to assume only complete fragments for now). However, fragments have nonzero area. This difference between the area of a fragment and the point sample representing it becomes very obvious in a common case of texturing.

As an example, imagine an object that is distant from the camera. Objects in a scene are generally textured at high detail. This is done to avoid the blurriness (such as the blurriness we saw in Figure 10.16b) that can occur when an object that is close to the camera has a low-resolution texture applied to it. As that same object and texture is moved into the distance (a common situation in a dynamic scene), this same, detailed texture will be mapped to smaller and smaller regions of the screen due to perspective scaling of the object. This is known as *minification* of a texture, as it is the inverse of magnification. This results in the same object and texture covering fewer and fewer fragments.

In an extreme (but actually quite common) case, the entire high-detail texture could be mapped in such a way that it maps to only a few fragments. Figure 10.17 provides such an example; in this case, note that if the object moves even slightly (even less than a pixel), the exact texel covering the fragment's center point can change drastically. In fact, such a point sample is almost random in the texture and can lead to the point-sampled color of the texture used for the fragment changing wildly from frame to frame as the object moves in tiny, subpixel amounts on the screen. This can lead to flickering over time, a distracting artifact in an animated, rendered image.

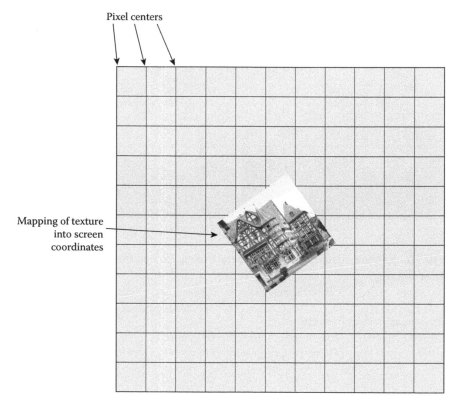

Figure 10.17. Extreme minification of a texture.

The problem lies in the fact that most of the texels in the texture have an almost equal "claim" to the fragment, as all of them are projected within the rectangular area of the fragment. The overall color of the fragment's texture sample *should* represent all of the texels that fall inside of it. One way of thinking of this is to map the square of a complete fragment on the projection plane onto the plane of the triangle, giving a (possibly skewed) quadrilateral, as seen in Figure 10.18. In order to evaluate the color of the texture for that fragment fairly, we need to compute a weighted average of the colors of all of the texels in this quadrilateral, based on the relative area of the quadrilateral covered by each texel. The more of the fragment that is covered by a given texel, the greater the contribution of that texel's color to the final color of the fragment's texture sample.

While an exact area-weighted-average method would give a correct fragment color and would avoid the issues seen with point sampling, in reality this is not an algorithm that is best suited for real-time rasterization. Depending on how the texture is mapped, a fragment could cover an almost unbounded number of texels. Finding and summing these texels on a per-fragment basis would require a potentially unbounded amount of per-fragment computation, which is well beyond the means of even hardware rasterization systems. A faster (preferably constant-time) method of approximating this texel averaging algorithm is required. For most modern graphics systems, a method known as *mipmapping* satisfies these requirements.

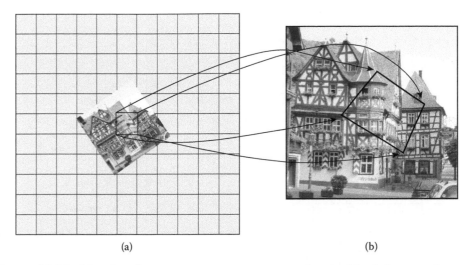

(a)   (b)

Figure 10.18. Mapping the square screen-space area of a pixel back into texel space: (a) screen space with pixel of interest highlighted and (b) texel-space back projection of pixel area.

10.7.3 Mipmapping

Source Code

Mipmapping

Mipmapping [157] is a texture-filtering method that avoids the per-fragment expense of computing the average of a large number of texels. It does so by precomputing and storing additional information with each texture, requiring some additional memory over standard texturing. This is a constant-time operation per texture sample and requires a fixed amount of extra storage per texture (in fact, it increases the number of texels that must be stored by approximately one-third). Mipmapping is a popular filtering algorithm in both hardware and software rasterizers and is relatively simple conceptually.

To understand the basic concept behind mipmapping, imagine a 2×2–texel texture. If we look at a case where the entire texture is mapped to a single fragment, we could replace the 2×2 texture with a 1×1 texture (a single color). One appropriate color would be the mean of the 4 texels in the 2×2 texture. We could use this new texture directly. If we precompute the 1×1–texel texture at load time of our application, we can simply choose between the two textures as needed (Figure 10.19).

When the given fragment maps in such a way that it only covers one of the 4 texels in the original 2×2–texel texture, we simply use a magnification method and the original 2×2 texture to determine the color. If the fragment covers the entire texture, we would use the 1×1 texture directly, again applying the magnification algorithm to it (although with a 1×1 texture, this is just the single texel color). The 1×1 texture adequately represents the overall color of the 2×2 texture in a single texel, but it does not include the detail of the original 2×2 texel texture. Each of these two versions of the texture has a useful feature that the other does not.

Mipmapping takes this method and generalizes it to any texture with power of two dimensions. For the purposes of this discussion, we assume that textures are square (the algorithm does not require this, as we shall see later in our discussion of mipmapping in practice). One approach to generating mipmap levels starts by taking the initial texture

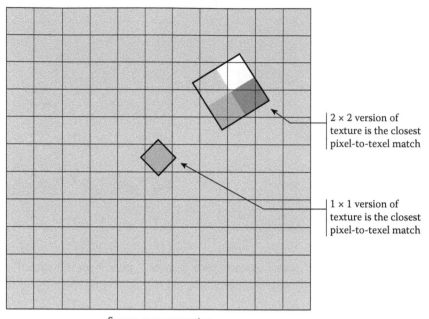

2 × 2 version of
texture is the closest
pixel-to-texel match

1 × 1 version of
texture is the closest
pixel-to-texel match

Screen-space geometry
(same mipmapped texture applied to both squares)

Figure 10.19. Choosing between two sizes of a texture.

image $Image_0$ (abbreviated I_0) of dimension $w_{texture} = h_{texture} = 2^L$ and generates a new version of the texture by averaging each square of four adjacent texels into a single texel. This generates a texture image $Image_1$ of size

$$\frac{1}{2}w_{texture} = \frac{1}{2}h_{texture} = 2^{L-1}$$

as follows:

$$Image_1(i,j) = \frac{I_0(2i,2j) + I_0(2i+1,2j) + I_0(2i,2j+1) + I_0(2i+1,2j+1)}{4}$$

where $0 \le i,j < \frac{1}{2}w_{texture}$. Each of the texels in $Image_1$ represents the overall color of a block of the corresponding 4 texels in $Image_0$ (Figure 10.20). Note that if we use the same original texture coordinates for both versions of the texture, $Image_1$ simply appears as a blurry version of $Image_0$ (with half the detail of $Image_0$). If a block of about four adjacent texels in $Image_0$ covers a fragment, then we can simply use $Image_1$ when texturing. But what about more extreme cases of minification? The algorithm can be continued recursively. For each image $Image_i$ whose dimensions are greater than 1, we can define $Image_{i+1}$, whose dimensions are half of $Image_i$, and average texels of $Image_i$ into $Image_{i+1}$. This generates an entire set of $L+1$ versions of the original texture, where the dimensions of $Image_i$ are equal to

$$\frac{w_{texture}}{2^i}$$

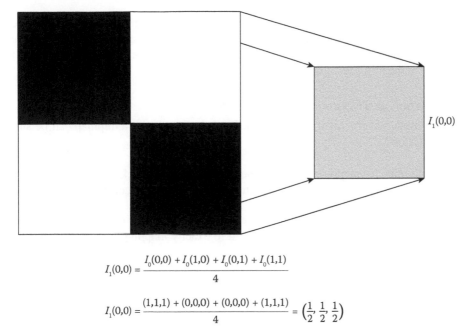

$$I_1(0,0) = \frac{I_0(0,0) + I_0(1,0) + I_0(0,1) + I_0(1,1)}{4}$$

$$I_1(0,0) = \frac{(1,1,1) + (0,0,0) + (0,0,0) + (1,1,1)}{4} = \left(\tfrac{1}{2}, \tfrac{1}{2}, \tfrac{1}{2}\right)$$

Figure 10.20. Texel-block-to-texel mapping between mipmap levels.

Figure 10.21. Mipmap-level size progression.

This forms a pyramid of images, each one-half the dimensions (and containing one-quarter the texels) of the previous image in the pyramid. Figure 10.21 provides an example of such a pyramid. We compute this pyramid for each texture in our scene once at load time or as an offline preprocess and store each entire pyramid in memory.

This simple method of computing the mipmap images is known as *box filtering* (as we are averaging a 2×2 "box" of texels into a single texel). Box filtering does not produce very high-quality mipmaps, as it tends to blur the images too much while still producing artifacts. Other, more complex methods are more often used to filter each mipmap level down to the next lower level. One good example is the Lanczos filter; see Turkowski [148] or

Wohlberg [159] for details of other image-filtering methods. One must also take care when generating each level to ensure the calculations are done on linear colors; if the original texture colors are in sRGB, convert to linear, do any computations, then convert back to sRGB to store the mipmap values.

10.7.3.1 Texturing a Fragment with a Mipmap

The most simple, general algorithm for texturing a fragment with a mipmap can be summarized as follows:

1. Determine the mapping of the fragment in screen space back into a quadrilateral in texture space by determining the texture coordinates at the corners of the fragment.

2. Having mapped the fragment square into a quadrilateral in texture space, select whichever mipmap level comes closest to exactly mapping the quadrilateral to a single texel.

3. Texture the fragment with the "best match" mipmap level selected in the previous step, using the desired magnification algorithm.

There are numerous common ways of determining the best-match mipmap level, and there are numerous methods of filtering this mipmap level into a final fragment texture value. We would like to avoid having to explicitly map the fragment's corners back into texture space, as this is expensive to compute. We can take advantage of information that other rasterization stages have already computed. As we saw in Sections 10.5.1 and 10.6.2, it is common in rasterization to compute the difference between the value of a fragment shader input (e.g., the texel coordinates) at a given fragment center and those of the fragment to the right and below the given fragment, for use in forward differencing. While we didn't explicitly say so, those differences can be expressed as derivatives. The listing that follows is designed to assign intuitive values to each of these four partial derivatives. For those unfamiliar with ∂, it is the symbol for a partial derivative, a basic concept of multivariable calculus. The ∂ operator represents how much one component of the output of a vector-valued function changes when you change one of the input components.

$$\frac{\partial u_{texel}}{\partial x_s} = \text{Change in } u_{texel} \text{ per horizontal pixel step}$$

$$\frac{\partial u_{texel}}{\partial y_s} = \text{Change in } u_{texel} \text{ per vertical pixel step}$$

$$\frac{\partial v_{texel}}{\partial x_s} = \text{Change in } v_{texel} \text{ per horizontal pixel step}$$

$$\frac{\partial v_{texel}}{\partial y_s} = \text{Change in } v_{texel} \text{ per vertical pixel step}$$

If a fragment maps to about 1 texel, then

$$\left(\frac{\partial u_{texel}}{\partial x_s}\right)^2 + \left(\frac{\partial v_{texel}}{\partial x_s}\right)^2 \approx 1, \text{ and } \left(\frac{\partial u_{texel}}{\partial y_s}\right)^2 + \left(\frac{\partial v_{texel}}{\partial y_s}\right)^2 \approx 1$$

In other words, even if the texture is rotated, if the fragment is about the same size as the texel mapped to it, then the overall change in texture coordinates over a single fragment has a length of about 1 texel. Note that all four of these differences are independent. These partials are dependent upon u_{texel} and v_{texel}, which are in turn dependent upon texture size. In fact, for each of these differentials, moving from $Image_i$ to $Image_{i+1}$ causes the differential to be halved. As we shall see, this is a useful property when computing mipmapping values.

A common formula that is used to turn these differentials into a metric of pixel–texel size ratio is described in Heckbert [74], which defines a formula for the *radius* of a pixel as mapped back into texture space. Note that this is actually the maximum of two radii, the radius of the pixel in u_{texel} and the radius in v_{texel}:

$$size = \max\left(\sqrt{\left(\frac{\partial u_{texel}}{\partial x_s}\right)^2 + \left(\frac{\partial v_{texel}}{\partial x_s}\right)^2}, \sqrt{\left(\frac{\partial u_{texel}}{\partial y_s}\right)^2 + \left(\frac{\partial v_{texel}}{\partial y_s}\right)^2}\right)$$

We can see (by substituting for the ∂) that this value is halved each time we move from $Image_i$ to $Image_{i+1}$ (as all of the ∂ values will halve). So, in order to find a mipmap level at which we map 1 texel to the complete fragment, we must compute the L such that

$$\frac{size}{2^L} \approx 1$$

where *size* is computed using the texel coordinates for $Image_0$. Solving for L,

$$L = \log_2 size$$

This value of L is the mipmap-level index we should use. Note that if we plug in partials that correspond to an exact one-to-one texture-to-screen mapping,

$$\frac{\partial u_{texel}}{\partial x_s} = 1, \frac{\partial v_{texel}}{\partial x_s} = 0, \frac{\partial u_{texel}}{\partial y_s} = 0, \frac{\partial v_{texel}}{\partial y_s} = 1$$

we get $size = 1$, which leads to $L = 0$, which corresponds to the original texture image as expected.

This gives us a closed-form method that can convert existing partials (used to interpolate the texture coordinates across a scan line) to a specific mipmap level L. The final formula is

$$L = \log_2\left(\max\left(\sqrt{\left(\frac{\partial u_{texel}}{\partial x_s}\right)^2 + \left(\frac{\partial v_{texel}}{\partial x_s}\right)^2}, \sqrt{\left(\frac{\partial u_{texel}}{\partial y_s}\right)^2 + \left(\frac{\partial v_{texel}}{\partial y_s}\right)^2}\right)\right)$$

$$= \log_2\left(\sqrt{\max\left(\left(\frac{\partial u_{texel}}{\partial x_s}\right)^2 + \left(\frac{\partial v_{texel}}{\partial x_s}\right)^2, \left(\frac{\partial u_{texel}}{\partial y_s}\right)^2 + \left(\frac{\partial v_{texel}}{\partial y_s}\right)^2\right)}\right)$$

$$= \frac{1}{2}\log_2\left(\max\left(\left(\frac{\partial u_{texel}}{\partial x_s}\right)^2 + \left(\frac{\partial v_{texel}}{\partial x_s}\right)^2, \left(\frac{\partial u_{texel}}{\partial y_s}\right)^2 + \left(\frac{\partial v_{texel}}{\partial y_s}\right)^2\right)\right)$$

Note that the value of L is real, not integer (we will discuss the methods of mapping this value into a discrete mipmap pyramid later). The preceding function is only one possible option for computing the mipmap level L. Graphics systems use numerous simplifications and approximations of this value (which is itself an approximation) or even other functions to determine the correct mipmap level. In fact, the particular approximations of L used by some hardware devices are so distinct that some experienced users of 3D hardware can actually recognize a particular piece of display hardware by looking at rendered, mipmapped images. Other pieces of 3D hardware allow the developer (or even the end user) to bias the L values used, as some users prefer "crisp" images (biasing L in the negative direction, selecting a larger, more detailed mipmap level and more texels per fragment), while others prefer "smooth" images (biasing L in the positive direction, tending toward a less detailed mipmap level and fewer texels per fragment). For a detailed derivation of one case of mipmap-level selection, see page 106 of Eberly [35].

Another method that has been used to lower the per-fragment expense of mipmapping is to select an L value, and thus a single mipmap level per triangle in each frame, and rasterize the entire triangle using that mipmap level. While this method does not require any per-fragment calculations of L, it can lead to serious visual artifacts, especially at the edges of triangles, where the mipmap level may change sharply. Software rasterizers that support mipmapping often use this method, known as *per-triangle mipmapping*.

Note that by its very nature, mipmapping tends to use smaller textures on distant objects. This means that mipmapping can actually *increase* performance for software rasterizers, because the smaller mipmap levels are more likely to fit in the processor's cache than the full-detail texture. This is true on most GPUs as well, due to the small, on-chip texture cache memories used to hold recently accessed texture image regions. As GPUs and software rasterizers are performance-bound to some degree by the memory bandwidth of reading textures, keeping a texture in the cache can decrease these bandwidth requirements significantly. Furthermore, if point sampling is used with a nonmipmapped texture, adjacent pixels may require reading widely separated parts of the texture. These large per-pixel strides through a texture can result in horrible cache behavior and can impede the performance of nonmipmapped rasterizers severely. These processor pipeline "stalls" or waits, caused by cache misses make the cost of computing mipmapping information (at least on a per-triangle basis) worthwhile, independent of the significant increase in visual quality.

10.7.3.2   Texture Filtering and Mipmaps

The methods described above work on the concept that there will be a single, "best" mipmap level for a given fragment. However, since each mipmap level is twice the size of the next mipmap level in each dimension, the closest mipmap level may not be an exact fragment-to-texel mapping. Rather than selecting a given mipmap level as the best, linear mipmap filtering uses a method similar to (bi)linear texture filtering. Basically, mipmap filtering uses the real-valued L to find the pair of adjacent mipmap levels that bound the given fragment-to-texel ratio, $\lfloor L \rfloor$ and $\lceil L \rceil$. The remaining fractional component $(L - \lfloor L \rfloor)$ is used to blend between texture colors found in the two mipmap levels.

Put together, there are now two independent filtering axes, each with two possible filtering modes, leading to four possible mipmap filtering modes as shown in Table 10.1. Of these methods, the most popular is *linear–bilinear*, which is also known as *trilinear interpolation* filtering, or *trilerp*, as it is the exact 3D analog to bilinear interpolation. It is the most expensive of these mipmap filtering operations, requiring the lookup of

Table 10.1. Mipmap Filtering Modes

Mipmap Filter	Texture Filter	Result
Nearest	Nearest	Select best mipmap level and then select closest texel from it
Nearest	Bilinear	Select best mipmap level and then interpolate 4 texels from it
Linear	Nearest	Select two bounding mipmap levels, select closest texel in each, and then interpolate between the 2 texels
Linear	Bilinear	Select two bounding mipmap levels, interpolate 4 texels from each, and then interpolate between the two results; also called trilerp

8 texels per fragment, as well as seven linear interpolations (three per each of the two mipmap levels, and one additional to interpolate between the levels), but it also produces the smoothest results. Filtering between mipmap levels also increases the amount of texture memory bandwidth used, as the two mipmap levels must be accessed per sample. Thus, multilevel mipmap filtering often counteracts the aforementioned performance benefits of mipmapping on hardware graphics devices.

A final, newer form of mipmap filtering is known as *anisotropic* filtering. The mipmap filtering methods discussed thus far implicitly assume that the pixel, when mapped into texture space, produces a quadrilateral that is fit quite closely by some circle—in other words, cases in which the quadrilateral in texture space is basically square. In practice, this is generally not the case. With polygons in extreme perspective, a complete fragment often maps to a very long, thin quadrilateral in texture space. The standard *isotropic* filtering modes can tend to look too blurry (having selected the mipmap level based on the long axis of the quad) or too sharp (having selected the mipmap level based on the short axis of the quad). Anisotropic texture filtering takes the aspect ratio of the texture-space quadrilateral into account when sampling the mipmap and is capable of filtering nonsquare regions in the mipmap to generate a result that accurately represents the tilted polygon's texturing.

10.7.3.3 Mipmapping in Practice

The default `CreateTexture` interface that we saw in Chapter 8 allocates only the base-level texture data, and no other mipmap levels. To create a texture with mipmaps, we use `CreateMipmappedTexture`, as follows:

```
IvResourceManager* manager;
// image data
const int numLevels = 5;
void* data[numLevels];
// ...

{
    IvTexture* texture = manager->CreateMipmappedTexture(kRGBA32TexFmt,
                            width, height,
                            data, numLevels, kImmutableUsage);
}
```

Table 10.2. Mipmap-Level Size Progression

Level	Width	Height
0	32	8
1	16	4
2	8	2
3	4	1
4	2	1
5	1	1

Notice that we are now passing in an array of image data—each array entry is a mipmap level. We must also specify the number of levels. We can also use the `IvTexture` functions `BeginLoadData` and `EndLoadData` if a mipmapped texture is created with dynamic or default usage. However, in the case of mipmaps, we use the argument to these functions, `unsigned int level` (previously defaulted to 0), which specifies the mipmap level. The mipmap level of the highest-resolution image is 0. Each subsequent level number (1, 2, 3, . . .) represents the mipmap pyramid image with half the dimensions of the previous level. Some APIs require that a "full" pyramid (all the way down to a 1 × 1 texel) be specified for mipmapping to work correctly. In practice, it is a good idea to provide a full pyramid for all mipmapped textures. The number of mipmap levels in a full pyramid is equal to

$$Levels = \log_2 \left(max(w_{texture}, h_{texture}) \right) + 1$$

Note that the number of mipmap levels is based on the larger dimension of the texture. Once a dimension falls to 1 texel, it stays at 1 texel while the larger dimension continues to decrease. So, for a 32 × 8–texel texture, the mipmap levels are shown in Table 10.2.

Note that the texels of the mipmap level images provided in the array passed to `CreateMipmappedTexture` or set in the array returned by `BeginLoadData` must be computed by the application. `Iv` simply accepts these images as the mipmap levels and uses them directly. Once all of the mipmap levels for a texture are specified, the texture may be used for mipmapped rendering by attaching the texture sampler as a shader uniform. An example of specifying an entire pyramid follows:

```
IvTexture* texture;
// ...

{
    for (unsigned int level = 0; level < texture->GetLevels(); level++) {
        unsigned int width = texture->GetWidth(level);
        unsigned int height = texture->GetHeight(level);
        IvTexColorRGBA* texels
            = (IvTexColorRGBA*)texture->BeginLoadData(level);

        for (unsigned int y = 0; y < height; y++) {
            for (unsigned int x = 0; x < width; x++) {
                IvTexColorRGBA& texel = texels[x + y * width];
```

```
                // Set the texel color, based on
                // filtering the previous level...
            }
        }
        texture->EndLoadData(level);
    }
```

In order to set the minification filter, the `IvTexture` function `SetMinFiltering` is used. `Iv` supports both nonmipmapped modes (bilinear filtering and nearest-neighbor selection) and all four mipmapped modes. The best-quality mipmapped mode (as described previously) is trilinear filtering, which is set using

```
IvTexture* texture;

// ...

    texture->SetMinFiltering(kBilerpMipmapLerpTexMinFilter);

    // ...
```

10.8   From Fragments to Pixels

Thus far, this chapter has discussed generating fragments, computing the per-fragment source values for a fragment's shader, and some details of the more complex aspects of evaluating a fragment's shader (texture lookups). However, the first few sections of the chapter outlined the real goal of all of this per-fragment work: to generate the final color of a pixel in a rendered view of a scene. Recall that pixels are the destination values that make up the rectangular gridded screen (or framebuffer). The pixels are "bins" into which we place pieces of surface that impinge upon the area of that pixel. Fragments represent these pixel-sized pieces of surface. In the end, we must take all of the fragments that fall into a given pixel's bin and convert them into a single color and depth for that pixel. We have made two important simplifying assumptions in the chapter so far:

- All fragments are opaque; that is, near fragments obscure more distant ones.

- All fragments are complete; that is, a fragment covers the entire pixel.

Put together, these two assumptions lead to an important overall simplification: the nearest fragment at a given pixel completely determines the color of that pixel. In such a system, all we need do is find the nearest fragment at a pixel, shade that fragment, and write the result to the framebuffer. This was a useful simplifying assumption when discussing visible surface determination and texturing. However, it limits the ability to represent some common types of surface materials. It can also cause jagged visual artifacts at the edges of objects on the screen. As a result, two additional features in modern graphics systems have removed these simplifying assumptions: pixel blending allows fragments to be partially transparent, and antialiasing handles pixels containing multiple partial fragments. We will close the chapter with a discussion of each.

10.8.1 Pixel Blending

Pixel blending is a per-fragment, nongeometric function that takes as its inputs the shaded color of the current fragment (which we will call C_{src}), the fragment's alpha value (which is properly a component of the fragment color, but which we will refer to as A_{src} for convenience), the current color of the pixel in the framebuffer (C_{dst}), and sometimes an existing alpha value in the framebuffer at that pixel (A_{dst}). These inputs, along with a pair of blending functions F_{src} and F_{dst}, define the resulting color (and potentially alpha value) that will be written to the pixel in the framebuffer, C_P. Note that C_P, once written, will become C_{dst} in later blending operations involving the same pixel. The general form of blending is

$$C_P = F_{src}C_{src} \oplus F_{dst}C_{dst}$$

where $\oplus$ can represent $+$, $-$, min (), or max (). We can also have a second pair of functions that affect only A. In most cases in games, however, we use the formula above with $\oplus$ set to $+$.

The alpha value, for both the source and the destination, is commonly interpreted as opacity (we'll see why when we discuss alpha blending, below). However, alpha can also be interpreted as fractional coverage of a pixel by a color—alpha in this case is the percentage of the pixel covered by the color. In general, this interpretation is not used much in games except possibly in interfaces; it is used more often when using pixel blending for 2D compositing or layering of images (also known as *alpha compositing*). We discuss pixel coverage in more detail in Section 10.8.2. For more information on alpha as coverage, see [123].

The simplest form of pixel blending is to disable blending entirely ("source replace" mode), in which the fragment replaces the existing pixel. This is equivalent to

$$F_{src} = 1$$
$$F_{dst} = 0$$
$$C_P = F_{src}C_{src} + F_{dst}C_{dst} = (1)C_{src} + (0)C_{dst} = C_{src}$$

Pixel blending is more commonly referred to by the name of its most common special case: alpha blending. Alpha blending involves using the source alpha value A_{src} as the opacity of the new fragment to linearly interpolate between C_{src} and C_{dst}:

$$F_{src} = A_{src}$$
$$F_{dst} = (1 - A_{src})$$
$$C_P = F_{src}C_{src} + F_{dst}C_{dst} = A_{src}C_{src} + (1 - A_{src})C_{dst}$$

Alpha blending requires C_{dst} as an operand. Because C_{dst} is the pixel color (generally stored in the framebuffer), alpha blending can (depending on the hardware) require that the pixel color be read from the framebuffer for each fragment blended. This increased memory bandwidth means that alpha blending can impact performance on some systems (in a manner analogous to depth buffering). In addition, alpha blending has several other properties that make its use somewhat challenging in practice.

Alpha blending is designed to compute a new pixel color based on the idea that the new fragment color represents a possibly translucent surface whose opacity is given by A_{src}.

Alpha blending only uses the fragment alpha value, not the alpha value of the destination pixel. The existing pixel color is assumed to represent the entirety of the existing scene at that pixel that is more distant than the current fragment, in front of which the translucent fragment is placed. For the following discussion, we will write alpha blending as

$$Blend(C_{src}, A_{src}, C_{dst}) = (A_{src})C_{src} + (1 - A_{src})C_{dst}$$

The result of multiple alpha blending operations is order dependent. Each alpha blending operation assumes that C_{dst} represents the final color of all objects more distant than the new fragment. If we view the blending of two possibly translucent fragments (C_1, A_1) and (C_2, A_2) onto a background color C_0 as a sequence of two blends, we can quickly see that, in general, changing the order of blending changes the result. For example, if we compare the two possible blending orders, set $A_1 = 1.0$, and expand the functions, we get

$$Blend(C_2, A_2, Blend(C_1, A_1, C_0)) \stackrel{?}{=} Blend(C_1, A_1, Blend(C_2, A_2, C_0))$$

$$Blend(C_2, A_2, Blend(C_1, 1.0, C_0)) \stackrel{?}{=} Blend(C_1, 1.0, Blend(C_2, A_2, C_0))$$

$$Blend(C_2, A_2, C_1) \stackrel{?}{=} C_1$$

These two sides are almost never equal; the two blending orders will generally produce different results. In most cases, alpha blending of two surfaces with a background color is order dependent.

10.8.1.1   Pixel Blending and Depth Buffering

In practice, this order dependence of alpha blending complicates depth buffering. The depth buffer is based on the assumption that a fragment at a given depth will completely obscure any fragment that is at a greater depth, which is only true for opaque objects. In the presence of alpha blending, we must compute the pixel color in a very specific ordering. We could depth sort all of the triangles, but as discussed above, this is expensive and has serious correctness issues with many datasets. Instead, one option is to use the assumption that for most scenes, the number of translucent triangles is much smaller than the number of opaque triangles. Given a set of triangles, one method of attempting to correctly compute the blended pixel color is as follows:

1. Collect the opaque triangles in the scene into a list, O.

2. Collect the translucent triangles in the scene into another list, T.

3. Render the triangles in O normally, using depth buffering.

4. Sort the triangles in T by depth into a far-to-near ordering.

5. Render the sorted list T with blending, using depth buffering.

This might seem to solve the problem. However, per-triangle depth sorting is still an expensive operation that has to be done on the host CPU in most cases. Also, per-triangle sorting cannot resolve all differences, as there are common configurations of triangles that cannot be correctly sorted back to front. Other methods have been suggested to avoid both of these issues. One such method is to depth sort at a per-object level to avoid gross-scale out-of-order blending, and then use more complex methods such as depth peeling [44], which uses advanced programmable shading and multiple renderings of objects to "peel away" closer surfaces (using the depth buffer) and generate depth-sorted colors. While quite complicated, the method works entirely on the GPU, and focuses on getting the closest layers correct, under the theory that deeper and deeper layers of transparency gain diminishing returns (as they contribute less and less to the final color).

Depth sorting or depth peeling of pixel-blended triangles can be avoided in some application-specific cases. Two other common pixel blending modes are commutative, and are thus order independent. The two blending modes are known as *add* and *modulate*. Additive blending creates the effect of "glowing" objects and is defined as follows:

$$F_{src} = 1$$
$$F_{dst} = 1$$
$$C_P = F_{src}C_{src} + F_{dst}C_{dst} = (1)C_{src} + (1)C_{dst} = C_{src} + C_{dst}$$

Modulate blending implements color filtering. It is defined as

$$F_{src} = 0$$
$$F_{dst} = C_{src}$$
$$C_P = F_{src}C_{src} + F_{dst}C_{dst} = (0)C_{src} + C_{src}C_{dst} = C_{src}C_{dst}$$

Note that neither of these effects involves the alpha component of the source or destination color. Both additive and modulate blending modes still require the opaque objects to be drawn first, followed by the blended objects, but neither requires the blended *objects* to be sorted into a depthwise ordering. As a result, these blending modes are very popular for particle system effects, in which many thousands of tiny, blended triangles are used to simulate smoke, steam, dust, or water. Other and more complex order-independent transparency solutions are possible; see [107] for one example.

Note that if depth buffering is used with unsorted, blended objects, the blended objects must be drawn with depth buffer *writing* disabled, or else any out-of-order (front-to-back) rendering of two blended objects will result in the more distant object not being drawn. In a sense, blended objects do not exist in the depth buffer, because they do not obscure other objects.

10.8.1.2   Premultiplied Alpha

In the above discussion, we have assumed that our colors are stored with straight RGB values and an associated alpha value. The RGB values represent our base color and the alpha represents its transparency or coverage; for example, (1, 0, 0, 1/2) represents semi-transparent red. However, as we mentioned in Chapter 8, a better format for blending is when we take the base *RGB* values and multiply them by the alpha value *A*, or

$$C' = CA$$

This is known as *premultiplied alpha*, and now our RGB values represent the contribution to the final result. The assumption here is that our alpha value lies in the range $[0, 1]$. If using 8-bit values for each channel you'll need to divide by 255 after the multiply. And when using sRGB be sure to apply the linear-to-sRGB conversion to the premultiplied color—don't apply it to the base color and then multiply by alpha.

Using this formulation, alpha blending becomes

$$Blend(C'_{src}, A_{src}, C'_{dst}) = (1)C'_{src} + (1 - A_{src})C'_{dst}$$

Since C'_{src} is $A_{src}C_{src}$ this doesn't appear to have gained us much, except perhaps saving a multiply. But premultiplied alpha has a number of significant advantages.

First of all, consider the case where we're using a texture with bilinear sampling. Suppose we have a solid $(A = 1)$ red texel right next to a transparent $(A = 0)$ green texel. Using standard colors, if we bilerp halfway between them, we get

$$\frac{1}{2}(1, 0, 0, 1) + \frac{1}{2}(0, 1, 0, 0) = \left(\frac{1}{2}, \frac{1}{2}, 0, \frac{1}{2}\right)$$

Despite the fact that we are interpolating from solid red to a completely transparent color, we have somehow ended up with a semitransparent color of yellow. If we use premultiplied alpha instead, the color values for the transparent color all become 0, so we have

$$\frac{1}{2}(1, 0, 0, 1) + \frac{1}{2}(0, 0, 0, 0) = \left(\frac{1}{2}, 0, 0, \frac{1}{2}\right)$$

which is the premultiplied alpha version of semitransparent red, which is what we want.

Even if we don't use a completely transparent color, we still get odd results, say

$$\frac{1}{2}(1, 0, 0, 1) + \frac{1}{2}\left(0, 1, 0, \frac{1}{4}\right) = \left(\frac{1}{2}, \frac{1}{2}, 0, \frac{5}{8}\right)$$

which is again much yellower than we'd expect. Using premultiplied alpha colors, we get

$$\frac{1}{2}(1, 0, 0, 1) + \frac{1}{2}\left(0, \frac{1}{4}, 0, \frac{1}{4}\right) = \left(\frac{1}{2}, \frac{1}{8}, 0, \frac{5}{8}\right)$$

and the contribution from the second color is appropriately reduced. So the first, and most important, advantage of premultiplied alpha is that it gives proper results from texture sampling.

The second advantage is that it allows us to expand our simple blending equation to a much larger set of blending operations, known as the Porter–Duff blending modes [123]. These are not often used in rendering 3D worlds, but they are very common in blending 2D elements, so knowing how to duplicate those effects for your in-game UI can be useful for certain effects. An example is the "Src In" operator, which replaces any contribution of the destination with a proportional fraction of the source, or

$$(A_{dst})C'_{src} + (0)C'_{dst}$$

This ends up becoming

$$(A_{dst})A_{src}C_{src}$$

which is not possible with the standard blending modes and straight colors.

The third advantage allows us to create transparent values with color values greater than 1, which is useful for creating lighting effects. For example, we could use a premultiplied alpha color of $(1, 1, 1, 1/2)$, which has the straight color equivalent of $(2, 2, 2, 1/2)$. The end result will have twice the contribution to the scene, creating an emissive effect. Forsyth [49] presents a great use for this in particle effects. Often we want particles to start out as additive (i.e., sparks) and then become alpha blended (soot and smoke). By using premultiplied alpha, we can create a single texture that has subareas representing the colors for the different particle types, and use it with a single blending mode. The spark particles can use a zero-alpha color with nonzero RGB values, or $(R, G, B, 0)$. The smoke particles can use a standard premultiplied alpha color, or (RA, GA, BA, A). By using these with the premultiplied alpha blend equation, the spark areas get added to the scene and the smoke areas will be alpha blended. For a single particle's lifetime, all we need to do is shift its texture coordinates to map from the different areas of the texture, and its visible representation will slowly change from spark to smoke. And all the particles will composite correctly with a single texture and a single draw call, without having to switch between additive and alpha blend modes.

Finally, when using straight colors, blending layers is not associative, so in order to get the correct result of blending layers A–D, you must blend A with B, then that result with C, then that result with D. However, there may be times when you want to blend B and C first, say for some sort of screen-based postprocessing effect. Premultiplied alpha allows you to do that and add the contributions A and D later. Note again that the order that is set for the blend operations must be the same—you can't blend C with B and expect to get the same result as blending B with C.

The only downside of premultiplied alpha is that when using it with 8-bit or smaller color channels, you end up losing precision. This is only an issue if you plan to use only the RGB colors in some operation that would scale them up. Otherwise, for best results, the use of premultiplied alpha is highly recommended.

10.8.1.3 Blending in Practice

Blending is enabled and controlled quite simply in most graphics systems, although there are many options beyond the modes supported by Iv. Setting the blending mode is done via the IvRenderer function SetBlendFunc, which sets F_{src}, F_{dst} and the operator $\oplus$ in a single function call. To use classic alpha blending (without premultiplied alpha), the function call is

```
renderer->SetBlendFunc(kSrcAlphaBlendFunc, kOneMinusSrcAlphaBlendFunc,
                       kAddBlendOp);
```

Additive mode is set using the call

```
renderer->SetBlendFunc(kOneBlendFunc, kOneBlendFunc, kAddBlendOp);
```

Modulate blending may be used via the call

```
renderer->SetBlendFunc(kZeroBlendFunc, kSrcColorBlendFunc, kAddBlendOp);
```

There are many more blending functions and operations available; see the source code for more details.

Recall that it is often useful to disable z-buffer writing while rendering blended objects. This is accomplished via depth buffer masking, described previously in the depth buffering section.

10.8.2 Antialiasing

The other simplifying rasterization assumption we made earlier, the idea that partial fragments are either ignored or "promoted" to complete fragments, induces its own set of issues. The idea of converting all fragments into all-or-nothing cases was to allow us to assume that a single fragment would "win" a pixel and determine its color. We used this assumption to reduce per-fragment computations to a single-point sample.

This is reasonable if we treat pixels as pure point samples, with no area. However, in our initial discussion of fragments and our detailed discussion of mipmapped textures, we saw that this is not the case; each pixel represents a rectangular region on the screen with a nonzero area. Because of this, more than one (partial) fragment may be visible inside of a pixel's rectangular region. Figure 10.22 provides an example of such a multifragment pixel.

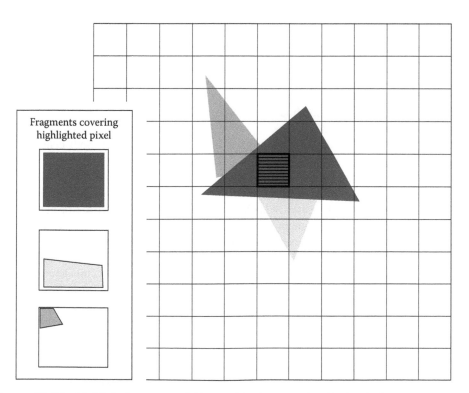

Figure 10.22. Multiple fragments falling inside the area of a single pixel.

Point samples of
partial fragments

Final on-screen color of pixels

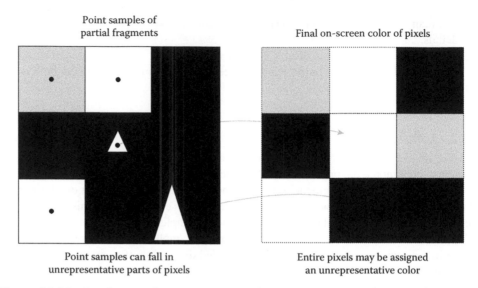

Point samples can fall in
unrepresentative parts of pixels

Entire pixels may be assigned
an unrepresentative color

Figure 10.23. A point sample may not accurately represent the overall color of a pixel.

Using the point-sampled methods discussed, we would select the color of a single fragment to represent the entire area of the pixel. However, as can be seen in Figure 10.23, this pixel center point sample may not represent the color of the pixel as a whole. In the figure, we see that most of the area of the pixel is dark gray, with only a very small square in the center being bright white. As a result, selecting a pixel color of bright white does not accurately represent the color of the pixel rectangle as a whole. Our perception of the color of the rectangle has to do with the relative areas of each color in the rectangle, something that the single-point sampling method cannot represent.

Figure 10.24 makes this even more apparent. In this situation, we see two examples of a pixel of interest (the center pixel in each 9-pixel 3×3 grid). In both center-pixel configurations (top and bottom of the left side of the figure), the vast majority of the surface area is dark gray. In each of the two cases, the center pixel contains a small, white fragment. The white fragments are the same size in both cases, but they are in slightly different positions relative to the center pixel in each of the two cases. In the first (top) example, the white fragment happens to contain the pixel center, while in the bottom case, the white fragment does not contain the pixel center. The right column shows the color that will be assigned to the center pixel in each case. Very different colors are assigned to these two pixels, even though their geometric configurations are almost identical. This demonstrates the fact that single-point sampling the color of a pixel can lead to somewhat arbitrary results. In fact, if we imagine that the white fragment were to move across the screen over time, an entire line of pixels would flash between white and gray as the white fragment moved through each pixel's center.

It is possible to determine a more accurate color for the 2 pixels in the figure. If the graphics system uses the relative areas of each fragment within the pixel's rectangle to weight the color of the pixel, the results will be much better. In Figure 10.25, we can see that the white fragment covers approximately 10 percent of the area of the pixel, leaving

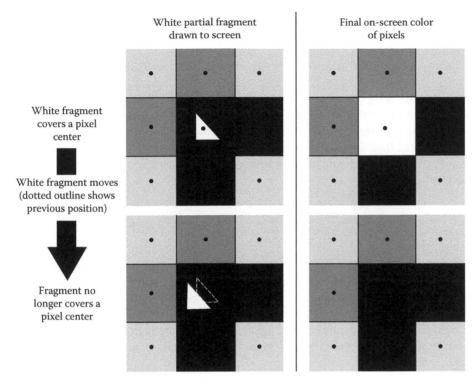

Figure 10.24. Subpixel motion causing a large change in point-sampled pixel color.

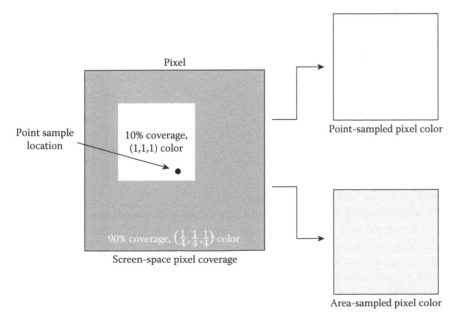

Figure 10.25. Area sampling of a pixel.

the other 90 percent as dark gray. Weighting the color by the relative areas, we get a pixel color of

$$C_{area} = 0.1 \times (1.0, 1.0, 1.0) + 0.9 \times (0.25, 0.25, 0.25) = (0.325, 0.325, 0.325)$$

Note that this computation is independent of *where* the white fragment falls within the pixel; only the size and color of the fragment matter. Such an area-based method avoids the point-sampling errors we have seen. This system can be extended to any number of different colored fragments within a given pixel. Given a pixel with area a_{pixel} and a set of n disjoint fragments, each with an area within the pixel a_i and a color C_i, the final color of the pixel is then

$$\frac{\sum_{i=1}^{n} a_i \times C_i}{a_{pixel}} = \sum_{i=1}^{n} \frac{a_i}{a_{pixel}} \times C_i = \sum_{i=1}^{n} F_i \times C_i$$

where F_i is the fraction of the pixel covered by the given fragment, or the fragment's coverage. This method is known as *area sampling*. In fact, this is really a special case of a more general definite integral. If we imagine that we have a screen-space function that represents the color of every position on the screen (independent of pixels or pixel centers) $C(x, y)$, then the color of a pixel defined as the region $l \leq x \leq r, t \leq y \leq b$ (the left, right, top, and bottom screen coordinates of the pixel), using this area-sampling method, is equivalent to

$$\frac{\int_t^b \int_l^r C(x, y) dx dy}{\int_t^b \int_l^r dx dy} = \frac{\int_t^b \int_l^r C(x, y) dx dy}{(b-t)(r-l)} = \frac{\int_t^b \int_l^r C(x, y) dx dy}{a_{pixel}} \qquad (10.5)$$

which is the integral of color over the pixel's area, divided by the total area of the pixel. The summation version of Equation 10.5 is a simplification of this more general integral, using the assumption that the pixel consists entirely of areas of piecewise constant color, namely, the fragments covering the pixel.

As a verification of this method, we shall assume that the pixel is entirely covered by a single, complete fragment with color $C(x, y) = C_T$, giving

$$\frac{\int_t^b \int_l^r C(x, y) dx dy}{a_{pixel}} = \frac{\int_t^b \int_l^r C_T dx dy}{a_{pixel}} = C_T \frac{\int_t^b \int_l^r dx dy}{a_{pixel}} = C_T \frac{a_{pixel}}{a_{pixel}} = C_T \qquad (10.6)$$

which is the color we would expect in this situation.

While area sampling does avoid completely missing or overemphasizing any single sample, it is not the only method used, nor is it the best at representing the realities of display devices (where the intensity of a physical pixel may not actually be constant within the pixel rectangle). The area sampling shown in Equation 10.5 implicitly weights all regions of the pixel equally, giving the center of the pixel weighting equal to that of the edges. As a result, it is often called *unweighted area sampling*. *Weighted area sampling*, on the other hand, adds a weighting function that can bias the importance of the colors in any region of the pixel as desired. If we simplify the original pixel boundaries and the functions associated with Equation 10.5 such that boundaries of the pixel are $0 \leq x, y \leq 1$, then Equation 10.5 becomes

$$\frac{\int_t^b \int_l^r C(x,y)dxdy}{\int_t^b \int_l^r dxdy} = \frac{\int_0^1 \int_0^1 C(x,y)dxdy}{1} \qquad (10.7)$$

Having simplified Equation 10.5 into Equation 10.7, we define a weighting function $W(x,y)$ that allows regions of the pixel to be weighted as desired:

$$\frac{\int_0^1 \int_0^1 W(x,y)C(x,y)dxdy}{\int_0^1 \int_0^1 W(x,y)dxdy} \qquad (10.8)$$

In this case, the denominator is designed to normalize according to the weighted area. A similar substitution to Equation 10.6 shows that constant colors across a pixel map to the given color. Note also that (unlike unweighted area sampling) the position of a primitive within the pixel now matters. From Equation 10.8, we can see that unweighted area sampling is simply a special case of weighted area sampling. With unweighted area sampling, $W(x,y) = 1$, giving

$$\frac{\int_0^1 \int_0^1 W(x,y)C(x,y)dxdy}{\int_0^1 \int_0^1 W(x,y)dxdy}$$
$$= \frac{\int_0^1 \int_0^1 (1)C(x,y)dxdy}{\int_0^1 \int_0^1 (1)dxdy}$$
$$= \frac{\int_0^1 \int_0^1 C(x,y)dxdy}{\int_0^1 \int_0^1 dxdy}$$
$$= \frac{\int_0^1 \int_0^1 C(x,y)dxdy}{1}$$

A full discussion of weighted area sampling, the theory behind it, and numerous common weighting functions is given in Hughes et al. [82]. For those desiring more depth, Glassner [54] and Wohlberg [159] detail a wide range of sampling theory.

10.8.2.1 Supersampled Antialiasing

The methods so far discussed show theoretical ways for computing area-based pixel colors. These methods require that pixel-coverage values be computed per fragment. Computing analytical (exact) pixel-coverage values for triangles can be complicated and expensive. In practice, the pure area-based methods do not lead directly to simple, fast hardware antialiasing implementations.

The conceptually simplest, most popular antialiasing method is known as oversampling, supersampling, or supersampled antialiasing (SSAA). In SSAA, area-based sampling is approximated by point sampling the scene at more than one point per pixel. In SSAA, fragments are generated not at the per-pixel level, but at the per-sample level. In a sense, SSAA is conceptually little more than rendering the entire scene to a larger (higher-resolution) framebuffer, and then filtering blocks of pixels in the higher-resolution framebuffer down to the resolution of the final framebuffer. For example, the supersampled framebuffer may be N times larger in width and height than the final destination framebuffer on-screen. In this case, every $N \times N$ block of pixels in the supersampled framebuffer will be filtered down to a single pixel in the on-screen framebuffer.

The supersamples are combined into a single pixel color via a weighted (or in some cases unweighted) average. The positions and weights used with weighted area versions of these sampling patterns differ by manufacturer; common examples of sample positions are shown in Figure 10.26. Note that the number of supersamples per pixel varies from as few as 2 to as many as 16. M-sample SSAA represents a pixel as an M-element piecewise constant function. Partial fragments will only cover some of the point samples in a pixel, and will thus have reduced weighting in the resulting pixel.

Some of the $N \times N$ sample grids also have rotated versions. The reason for this is that horizontal and vertical lines happen with high frequency and are also correlated with the pixel layout itself. By rotating the samples at the correct angle, all N^2 samples are located at distinct horizontal and vertical positions. Thus, a horizontal or vertical edge moving slowly from left to right or top to bottom through a pixel will intersect each sample individually and will thus have a coverage value that changes in $1/N^2$ increments. With screen-aligned $N \times N$ sample patterns, the same moving horizontal and vertical edges would intersect entire rows or columns of samples at once, leading to coverage values that changed in $1/N$ increments. The rotated patterns can take better advantage of the number of available samples.

M-sample SSAA generates M times (as mentioned above, generally 2–16 times) as many fragments per pixel. Each such (smaller) fragment has its own color computed by evaluating per-vertex attributes, texture values, and the fragment shader itself as many as M times more frequently per frame than normal rendering. This per-sample full rendering

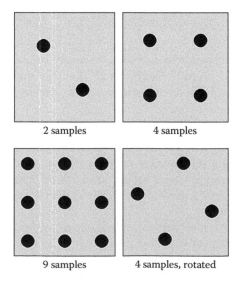

Figure 10.26. Common sample-point distributions for multisample-based antialiasing.

pipeline is very powerful, since each sample truly represents the color of the geometry at that sample. It is also extremely expensive, requiring the entire rasterization pipeline to be invoked per sample, and thus increasing rasterization overhead by 2–16 times. For even powerful 3D hardware systems, this can simply be too expensive.

10.8.2.2   Multisampled Antialiasing

The most expensive aspect of supersampled antialiasing is the creation of individual fragments per sample and the resulting texturing and fragment shading per sample. Another form of antialiasing recognizes the fact that the most likely causes of aliasing in 3D rendering are partial fragments at the edges of objects, where pixels will contain multiple partial fragments from different objects, often with very different colors. Multisampled antialiasing (MSAA) attempts to fix this issue without raising the cost of rendering as much as does SSAA. MSAA works like normal rendering in that it generates fragments (including partial fragments) at the final pixel size. It only evaluates the fragment shader once per fragment, so the number of fragment shader invocations is reduced significantly when compared to SSAA.

The information that MSAA does add is per-sample fragment coverage. When a fragment is rendered, its color is evaluated once, but then that same color is stored for each visible sample that the fragment covers. The existing color at a sample (from an earlier fragment) may be replaced with the new fragment's color. But this is done at a per-sample level. At the end of the frame, a "resolve" is still needed to compute the final color of the pixel from the multiple samples. However, only a coverage value (a simple geometric operation) and possibly a depth value is computed per sample, per fragment. The expensive steps of computing a fragment color are still done once per fragment. This greatly reduces the expense of MSAA when compared to SSAA.

There are two subtleties to MSAA worth mentioning. First, since MSAA is coverage based, no antialiasing is computed on complete fragments. The complete fragment is rendered as if no antialiasing was used. SSAA, on the other hand, antialiases every pixel by invoking the fragment's shader several times per pixel. A key observation is that perhaps the most likely item to cause aliasing in single-sampled complete fragments is texturing (since it is the highest-frequency value across a fragment). Texturing already has a form of antialiasing applied: mipmapping. Thus, this is not a problem for MSAA in most cases.

The other issue is the question of selecting the position in the pixel at which to evaluate a shader on a partial fragment. Normally, we evaluate the fragment shader at the pixel center. However, a partial fragment may not even cover the pixel center. If we sample the fragment shader at the pixel center, we actually will be extrapolating the vertex attributes beyond the intended values. This is particularly noticeable with textures, as we will read the texture at a location that may not have been mapped in the triangle. This can lead to glaring visual artifacts. The solution in most 3D MSAA hardware is to select the centroid of the samples covered by a fragment. Since fragments are convex, the centroid will always fall inside of the fragment. This does add some complexity to the system, but the number of possible configurations of a fragment that does not include the pixel center is limited. The convexity and the fact that the central sample is not touched means that there are a very limited set of covered-sample configurations possible. The set of possible positions can be precalculated before the hardware is even built. However, centroid sampling must be requested on a per-attribute basis. Otherwise, the hardware will default to using pixel center sampling.

10.8.3   Antialiasing in Practice

For most rendering APIs, the most important step in using MSAA is to create a framebuffer for rendering that is compatible with the technique. Whereas depth buffering required an additional buffer alongside the framebuffer to store the depth values, MSAA requires a special framebuffer format that includes the additional color, depth, and coverage values per sample within each pixel. Different rendering APIs and even different rendering hardware on the same APIs often have different methods for explicitly requesting MSAA-compatible framebuffers. Some rendering APIs allow the application to specify the number and event layout of samples in the pixel format, while others simply use a single flag for enabling a single (unspecified) level of MSAA.

Finally, some rendering APIs can require special flags or restrictions when presenting an MSAA framebuffer to the screen. For example, sometimes MSAA framebuffers must be presented to the screen using a special mode that marks the framebuffer's contents as invalid after presentation. This takes into account the fact that the framebuffer must be "resolved" from its multisample-per-pixel format into a single color per pixel during presentation, destroying the multisample information in the process.

10.9   Chapter Summary

This chapter concludes the discussion of the rendering pipeline. Rasterization provides us with some of the lowest-level yet most mathematically interesting concepts in the entire pipeline. We have discussed the connections between mathematical concepts, such as projective transforms, and rendering methods, such as perspective-correct texturing. In addition, we addressed issues of mathematical precision in our discussion of the depth buffer. Finally, the concept of point sampling versus area sampling appeared twice, relating to both mipmapping and antialiasing. Whether it is implemented in hardware, software, or a mixture of the two, the entire graphics pipeline is ultimately designed only to feed a rasterizer, making the rasterizer one of the most important, yet least understood, pieces of rendering technology.

Thanks to the availability of high-quality, low-cost 3D hardware on a wide range of platforms, the percentage of readers who will ever have to implement their own rasterizer is now vanishingly small. However, an understanding of how rasterizers function is important even to those who will never need to write one. For example, even a basic practical understanding of the depth buffering system can help a programmer build a scene that avoids visual artifacts during visible surface determination. Understanding the inner workings of rasterizers can help a 3D programmer quickly debug problems in the geometry pipeline. Finally, this knowledge can guide programmers to better optimize their geometry pipeline, "feeding" their rasterizer with high-performance datasets.

⑪ Random Numbers

11.1 Introduction

Now that we've spent some time in the deterministic worlds of pure mathematics, graphics, and interpolation, it's time to look at some techniques that can make our world look less structured and more organic. We'll begin in this chapter by considering randomness and generating random numbers in the computer.

So why do we need random numbers in games? We can break down our needs into a few categories: the basic randomness needed for games of chance, as in simulating cards and dice; randomness for generating behavior for intelligent agents, such as enemies and nonplayer allies; turbulence and distortion for procedural textures; and randomly spreading particles, such as explosions and gunshots, in particle systems.

In this chapter we'll begin by covering some basic concepts in probability and statistics that will help us build our random processes. We'll then move to techniques for measuring random data and then basic algorithms for generating random numbers. Finally, we'll close by looking at some applications of our random number generators (RNGs).

11.2 Probability

Probability theory is the mathematics of measuring the likelihood of unpredicable behavior. It was originally applied to games of chance such as dice and cards. In fact, Blaise Pascal and Pierre de Fermat worked out the basics of probability to solve a problem posed by a famous gambler, the Chevalier de Mere. His question was, which is more likely, rolling at least one 6 in 4 throws of a single die, or at least one double 6 in 24 throws of a pair of dice? (We'll answer this question at the end of the next section.)

These days probability can be used to predict the likelihood of other events such as the weather (i.e., the chance of rain is 60 percent) and even human behavior. In the following section we will summarize some elements of probability, enough for simple applications.

11.2.1 Basic Probability

The basis of probability is the *random experiment*, which is an experiment with a nondetermined outcome that can be observed and reobserved under the same conditions. Each time we run this experiment we call it a *random trial*, or just a trial. We call any of the particular outcomes of this experiment an *elementary outcome*, and the set of all elementary outcomes the *sample space*. Often we are interested in a particular set of outcomes, which we call the *favorable outcomes* or an *event*.

We define the *probability* of a particular event as a real number from 0 to 1, where 0 represents that the event will never happen, and 1 represents that the event will always happen. This value can also be represented as a percentage, from 0 to 100 percent. For a particular outcome ω_i, we can represent the probability as $P(\omega_i)$.

The classical computation of probability assumes that all outcomes are equally likely. In this case, the probability of an event is the number of favorable outcomes for that event divided by the total number of elementary outcomes. As an example, suppose we roll a fair (i.e., not loaded) six-sided die. This is our random experiment. The sample space Ω for our experiment is all the possible values on each side, so $\Omega = \{1, 2, 3, 4, 5, 6\}$. The event we're interested in is, how likely is it for a 3 or 4 to come up? Or, what is P(3 or 4)? The number of favorable outcomes is two (either a 3 or a 4) and the number of all elementary outcomes is six, so the probability is 2 over 6, or $1/3$.

Another classic example is drawing a colored ball out of a jar. If we have 3 red balls, 2 blue balls, and 5 yellow balls, the probability of drawing a red ball out is $3/(3+2+5)$, or $3/10$, the probability of drawing a blue ball is $2/10 = 1/5$, and the probability of drawing a yellow ball is $5/10 = 1/2$.

However, it's not always the case that each outcome is equally likely (life is not necessarily fair). Because of this, there are two additional approaches to computing probabilities. The first is the frequentist approach, which has as its central tenet that if we perform a large number of trials, the number of observed favorable outcomes over the number of trials will approach the probability of the event. This also is known as the *law of large numbers*. The second is the Bayesian approach, which is more philosophical and is based on the fact that many events are not in practice repeatable. The probability of such events is based on a personal assessment of likelihood. Both have their applications, but for the purposes of this chapter, we will be focusing on the frequentist definition.

As an example of the law of large numbers, look at Figure 11.1. Figure 11.1a shows the result of a computer simulation of rolling a fair die 1,000 times. Each column represents the number of times each side came up. As we can see, while the columns are not equal, they are pretty close. For example, if we divide the number of 3s generated by the total number of rolls, we get 0.164—pretty close to the actual answer of $1/6$.

Figure 11.1b, on the other hand, shows the result of rolling a loaded die, where 6s come up more often. As we'd expect, the 6 column is much higher than the rest, and dividing the number of 6s generated by the total gives us 0.286—not at all close to the expected probability. Clearly something nefarious is going on. While we never can be exact about

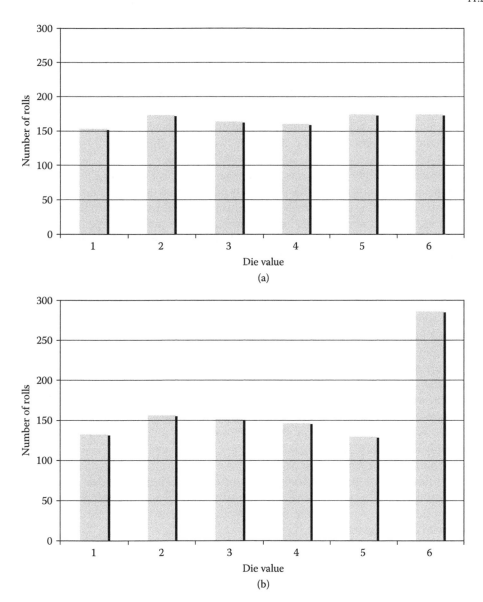

Figure 11.1. (a) Simulation results for rolling a fair die 1,000 times and (b) simulation results for rolling a loaded die 1,000 times.

whether observed results match expected behavior (this is probability, after all), we'll talk later about a way to measure whether our observed outcomes match the expected outcomes.

We often consider the probability of more than one trial at a time. If performing the experiment has no effect on the probability of future trials, we call these *independent events* or *independent trials*. For example, each instance of rolling a die is an independent trial. Drawing a ball out of the jar and not putting it back is not; future trials are affected by what happens. For example, if we draw a red ball out of the jar and don't replace it, the probability of drawing another red ball is 2/9, as there are now only two red balls and nine balls total in the jar. These are known as *dependent events* or *dependent trials*.

A few algebraic rules for probability may prove useful for game development. First of all, the probability of an event *not* happening is 1 minus the probability of the event, or $P(\text{not } E) = 1 - P(E)$. For example, the probability of not rolling a 6 on a fair die is $1 - 1/6 = 5/6$.

Secondly, the probability of two independent events E and F occurring is $P(E) \cdot P(F)$. So, for example, the probability of rolling a die twice and rolling a 1 or 2 on the first roll and a 3, 4, or 6 on the second roll is $2/6 \cdot 3/6 = 6/36 = 1/6$.

Finally, the probability of one event E *or* another event F is $P(E) + P(F) - P(E \text{ and } F)$. An example of this is considering the probability of rolling an odd number or a 1 on a die. The probability of rolling an odd number *and* a 1 is just the probability of rolling a 1, or 1/6 (we can't use the multiplicative rule here because the events are not independent). So, the result is $3/6 + 1/6 - 1/6 = 1/2$—just the probability of rolling an odd number.

With these rules we can answer Chevalier de Mere's question. The first part of the question is, what is the probability of rolling at least one 6 in four throws of a single die? We'll represent this as $P(E)$. It's a little easier to turn this around and ask, what is the probability of *not* rolling a 6 in four throws of one die? We can call the event of not throwing a 6 on the ith roll A_i, and the probability of this event is $P(A_i)$. Then the probability of all 4 is $P(A_1 \text{ and } A_2 \text{ and } A_3 \text{ and } A_4)$. As each roll is an independent event, we can just multiply the four probabilities together to get a probability of $(5/6)^4$. But this probability is $P(\text{not } E)$, so we must use the "not" rule and subtract the result from 1 to get $P(E) = 1 - (5/6)^4$, or 0.518.

The other half of the question is, what is the probability of rolling at least one double 6 in 24 throws of a pair of die? This can be answered similarly. We represent this as $P(F)$. Again, we turn the question around and compute the probability of the negative: rolling *no* double 6s. For a given roll i, the probability of not rolling a double 6 is $P(B_i) = 35/36$. We multiply the results together to get $P(\text{not } F) = (35/36)^{24}$ and so $P(F) = 1 - (35/36)^{24}$, or 0.491. So, the first event is more likely.

This is just a basic example of computing probabilities. Those interested in computing the probability of more complex examples are advised to look to the references noted at the end of the chapter—it can get more complicated than one expects, particularly when dealing with dependent trials.

11.2.2 Random Variables

As we saw with vectors, mathematicians like abstractions so they can wrap an algebra around a concept and perform symbolic operations on it. The abstraction in this case is the *random variable*. Suppose we have a random experiment that generates values (if not, we can assign a value to each outcome of our experiment). We call the values generated by this process a random variable, usually represented by X. Note that X represents all possible values; a particular result of a random experiment is represented by X_i, and a particular value is represented by x.

If the set of all random values for our given problem has a fixed size,[1] as in the examples above, then we say it is a *discrete random variable*. In this case, we're interested in the

---

[1] Or, is countably infinite, though in games this is rarely considered, if ever.

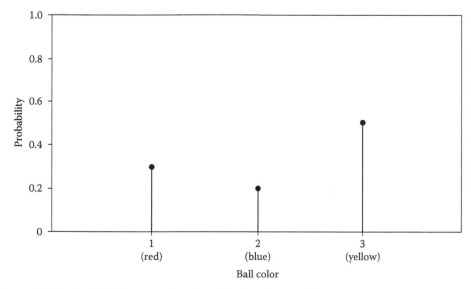

Figure 11.2. Probability mass function for drawing one ball out of a jar with three red balls, two blue balls, and five yellow balls.

probability of a particular outcome x. We can represent this as a function $m(x)$, where the function's domain is the sample space Ω. As an example, suppose we create such a function for our jar experiment. We'll say that red $= 1$, blue $= 2$, and yellow $= 3$. The sample space of our random variable is now $\Omega = 1, 2,$ or 3. The value of $m(x)$ for each possible x is the probability that x is the result of the draw out of the jar. The resulting graph can be seen in Figure 11.2. Notice that $m(x)$ only has a value at $1, 2,$ or 3, and is 0 everywhere else. This is known as a *probability mass function*, or sometimes a *probability distribution function*. This function has three important properties: its domain is the sample space of a random variable; for all values x, $m(x) \geq 0$ (i.e., there are no negative probabilities); and the sum of the probabilities of all outcomes is 1, or

$$\sum_{i=0}^{n-1} m(x_i) = 1$$

where n is the number of elements in Ω.

Now, suppose that our sample space has an uncountably infinite number of outcomes. One example of this is spinning a disc with a pointer: its angle relative to a fixed mark has an infinite number of possible values. This is known as a *continuous random variable*. Another example of a continuous random variable is randomly choosing a value from all real values in the range [0, 1]. Assuming all numbers have an equal probability, this is known as a *uniform variate*, or sometimes as the *canonical random variable* ξ [121].

One interesting thing about a continuous random variable is that the probability of a given outcome x is 0, since the number of possible outcomes we're dividing by is infinite. However, we can still measure probabilities by considering ranges of values and use a special kind of function to encapsulate this. Figure 11.3 shows one such function over the

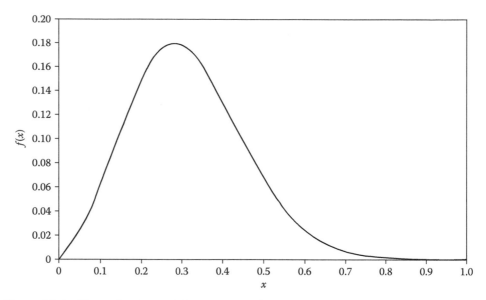

Figure 11.3. Example of a probability density function.

canonical random variable. This function $f(x)$ is known as a *probability density function* (PDF). It has characteristics similar to the probability mass function for the discrete case: all values $f(x)$ are greater than or equal to 0 and the area under the curve is equal to 1. As with the discrete case, the second characteristic indicates that the sum of the probabilities for all outcomes is 1 and can be represented by the integral:

$$\int_{-\infty}^{\infty} f(x)dx = 1$$

We can also find the probability of a series of random events, say from a to b. In the discrete case, all we need to do is take the sum across that interval:

$$P(a \leq \omega \leq b) = \sum_{x=a}^{b} m(x)$$

In the continuous case, again we take the integral:

$$P(a \leq x \leq b) = \int_{a}^{b} f(x)dx$$

Sometimes we want to know the probability of a random value being less than or equal to some value y. Using the mass function, we can compute this in the discrete case as

$$F(y) = \sum_{x=x_0}^{y} m(x)$$

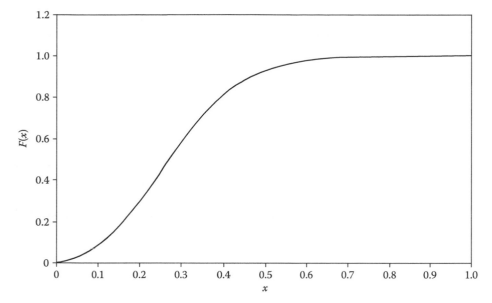

Figure 11.4. Corresponding cumulative distribution function for the probability density function in Figure 11.3.

or in the continuous case using the density function as

$$F(y) = \int_{-\infty}^{y} f(x)dx$$

This function $F(x)$ is known as the *cumulative distribution function* (CDF). We can think of this as a cumulative sum across the domain. Note that because the CDF is the integral of the PDF in the continuous realm, the PDF is actually the derivative of the CDF.

Figure 11.4 shows the cumulative distribution function for the continuous PDF in Figure 11.3. Note that it starts at a value of 0 for the minimum in the domain and increases to a maximum value of 1: all cumulative distribution functions have this property. We'll be making use of cumulative distribution functions when we discuss the chi-square method below.

11.2.3 Mean and Standard Deviation

Suppose we conduct N random trials with the random variable X, giving us results (or *samples*) $X_0, X_1, \ldots, X_{N-1}$. If we take what is commonly known as the average of the values, we get the *sample mean*

$$\bar{X} = \frac{1}{N} \sum_{i=0}^{N-1} X_i$$

We can think of this as representing the center of the values produced. We can get some sense of spread of the values from the center by computing the *sample variance* s^2 as

$$s^2 = \frac{1}{N-1} \sum_{i=0}^{N-1} (X_i - \bar{X})^2$$

The larger the sample variance, the more the values spread out from the mean. The smaller the variance, the closer they cluster to the mean. The square root s of this is known as the *standard deviation of the sample*.

Note that these values are computed for the samples we record. We can compute similar values for the mass or density function for X as well, dropping the reference to "sample" in the definitions.

The *expected value* or *mean* of a discrete random variable X with sample space Ω of size n and mass function $m(x)$ is

$$E(X) = \sum_{i=0}^{n-1} x_i m(x_i)$$

And for a continuous random variable, it is

$$E(X) = \int_{-\infty}^{\infty} x f(x) dx$$

Both are often represented as μ for short. Similar to the sample mean, these represent the centers of the probability mass and density functions, respectively.

The corresponding spread from the mean is the variance, which is computed in the discrete case as

$$\sigma^2 = \sum_{i=0}^{n-1} (x_i - \mu)^2 m(x)$$

and in the continuous case as

$$\sigma^2 = \int_{-\infty}^{\infty} (x - \mu)^2 f(x) dx$$

As before, the square root of the variance, or σ, is called the *standard deviation*.

We'll be making use of these quantities below, when we discuss the normal distribution and the central limit theorem.

11.2.4   Special Probability Distributions

There are a few specific probability mass functions and probability density functions that are good to be aware of. The first is the uniform distribution. A uniform probability mass function for n discrete random variables has $m(x_i) = 1/n$ for all x_i. Similarly, a uniform probability density function over the interval $[a, b]$ has $f(x) = 1/(b-a)$ for all $a \leq x \leq b$ and $f(x) = 0$ everywhere else. Examples of uniform probability distributions are rolling a fair die or drawing a card. On the other hand, the distribution of a loaded die is nonuniform.

Similarly, our PDF in Figure 11.3 has a nonuniform distribution. Our immediate goal in building a random number generator is simulating a uniformly distributed random variable, but as the large majority of situations we deal with will have nonuniform distributions, simulating those also will be important.

There are two other distributions that are of general interest. The first is a discrete distribution known as the *binomial distribution*. Suppose we have a random experiment where there are only two possible outcomes: success or failure. How we measure success depends on the experiment: it could be rolling a 2 or 3 on a single die roll, or flipping a coin so it lands heads, or picking out a red ball. Each time we perform the experiment it must not affect any other time (i.e., it is independent), and the probabilities must remain the same each time. Now we repeat this experiment n times, and ask the question, how many successes will we have? This is another random variable, called the *binomial random variable*.

In general, we're more interested in the probability that we will have k successes, which is

$$Pr(X = k) = \binom{n}{k} p^k (1 - p)^{n-k}$$

where

$$\binom{n}{k} = \frac{n!}{k!(n-k)!}$$

This is known as the *binomial coefficient*. If we graph the result for $n = 8, p = 2/3$, and all values of k from 1 to n, we get a lopsided pyramid shape (Figure 11.5). Note that the mean lies near the peak of the pyramid. It will only lie at the peak if the result is symmetric, which only happens if the probability $p = 1/2$.

This discrete distribution can lead to a continuous density function. Suppose that n gets larger and larger. As n approaches ∞, the discrete distribution will start to approximate a continuous density function; oddly, this function also becomes symmetric. Now we take

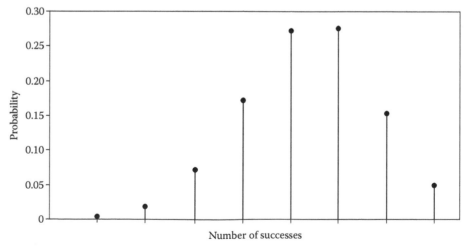

Figure 11.5. Binomial distribution for $n = 8$ and $p = 2/3$.

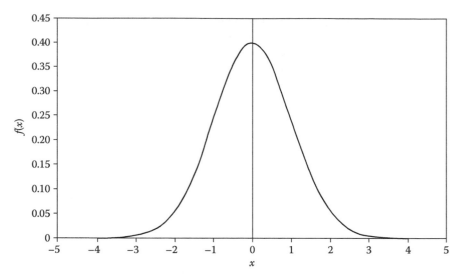

Figure 11.6. The standard normal distribution.

this continuous function and translate it so that the mean lies on 0, and scale it so that the standard deviation is 1, while maintaining an area of 1 under the curve. What we end up with is seen in Figure 11.6: the *standard normal distribution*. This can be represented by the function

$$f(x) = \frac{1}{\sqrt{2\pi}} e^{-x^2/2}$$

We can also have a general normal distribution where we can specify mean and standard deviation, also known as a *Gaussian distribution* or a *bell curve*:

$$f(x) = \frac{1}{\sigma\sqrt{2\pi}} e^{-(x-\mu)^2/2\sigma^2}$$

Note that the Gaussian distribution is also the same one used (albeit in 2D) when applying a blur filter to an image or to generate a mipmap.

Figure 11.7 shows a general normal distribution with a mean of 3.75 and a standard deviation of 2.4. For any value of p, the binomial distribution of n trials can be approximated by a normal distribution with $\mu = np$ and $\sigma = np(1 - p)$. Also, for a further intuitive sense of standard deviation it's helpful to note that in the normal distribution 68 percent of results are within 1 standard deviation around the mean, and 95 percent are within 1.96 standard deviations.

The interesting thing about the normal distribution is that it can be applied to all sorts of natural phenomena. Test values for a large group of students will fall in a normal distribution. Or measurements taken by a large group, say length or temperature, will also fall in a normal distribution.

With the introduction of the normal distribution we can also draw a better relationship between the mean and the sample mean. Suppose we take N random samples using a probability distribution with mean μ and standard deviation σ. Due to a theorem known as

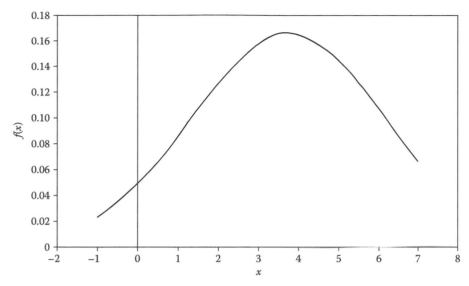

Figure 11.7. General normal distribution with mean of 3.75 and standard deviation of 2.4.

the central limit theorem, it can be shown that the sample mean $\bar{X}$ of our samples should be normally distributed around the mean μ, and that the standard deviation of $\bar{X}$ is $\sigma/\sqrt{N}$. So, the average of random samples from a normal distribution is also normally distributed, and the larger N is, the smaller $\sigma/\sqrt{N}$ will be. So, what this is saying is that for very large N, the mean μ and the sample mean $\bar{X}$ should be nearly equal. We'll be making use of this when we discuss hypothesis testing in the next section.

11.3 Determining Randomness

Up to this point we have been talking about random variables and probabilities while dancing around the primary topic of this chapter—randomness. What does it mean for a variable to be random? How can we determine that our method for generating a random variable is, in fact, random? Unfortunately, as we'll see, there is no definitive test for randomness, but we can get a general sense of what randomness is.

We use the term *random* loosely to convey a sense of nondeterminism and unpredictability. Note that human beings are notoriously bad at generating random numbers. Ask a large group of people for a number between 1 and 10, and the majority of the people will pick 7. The reason is that they are consciously trying to be random—trying to avoid creating a pattern, as it were—and by doing so they create a new pattern. The same can happen if you ask someone to generate a random sequence of numbers. They will tend to mix things up, placing large numbers after small ones, and avoiding "patterns," such as having the same number twice in a row. The problem is that a true random process will generate such results—streaks happen. So again, by trying to avoid patterns, a new and more subtle pattern is generated.

This gives us a clue as to how we might define a sequence of random numbers: a sequence with no discernable pattern. Statistically, when we say a process is random, we mean that it lacks *bias* and *correlation*. A biased process will tend toward a single value or set of values, such as rolling a loaded die. Informally, correlation implies that values within the sequence are related to each other by a pattern, usually some form of linear equation. As we will see, when generating random numbers on a computer we can't completely remove correlation, but we can minimize it enough so that it doesn't affect any random process we're trying to simulate.

11.3.1   Chi-Square Test

In order to test for bias and somewhat for correlation, we will perform a series of random experiments with known probabilities and compare the results of the experiments with their expected distribution. For this comparison, we'll use a common statistical technique known as *hypothesis testing*. The way we'll use it is to take a set of observed values generated by some sort of random process (we hope), compare against an expected distribution of values, and determine the probability that the result is suitably random. Most of the tests we'll see below pick a particularly nasty test case and then use hypothesis testing to measure how well a random number generator does with that case.

The first step of hypothesis testing is to declare a *null hypothesis*, which, in this case, is that the random number generator is a good one and our samples approximate the probability distribution for our particular experiment. Our *alternate hypothesis* is that the results are not due to chance—that something else is biasing the experiment.

The second step is to declare a test statistic against which we'll measure our results. In our case, the test statistic will be the particular probability distribution for our experiment.

The third step is to compute a p value comparing our test statistic to our samples. This is another random variable that measures the probability that our observed results match the expected results. The lower this probability, the more likely that the null hypothesis is not true for our results. Finally, we compare this p value to a fixed significance level α. If the p value is less than or equal to α, then we agree that the null hypothesis is highly unlikely and we accept the alternate hypothesis.

One possibility for our p value is to compare the sample mean for our results with the mean for our probability distribution. From the central limit theorem, we know that the sample mean is normally distributed, and the probability of the sample mean lying outside of 1.96 standard deviations from the mean is around 5 percent. So, one choice is to let the p value be the probability of our deviation from the sample mean, and our significance level 5 percent (i.e., if we lie outside two standard deviations we fail the null hypothesis).

However, in our case we're going to use a different technique known as *Pearson's chi-square test*, or more generally the chi-square (or χ^2) test. Chi-square in this case indicates a certain probability distribution, so there can be other chi-square tests, which we won't be concerned with in this text.

To see how the chi-square test works, let's work through an example. Suppose we want to simulate the roll of two dice, summed together. The probabilities of each value are as follows:

Die Value	2	3	4	5	6	7	8	9	10	11	12
Probability	1/36	1/18	1/12	1/9	5/36	1/6	5/36	1/9	1/12	1/18	1/36

So, if we were to perform, say, 360 rolls of the dice, we'd expect that the dice would come up the following number of times:

Die Value	2	3	4	5	6	7	8	9	10	11	12
Frequency	10	20	30	40	50	60	50	40	30	20	10

These are the *theoretical frequencies* for our sample trial. Our null hypothesis is that our random number generator will simulate this distribution. The alternate hypothesis is that there is some bias in our random number generator. Our test statistic is, as we'd expect, this particular distribution. In addition, note that we need a large number of samples in order for our chi-square test to be valid.

Now take a look at some counts generated from two different random number generators.

Die Value	2	3	4	5	6	7	8	9	10	11	12
Experiment 1	9	21	29	43	52	59	47	38	31	19	12
Experiment 2	17	24	28	29	35	76	46	35	32	23	15

First of all, note that neither matches the theoretical frequencies exactly. This is actually what we want. If one set matched exactly, it would not be very random, and its behavior would be very predicable. On the other hand, we don't want our random number generator to favor one number too much over the others. That may indicate that our dice are loaded, which is also not very random.

The first step in determining our p value is computing the chi-square value. What we want to end up with is a value that straddles the two extremes—neither too high nor too low. Computing it is very simple: for each entry, we just subtract the theoretical value from e_i, the observed value o_i, square the result, and divide by the theoretical value. Sum all these up and you have the chi-square value. In equation form, this is

$$V = \sum_{i=0}^{n} \frac{(e_i - o_i)^2}{e_i}$$

Using this, we can now compute the chi-square values for our two trials. For the first we get 1.269, and for the second we get 21.65.

Now that we have a chi-square value, we can compute a p value. To do that, we compare our result against the chi-square distribution. Or more accurately, we compare against the cumulative distribution function of a particular chi-square distribution.

To understand the chi-square distribution, suppose we have a random process that generates values with a standard normal distribution (i.e., a mean of 0 and a standard deviation of 1). Now let's take k random values and compute the following function:

$$\chi^2 = \sum_{i=1}^{k} x_i^2$$

The chi-square distribution indicates how the results from this function will be distributed. Figure 11.8 shows the probability density function and cumulative density function for various values of k.

In order to know which chi-square distribution to use, we need to know the degrees of freedom k in our experiment. This is equal to the number of possible outcomes minus 1. In our example above, the k value is $11 - 1 = 10$. If we now substitute our computed chi-square value into the appropriate chi-square cumulative density function, that gives us the probability that we will get this chi-square value or less. This is the p value we're looking for. If the resulting p value is very low, say from 0 to 0.1, then our numbers aren't very random, because they're too close to the theoretical results. If the p value lies in the higher probability range, say from 0.9 to 1.0, then we know that our numbers aren't random because one or more values are being emphasized over the others. What we want is a p value that lies in the sweet spot of the middle. This is a slightly different approach to hypothesis testing, because we're trying to check two conditions here instead of one.

So, how do we calculate the p value? This can be calculated directly, but the process is fairly complex. Fortunately, tables of pregenerated values are available (e.g., Table 11.1), and looking up the closest value in a table is good enough for our purposes.

For the particular row that corresponds to our number of degrees of freedom, we find the entry closest to our value V. The column for that entry gives us the p value. Looking at the $k = 10$ column, we see that the chi-square value of 1.269 for experiment 1 produces a p value of at most 0.01, and the chi-square value of 21.65 for experiment 2 produces a value between 0.95 and 0.99. So experiment 1 is too close to the expected probability distribution, and experiment 2 is far away. This fits the way they were generated. The first set of random numbers we simply chose to be very close to the expected value. The second set were weighted so that 1 would be more likely to come up on one die and 6 more likely on the other.

An alternative to looking up the result in a table is to use a statistical package to compute this value for us. Microsoft Excel has a surprising amount of statistical calculations available, and the chi-square test is one of those. A quick online search for "chi-square calculator" also finds a number of Web applications that perform this operation. Note that Excel and most tables reverse the sense of the p value; that is, rather than compute the probability that the chi-square value is less than or equal to our computed value, they compute the probability it will exceed that value. This allows them to use the standard approach to using p values, where a low p value means that our experiment is biased. Therefore, when using these packages keep this in mind.

This procedure gives us the basic core of what we need to test our random number generators: we create a test with random elements and then determine the theoretical frequencies for our test. We then perform a set of random trials using our random number generator and compare our results to the theoretical ones using the chi-square test. If the p value generated is acceptable, we move on; otherwise, the random number generator has failed. Note that if a generator passes the test, it only means that the random number generator produces good results for that statistic. If the statistic is one we might use in our game, that might be good enough. If it fails, it may require more testing, since we might have gotten bad results for that one run. With this in place, we can now talk about a few of the most basic tests.

The most basic test we can perform is the *equidistribution test*, which determines whether our presumably uniform random number generator produces a uniform sequence. Our test

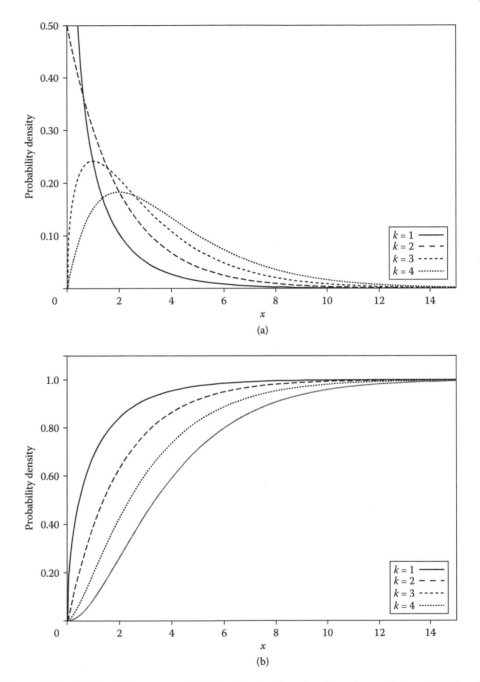

Figure 11.8. (a) The chi-square probability density function for values of k from 1 to 4 and (b) the chi-square cumulative density function for values of k from 1 to 4.

Table 11.1. Chi-Square CDF Values for Various Degrees of Freedom k

	$p = 0.01$	$p = 0.05$	$p = 0.1$	$p = 0.9$	$p = 0.95$	$p = 0.99$
$k = 1$	0.00016	0.00393	0.01579	2.70554	3.84146	6.63489
$k = 2$	0.02010	0.10259	0.21072	4.60518	5.99148	9.21035
$k = 3$	0.1148	0.35184	0.58438	6.25139	7.81472	11.3449
$k = 4$	0.29710	0.71072	1.06362	7.77943	9.48772	13.2767
$k = 5$	0.55430	1.14548	1.61031	9.23635	11.0704	15.0863
$k = 6$	0.8720	1.63538	2.20413	10.6446	12.5916	16.811
$k = 7$	1.23903	2.16734	2.83311	12.0170	14.0671	18.4753
$k = 8$	1.6465	2.73263	3.48954	13.3616	15.5073	20.0901
$k = 9$	2.08789	3.3251	4.16816	14.6837	16.9190	21.6660
$k = 10$	2.55820	3.94030	4.86518	15.9871	18.3070	23.2092
$k = 11$	3.0534	4.57480	5.57779	17.275	19.6751	24.7250
$k = 12$	3.57055	5.22602	6.30380	18.5493	21.0260	26.2170
$k = 13$	4.10690	5.8919	7.04150	19.8119	22.3620	27.6881
$k = 14$	4.66041	6.5706	7.78954	21.064	23.6848	29.1411
$k = 15$	5.22936	7.26093	8.54675	22.3071	24.9958	30.5780

statistic is that the counts will be the same for all groups. Ideally, we set one bucket for each possible value, but given that we can have thousands of values, that's not often practical. Usually, values are grouped into sequential groups; that is, we might shift a 32-bit random number right by 24 and count values in 256 possible groups.

The *serial test* follows onto the equidistribution test by considering sequences of random numbers. In this case, we generate pairs of numbers (e.g., $(x_0, x_1), (x_2, x_3), \ldots,$) and count how many times each pair appears. Our test statistic is that we expect the count for each particular pair to be uniformly distributed. The same is true for triples, quadruples, and so on up, although managing any size larger than quadruples gets unwieldy, and so something like the poker hand test, below, is recommended.

The *poker hand test* consists of building hands of cards, ignoring suits, and counting the number of poker hands, which Knuth [90] represents as follows:

All different	*abcde*
Pair	*aabcd*
Two pair	*aabbc*
Three of a kind	*aaabc*
Full house	*aaabb*
Four of a kind	*aaaab*
Five of a kind	*aaaaa*

Each of these outcomes have different probabilities. We generate numbers between, say, 2 and 13, and track the number of poker hands of each type. Then, as before, we compare the results with the expected probabilities by performing the chi-square test.

There is a simplification of this, where we only count the number of different values in the poker hand. This becomes

5 values   All different
4 values   One pair
3 values   Two pair, three of a kind
2 values   Full house, four of a kind
1 value   Five of a kind

This is easier to count, and the probabilities are easier to compute. In general, if we're generating numbers from 0 to $d - 1$, with a poker hand of size k, Knuth gives the probability of r different values as

$$p_r = \frac{d(d-1)\dots(d-r+1)}{d^k} \left\{ \begin{matrix} k \\ r \end{matrix} \right\}$$

where

$$\left\{ \begin{matrix} k \\ r \end{matrix} \right\} = \frac{1}{r!} \sum_{j=0}^{r} (-1)^{r-j} \binom{r}{j} j^k$$

This last term is known as a *Stirling number of the second kind*, and counts the number of ways to partition k elements into r subsets.

These three are just a few of the possibilities. There are other tests, many with colorful names, such as the birthday spacing test or the monkey test. For those who want to create their own random number generators and need to run them through a series of tests, a few open-source libraries are available. The first is DIEHARD, created by George Marsaglia, and so named because a non-English speaker misunderstood the notion of a "battery" of tests. However, the name is appropriate, as the tests are very thorough. DIEHARD is no longer maintained, but is available online. For a regularly updated library, there is DieHarder, which was created by Robert G. Brown of Duke University. In addition to regular maintenance, this one adds some additional tests suggested by the National Institute of Standards and Technology, and is released under the GNU Public License. It is also available online and installable on Linux as a package.

A more recent and rigorous set of tests is TESTU01 [94], which has three increasingly more stringent batteries of tests called Small Crush, Crush, and Big Crush. It's also available online as an open-source package. Any random number generator that passes Crush or even Small Crush is certainly suitable for games.

Finally, Marsaglia and Tsang created Tuftest [102], a set of three tests that purport to cover most of the cases necessary for a good random number generator. It's arguable whether these three tests are sufficient, but even if you'd rather use Big Crush instead, they still serve well as a reasonably quick unit test.

In general, however, we will not be creating our own random number generator. In those cases, a chi-square test is more useful for verifying that your use of a random number generator matches your expected behavior. For example, suppose you were trying to generate a particular probability distribution that a designer has created. If your results in-game don't match this distribution, you know you've done something wrong. The chi-square test allows you to verify this.

11.3.2 Spectral Test

There is one test of random number generators that falls outside of the standard chi-square-based or other statistical tests, and that is the *spectral test*. The spectral test is derived from the fact that researchers noticed that if they constructed points in space using certain RNGs, those points would align along a fixed number of planes (a statistician would say that the data are linearly correlated). This means that no point could be generated in the space between these planes—not very random. For many bad RNGs, this can be seen by doing a two-dimensional (2D) plot; for others, a three-dimensional (3D) plot is necessary. Some extreme examples can be seen in Figure 11.9.

In fact, Marsaglia [98] showed that for certain classes of RNGs (the linear congruential generators, which we'll cover below) this alignment is impossible to avoid. For a given dimension k, the results will lie "mainly in the planes," to quote the title of the article.

The spectral test was created to test for these cases. It takes d-tuples $(x_i, x_{i+1}, \ldots, x_{i+d-1})$ of a random sequence and looks for the spacings between them that lie along a d-dimensional hyperplane. For our purposes, we are not going to implement the spectral test. It mostly applies to a single class of RNGs, and as we'll see, a great deal of research has been done on determining good RNGs, so it's unlikely that we'll need a spectral test. Also, if the spacing between the planes is small enough, it's unlikely that it will significantly affect the sort of random data that are generated for games. However, this property of some RNGs is something to be aware of.

11.4 Random Number Generators

Now that we've covered some basic probability and some means of testing randomness, we can talk about how we generate random numbers. True random generators for computers are only possible by creating circuitry that depends on some physical phenomenon. One example is a generator that took video of lava lamps and used that to generate random numbers over time. Alternatively, we could track the particles generated by a radioactive isotope. Usually, however, a circuit is built that takes advantage of the fact that power to the computer has a certain amount of unpredictable noise in it. This noise is amplified and used to generate random values.

In our case, we can't assume access to such hardware. Instead, we'll have to make use of what is called a pseudorandom number generator. We will start with a set of one or more numbers and use a deterministic algorithm to generate a sequence of numbers that appear random. That is, our process is completely predetermined, but the numbers generated fulfill certain characteristics that make them suitable for simulating actual random processes. Because of this, pseudorandom number generators are just referred to as *random number generators*.

There is another class of RNGs known as quasi-random number generators. These generate numbers in a way that avoids streaks and clumping, and are primarily used for a numerical integration technique known as Monte Carlo integration. However, we won't be considering those as they tend to be more expensive and we don't require that kind of precision.

Why study random number algorithms when most languages these days come with a built-in RNG? The reason is that these built-in RNGs are usually not very random. Understanding why they are flawed is important if we intend on using them and working around

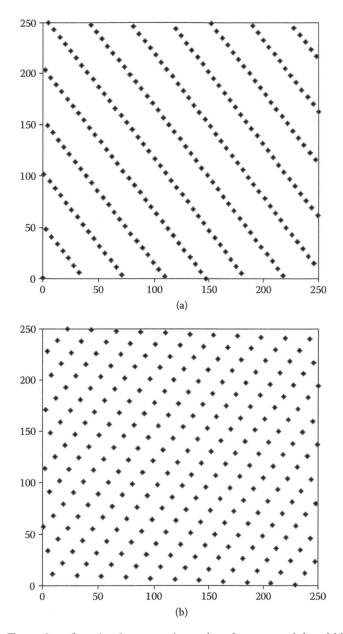

Figure 11.9. Examples of randomly generating points that stay mainly within the planes.

their flaws, and understanding what makes a good generator is important if we want to create our own.

Our goal in building an RNG is to generate a series or *stream* of numbers with properties close to those of actual random events. Because this series of numbers is usually very large, all of the RNGs that we're going to discuss can be described by a special type of function known as a *recurrence relation*. Those with experience in recursion should be familiar with the concept: the value at a given step n is dependent on values from previous steps (in many cases, only the immediately previous step). For example, here is the recurrence relation for the Fibonacci series:

$$x_n = x_{n-1} + x_{n-2}$$

To start things off, one or more *seed* values are set, and these control how the sequence of numbers will proceed. Again, using our Fibonacci example, using seed values $x_0 = 0$ and $x_1 = 1$, we get the series

$$0, 1, 1, 2, 3, 5, 8, 13, 21, \ldots$$

The process alone doesn't produce our sequence—the seed also plays a part. For example, if we use seed values $x_0 = 2$ and $x_1 = 1$, we get the Lucas numbers:

$$2, 1, 3, 4, 7, 11, 18, 29, \ldots$$

So, choosing the proper seed value is very important. If we use the same seed all the time, we'll always get the same sequence every time. This can be useful for debugging, so that we get the same results during each debugging pass, but in the final game we'll probably want to randomize this seed value somehow. One common method is to use the operating system clock value. Another uses the frequency of the user's keystrokes, mouse movement, or joystick movement at start-up time to compute a random value for the rest of the game.

The Fibonacci series is infinite, since the values get progressively larger and larger. However, we will need to limit our results to fit within calculable values on the computer, so we will take a modulus of anything we compute to ensure that it stays within bounds. Doing this with Fibonacci gives us

$$x_n = (x_{n-1} + x_{n-2}) \mod m$$

The value m is often one more than the largest representable number, although as we'll see below, other values work better with certain algorithms.

Another final concept we need to discuss before diving in is the *period* of a random number sequence. Because of the modulus, eventually all generators will repeat their values; you will end up generating your original seed values and the sequence will start again. For example, take this (very poor) RNG (please):

$$x_n = (x_{n-1} + 2) \mod 4$$

Given a seed value of 0, this will generate the sequence

$$0, 2, 0, 2, 0, 2, 0, 2, \ldots$$

This is a poor RNG for two reasons. First, as we can see, the values are very regular. But also, it has a very small period of 2. We want this period to be as large as possible; at the very least, it should encompass all values $(0, \ldots, m - 1)$, and ideally be much larger than that so that we can get streaks of numbers and handle large sets of permutations. For example, a deck of cards has 2^{226} possible permutations; to have the possibility of generating them all, we would need a random number generator with a period at least that large.

This should give some general sense of the structure of the algorithms we'll be discussing. Note that this is by no means an exhaustive list. We are merely trying to present some standard algorithms to demonstrate the wide variety of possibilities. A few of the generators we'll discuss are not very good. This is mainly to show what can go wrong in case you are tempted to create your own. Also note that when discussing generators in this section, we'll only be constructing those that generate unsigned integers. We'll cover how to create signed integers, smaller than full integer ranges, and floating-point numbers in Section 11.5.

11.4.1 Linear Congruential Methods

11.4.1.1 Definition

The *linear congruential generator* (LCG) is a very popular random number generator. It was first introduced by D. H. Lehmer in 1949 and is introduced in most algorithm classes and implemented in most standard libraries. The LCG is represented by the following equation:

$$x_n = (ax_{n-1} + c) \bmod m$$

where

$$0 \leq m$$
$$0 \leq a < m$$
$$0 \leq c < m$$

In this case, m is called the *modulus*, a is called the *multiplier*, and c is called the *increment*. If c is 0, this is called a *multiplicative congruential method*; otherwise, it is a *mixed congruential method*.

Note that no matter what the values are, the maximum period is m. This makes sense; because we're only tracking one variable, if we ever repeat a value, the sequence will begin again from that point. So, the maximum we can possibly do is to run through all of the values between 0 and $m - 1$ and then start again from the seed. Of course, this is only possible if $c \neq 0$. If $c = 0$, then if our sequence generates 0, we'll end up with something like

$$\ldots, 4, 24, 6, 0, 0, 0, 0, \ldots$$

This is because once x is 0, $ax \bmod m$ will always be 0. So if $c = 0$, we can only use values between 1 and $m - 1$. In this case, if we need 0, we can always subtract 1 from the result.

11.4.1.2 Choosing the Modulus

The first question when constructing an LCG is what the value of m should be. The most obvious choice, as we stated above, is to use one more than the largest representable integer value, or, if our word size is w bits, 2^w. As most adders will wrap values when overflow occurs, they are automatically performing a modulus 2^w, which makes our computation very efficient.

However, there are two problems when using a power of 2 for m. First, it can be shown that if $c = 0$, then the maximum period is only $m/4$, and this in turn can only happen if a mod 8 is 5 and the initial seed is odd. Since we're multiplying an odd number by an odd number, the result will be an odd number, and in fact, we'll get half of the odd numbers between 1 and $2^w - 1$. To avoid this, most generators that use $m = 2^w$ use an odd value for c, which allows the resulting value to alternate between even and odd, and provides the full period of m.

However, when our numbers alternate between even and odd, the least significant bit will alternate between 0 and 1—not very random. This signifies another problem, as pointed out by Knuth [90]. Suppose that d is a divisor of m (i.e., m mod $d = 0$) and

$$y_{n-1} = x_{n-1} \bmod d$$

If $d = 2^e$, we can think of this as representing the eth least significant bits of x_{n-1}. It can be shown that

$$y_n = (ay_{n-1} + c) \bmod d$$

In other words, while our random sequence may have a maximum period of m, its least significant bits have a maximum period of d—they are much less random than the most significant bits.

This really only comes into play if we're using our RNG for small value simulations such as rolling dice. One solution is to shift the result from the RNG to the right so that the more random bits in the middle word become the new least significant bits and then take our modulus, which is what we have done in ours.

However, because of these problems, most researchers recommend using a large prime number for m instead. There is some debate on what that suitable prime number is. One popular choice for 32-bit numbers is $2^{31} - 1$, because it is close to our maximum value of $2^{32} - 1$ (which is not prime, by the way). Marsaglia [99] also suggests $2^{32} - 5$ and $2^{32} - 2$. However, in general, we simply want a large enough prime for our purposes—the larger the prime, the larger the period.

The big advantage of the prime value is that it guarantees to give us a full period without having to use an increment. However, using a prime value has some consequences. First of all, we need to be sure to store the maximum possible value somewhere, so that we can compute the floating-point equivalents (see below). We also lose the convenience and efficiency of letting the hardware magically handle our modulus for us. In fact, rather than taking advantage of overflow, we have to be careful that it doesn't take advantage of us. For example, suppose we have a 4-bit architecture, and are computing the following LCG:

$$x_n = (3x_{n-1}) \bmod 13$$

If x_{n-1} is 12, then $3x_{n-1} = 36$. But this value doesn't fit into the word, so it is truncated to 36 mod $16 = 4$. This mod 13 gives a final value of 4. But the actual result should be 36 mod $13 = 10$.

To solve this problem, Park and Miller [118] recommend the following replacement formula, based on Schrage [134], for ax mod m:

$$ax \bmod m = a(x \bmod q) - r(x \div q) + m\delta(x)$$

where $\div$ is integer division, $q = m \div a$, $r = m \bmod a$, and $\delta(x) = (z \div q) - (az \div m)$. The value of $\delta(x)$ will be either 0 or 1, and only 1 if $a(x \bmod q) - r(z \div q)$ is less than 0. We can represent this in code as follows:

```
x = a*(x%q) - r*(x/q);
if (x <= 0)
    x += m;
```

Note that this only works if $r < q$; otherwise, overflow will still occur.

Of course, a simpler solution is to do our calculations in a larger word size and truncate down to our desired, smaller word size (i.e., compute in 64-bit integers for a 32-bit result), but that assumes this option is available to us.

11.4.1.3   Choosing the Multiplier

So these are our two logical possibilities for a modulus: either a power of 2 or a large prime number. We've already noted that $c > 0$ is only necessary when using a modulus that is a power of 2, and in that case any odd number will do (1 is a popular choice). So, the remaining question is, what do we choose for a, our multiplier?

We want to make our choice to maximize two things: the period of the random sequence and the randomness of the resulting numbers within that sequence. The most common measure of this randomness is to use the spectral test, as LCGs are particularly susceptible to obviously regular patterns unless the values are chosen appropriately.

Let's consider the maximum period first. We've already noted that if $m = 2^w$, we want $a \bmod 8 = 5$ and c to be odd to get the full period. In general, however, we want the following. Suppose we begin with seed x_0, multiplier a, modulus m, and $c = 0$. Then we can find the value at the ith step by calculating

$$x_i = a^i x_0 \bmod m$$

Without loss of generality, let's assume that $x_0 = a$. If we have a full period, eventually we'll generate a as our current random number, and that starting point is as good as any other. So, this formula becomes

$$x_i = a^{i+1} \bmod m$$

In order for us to get a full cycle, $a \bmod m, a^2 \bmod m, a^3 \bmod m, \ldots, a^{m-1} \bmod m$ must all be distinct values. How do we find this a? Fortunately, there is a mathematical entity known as a *primitive element*, which has just this property when a and m are relatively prime (that is, their only common factor is 1). So, for our particular m, we just need to generate all the primitive elements, and that will give us good values for a.

This gives us a starting point for both cases. However, the number of possible values is still quite large, and narrowing this down requires the use of the spectral test. Fortunately for us, many people have already done studies of the primitive elements for specific values of m. A fairly recent work by L'Ecuyer [93] in particular has laid out tables of possible values for all of the cases we're interested in, including the power of 2 cases with no addition. Using a value from these tables will guarantee excellent results. For our generators, we have chosen default values of $a = 2,862,933,555,777,941,757$ for the 2^{64} generator and m of $2^{32} - 5$

and a of 93,167 for the prime generator. While those values will produce good results, we of course let users set their own values if they want.

11.4.1.4 Summary

In summary, the LCG is the most commonly taught and used RNG today. It is usually the basis for RNGs in most math libraries. Although it's not the best generator, when the values for a and m are chosen carefully, it can produce results good enough for most games. However, when using it, one needs to be wary of the limited period, the randomness of the least significant bits, and the problems with the spectral test. Because of this, we'll be looking at some other possibilities for RNGs.

11.4.2 Lagged Fibonacci Methods

The linear congruential methods are reasonable RNGs, but they do have their flaws. The most major flaw is that when performing the spectral test for k-dimensional points, the best we can do is to limit the number of distinct hyperplanes that points will fall on to $m^{1/k}$. In other words, there will always be points in space that we cannot randomly generate. The other problem is that the maximum period we can expect is m. Can we do better?

One thing we note about LCGs is that they only make use of the last value—perhaps we can do better by looking at more than one previous value. We've already mentioned the Fibonacci method, where we take values from the previous two steps. Recall that this has the recurrence relation

$$x_n = (x_{n-1} + x_{n-2}) \bmod m$$

and requires two seed values. For the traditional Fibonacci sequence, this would be $x_0 = 1$ and $x_1 = 1$.

Unfortunately, while the standard Fibonacci method has a large period, it has been shown to not produce very random numbers. There is actually a hidden pattern, where the ratio between one value and the previous value is approximately the golden ratio: $(\sqrt{5} + 1)/2$.

A better approach is to use a *lagged Fibonacci generator*, where we look further back into the sequence for values, and they are not necessarily one after the other. This can be generalized as

$$x_n = (x_{n-j} \star x_{n-k}) \bmod m$$

where $\star$ is any binary operation (addition, subtraction, multiplication, and exclusive-OR are common choices) and $0 < j < k$. Assuming that $m = 2^w$ and addition, subtraction, or exclusive-OR is used, the maximum possible period for lagged Fibonacci generators is $2^{w-1}(2^k - 1)$. For multiplication, this drops to $2^{w-3}(2^k - 1)$. However, multiplication has been shown to mix bits better. In any case, assuming k is large enough, this period can be much larger than a standard LCG.

There are two decisions that we have to make when dealing with lagged Fibonacci generators: What are the values of j and k, and how do we initialize the starting k values? As far as the first question, which will determine the actual maximum period, tables of good values can be found in Knuth [90]. The choice of our initial values can be even more critical, as choosing poorly will seriously affect the randomness of the resulting sequence (e.g., consider what would happen if all the initial values were the same). One good possibility presented by Mitchell and Moore is to use addition with $j = 24$ and $k = 55$. The values $x_0 \ldots x_n$ are initialized with arbitrary integers, but guaranteed to be noneven. This gives us a period of $2^{w-1}(2^{55} - 1)$, which is quite respectable. In general, however, the problem of choosing good starting values haunts the Fibonacci generator, so again we will look for other solutions.

11.4.3   Carry Methods

Source Code
Library
IvRandom
Filename
IvCarryMultiply

One of the flaws of the LCG is that it works best with a value m that is prime. This is bad for two reasons: it's not cheap to compute and computer word sizes are powers of 2 (which are definitely not prime). For those values that are a power of 2, the least significant bits have a lower period than the entire sequence.

One solution was presented by George Marsaglia and Arif Zaman [103], who noted that with the LCG, the most significant bits get mixed better than the least significant bits. To mix the least significant bits, they proposed a carry or borrow operation, which takes part of the result from the previous stage and carries it forward to be applied to the least significant bits in the next stage. The standard formula for an add-with-carry generator is

$$x_n = (x_{n-k} + x_{n-r} + c_{n-1}) \bmod m$$
$$c_n = (x_{n-k} + x_{n-r} + c_{n-1}) \div m$$

Again, $\div$ represents an integer divide. As we can see, the bits that would normally be cast out from the modulus operation are added to the next stage, thereby mixing the lower bits. Something similar can be done with the subtract-with-borrow generator:

$$x_n = (x_{n-k} - x_{n-r} - c_{n-1}) \bmod m$$
$$c_n = (x_{n-k} - x_{n-r} - c_{n-1}) \div m$$

While these generators have large periods, solve the least significant bit issue, and otherwise show some promise, it was shown that they also fall prey to the same problem of falling mainly in the planes that linear congruent generators do.

In 1992, George Marsaglia posted a modification of this technique [100], which he called the mother of all random number generators, also known as the multiply-with-carry technique. This algorithm also works much better with values of m that are powers of 2. As before, the idea is to add the high-entropy bits (i.e., those that are changing a lot) to the low-entropy bits (i.e., those that don't normally change all that much). If we represent a 32-bit integer x_i as two 16-bit integers $a_i b_i$ (i.e., a is the high bits and b is the low bits), then a basic example of his algorithm is

$$x_n = 30{,}903 \, b_{n-1} + a_{n-1}$$

Instead of returning x_n as the result, we return the low bits b_n as a 16-bit integer.

This can easily be represented in C as follows:

```
k=30903*(k&65535)+(k>>16);
return(k&65535);
```

In this case we're doing a carry mod 2^{16}.

This has a period of 2^{59} 16-bit numbers. We can generate 32-bit numbers with the same period by concatenating two results together.

```
k=30903*(k&65535)+(k>>16);
j=18000*(j&65535)+(j>>16);
return ((k << 16) + j);
```

The multipliers for j and k are chosen to give good results for a modulus of 2^{16}.

This can be extended further, giving a period of 2^{118}.

```
k=30903*(k&65535)+(k>>16);
j=18000*(j&65535)+(j>>16);
i=29013*(i&65535)+(i>>16);
l=30345*(l&65535)+(l>>16);
m=30903*(m&65535)+(m>>16);
n=31083*(n&65535)+(n>>16);
return((k+i+m)<<16)+j+l+n);
```

This is a considerable improvement over the previous two methods: it gives us very large periods, it does a good job of randomizing the bits, it works well with computer word sizes and so is good for both floating-point numbers and integers, and it's very fast. It also only requires six starting values, as opposed to the large table needed for the lagged Fibonacci methods. However, it still doesn't pass all the TestU01 batteries, so while an improvement, it's not ideal.

Source Code

Library
IvRandom
Filename
IvXorshift

11.4.4 Xorshift

A more recent set of methods is the xorshift generators, again first presented by George Marsaglia [101]. The base form is to xor a value with a shifted version of itself a number of times, for example,

```
UInt32 x = kNonZeroNumber;
UInt32 xorshift32()
{
    x ^= (x << a);
    x ^= (x >> b);
    x ^= (x << c);
    return x;
}
```

where a, b, and c are chosen appropriately to guarantee the maximum period, which in this case will be $2^{32} - 1$ (the value 0 is not possible, unless all you want to generate is a sequence of 0s). We can increase the period by adding more state variables; for example, the following provides a period of $2^{128} - 1$:

```
UInt32 x, y, z, w;    // state variables
UInt32 xorshift128()
{
    UInt32 t;
    t = (x << a);
```

```
    x = y; y = z; z = w;
    w ^ = (w >> b) ^ t ^ (t << c);
}
```

These methods are very fast, but still not ideal as they in general fail the Small Crush suite.

An improvement, suggested by Marsaglia and tested by Sebastiano Vigna, is to multiply the result of the xorshift by a factor to permute the result. This is known as the xorshift* algorithm [152], and it produces much better results. For example, the following $2^{1024} - 1$ period algorithm is much better behaved, quite fast, and systematically passes all the TestU01 tests (for certain state values it can fail Big Crush, but for most states it passes).

```
UInt64 s[ 16 ];    // state variables
int p;
UInt64 xorshiftmul1024()
{
    UInt64 s0 = s[p];
    UInt64 s1 = s[p = (p + 1) & 15];
    s1 ^= s1 << 31;
    s1 ^= s1 >> 11;
    s0 ^= s0 >> 30;
    return (s[p] = s0 ^ s1)*1181783497276652981LL;
}
```

If space is an issue, then an alternative is the xorshift+ algorithm [153], which adds the last returned value to the xorshift result:

```
UInt64 s[2];
UInt64 xorshiftadd128(void) {
    UInt64 s1 = s[0];
    const UInt64 s0 = s[1];
    s[0] = s0;
    s1 ^= s1 << 23; // a
    return (s[1] = (s1 ^ s0 ^ (s1 >> 17) ^ (s0 >> 26))) + s0; // b, c
}
```

This has a period of $2^{128} - 1$, which Vigna states is not suitable for large-scale parallel simulations, but should be more than fine for many games (an exception would be card games—recall that a deck of cards has 2^{226} possible permutations). The main downside of xorshift* or xorshift+ is that they are designed for 64-bit integers, which are not supported by older mobile processors. An already mentioned minor issue is that 0 is not part of the sequence—this can be solved by either subtracting one from the result (and using $2^{64} - 2$ as the divisor for computing [0, 1] floating-point ranges) or, if you only need 32-bit values, using the top 32 bits.

11.4.5   Other Alternatives

Source Code
Library
IvRandom
Filename
IvMersenne

For some time, the Mersenne Twister [106] was considered the pinnacle of RNGs. Its creators' (Matsumoto and Nishimura) goal was to create a generator with a large period that passes a large battery of tests, and still is fast enough for practical use. Even today it is still widely used in many languages and libraries as the standard random number generator. However, the Mersenne Twister has some issues. While it passes the older DIEHARD tests, it fails a number of the newer Crush and Big Crush tests. The standard algorithm requires a starting table of 624 values, which is quite large compared to other methods, and stresses the CPU cache. It's period of $2^{19,937} - 1$, while impressive, is unnecessarily excessive for games. Because of this, while we've included an implementation of the Mersenne Twister for those who want to use it, it's no longer recommended.

A reasonable alternative for 32-bit processors is the Keep It Simple, Stupid, or KISS generator. The original was again devised by George Marsaglia with Arif Zaman [104] and combines a linear congruential generator, a multiply-with-carry generator, and an xorshift method. By using all three in combination, the intent is to allow the strengths of each algorithm to compensate for the shortcomings of the others. However, the original algorithm does not pass all the TestU01 suite, and in any case the multiply-with-carry generator assumes 64-bit arithmetic (to manage the carry bits). A 32-bit alternative was proposed by Jones [86], and replaces the multiply-with-carry generator with an add-with-carry:

```
static unsigned int x, y, z, w, c=0;

unsigned int JKISS32()
{
    int t;

    y ^= (y<<5); y ^= (y>>7); y ^= (y<<22);
    t = z+w+c; z = w; c = t < 0; w = t & 2147483647;
    x += 1411392427;

    return x + y + w;
}
```

This has a period of 2^{121}, and passes all of TestU01, which is certainly suitable for games. As with other methods, x, y, z, and w must be seeded appropriately.

11.4.6 Setting Initial State

This last point is an important one, especially in algorithms with a large set of state: if we don't set up our initial variables well, we won't get the full period until the algorithm runs for a while, or "warms up." And setting the state variables using one seed and feeding it into another random number generator won't entirely work. If we start with a 2^{32} bit seed, that still only gives us a total of 2^{32} possible state values, which for generators with larger periods will not cover all of the possible starting states.

One solution, assuming that the system supports it, is to use /dev/random. This is a device on Unix and related platforms (including iOS and Android) that uses the internal state of various drivers and other operating system-level data to generate random numbers for cryptological applications. However, reading from it will block until a suitable level of randomness is achieved—because of this, it's better to use the nonblocking /dev/urandom, as we don't need a cryptological level of randomness. It's not fast enough for a general random number generator, but it's certainly suitable for seeding our state. The Windows equivalent is the system function rand_s.

That said, it's still useful to have a way to set an initial state deterministically, so that results can be duplicated during testing and debugging.

11.4.7 Conclusions

This concludes our discussion of basic random number generators. The question remains: Which one to use? Obviously, in the best possible cases, we would use xorshift*1,024. However, as mentioned, this assumes that we have 64-bit operations available. In that case, the multiply–carry or JKISS32 method might be good enough. And if space and speed are truly at a premium, a linear congruential method may do the trick—but be careful to choose one with a good modulus and multiplier.

So to summarize: whenever possible, use a good method that passes a wide battery of tests; be sure to pick good random seeds; and if you can ever help it, don't use the default generator—create your own or use a well-vetted library.

11.5 Special Applications

Up to this point, we've been discussing only how to randomly generate uniformly distributed unsigned integers. However, randomness in a computer game extends beyond this. In this section we'll discuss a few of the more common applications and how we can use our uniform generator to construct them.

11.5.1 Integers and Ranges of Integers

In addition to unsigned integers, it is useful to be able to generate other types of values, and in various ranges. In this section we'll discuss some of the possibilities and how to generate them. For the sake of this discussion, we will assume that we are generating values from 0 to $m - 1$: if 0 is not possible with our generator, we can simply subtract 1 from the result and substitute $m - 2$ for $m - 1$.

If our generator has $m = 2^w$, then generating signed integers is simply a case of recasting the unsigned result as signed. The alternative is to do a scale and translate transformation, so $y = 2x - m$. This assumes that $m < 2^{w-1}$; otherwise, we'll end up overflowing.

Another common case is generating a range of integers, say from a to b. If x is the result of our RNG, we could do

$$y = x \bmod (b - a + 1) + a$$

The problem is if the range $r = b - a$ is small and the RNG has poor mixing of the least significant bits, the result will not be very random. One solution is to shift x to the right before performing the modulus. Another, of course, is to use a different generator. But if r is large, then again we need to worry about overflow.

A better solution is to only use the lower k bits of x, where $k = \lceil \log_2 (r) \rceil$, throw out any values greater than r, and then add a:

```
unsigned int range = b - a;
unsigned int kmask = NextPowerOf2(range)-1;
unsigned int y;
do
{
    y = (x & kmask);
} while (y > range);
return (int)(y + a);
```

This again assumes that you are using a generator with good randomness in the lower bits, and that $b > a$. If you're not sure of the second, you can check for the condition and swap a and b accordingly.

11.5.2 Floating-Point Numbers

Usually when generating floating-point numbers, we want the range $[0, 1]$. Commonly, this is computed as

```
float f = float(random())*RECIP_MAX_RAND;
```

where `RECIP_MAX_RAND` is the floating-point representation of 1 over the maximum possible random number.

An alternative is to set the exponent of the floating-point number to $bias + 1$ (see Chapter 1 for the definition of floating-point bias) and take random bits from the integer to fill the mantissa. This gives a value in the interval $[1, 2)$. Subtracting 1 gives us an interval of $[0, 1)$. For a single-precision floating point this can be computed as

```
union
{
    unsigned int i;
    float f;
} floatConv;
floatConv.i = 0x3f80000 | random() >> 9;
return floatConv.f - 1.0f;
```

If the pointer cast does not compile efficiently, we can use a union to do the bit conversion.

Now that we have values from 0 to 1, computing a general random interval $[a, b)$ is simple:

```
y = (b-a)u + a;
```

As mentioned above, we can also use this to generate intervals for integers via casting.

11.5.3 Shuffling

Knowing how to efficiently shuffle a fixed set of data is useful for two reasons. First of all, it can be used to generate random permutations, which is good for the obvious cases (decks of cards) and the nonobvious cases (a randomized cycle of idle or fight behaviors for an AI). Secondly, it often comes up during technical interviews.

The answer (called the Fisher–Yates shuffle [46]) is quite simple, and can be thought of as a variation on selection sort. We work our way through the list, choosing a random element to the right of or including the current element, and swapping them:

```
for (int i = 0; i < n; ++i)
{
    int randpos = randRange(i, n-1);
    swap(a[randpos], a[i]);
}
```

We include the current element because doing nothing at all is a suitable random event.

Note that to produce all the possible permutations, we need a random number generator with at least that period.

11.5.4   Nonuniform Distributions

Source Code
Demo
SphereDisc

Up until now, we've only been considering uniform random numbers. However, as we've seen, a large class of random events have nonuniform distributions. How then do we calculate these?

If we have a discrete random variable and its distribution, then we can create a discrete CDF and store the results in a table. If we then roll a uniform floating-point number in the interval [0, 1), we can then find the minimum entry that is greater than that value, and generate that. Let's take our ball-drawing problem as an example. Figure 11.10 shows the CDF for the probability distribution in Figure 11.2. Notice that due to the discrete nature of the distribution the CDF is a step function. If, for example, we randomly generate the value 0.43, we find that value in the y-axis, and then trace along horizontally until we hit a step. Sliding down to the x-axis, we see that step begins at 2, which represents the color blue, so that is the result of our random variable.

If we have a continuous random variable, this is not as simple. However, we can observe that what we're doing with the discrete case is just inverting the CDF. Assuming that there is an inverse, we can do the same with the continuous CDF, plug in our uniform value, and take the result as our nonuniform random variable.

If there is no inverse or we don't know the exact function for either the PDF or the CDF, then there is one other technique we can try: the rejection method or *rejection sampling*. The idea is that we generate values using a PDF that is close to our unknown one, and then throw out those that don't match. One example is our solution for computing a range

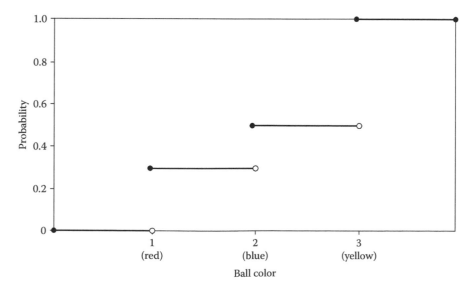

Figure 11.10. Discrete cumulative distribution function for the probability mass function in Figure 11.2.

of integer values, above, and we'll see more examples of this in the next two sections. However, this may not be the most efficient method, and in games it can be better to find an approximation of our distribution and use that.

11.5.5 Spherical Sampling

One common example of randomness in a game is generating the initial random direction for a particle. The most commonly used particle system of this type is spherical, where all the particles expand from a common point. We can compute the direction vector for this easily by generating a random point on a unit sphere.

One possible (but wrong) solution for this is to generate random components (v_0, v_1, v_2), where each v_i is a floating-point value in the range $[-1, 1]$, and then normalize the result. This will produce random points on the sphere, but the result will not be evenly distributed across the surface of the sphere. If we look at Figure 11.11, we can see the result. Because the initial random numbers generated are within a cube, the result on the sphere is biased toward the locations closest to the corners.

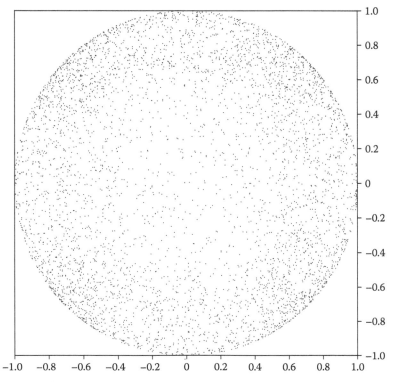

Figure 11.11. Spherical sampling, looking down along z. Result of normalizing random cube values; the points tend to collect near the "corners."

Another possibility is to use rejection sampling. We again generate our three values, but then test to see if $v_0^2 + v_1^2 + v_2^2$ is less than 1. It works fine, but can require a large number of RNG evaluations, so we'll consider one other option.

Rather than using Cartesian coordinates, let's look at spherical coordinates, which may be a little more natural to use on (say) a sphere. Recall that ϕ is the angle from the z-axis down, from 0 to π radians, and θ is the angle from the x-axis, from 0 to 2π radians. Since we're talking about a unit sphere, our radius ρ in this case is 1. So, we could generate two values ξ_0 and ξ_1 in the interval $[0, 1]$, and compute $\phi = \xi_0\pi$ and $\theta = 2\xi_1\pi$. From there we can compute x, y, and z as

$$x = \sin\phi\cos\theta$$
$$y = \sin\phi\sin\theta$$
$$z = \cos\phi$$

However, again we don't quite get the distribution that we expect. In Figure 11.12, we see that the points are now clustered around the poles of the sphere. The solution is to note that we want a latitude–longitude distribution, where z is our latitude and is uniformly distributed, and θ is longitude and also uniformly distributed. The radius at our latitude line

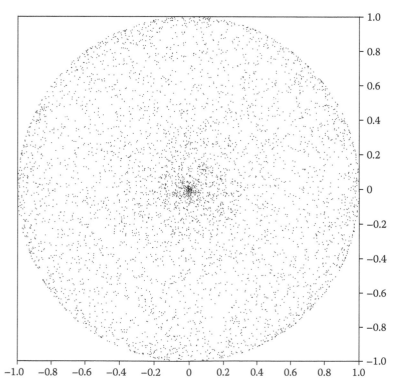

Figure 11.12. Spherical sampling, looking down along z. Result of randomizing spherical coordinates; the points tend to collect near the poles.

will depend on z—we want to guarantee that $x^2 + y^2 + z^2 = 1$. The following calculation handles this:

$$z = 1 - 2\xi$$
$$r = \sqrt{1 - z^2}$$
$$\theta = 2\pi\xi_1$$
$$x = r\cos\theta$$
$$y = r\sin\theta$$

The final result can be seen in Figure 11.13.

A similar calculation can be done if we want to generate points on a hemisphere. Instead of calculating $z = 1 - 2\xi$, we want z to vary from 0 to 1, which is just ξ_0.

Whether this spherical coordinate method or rejection sampling is faster depends on the cost of the square root and trigonometric functions. On one system, for example, four to five uniform variates can be generated in the time of a single trigonometric call. Profiling will be required in your particular application to determine which is best.

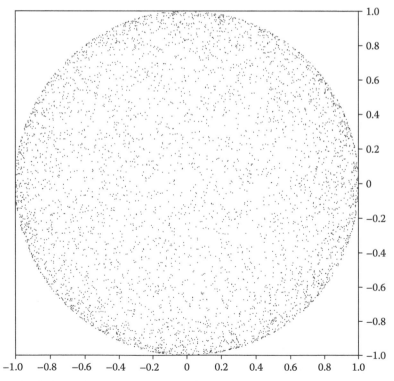

Figure 11.13. Spherical sampling, looking down along z. Result of randomizing latitude and longitude; the result is correct.

11.5.6   Disc Sampling

Another particle or ray casting shape that we might use is a cone. We can simulate a cone by using the cone tip as the source of our ray or particle and randomly selecting a point on the disc at the other end of the cone. This can be generalized by selecting a point on a unit disc. Afterwards we can scale the result by the radius of the cone, and then rotate it to be normal with the cone direction.

To select a point on a unit disc we could use rejection sampling again. The rejection sampling approach is similar to the 3D case: we generate two random numbers with range from 0 to 1. This time if the vector generated has length *greater* than 1, we try again; otherwise, we proceed as before.

Alternatively, we can generate a value using polar coordinates. The naive approach is to generate two values ξ_0 and ξ_1 as in the spherical case. This time we want the radius to vary from 0 to 1, and θ to vary from 0 to 2π, and so

$$r = \xi_0$$
$$\theta = 2\pi \xi_1$$
$$x = r \cos \theta$$
$$y = r \sin \theta$$

However, we find that we get clustering in the center, as we did in the spherical case (Figure 11.14). This may be close to what we want if we're calculating bullet trajectories,

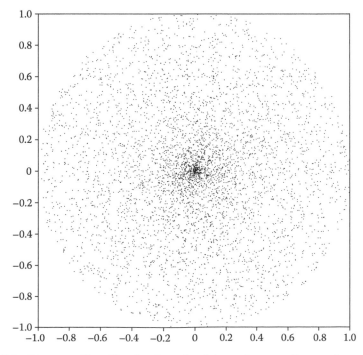

Figure 11.14. Disc sampling. Result of randomizing polar coordinates; the points tend to collect at the center.

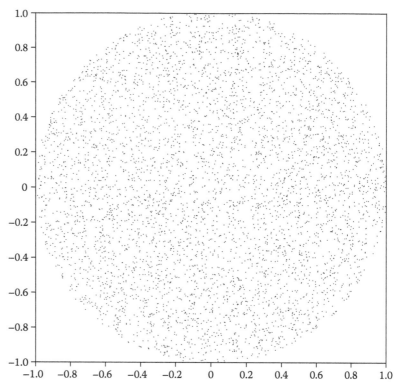

Figure 11.15. Disc sampling. Result of randomizing polar coordinates with radius correction; the result is correct.

where we want them to cluster around the aim direction. However, let's assume this is undesirable. The insight here is to set $r = \sqrt{\xi_0}$. This pushes the values back to the edges of the disc and gives us uniform sampling across the area of the disc (Figure 11.15).

Again, profiling will be needed to determine if this method or rejection sampling will be more efficient.

11.5.7 Noise and Turbulence

Source Code
Demo
Perlin

We will conclude our discussion of random numbers by briefly looking at some common noise functions and how they can be used to generate procedural textures. The first question is, why do we want to add randomness to our procedural textures? The main reason is that the world itself is random. Random bumps against the wall create scuff marks and divits. The way trees grow depends on rain, wind, and sun. Clouds in turn are dependent on humidity and wind. So, by adding random elements to textures that simulate natural features, we make them look less synthetic and more organic.

The common way to apply noise to textures is to build a noise lattice. In this case, we place random values at regular intervals in the texture space, and then interpolate between them to obtain the intermediary values. By using an appropriate interpolation function (usually cubic), we can guarantee that our noise function is continuous and smooth, which produces much better visual results. This lattice can be 1D, 2D, 3D, or even 4D. Higher lattices are not usually used because generation cost gets quite high.

The random function has a couple of competing requirements: it must not produce any obvious pattern, but it also must not vary every time we generate it, because we need the noise to be repeatable. The latter requirement is because otherwise the resulting textures will appear to flicker and move across the surfaces they're applied to. While this may be desirable in some cases, we'd like to control the situation. To manage this, most noise systems pregenerate a table of random values and then hash into the table, where the hash is usually based on the lattice coordinates. We also want these random values to be bounded—the most common interval is $[-1, 1]$.

The most basic lattice noise is known as *value noise*. In this case, we generate random values at each lattice point and then interpolate between them. An alternative is Ken Perlin's original noise function [119], also known as *gradient noise*. In this case, the position at each lattice point is set to zero, but the tangent vector is randomized. This can be done generating a random point on a unit sphere, as we did above. Value noise tends to be lower frequency (more smooth) and gradient noisc tends to be high frequency (more jaggy). Because of this, it's also common to combine them to create *value–gradient noise*.

To create interesting effects, we combine noise functions together. Often, we use the same noise function, but vary the spacing of the lattice. By doubling the frequency of lattice points we get what is called a new noise *octave*. This gives us a higher level of detail, which we can either use alone or combine with other octaves to get a more naturalistic effect: the lower octaves provide the broad strokes, while the higher octaves add the fiddly bits. For example, combining four octaves of gradient noise together gives us a *turbulence* function, which is very useful for producing cloud and marble effects. Usually the higher octaves are divided by their relative frequency before adding to help blend their effect into the lower-frequency base.

Let's look at a couple of examples using fragment shaders. Both of these are simplified from the *OpenGL Shading Language* text [132]. The first example generates a cloud texture on our object.

```
in vec3 localPos;
out vec4 fragColor;
void main()
{
    vec3 sky = vec3(0.0, 0.3, 0.8);
    vec3 cloud = vec3(0.8, 0.8, 0.8);
    float turb = (noise1(localPos) + noise1(2.0*localPos)*0.5
        + noise1(4.0*localPos)*0.25 + noise1(8.0*localPos)*0.125);
    vec3 color = mix(sky, cloud, turb);
    fragColor = vec4(color, 1.0);
}
```

Here we see the turbulence calculation. We use our nontransformed position (sent via a varying variable from the vertex shader) as the hash into our noise function and scale it to get different frequencies. We're using 2.0 as our frequency increment here to show the ideal behavior, but it's usually recommended to use a nonintegral value to decrease some of the gridlike behavior often seen with lattice noise. Once the turbulence value is calculated,

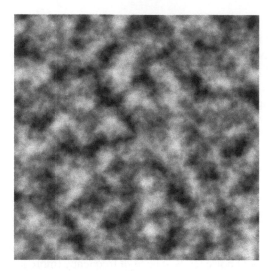

Figure 11.16. Sky texture generated using Perlin noise in a turbulence function.

we use it as a blending factor between our sky and cloud colors. Figure 11.16 shows the result.

We can do something similar to generate a marble texture. The base interpolant for the marble is the sine of the local y coordinate. We then perturb the base position by our turbulence to remove the regularity of the sine function as follows:

```
in vec3 localPos;
out vec4 fragColor;
void main()
{
    vec3 light = vec3(0.7, 0.7, 0.7);
    vec3 dark = vec3(0.0, 0.0, 0.0);
    float turb = (noise1(localPos) + noise1(2.0*localPos)*0.5
        + noise1(4.0*localPos)*0.25 + noise1(8.0*localPos)*0.125);
    float interp = sin(6.0*MCposition.y + 8.0*turb)*0.65;
    vec3 color = mix(light, dark, interp);
    fragColor = vec4(color, 1.0);
}
```

Figure 11.17 shows the result.

For both of these cases, we have used the built-in noise function in GLSL. Similar noise functions are available in HLSL, Cg, and other shading languages. Whether you use this function or not depends on the speed of your graphics processing unit. In these fragment shaders we are doing four function calls, which can get rather expensive. Because of this, graphics engineers often will generate a texture with different noise octave values in each color component and then do a lookup into that texture.

These examples give just a taste of what is available by making use of noise functions. Noise is used for generating wood textures, turbulence in fire texture, terrain, and many other cases. More detail on noise and other procedural generation can be found in Ebert et al. [39].

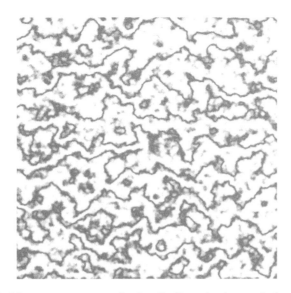

Figure 11.17. Marble texture generated using Perlin noise in a turbulence function.

11.6 Chapter Summary

In this chapter we discussed some basic probability and statistics that will help us build our random processes. We used some of these statistic measures to create basic techniques for measuring random data. We also surveyed the most common random number generators, in particular, the linear congruential generator and the Mersenne Twister. Finally, we wrapped things up by looking at some examples of using these random number generators, from simulating arbitrary distributions to building turbulence functions for computer graphics.

Further reading in random numbers is wide and varied. Gonick and Smith [59] is a very approachable guide to probability and statistics; Grinstead and Snell [65] is recommended as a more thorough and formal, yet still readable, text. While slightly out of date now, a standard survey of random number techniques can be found in Knuth [90]. A great deal of detail is given in this text to demonstrating the correctness of random algorithms and discussing techniques for measuring randomness. For those interested in unusual random distributions, particularly for graphics, Pharr and Humphries [121] is an excellent text. Finally, Ebert et al. [39] is the standard book for studying procedural algorithms.

12 Intersection Testing

12.1 Introduction

In the previous chapters we have been primarily focused on manipulating and displaying our game objects in isolation. Whether we are rendering an object or animating it, we haven't been concerned with how it might be interacting with other objects in our scene. This is neither realistic nor interesting. For example, you are manipulating an object right now: this book. You can hold it in your hand, turn its pages, or drop it on the floor. In the latter case, it stops reacting to you and starts reacting to the floor. If good game play derives from interesting interactions, then we need some way to detect when two game objects should be affecting one another and respond accordingly.

In this chapter we'll be concerned with a very straightforward question: How do we tell when two geometric entities are intersecting? This knowledge proves useful in many cases throughout a game engine. The most obvious is collision detection and response. Rather than have game objects pass through each other, we want them to push against each other and respond realistically. In the real world, this is a simple problem. Solid objects are solid; due to their physical properties, they just don't interpenetrate. But in the virtual world we have to create these constraints ourselves. Despite the fact that we have completely defined the geometry of our game objects, we still need to provide methods to detect when they interpenetrate. Only when we have a way to handle this can we write the code to perform the proper response.

Another time when we want to detect when two geometric entities interpenetrate is when we want to cast a ray and see what objects it intersects. One example of this we have seen already: detecting the object we've clicked on by generating a pick ray from a screen-space mouse click, and determining the first object we hit with that ray. Another way this is used is in artificial intelligence (AI). In order to simulate whether one AI agent can see another, we cast a ray from the first to the second and see if it intersects any objects. If not, then we can say that the first agent's target is in sight.

We have also mentioned a third use of object intersection before: determining which objects are visible in a view frustum so that we can do quick visibility culling. If they interpenetrate or are inside the frustum, then we go ahead to the rendering step; otherwise, they get skipped. This can considerably speed up our rendering.

Due to the variety of shapes and primitives used in a standard game engine, finding intersections between all of the cases can get quite complex; a single chapter is not enough to cover everything. Instead, we'll cover five basic objects, some methods for improving performance and accuracy, and directions for improvement. We will also briefly discuss how to use these methods in a simple collision detection system, and how we can apply similar techniques to our ray casting and frustum culling problems. Details on more complex systems can be found in the recommended reading at the end of the chapter.

12.2   Closest Point and Distance Tests

As we'll find, object intersection tests often can be described more easily in terms of a distance computation between two primitives, such as a point and a line. In particular, we'll often want to know if the distance between two primitives is less than some value, such as a radius. So, before we begin our discussion of determining intersections between bounding objects, we will cover a selection of useful methods for testing distances between certain geometric primitives.

Related to that topic is determining the closest points of approach between those same primitives; if we can find the closest points, the distance between the two primitives is the distance between those points. Because of this, we'll first consider closest point problems followed by how to calculate the distance between the same two primitives.

12.2.1   Closest Point on Line to Point

Our first problem is illustrated in Figure 12.1: Given a point Q, and a line L defined by a point P and a vector $\mathbf{v}$, how do we find the point Q' on the line that is closest to Q? We approach this by examining the geometric relationships between the point and line. In particular, we notice that the dotted line segment between Q and Q' is orthogonal to the line. This line segment corresponds to a line of projection: to find Q', we need to project Q onto the line.

To do this, we begin by computing the difference vector $\mathbf{w}$ between Q and P, or $\mathbf{w} = Q - P$. Then we project this onto $\mathbf{v}$, to get the component of $\mathbf{w}$ that points along $\mathbf{v}$. Recall that this is

$$\text{proj}_{\mathbf{v}}\mathbf{w} = \frac{\mathbf{w} \bullet \mathbf{v}}{\|\mathbf{v}\|^2}\mathbf{v}$$

We add this to the line point P to get our projected point Q', or

$$Q' = P + \frac{\mathbf{w} \bullet \mathbf{v}}{\|\mathbf{v}\|^2}\mathbf{v}$$

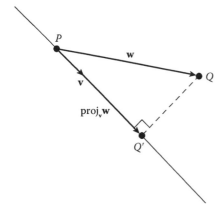

Figure 12.1. Closest point on a line.

The equivalent code is as follows:

```
IvVector3 IvLine3::ClosestPoint(const IvVector3& point)
{
    IvVector3 w = point - mOrigin;
    float vsq = mDirection.Dot(mDirection);
    float proj = w.Dot(mDirection);

    return mOrigin + (proj/vsq)*mDirection;
}
```

12.2.2 Line–Point Distance

Source Code
Library
Filename
IvLine3

As before, we're given a point Q and a line L defined by a point P and a vector $\mathbf{v}$. In this case, we want to find the distance between the point and the line. One way is to compute the closest point on the line and compute the distance between that and Q. A more direct approach is to use the Pythagorean theorem (Figure 12.2).

We note that $\mathbf{w} = Q - P$ can be represented as the sum of two vectors, one parallel to $\mathbf{v}(\mathbf{w}_{\parallel})$ and one perpendicular ($\mathbf{w}_{\perp}$). These form a right triangle, so from Pythagoras, $\|\mathbf{w}\|^2 = \|\mathbf{w}_{\parallel}\|^2 + \|\mathbf{w}_{\perp}\|^2$. We want to know the length of $\mathbf{w}_{\perp}$, so we can rewrite this as

$$\|\mathbf{w}_{\perp}\|^2 = \|\mathbf{w}\|^2 - \|\mathbf{w}_{\parallel}\|^2$$
$$= \mathbf{w} \cdot \mathbf{w} - \left\|\frac{\mathbf{w} \cdot \mathbf{v}}{\mathbf{v} \cdot \mathbf{v}}\mathbf{v}\right\|^2$$
$$= \mathbf{w} \cdot \mathbf{w} - \left(\frac{\mathbf{w} \cdot \mathbf{v}}{\mathbf{v} \cdot \mathbf{v}}\right)^2 \mathbf{v} \cdot \mathbf{v}$$
$$= \mathbf{w} \cdot \mathbf{w} - \frac{(\mathbf{w} \cdot \mathbf{v})^2}{\mathbf{v} \cdot \mathbf{v}}$$

Taking the square root of both sides will give us the distance between the point and the line.

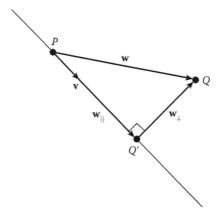

Figure 12.2. Computing distance from point to line, using a right triangle.

The equivalent code is as follows:

```
float IvLine3::DistanceSquared(const IvVector3& point)
{
    IvVector3 w = point - mOrigin;
    float vsq = mDirection.Dot(mDirection);
    float wsq = w.Dot(w);
    float proj = w.Dot(mDirection);

    return wsq - proj*proj/vsq;
}
```

Note that in this case we're computing the squared distance. In most cases we'll be using this to avoid computing a square root. Another optimization is possible if we can guarantee that $\mathbf{v}$ is normalized; in that case, we can avoid calculating and dividing by $\mathbf{v} \cdot \mathbf{v}$, since its value is 1.

Source Code
Library
IvMath
Filename
IvLineSegment3

12.2.3   Closest Point on Line Segment to Point

Recall that a line segment can be defined as the convex combination of two points P_0 and P_1, or

$$S(t) = (1 - t)P_0 + tP_1$$

where $0 \leq t \leq 1$. We can rewrite this as

$$S(t) = P_0 + t(P_1 - P_0)$$

or

$$S(t) = P + t\mathbf{v}$$

where as before $0 \leq t \leq 1$. In this case, $\mathbf{v}$ should not be normalized, as its length is the length of our line segment, and the endpoints are P and $P + \mathbf{v}$.

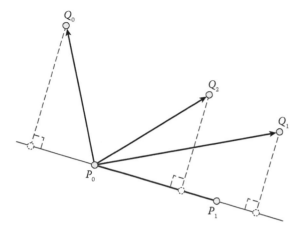

Figure 12.3. Three cases when projecting a point onto a line segment.

In the problem of finding the closest point on a line, we computed the projection of the point onto the line. Doing the same for a line segment gives us three cases (Figure 12.3). In the first case, the result of projecting Q_0 lies outside the segment but closest to P_0. In the second case, the result of projecting Q_1 lies outside the segment but closest to P_1. In the third case, the projected Q_2 lies on the segment, and we can use the same projection calculations that we used with a line.

To determine which case we're in, we begin by noting that

$$t = \frac{\mathbf{w} \cdot \mathbf{v}}{\mathbf{v} \cdot \mathbf{v}}$$

is acting as our parameter t for the projected point, where again $\mathbf{w} = Q - P$. If $t < 0$, then the projected point lies beyond P_0, and the closest point is P_0. Similarly, if $t > 1$, then the closest point is P_1.

Testing t directly requires a floating-point division. By modifying our test we can defer the division to be performed only when we truly need it, that is, when the point lies on the segment. Since $\mathbf{v} \cdot \mathbf{v} > 0$, then $\mathbf{w} \cdot \mathbf{v} < 0$ in order for $t < 0$. And in order for $t > 1$, then $\mathbf{w} \cdot \mathbf{v} > \mathbf{v} \cdot \mathbf{v}$.

The equivalent code is as follows:

```
IvVector3 IvLineSegment3::ClosestPoint(const IvVector3& point)
{
    IvVector3 w = point - mOrigin;

    float proj = w.Dot(mDirection);
    if ( proj <= 0 )
        return mOrigin;
    else
    {
        float vsq = mDirection.Dot(mDirection);
        if ( proj >= vsq )
```

```
                    return mOrigin + mDirection;
            else
                    return mOrigin + (proj/vsq)*mDirection;
        }
    }
```

12.2.4 Line Segment–Point Distance

As with lines, we can compute the distance to the line segment by computing the distance to the closest point on the line segment. If we recall, there are three cases: the closest point is P_0, P_1, or a point somewhere else on the segment, which we'll calculate.

If the closest point is P_0, then we can compute the distance as $\|Q - P_0\|$. Since $\mathbf{w} = Q - P_0$, then the squared distance is equal to $\mathbf{w} \cdot \mathbf{w}$.

If the closest point is P_1, then the squared distance is $(Q - P_1) \cdot (Q - P_1)$. However, we're representing our endpoint as $P_1 = P_0 + \mathbf{v}$, so this becomes $(Q - P_0 - \mathbf{v}) \cdot (Q - P_0 - \mathbf{v})$. We can rewrite this as

$$\begin{aligned}
\mathrm{distsq}(Q, P_1) &= ((Q - P_0) - \mathbf{v}) \cdot ((Q - P_0) - \mathbf{v}) \\
&= (\mathbf{w} - \mathbf{v}) \cdot (\mathbf{w} - \mathbf{v}) \\
&= \mathbf{w} \cdot \mathbf{w} - 2\mathbf{w} \cdot \mathbf{v} + \mathbf{v} \cdot \mathbf{v}
\end{aligned}$$

We've already calculated most of these dot products when determining whether we're closest to P_1, so all we need to compute is $\mathbf{w} \cdot \mathbf{w}$ and add. If the closest point lies elsewhere on the segment, then we use the line distance calculation just given.

The final code is as follows:

```
float IvLineSegment3::DistanceSquared(const IvVector3& point)
{
    IvVector3 w = point - mOrigin;

    float proj = w.Dot(mDirection);
    if ( proj <= 0 )
    {
        return w.Dot(w);
    }
    else
    {
        float vsq = mDirection.Dot(mDirection);
        if ( proj >= vsq )
        {
            return w.Dot(w) - 2.0f*proj + vsq;
        }
        else
        {
            return w.Dot(w) - proj*proj/vsq;
        }
    }
}
```

12.2.5 Closest Points between Two Lines

Source Code
Library
IvMath
Filename
IvLine3

Sunday [141] provides the following construction for finding the closest points between two lines. Note that in this case there are two closest points, one on each line, since there are two degrees of freedom. The situation is shown in Figure 12.4. Line L_1 is described by the point P_0 and the vector $\mathbf{u}$. Correspondingly, line L_2 is described by the point Q_0 and the vector $\mathbf{v}$, or

$$L_1(s) = P_0 + s\mathbf{u}$$
$$L_2(t) = Q_0 + t\mathbf{v}$$

Vectors $\mathbf{u}$ and $\mathbf{v}$ are not necessarily normalized.

We'll define the two closest points that we're looking for as lying at parameters s_c and t_c on the lines, and call them $L_1(s_c)$ and $L_2(t_c)$, respectively. We'll refer to the vector from $L_2(t_c)$ to $L_1(s_c)$ as $\mathbf{w}_c$.

Expanding $\mathbf{w}_c$, we have

$$\mathbf{w}_c = L_1(s_c) - L_2(t_c)$$
$$= P_0 + s_c\mathbf{u} - Q_0 - t_c\mathbf{v}$$
$$= (P_0 - Q_0) + s_c\mathbf{u} - t_c\mathbf{v}$$

We'll use $\mathbf{w}_0$ to represent the difference vector $P_0 - Q_0$, so

$$\mathbf{w}_c = \mathbf{w}_0 + s_c\mathbf{u} - t_c\mathbf{v} \tag{12.1}$$

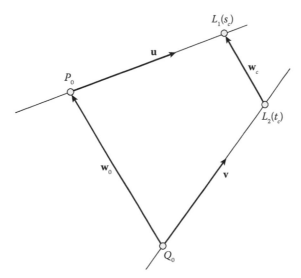

Figure 12.4. Finding the closest points between two lines.

In order for $\mathbf{w}_c$ to represent the vector of closest distance, it needs to be perpendicular to both L_1 and L_2. This means that

$$\mathbf{w}_c \bullet \mathbf{u} = 0$$
$$\mathbf{w}_c \bullet \mathbf{v} = 0$$

Substituting in Equation 12.1 and expanding, we get

$$0 = \mathbf{w}_0 \bullet \mathbf{u} + s_c \mathbf{u} \bullet \mathbf{u} - t_c \mathbf{u} \bullet \mathbf{v} \qquad (12.2)$$
$$0 = \mathbf{w}_0 \bullet \mathbf{v} + s_c \mathbf{u} \bullet \mathbf{v} - t_c \mathbf{v} \bullet \mathbf{v} \qquad (12.3)$$

We have two equations and two unknowns s_c and t_c, so we can solve for this system of equations. Doing so, we get the result that

$$s_c = \frac{be - cd}{ac - b^2} \qquad (12.4)$$
$$t_c = \frac{ae - bd}{ac - b^2} \qquad (12.5)$$

where

$$a = \mathbf{u} \bullet \mathbf{u}$$
$$b = \mathbf{u} \bullet \mathbf{v}$$
$$c = \mathbf{v} \bullet \mathbf{v}$$
$$d = \mathbf{u} \bullet \mathbf{w}_0$$
$$e = \mathbf{v} \bullet \mathbf{w}_0$$

There is one case where we need to be careful. If the two lines are parallel, then $\mathbf{u}$ and $\mathbf{v}$ are parallel, so $|\mathbf{u} \bullet \mathbf{v}| = \|\mathbf{u}\|\|\mathbf{v}\|$. Then the denominator $ac - b^2$ equals

$$ac - b^2 = (\mathbf{u} \bullet \mathbf{u})(\mathbf{v} \bullet \mathbf{v}) - (\mathbf{u} \bullet \mathbf{v})^2$$
$$= \|\mathbf{u}\|^2\|\mathbf{v}\|^2 - (\|\mathbf{u}\|\|\mathbf{v}\|)^2$$
$$= 0$$

This leads to a division by 0. The problem is that there are an infinite number of pairs of closest points spaced along each line. In this case, we'll just find the closest point Q' on L_2 to the origin P_0 of line L_1 and return P_0 and Q'.

```
void ClosestPoints( IvVector3& point1,
                    IvVector3& point2,
                    const IvLine3& line1,
                    const IvLine3& line2 )
{
    IvVector3 w0 = line1.mOrigin - line2.mOrigin;
    float a = line1.mDirection.Dot( line1.mDirection );
    float b = line1.mDirection.Dot( line2.mDirection );
    float c = line2.mDirection.Dot( line2.mDirection );
```

```
    float d = line1.mDirection.Dot( w0 );
    float e = line2.mDirection.Dot( w0 );
    float denom = a*c - b*b;
    if ( IsZero(denom) )
    {
        point1 = line1.mOrigin;
        point2 = line2.mOrigin + (e/c)*line2.mDirection;
    }
    else
    {
        point1 = line1.mOrigin + ((b*e - c*d)/denom)*line1.mDirection;
        point2 = line2.mOrigin + ((a*e - b*d)/denom)*line2.mDirection;
    }
}
```

12.2.6 Line–Line Distance

Source Code
Library
IvMath
Filename
IvLine3

From the calculation of closest points between two lines, we know that $\mathbf{w}_c$ is the vector of closest distance. Therefore, its length equals the distance between the two lines. Rather than compute the closest points directly, we can substitute the values of s_c and t_c into Equation 12.1 and compute the length of $\mathbf{w}_c$. As before, to avoid the square root, we can use $\|\mathbf{w}_c\|^2 = \mathbf{w}_c \cdot \mathbf{w}_c$ instead.

The code is as follows:

```
float DistanceSquared( const IvLine3& line1, const IvLine3& line2 )
{
    // compute parameters
    IvVector3 w0 = line1.mOrigin - line2.mOrigin;
    float a = line1.mDirection.Dot( line1.mDirection );
    float b = line1.mDirection.Dot( line2.mDirection );
    float c = line2.mDirection.Dot( line2.mDirection );
    float d = line1.mDirection.Dot( w0 );
    float e = line2.mDirection.Dot( w0 );
    float denom = a*c - b*b;
    // if lines parallel
    if ( IsZero(denom) )
    {
        IvVector3 wc = w0 - (e/c)*line2.mDirection;
        return wc.Dot(wc);
    }
    // otherwise
    else
    {
        IvVector3 wc = w0 + ((b*e - c*d)/denom)*line1.mDirection
                          - ((a*e - b*d)/denom)*line2.mDirection;
        return wc.Dot(wc);
    }
}
```

12.2.7 Closest Points between Two Line Segments

Source Code
Library
IvMath
Filename
IvLineSegment3

Finding the closest points between two line segments follows from finding the closest points between two lines. We compute s_c and t_c, as we've done, but then need to clamp the results to the ranges of s and t defined by the endpoints of the two line segments. As before, we'll define our line segments as starting at the source point of the line and ending at that source point plus the line vector. So for line L_1, the two points are P_0 and $P_0 + \mathbf{u}$, and for line L_2, the two points are Q_0 and $Q_0 + \mathbf{v}$. This gives us parameters 0 and 1 for the locations of the two endpoints. If our results s_c and t_c lie between the values 0 and 1, then our closest points lie on the two segments, and we're done.

Otherwise, we need to clamp our parameters to each of the endpoint parameters and try again. To see how to do that, let's take a look at the $s = 0$ endpoint. Remember that what we want to do is find the smallest possible distance between the two points while not sliding off the end of the segment; namely, we want to minimize the length of $\mathbf{w}_c$ while maintaining $s = 0$. Since length is always increasing, we'll use $\|\mathbf{w}_c\|^2$, which will be much easier to minimize. Remember that

$$\mathbf{w}_c = \mathbf{w}_0 + s_c\mathbf{u} - t_c\mathbf{v}$$

Since we're clamping s_c to 0, this becomes

$$\mathbf{w}_c = \mathbf{w}_0 - t_c\mathbf{v}$$

Therefore, for this endpoint we try to find the minimum value for

$$\mathbf{w}_c \bullet \mathbf{w}_c = (\mathbf{w}_0 - t_c\mathbf{v}) \bullet (\mathbf{w}_0 - t_c\mathbf{v}) \tag{12.6}$$

To do this, we return to calculus. To find a minimum value (in this case, there is only one) for a function, we find a place where the derivative is 0. Taking the derivative of Equation 12.6 in terms of t_c, we get the result

$$0 = -2\mathbf{v} \bullet (\mathbf{w}_0 - t_c\mathbf{v})$$

Solving for t_c, we get

$$t_c = \frac{\mathbf{v} \bullet \mathbf{w}_0}{\mathbf{v} \bullet \mathbf{v}} \tag{12.7}$$

So, for the fixed point on line L_1 at $s = 0$, this gives us the parameter of the closest point on line L_2. As we can see, this is equivalent to computing the closest point between a line and a point, where the line is L_2 and the point is P_0.

For the $s = 1$ endpoint, we follow a similar process. Our minimization function is

$$\mathbf{w}_c \bullet \mathbf{w}_c = (\mathbf{w}_0 + \mathbf{u} - t_c\mathbf{v}) \bullet (\mathbf{w}_0 + \mathbf{u} - t_c\mathbf{v}) \tag{12.8}$$

The corresponding zero derivative function is

$$0 = -2\mathbf{v} \bullet (\mathbf{w}_0 + \mathbf{u} - t_c\mathbf{v})$$

Solving for t_c gives us

$$t_c = \frac{\mathbf{v} \bullet \mathbf{w}_0 + \mathbf{u} \bullet \mathbf{v}}{\mathbf{v} \bullet \mathbf{v}}$$

Again, this is equivalent to computing the closest point between a line and a point, where the line is L_2 and the point is $P_0 + \mathbf{v}$. The solutions for s_c when clamping to $t = 0$ or $t = 1$ are similar.

One nice thing about these functions is that they use the a through e values that we've already calculated for the basic line–line distance calculation. So, Equation 12.7 becomes

$$t_c = \frac{e}{c}$$

So, which endpoints do we check? Well, if the parameter s_c is less than 0, then the closest segment point to line L_2 will be the $s = 0$ endpoint. And if s_c is greater than 1, then the closest segment point will be at $s = 1$. Choosing one or the other, we resolve for t_c and check that it lies between 0 and 1. If not, we perform the same process to clamp t_c to either the $t = 0$ or $t = 1$ endpoint and recalculate s_c accordingly (with some minor adjustments to ensure that we keep s_c within 0 and 1).

Once again, there is a trick we can do to avoid multiple floating-point divisions. Instead of computing, say, s_c directly and testing against 0 and 1, we can compute the numerator s_N and denominator s_D. The initial s_D is always greater than 0, so we know that if s_N is less than 0, s_c is less than 0 and we clamp to $s = 0$ accordingly. Similarly, if s_N is greater than s_D, we know that $s_c > 1$, and we clamp to $s = 1$. The same can be done for the t values. Using this, we can recalculate the numerator and denominator when necessary, and do the floating-point divides only after all the clamping has been done.

For example, the following code snippet calculates the s values:

```
// clamp s_c to 0
if (sN < 0.0f)
{
    sN = 0.0f;
    tN = e;
    tD = c;
}
// clamp s_c to 1
else if (sN > sD)
{
    sN = sD;
    tN = e + b;
    tD = c;
}
```

The full code is too long to contain here, but can be found at www.essentialmath.com.

12.2.8 Line Segment–Line Segment Distance

Finding the segment-to-segment squared distance is similar to line-to-line distance: we follow the procedure for closest points between line segments, calculate $\mathbf{w}_c$ directly from the final s_c and t_c, and then compute its length. The full code can be found at www.essentialmath. com in the `IvLineSegment3` friend function `DistanceSquared()`.

Source Code
Library
IvMath
Filename
IvLineSegment3

12.2.9 General Linear Components

Testing ray versus ray or line versus line segment is actually a simplification of the segment–segment closest point and distance determination. Instead of clamping against both components, we need only clamp against those endpoints that are necessary. So, for

Source Code
Library
IvMath
Filename
IvLine3
IvRay3
IvLineSegment3

example, if we treat $P_0 + s\mathbf{u}$ as the parameterization of a line segment, and $Q_0 + t\mathbf{v}$ as a line, then we need only to ensure that s_c is between 0 and 1, clamp to the appropriate endpoint, and adjust t_c accordingly. Similarly, if we're working with rays, we need only to clamp s_c or t_c to 0.

Implementations of these algorithms can be found in the appropriate classes.

12.3 Object Intersection

Now that we've covered some methods for measuring distance between primitives, we can talk about object intersection. The most direct, and naive, approach to determine whether two objects are intersecting is to work directly from raw object data. We could start with a triangle in object A and a triangle in object B and see if they are intersecting. Then we move to the next triangle in object A and test again. While ultimately this may work (the exception is if one object is inside the other), it will take a while to do, and most of the time performing all those tests isn't even necessary. Take the two objects in Figure 12.5. They are clearly not intersecting—we can tell that in an instant. But our minds are not considering each object as a collection of lines and doing individual tests. Rather, we are comparing them as a whole, as two rough blobs, and determining that the blobs aren't intersecting. By using a similar process in our intersection routines, we can save ourselves a lot of time.

For instance, suppose we surround each object with a sphere (Figure 12.6). We can begin by testing for intersection between the spheres. If the two spheres aren't intersecting, we know the objects aren't either. If the spheres are intersecting, we can try comparing another simplified version of our object—say, two boxes. The boxes fit the shape of our objects better, but are still a simpler test than our full triangle–triangle comparison. If the boxes intersect, only then do we perform our complex collision detection routine.

This technique of using simplified objects to test intersections before performing more expensive operations is commonly used in game engines, and is necessary to get collision detection and other intersection-based systems running in real time. The simplified objects

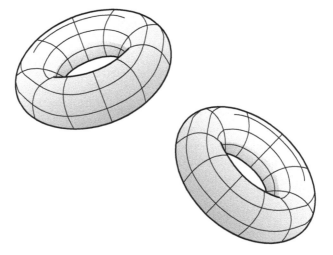

Figure 12.5. Nonintersecting objects.

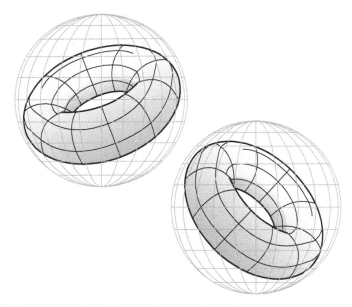

Figure 12.6. Nonintersecting objects with bounding sphere.

are known as *bounding objects* and are named specifically after the basic primitives we
used to approximate the object: bounding spheres and bounding boxes. In games, we can
often get away with ignoring the underlying geometry completely and only using bounding
objects to determine intersections. For example, when handling collisions in this way, either
the action happens so fast that we don't notice any overlapping objects or objects reacting
to collision when they appear separated, or the error is so slight that it doesn't matter. In any
case, choosing the side of making the simulation run faster for a better play experience is
usually a good decision.

One thing to note with the following algorithms is that their performance is often
dependent on the platform that they are run on. For example, some systems don't have
predictive branching, so conditionals are quite slow. So, on such a platform, an algorithm
that calculates unnecessary data may actually turn out to be faster than one that attempts
to avoid this using if-then-else clauses. Even on relatively similar architectures there can
be surprising differences in relative performance. This is shown strikingly by Löfstedt and
Akenine-Möller [105].

To keep things concise, we have chosen a few algorithms that are commonly used and
are relatively fast on a broad variety of architectures. Other books are more detailed, covering
many different polytopes (the 3D equivalent of polygons) and interactions between all sorts
of bounding objects. In our case, we'll focus on a few simple shapes, beginning with
the simplest objects and moving on to the most complex, or most expensive, to compute.
However, the reader should be aware of the issues above and may need to explore alternatives
for his or her particular application.

Within each section we'll only consider three cases of intersection. We'll first look
at intersections between objects of the same bounding type, which is useful in collision
detection. Second, we'll cover intersections between a ray and the particular bounding

object, which we'll need for picking and visibility testing for AI. Finally, we'll discuss how to determine intersection between a plane and the bounding object, which can be used for both culling against frustum planes and collisions with essential planar objects like walls. In all cases, we aren't concerned with the exact point of intersection, just whether the items intersect.

12.3.1 Spheres

12.3.1.1 Definition

The simplest possible bounding object is a sphere. It also has the most compact representation: a center point C and a radius r (Figure 12.7). When bounding a rigid object, a sphere is also independent of the object's orientation. This allows us to update a sphere quickly—when an object moves, we need only to update the sphere's position. If the object is scaled, we can scale the radius accordingly. The combination of low memory usage, fast update time, and fast intersection tests makes bounding spheres a first choice in any real-time system.

The surface of the sphere is defined as all points P such that the length of the vector from C to P is equal to the radius:

$$\sqrt{(P_x - C_x)^2 + (P_y - C_y)^2 + (P_z - C_z)^2} = r$$

or

$$\sqrt{(P - C) \bullet (P - C)} = r$$

Ideally, we'll want to choose the smallest possible sphere that encompasses the entire object. Too small a sphere, and we may skip two objects that are actually intersecting. Too large, and we'll be unnecessarily performing our more expensive tests for objects that are

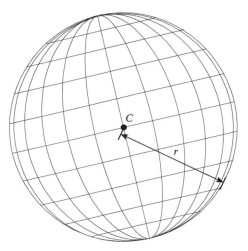

Figure 12.7. Bounding sphere.

clearly separate. Unfortunately, the most obvious methods for choosing a bounding sphere will not always generate as tight a fit as we might like.

One such method is to take the local origin of the object as our center C, and compute r by taking the maximum distance from that to all the vertices in the object. There are many problems with this. The most common is that the local origin could be considerably offset from the most desirable center point for the object (Figure 12.8a). This could happen if you have a character whose origin is at its feet, so it can be placed on the ground properly. An alternate but equivalent situation is where the origin is at a reasonable center point for the majority of the object's vertices, but there are one or two outlying vertices that cause problems (Figure 12.8b).

Eberly [35] provides a number of methods for finding a better fit. One is to average all the vertex locations to get the centroid and use that as our center. This works well for the case of a noncentered origin, but still is a problem for an object with outlying points (Figure 12.8c). The reason is that the majority of the points lie within a small area and thus weight the centroid in that direction, pulling it away from the extrema.

We could also take an axis-aligned bounding box in the object's local space and use its endpoints to compute our sphere position and radius (Figure 12.8d). This tends to center the sphere better but leads to a looser fit. A compromise method uses the center of the bounding box as our sphere position, and computes the radius as the maximum distance from the center to our points. This gives a slightly better result. The code for this last method is as follows:

```
void
IvBoundingSphere::Set( const IvPoint3* points, unsigned int numPoints )
  // compute minimal and maximal bounds
  IvVector3 min(points[0]), max(points[0]);
  for ( unsigned int i = 1; i < numPoints; ++i )
  {
    if (points[i].x < min.x)
      min.x = points[i].x;
    else if (points[i].x > max.x )
      max.x = points[i].x;
    if (points[i].y < min.y)
      min.y = points[i].y;
    else if (points[i].y > max.y )
      max.y = points[i].y;
    if (points[i].z < min.z)
      min.z = points[i].z;
    else if (points[i].z > max.z )
      max.z = points[i].z;
  }

  // compute center and radius
  mCenter = 0.5f*(min + max);
  float maxDistance = ::DistanceSquared( mCenter, points[0] );
  for ( unsigned int i = 1; i < numPoints; ++i )
  {
    float dist = ::DistanceSquared( mCenter, points[i] );
    if (dist > maxDistance)
      maxDistance = dist;
  }
  mRadius = ::IvSqrt( maxDistance );
}
```

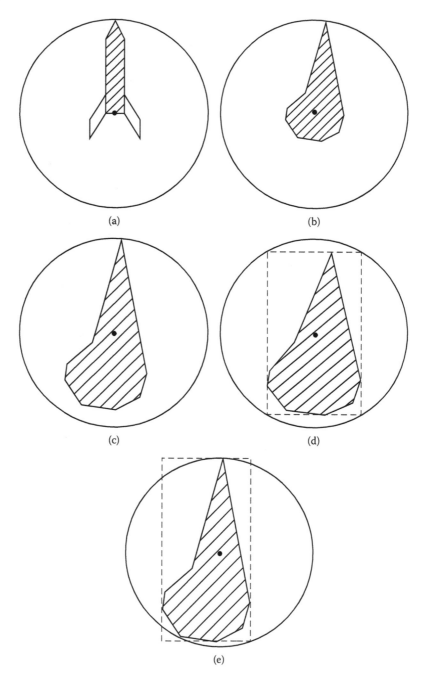

Figure 12.8. (a) Bounding sphere, offset origin; (b) bounding sphere, outlying point; (c) bounding sphere, using centroid, object vertices; (d) bounding sphere, using box center, box vertices; and (e) bounding sphere, smallest possible.

It should be noted that none of these methods is guaranteed to find the smallest bounding sphere. The standard algorithm for this is by Welzl [156], who showed that linear programming can be used to find the optimally smallest sphere surrounding a set of points (Figure 12.8e). Many implementations are readily available online: one by Bernd Gaertner is provided under the GNU General Public License.

While we don't want to be cavalier about using ridiculously large bounding spheres, in some cases having the tightest possible fit isn't that much of an issue. Our objects will not be generally spherical, and so we'll be using something more complex for our final intersection test. As long as our spheres are reasonably close to a good fit, they will act to cull a great number of obvious cases, which is all we can ask for.

12.3.1.2 Sphere–Sphere Intersection

Determining whether two spheres are intersecting is as simple as their representation. We need only to determine whether the distance between their centers is less than the sum of their two radii (Figure 12.9), or

$$\sqrt{(C_1 - C_2) \bullet (C_1 - C_2)} <= r_1 + r_2 \qquad (12.9)$$

The square root operation is expensive, and in any case, it is unnecessary. Since we're not looking for the absolute difference, just a relation, we can use

$$(C_1 - C_2) \bullet (C_1 - C_2) <= (r_1 + r_2)^2 \qquad (12.10)$$

As promised, this gives us an extremely cheap test for culling large numbers of intersections. This is why bounding spheres are used everywhere in computer graphics and simulation; we perform an initial fast check with a bounding sphere first before even considering the more complex cases.

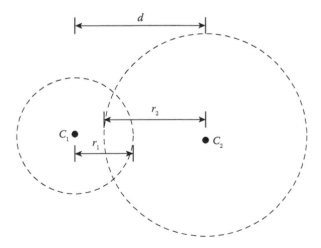

Figure 12.9. Sphere–sphere intersection.

The code is as follows:

```
bool
IvBoundingSphere::Intersect( const IvBoundingSphere& other )
{
    IvVector3 centerDiff = mCenter - other.mCenter;
    float radiusSum = mRadius + other.mRadius;
    return ( centerDiff.Dot(centerDiff) <= radiusSum*radiusSum );
}
```

12.3.1.3 Sphere–Ray Intersection

Intersection between a sphere and a ray is nearly as simple. Instead of testing two centers and comparing the distance with the sum of two radii, we test the distance between a single sphere center and a ray. If the distance is less than or equal to the sphere's radius, then the ray intersects the sphere (Figure 12.10).

We can use the line–point distance measurement described as the basis for this test. The code is as follows (it assumes an initial nonzero, nonnormalized **v**):

```
bool
IvBoundingSphere::Intersect( const IvRay3& ray )
{
    // compute intermediate values
    IvVector3 w = mCenter - ray.mOrigin;
    float wsq = w.Dot(w);
    float proj = w.Dot(ray.mDirection);
    float rsq = mRadius*mRadius;

    // if sphere behind ray, no intersection
    if ( proj < 0.0f && wsq > rsq )
       return false;

    float vsq = ray.mDirection.Dot(ray.mDirection);

    // test length of difference vs. radius
    return ( vsq*wsq - proj*proj <= vsq*mRadius*mRadius );
}
```

An additional check has been added since we're using a ray. If the sphere lies behind the origin of the ray, then there is no intersection. This is true if the angle between the difference vector **w** and the line direction is greater than 90 degrees (`proj < 0.0f`) and the line origin lies outside of the sphere (`wsq > rsq`).

We also remove the need for a floating-point divide by multiplying through by `vsq`. This adds two multiplications, but this still should be faster on most floating-point processors. As before, if we can guarantee that the ray direction vector is normalized, then we can remove the need for `vsq` altogether.

12.3.1.4 Sphere–Plane Intersection

Testing whether a sphere lies entirely on one side of a plane can be done quite efficiently. Recall that we can determine the distance between a point and such a plane by taking the

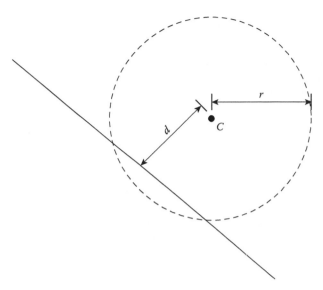

Figure 12.10. Line–sphere intersection.

absolute value of the result of the plane equation. If the result is positive and the distance is greater than the radius, then the sphere lies on the inside of the plane. If the result is negative, and the distance is greater than the sphere's radius, then the sphere lies outside of the plane. Otherwise, the sphere intersects the plane.

The code for this test is as follows:

```
float
IvBoundingSphere::Classify( const IvPlane& plane )
{
    float distance = plane.test(mCenter):
    if ( distance > radius)
    {
        return distance-radius;
    }
    else if ( distance < -radius )
    {
        return distance+radius;
    }
    else
    {
        return 0.0f;
    }
}
```

Here we're returning a signed distance, like the standard plane test. If the sphere intersects, we return zero. Otherwise, we return the signed distance minus the signed distance of the radius.

12.3.2 Axis-Aligned Bounding Boxes

Source Code
Library
IvCollision
Filename
IvAABB

12.3.2.1 Definition

Spheres work well as either cheap culling objects or bounding objects for a small class of models (i.e., if you're tossing grenades or writing a billiards game). For more angular objects, we need a better-fitting bounding surface. One possibility is the bounding box. Just like the bounding sphere, the ideal bounding box is the smallest possible box that encloses an object.

The first type we'll consider is the AABB, or axis-aligned bounding box, so called because the box edges are aligned to the *world* axes. This makes representation of the box simple: we use two points, one each for the minimum and maximum *xyz* positions (Figure 12.11). When the object is translated, to update the box we translate the minimum and maximum points. Similarly, if the object is scaled, we scale the two points relative to the box center. However, because the box is aligned to the world axes, any rotation of the object means that we have to recalculate the minimum and maximum points from the object vertices' new positions in world space.

The other disadvantage AABBs have is that in many cases, like spheres, they still aren't a very close fit to the object they are trying to approximate (Figure 12.12). And for rounded objects like submarines or organic objects like humans, the fact that they have corners is a disadvantage as well. However, they are relatively cheap to compute and cheap to test as well, so they continue to prove useful.

One advantage that world axis-aligned boxes have over a box oriented to the object's local space is that we need only recompute them once per frame, and then we can compare them directly without further transformation, since they are all in the same coordinate frame. So, while AABBs have a high per-frame overhead (since we have to recalculate them each time an object reorients), they are *extremely* cheap to test against one another. As we'll see,

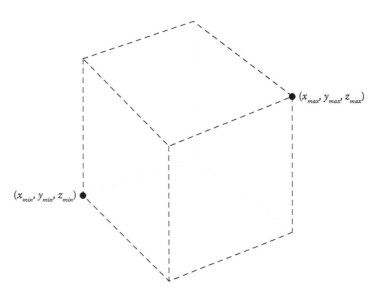

Figure 12.11. Axis-aligned bounding box.

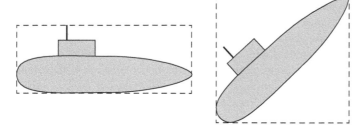

Figure 12.12. Fitting axis-aligned bounding box.

there is a lot more overhead for determining intersection between oriented boxes. Oriented boxes are generally cheap per frame (they move with the transforms of the object), but are more expensive to test against one another.

To compute an AABB, we first transform the object into world space. Then we set the minimum and maximum points to be equal to the first point (in world space, remember) in the object. Starting with the second point, we compare the *xyz* values of each point with those in the minimum and maximum. If any coordinate is less than that in the minimum, set the minimum coordinate to that value, and the same for the maximum, except use greater than. When done, this will give you the axis-aligned extrema for your box.

```
void
IvAABB::Set( const IvPoint3* points, unsigned int numPoints )
{
    ASSERT( points );

    // compute minimal and maximal bounds
    mMinima.Set(points[0]);
    mMaxima.Set(points[0]);
    for ( unsigned int i = 1; i < numPoints; ++i )
    {
        if (points[i].x < mMinima.x)
            mMinima.x = points[i].x;
        else if (points[i].x > mMaxima.x )
            mMaxima.x = points[i].x;
        if (points[i].y < mMinima.y)
            mMinima.y = points[i].y;
        else if (points[i].y > mMaxima.y )
            mMaxima.y = points[i].y;
        if (points[i].z < mMinima.z)
            mMinima.z = points[i].z;
        else if (points[i].z > mMaxima.z )
            mMaxima.z = points[i].z;
    }
}
```

12.3.2.2 AABB–AABB Intersection

In order to understand how we find intersections between two axis-aligned boxes, we introduce the notion of a *separating plane*. The general idea is this: we check the boxes in each of the coordinate directions in world space. If we can find a plane that separates the two boxes in any of the coordinate directions, then the two boxes are not intersecting. If we fail all three separating plane tests, then they are intersecting and we handle it appropriately.

Let's look at the process of finding a separating plane between two boxes in the x direction. Since the boxes are axis aligned, this becomes a one-dimensional (1D) problem on a number line. The minimum and maximum values of the two boxes become the extrema of two intervals on the line. If the two intervals are separate, then there is a separating plane and the two boxes are separate along the x direction. This is the case only if the maximum value of one interval is less than the minimum value of the other interval (Figure 12.13). Expressing this for all three axes:

```
bool
IvAABB::Intersect( const IvAABB& other )
{
  // if separated in x direction
  if (mMinima.x > other.mMaxima.x || other.mMinima.x > mMaxima.x )
    return false;

  // if separated in y direction
  if (mMinima.y > other.mMaxima.y || other.mMinima.y > mMaxima.y )
    return false;

  // if separated in z direction
  if (mMinima.z > other.mMaxima.z || other.mMinima.z > mMaxima.z )
    return false;

  // no separation, must be intersecting
  return true;
}
```

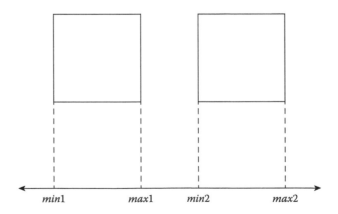

Figure 12.13. Axis-aligned box–box separation test.

Examining this code makes another advantage of AABBs clear. If we're using three-dimensional (3D) objects in an essentially two-dimensional (2D) game, we can ignore the z-axis and so save a step in our computations. This is not always possible with boxes aligned to the local axes of an object.

12.3.2.3 AABB–Ray Intersection

Determining intersection between a ray and an axis-aligned box is similar to determining intersection between two boxes. We check one axis direction at a time as before, except that in this case, there is a little more interaction between steps.

Figure 12.14 shows a 2D cross section of the situation. The ray R shown intersects the minimum and maximum x planes of the box at $R(s_x)$ and $R(t_x)$, respectively, and the minimum and maximum y planes at $R(s_y)$ and $R(t_y)$. Instead of testing for extrema overlaps in the box axes directions, we'll test whether there is overlap between the line segment from $R(s_x)$ to $R(t_x)$ and the line segment from $R(s_y)$ to $R(t_y)$. This is the same as testing whether the intervals of the line parameters $[s_x, t_x]$ and $[s_y, t_y]$ overlap.

If the ray misses the box, as in the figure, then the $[s_x, t_x]$ interval doesn't overlap the $[s_y, t_y]$ interval, just like the preceding box–box intersection. So, if there's no overlap (if $t_x < s_y$, or vice versa), then there's no intersection, and we stop. If they do overlap, then we test that overlap interval against the z intersections. If there's overlap there as well, then we know that the ray intersects the box.

For each axis, we begin by computing the parameters where the ray (represented by the point P and vector $\mathbf{v}$) crosses the minimum and maximum planes. So, for example, in the x direction we'll calculate intersections with the $x = x_{min}$ and $x = x_{max}$ planes. To do this, we need to solve the following equations:

$$P_x + s_x v_x = x_{min}$$
$$P_x + t_x v_x = x_{max}$$

Solving for s_x and t_x, we get

$$s_x = \frac{x_{min} - P_x}{v_x}$$
$$t_x = \frac{x_{max} - P_x}{v_x}$$

To simplify adjustment of our overlap interval, we want to ensure that $s_x < t_x$. This can be handled by checking whether $1/v_x < 0$; if so, we'll swap the x_{min} and x_{max} terms.

We'll track our parameter overlap interval by using two values s_{max} and t_{min}, initialized to the maximum interval. For a ray this is $[0, \infty]$; for a line this would be $[-\infty, \infty]$; for a segment it would be $[0, s]$, where s is the length of the segment. These represent the maximum s and minimum t values seen so far. As we calculate intersection parameters for each axis, we'll sort them so that $s < t$, and then update s_{max} and t_{min} if $s > s_{max}$ or $t < t_{min}$. We know that the ray misses the box if we ever find that $s_{max} > t_{min}$. For example, looking at Figure 12.14, after doing the x-axis calculations we see that $s_{max} = s_x$ and $t_{min} = t_x$. After the y-axis parameters are computed, t_{min} is updated to t_y, and s_{max} remains s_x. But $s_x > t_y$, so there is no intersection.

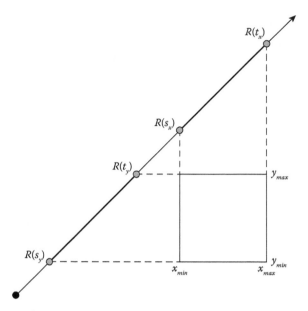

Figure 12.14. Axis-aligned box–ray separation test.

The code, abbreviated for space, is as follows:

```
bool
IvAABB::Intersect( const IvRay3& ray )
{
    float maxS = 0.0f;   // for line, use -FLT_MAX
    float minT = FLT_MAX;   // for line segment, use length

    // do x coordinate test (yz planes)

    // compute sorted intersection parameters
    float s, t;
    float recipX = 1.0f/ray.mDirection.x;
    if ( recipX >= 0.0f )
    {
        s = (mMin.x - ray.mOrigin.x)*recipX;
        t = (mMax.x - ray.mOrigin.x)*recipX;
    }
    else
    {
        s = (mMax.x - ray.mOrigin.x)*recipX;
        t = (mMin.x - ray.mOrigin.x)*recipX;
    }

    // adjust min and max values
    if ( s > maxS )
        maxS = s;
    if ( t < minT )
        minT = t;
```

```
    // check for intersection failure
    if ( maxS > minT )
        return false;

    // do y and z coordinate tests (xz & xy planes)
    ...

    // done, have intersection
    return true;
}
```

There's one special case that is implicitly handled: clearly if v_x is 0, then there are no solutions for s_x and t_x; the ray is parallel to the minimum and maximum planes. Normally in this case we'd need to test whether P_x lies between x_{min} and x_{max}. If not, the ray misses the box and there is no intersection. However, when using the IEEE floating-point standard, division by zero will return $-\infty$ for a negative numerator, and ∞ for a positive numerator. Hence, if the ray misses the box, the resulting interval will be either $[-\infty, -\infty]$ or $[\infty, \infty]$, which will lead to intersection failure. The only odd case is when the origin of the ray lies on one of the box planes, so s or t will end up being $0/\infty = \text{NaN}$. This still isn't a problem, as the subsequent comparisons will fail (because any comparison involving NaN returns false) and imply intersection, which is correct. More detail can be found in [5].

12.3.2.4 AABB–Plane Intersection

The most naive test to determine whether a box intersects a plane is to see whether a single box edge crosses the plane. That is, if two neighboring vertices lie on either side of the plane, there is an intersection. There are 12 edges, so this requires 24 plane tests. There are two improvements we can make to this. The first is to note that we need to test only *opposing* corners of the box, that is, two vertices that lie at either end of a diagonal that passes through the box center. This cuts the number of "edges" to be checked down to four. The second improvement is provided by Akenine-Möller et al. [1], who note that we really need to test only one: the diagonal most closely aligned with the plane normal. Figure 12.15 shows a cross section of the situation.

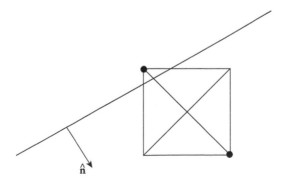

Figure 12.15. Axis-aligned box–plane separation test.

Code to manage this is as follows. As before, we return zero if there is an intersection, and the signed distance otherwise.

```
float
IvAABB::Classify( const IvPlane& plane )
{
    IvVector3 diagMin, diagMax;
    // set min/max values for x direction
    if ( plane.mNormal.x >= 0)
    {
        diagMin.x = mMin.x;
        diagMax.x = mMax.x;
    }
    else
    {
        diagMin.x = mMax.x;
        diagMax.x = mMin.x;
    }

    // ditto for y and z directions
    ...
    // minimum on positive side of plane, box on positive side
    float test = plane.mNormal.Dot( diagMin ) + plane.mD;
    if ( test > 0.0f )
        return test;

    test = plane.mNormal.Dot ( diagMax ) + plane.mD;
    // min on nonpositive side, max on nonnegative side, intersection
    if ( test >= 0.0f )
        return 0.0f;
    // max on negative side, box on negative side
    else
        return test;
}
```

A further optimization is to remove the conditionals for generating the maximal diagonal by computing a central point and half-extent vectors for the AABB, and using the algorithm for object-oriented bounding boxes; see Section 12.3.4 for more details.

12.3.3 Swept Spheres

Source Code
Library
IvCollision
Filename
IvCapsule

12.3.3.1 Definition

The bounding sphere and the axis-aligned bounding box have one problem: there is no real sense of orientation. The sphere is symmetric across all axes and the AABB is always aligned to the world axes. For objects that have definite long and short axes (e.g., a human), this doesn't provide for an ideal approximation. The next two bounding objects we'll consider are not tied to the world axes at all, which makes them much more suitable for general models.

The simplest of such bounding regions are the swept spheres. If we consider the sphere as a region enclosed by a radius around a point, or a zero-dimensional center, the swept spheres use higher-dimensional centers. One example is the *capsule*, which is a line segment

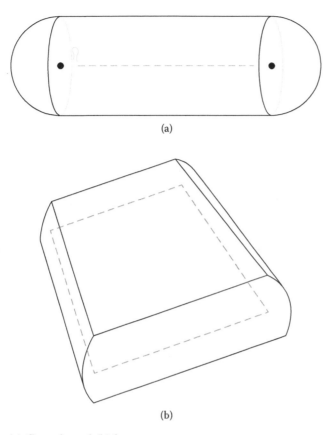

(a)

(b)

Figure 12.16. (a) Capsule and (b) lozenge.

surrounded by a radius (Figure 12.16a). Another possibility is the lozenge, which has a quadrilateral center (Figure 12.16b). For our purposes, we'll concentrate on capsules (Eberly [35] provides more information on lozenges and other swept spheres).

Computing the capsule in local space for a set of points is fairly straightforward, but not as simple as spheres or bounding boxes. Our first step is to compute a bounding box for the points. If the object is generally axis aligned (not unreasonable considering that the artists usually build objects in this way), we can use an axis-aligned bounding box. Otherwise, we may need an oriented bounding box (see Section 12.3.4 on how to compute this). We then find the longest side. The line that we will use for our baseline segment runs through the middle of the box. We'll use the center of one end of the box as our line point A, and the box axis $\mathbf{w}$ as our line vector. We could use the local origin and a coordinate axis for our line, but while we're willing to assume axis alignment, we're not so optimistic as to assume that the object is centered on a coordinate axis.

Now we need to compute the radius r of the capsule. For each point in the object, we compute the distance from the point to the line. The maximum distance becomes our radius. The line combined with the radius gives us a tube with radius r and ends extending to infinity. All the points in the object just fit inside the tube.

The final part to building the capsule is capping the tube with two hemispheres that just contain any points near the end of the object. Eberly [35] describes a method for doing this. The center of each hemisphere is one of the two endpoints of the line segment, so finding the hemisphere allows us to define the line segment. Let's consider the endpoint with the smaller t value—call it $L(\xi_0)$—shown in Figure 12.17. We want to find the leftmost hemisphere (i.e., the one with the smallest ξ_0) so that all points in the model lie either on the hemisphere (such as point P_0) or to the right of it (point P_1). Another way to think of this is that for each point we'll compute a hemisphere centered on the line that exactly contains that point and choose the hemisphere with the smallest ξ_0 value. If we do the same at the other end, with hemispheres oriented the other way and choosing the one with largest parameter value ξ_1, then all points will be tightly enclosed by the capsule.

To set this up, we first need to transform our points from the local space of the object to the local space of the line. We'll build a coordinate frame consisting of the line point A, normalized line vector $\hat{\mathbf{w}}$, and two vectors perpendicular to $\hat{\mathbf{w}}$: $\hat{\mathbf{u}}$ and $\hat{\mathbf{v}}$. Subtracting the line point from the object point and multiplying by a 3×3 matrix formed from $\hat{\mathbf{u}}$, $\hat{\mathbf{v}}$, and $\hat{\mathbf{w}}$ transforms the object-space point P to a line-space point P' with line-space coordinates (u, v, w). Since $\hat{\mathbf{w}}$ is normalized, a point $L(\xi_0)$ on the line equals $(0, 0, \xi_0)$ in line space.

If P' lies on a hemisphere with radius r and center X_0 on the line, the length of a vector $\mathbf{d}$ from X_0 to P_i' should be equal to the radius r (Figure 12.18). Given this and the other parameters, we should be able to solve for X_0, and hence ξ_0.

The vector $\mathbf{d} = P' - X_0$. In line space, $\mathbf{d} = (u, v, w) - (0, 0, \xi) = (u, v, w - \xi)$. Ensuring that $\|\mathbf{d}\| = r$ means that

$$u^2 + v^2 + (w - \xi_0)^2 = r^2$$

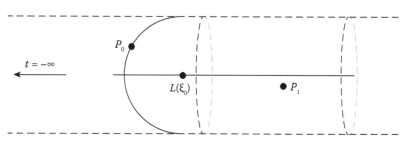

Figure 12.17. Capsule endcap fitting.

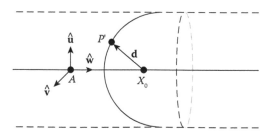

Figure 12.18. Determining hemisphere center X_0 for given point P'.

Solving for ξ_0, we get

$$\xi_0 = w - \left(\pm \sqrt{r^2 - (u^2 + v^2)} \right)$$

Since this is a hemisphere, we want X_0 to be to the right of P, so $w \geq \xi_0$, and this becomes

$$\xi_0 = w + \sqrt{r^2 - (u^2 + v^2)}$$

Computing this for every point P in our model and finding the minimum ξ_0 gives us our first endpoint. Similarly, the second endpoint is found by finding the maximum value of

$$\xi_1 = w - \sqrt{r^2 - (u^2 + v^2)}$$

12.3.3.2 Capsule–Capsule Intersection

Handling capsule–capsule intersection is very similar to sphere–sphere intersection. Instead of calculating the distance between two points, and determining whether that is less than the sum of the two radii, we calculate the distance between two line segments and check against the radii. As before, if the distance is less than the sum of the two radii, we have intersecting capsules.

```
bool
IvCapsule::Intersect( const IvCapsule& other )
{
    float radiusSum = mRadius + other.mRadius;
    return ( mSegment.DistanceSquared( other.mSegment )
                                    <= radiusSum*radiusSum );
}
```

12.3.3.3 Capsule–Ray Intersection

Capsule–ray intersection follows from capsule–capsule collision. Instead of finding the distance between two line segments, we need to find the distance between a ray and a line segment and compare it to the radius of the capsule, as follows:

```
bool
IvCapsule::Intersect( const IvRay3& ray )
{
    // test distance between line and segment vs. radius
    return ( ray.DistanceSquared( mSegment ) <= mRadius*mRadius );
}
```

12.3.3.4 Capsule–Plane Intersection

There are two tests necessary to determine whether a capsule intersects a plane. First of all, if the two endpoints of the line segment defining the capsule lie on either side of the plane, then clearly the capsule intersects the plane. However, even if the line segment lies on one side of the plane, the distance between one of the endpoints and the plane may be less than the radius. In this case, the capsule and plane would also intersect. Both cases are easy to test; we already have the pieces in place.

The code is as follows:

```
float
IvCapsule::Classify( const IvPlane& plane )
{
    float s0 = plane.Test( mSegment.GetEndpoint0() );
    float s1 = plane.Test( mSegment.GetEndpoint1() );

    // points on opposite sides or intersecting plane
    if (s0*s1 <= 0.0f)
        return 0.0f;

    // intersect if either endpoint is within radius distance of plane
    if( IvAbs(s0) <= mRadius || IvAbs(s1) <= mRadius )
        return 0.0f;

    // return signed distance
    return ( IvAbs(s0) < IvAbs(s1) ? s0 : s1 );
}
```

12.3.4 Object-Oriented Boxes

12.3.4.1 Definition

World axis-aligned boxes are easy to create and fast to use for detecting intersections, but are not a very tight fit around objects that are not themselves generally aligned to the world axes (Figure 12.12). A more accurate approach is to create an initial bounding box that is a tight fit around the object in local space, and then rotate and translate the box as well as the object (Figure 12.19). These are known as object-oriented bounding boxes, or OBBs. This has another advantage in that we don't have to recalculate the box every time the object moves, but just transform the initial one. Also, for rigid objects with a large number of vertices, recomputing the AABB every frame may be too expensive. The disadvantage is that testing intersections between two object-oriented boxes is more complicated. In the axis-aligned case, we could simplify our cases down to three tests because of the alignment.

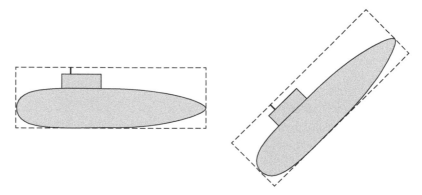

Figure 12.19. Object-oriented bounding boxes.

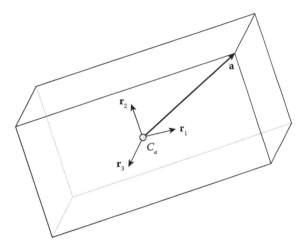

Figure 12.20. Properties of OBBs.

In the OBB case, the two can be at any relative orientation to each other, which complicates the issue considerably.

The representation for an OBB A consists of the center point C_a, an orientation matrix $\mathbf{R}_a$, and an extent vector $\mathbf{a}$ (Figure 12.20). The extent vector represents the difference from the center point to the point of maximum x, y, and z on the box. Note that the center of the box is not necessarily the same as the local origin of the object, nor does the orientation of the box have to match the orientation of the object. If either is the case, some adjusting of the object's local-to-world transformation will have to be done to generate the box axes and center location in world space. If the transformation of the box to object-space orientation is $\mathbf{R}_{box \rightarrow object}$ and the object's orientation is $\mathbf{R}_{object \rightarrow world}$, then the box's local-to-world rotation is

$$\mathbf{R}_{box \rightarrow world} = \mathbf{R}_{object \rightarrow world}\, \mathbf{R}_{box \rightarrow object}$$

To simplify our life, however, we can use boxes aligned to the object's local coordinates, with a vector $\mathbf{d}$ in object space indicating the box center relative to the object center (as mentioned in Chapter 4, it's not usually practical to build objects with their bounding box center as their origin). In either case, any time we need the box center $\mathbf{c}$ in world space we can use

$$\mathbf{c} = \mathbf{R}_{object \rightarrow world}\mathbf{d} + \mathbf{t}$$

If we're simply simulating an object using an OBB, aligning it to the local axes may produce the results we want. However, when using an OBB for culling, we often want a tighter fit than that. Ideally, we want to find the set of box axes that produce the minimum volume box, and there are a number of techniques that do just that. The most commonly used method approximates this by taking a statistical measure of the object known as the *covariance matrix* [150].

The covariance matrix is a 3×3 array represented as

$$\mathbf{C} = \begin{bmatrix} C_{xx} & C_{xy} & C_{xz} \\ C_{yx} & C_{yy} & C_{yz} \\ C_{zx} & C_{zy} & C_{zz} \end{bmatrix}$$

where C_{xy}, for example, is

$$C_{xy} = \frac{1}{n} \sum_{i=1}^{n} (x_i - \bar{x})(y_i - \bar{y})$$

The value $\bar{x}$ is the mean of x values of the n points, and $\bar{y}$ is the corresponding mean of the y values.

By computing the eigenvectors (called the principal axes) of this matrix, we can determine the direction of greatest variance, or where the points are most spread out, which will become our long axis. The other eigenvectors become the directions of the remaining axes for our OBB. The mean of the points becomes the center of the box, and from there we can project the points onto the axes to determine the maximum extent along each axis. Code for this computation can be found in the accompanying example code.

It should be noted that while the principal axes method works reasonably well for general point clouds, it's not always optimal, or even close to optimal. For example, a cube centered on the origin is statistically symmetric, no matter how it's rotated. So the resulting covariance matrix will have an infinite number of eigenvectors, and the ones chosen may not be orthogonal to the faces of the cube, which could give you a very poor fit. In 2D, it's possible to generate a minimum enclosing box in $O(n \log (n))$ time by computing a convex polygon enclosing the points (known as a *convex hull*), producing bounding rectangles coincident to each face of the polygon, and choosing the rectangle with the smallest area [50]. This is known as the rotating caliper method. O'Rourke [114] has extended it to 3D with an $O(n^3)$ algorithm, but in practice most people do an approximation by choosing a reasonably long axis, projecting the points onto an orthogonal plane, and running the 2D algorithm to get the best fit in the remaining two directions.

12.3.4.2   OBB–OBB Intersection

There have been many methods for testing intersections between two arbitrarily oriented boxes, including linear programming techniques and closest-feature tracking. The most efficient technique known to date, however, uses the concept of separating axes and is due to Gottschalk et al. [60]. The following discussion is heavily drawn from this paper, with some additional concepts due to Eberly [35] and van den Bergen [149].

Recall that to test whether two axis-aligned boxes were intersecting, we did three tests, one for each axis x, y, and z. For each test, we checked the extents of each box along each of the axis directions. This is equivalent to projecting the box along the basis vectors $\mathbf{i}$, $\mathbf{j}$, and $\mathbf{k}$. If the intervals of a given projection don't overlap, then there is a separating plane normal to the test vector and therefore no intersection. The corresponding axis is known as a *separating axis*.

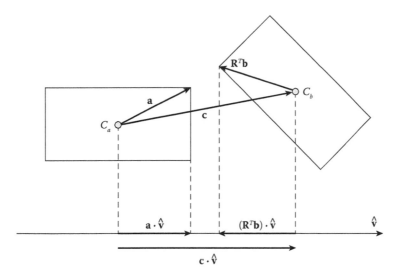

Figure 12.21. Example of OBB separation test.

This works well for axis-aligned boxes, but we need a slightly different test for oriented boxes. To simplify our equations and improve performance, we'll use transformations relative to box A. We end up with a single translation vector $\mathbf{c}$ from A to B, where $\mathbf{c} = \mathbf{R}_a^T \bullet (C_b - C_a)$, and a relative rotation matrix $\mathbf{R} = \mathbf{R}_b^T \mathbf{R}_a$. A's extent vector remains the same, since it's relative to its local space. B's extent vector becomes $\mathbf{R}^T \mathbf{b}$.

Now suppose we have a potential separating axis direction $\mathbf{v}$. We want to perform the same test we did with the AABBs: project each box onto the vector and check to see whether the projections are separate or not. Another way of representing this is to project the box centers onto the vector as endpoints, and then project the extent vectors closest to the center onto the vector as well (Figure 12.21). If the distance between the projected box centers is less than the sum of the lengths of the projected extents, then there is no intersection. Expressed mathematically, there is no intersection if

$$|\mathbf{c} \bullet \mathbf{v}| > |\mathbf{a} \bullet \mathbf{v} + (\mathbf{R}^T \mathbf{b}) \bullet \mathbf{v}|$$

This works if the extent vectors are aligned appropriately to give us the maximum projected length, but we can't make that assumption. Instead, we'll use a pseudo–dot product that forces maximum length, so the equivalent to $\mathbf{a} \bullet \mathbf{v}$ is

$$|a_x v_x| + |a_y v_y| + |a_z v_z|$$

This is legal because the extents can be taken from any of the eight octants, so we can get any sign we want for any term.

An equivalent equation can be found for $(\mathbf{R}^T \mathbf{b}) \bullet \mathbf{v}$. The final separating axis equation is

$$|\mathbf{c} \bullet \mathbf{v}| > \sum_i |a_i v_i| + \sum_i |(\mathbf{R}^T \mathbf{b})_i v_i| \qquad (12.11)$$

While this gives us our test, there is an infinite number of choices for $\mathbf{v}$, which is not practical. Gottschalk et al. [60] demonstrate that any separating plane will be either parallel to one of the box faces or parallel to an edge from each box. This means that a maximum of 15 separating axis tests are necessary: 3 against the axes of box A, 3 against the axes of box B, and 9 cross products using one axis from A and one from B.

The nice thing about this result is that it allows us to simplify our equations considerably. For example, let's use the cross product of the local x-axis from A and the local y-axis from B. In A's local space, the x-axis of A is $\mathbf{i} = (1,0,0)$. If we represent the matrix $\mathbf{R}$ as the three row vectors $(\mathbf{r}_0^T, \mathbf{r}_1^T, \mathbf{r}_2^T)$, then the y-axis of B in A's space is (r_{10}, r_{11}, r_{12}). Performing the cross product $\mathbf{i} \times \mathbf{r}_1$, we get

$$\mathbf{v} = (0, -r_{12}, r_{11}) \tag{12.12}$$

For the B terms, it's convenient to transform $\mathbf{v}$ to be relative to B's basis via $\mathbf{R}^T$:

$$\mathbf{R}^T(\mathbf{i} \times \mathbf{r}_1) = \begin{bmatrix} \mathbf{r}_0 \cdot (\mathbf{i} \times \mathbf{r}_1) \\ \mathbf{r}_1 \cdot (\mathbf{i} \times \mathbf{r}_1) \\ \mathbf{r}_2 \cdot (\mathbf{i} \times \mathbf{r}_1) \end{bmatrix} = \begin{bmatrix} \mathbf{i} \cdot (\mathbf{r}_1 \times \mathbf{r}_0) \\ \mathbf{i} \cdot (\mathbf{r}_1 \times \mathbf{r}_1) \\ \mathbf{i} \cdot (\mathbf{r}_1 \times \mathbf{r}_2) \end{bmatrix} = \begin{bmatrix} \mathbf{i} \cdot (-\mathbf{r}_2) \\ \mathbf{i} \cdot \mathbf{0} \\ \mathbf{i} \cdot \mathbf{r}_0 \end{bmatrix}$$

So, $\mathbf{v}$ in B space is

$$\mathbf{R}^T\mathbf{v} = (-r_{20}, 0, r_{00}) \tag{12.13}$$

Substituting Equations 12.12 and 12.13 into Equation 12.11 and multiplying out the terms, the final axis test is

$$|c_2 r_{11} - c_1 r_{12}| > a_1 |r_{12}| + a_2 |r_{11}| + b_0 |r_{20}| + b_2 |r_{00}|$$

The test for other axes can be derived similarly. All use the absolute value of elements from the matrix $\mathbf{R}$, so it is far more efficient to precompute them and then perform the axis tests. If this is done, the algorithm takes about 200 operations. It can be found in `IvOBB::Intersect()`.

One caveat: any implementation of this algorithm needs to take steps to avoid numerical problems with floating-point precision. In particular, if two edges, one from each box, are nearly parallel, the resulting cross product will be near zero. This will lead to invalid results for the separation test. The solution is to detect the condition, and only test against the six main axes of the boxes. Even in this case, care must be taken, as numeric error can lead to false negatives.

12.3.4.3   OBB–Ray Intersection

Detecting intersection between a linear component and an oriented box is much simpler than detecting intersection between two boxes. One method is to transform the ray into the box's local space and perform a standard AABB intersection test. To transform the linear component, the origin point is transformed by the inverse of the box's world transform matrix, and the direction vector by the inverse rotation of the box's transformation matrix. The newly transformed line, ray, or line segment can be passed into the appropriate AABB routine.

An alternative is to use a modified version of the AABB algorithm, as described by Akenine-Möller et al. [1]. In this case, instead of using planes normal to the three world axes, we'll use planes normal to the three box axes. Recall that these axes are specified as the three column vectors in our rotation matrix.

Each axis has two parallel planes associated with it. If we treat the box's center as the origin of our frame, the extent vector $\mathbf{a}$ contains the magnitude of our d values for these planes. For example, two of the parallel box planes are $r_{00}x + r_{10}y + r_{20}z + a_x = 0$ and $r_{00}x + r_{10}y + r_{20}z - a_x = 0$.

If we translate our ray so that its origin is relative to the box origin, we can determine s and t parameters for the intersections with these planes, just as we did with the axis-aligned box. In this case, the formulas for s and t for each axis (including the translation) are

$$s = \frac{\mathbf{r}_i \bullet (C - P) - \mathbf{a}_i}{\mathbf{r}_i \bullet \mathbf{v}} \quad t = \frac{\mathbf{r}_i \bullet (C - P) + \mathbf{a}_i}{\mathbf{r}_i \bullet \mathbf{v}}$$

We also need to modify our test to determine whether the ray is parallel to the current pair of planes we're testing. This is easily done by taking the dot product of the direction vector $\mathbf{v}$ and the plane normal and seeing if it is close to zero. If so, the ray is parallel to the plane, and we need to project the vector $C - P$ onto the current axis, and see if the result lies outside the extents.

The modified code is as follows:

```
bool
IvOBB::Intersect( const IvRay3& ray )
{
    float maxS = -FLT_MAX;
    float minT = FLT_MAX;

    // compute difference vector
    IvVector3 diff = mCenter - ray.mOrigin;

    // for each axis do
    for (int i = 0; i < 3; ++i)
    {
        // get axis i
        IvVector3 axis = mRotation.GetColumn( i );
        // project relative vector onto axis
        float e = axis.Dot( diff );
        float f = ray.mDirection.Dot( axis );

        // ray is parallel to plane
        if ( IsZero( f ) )
        {
            // ray passes by box
            if ( -e - mA[i] > 0 || -e + mA[i] > 0 )
                return false;
            continue;
        }

        float s = (e - mA[i])/f;
        float t = (e + mA[i])/f;

        // fix order
        ...
```

```
        // adjust min and max values
        ...
        // check for intersection failure
        ...
    }

    // done, have intersection
    return true;
}
```

Performance can be improved here by storing the rotation matrix as an array of three vectors instead of an `IvMatrix33`.

12.3.4.4 OBB–Plane Intersection

As we did with OBB–ray intersection, we can classify the intersection between an OBB and a plane by transforming the plane to the OBB's frame and using the AABB–plane classification algorithm. Since the transformation is just a pure rotation and a translation, we can find the transformed normal by

$$\hat{\mathbf{n}}' = \mathbf{R}^T \hat{\mathbf{n}}$$

We apply the transpose since we're going from world space into box space. The minimal and maximal points for the AABB in this case are the extent vector and its negative, $\mathbf{a}$ and $-\mathbf{a}$, respectively.

An alternative, presented by Akenine-Möller et al. [1], is to use the principle of separating planes again. This time, our test vector will be the plane normal, and we'll project the box diagonal on to it. To ensure we get maximum extent, we'll add the absolute values of the elements together, similar to what we did before:

$$r = |(a_0\mathbf{r}_0) \bullet \mathbf{n}| + |(a_1\mathbf{r}_1) \bullet \mathbf{n}| + |(a_2\mathbf{r}_2) \bullet \mathbf{n}|$$

Here, each $\mathbf{r}_i$ represents a column of the rotation matrix. The box intersects the plane if the distance between the box center and the plane is less than r.

The resulting code is as follows:

```
float IvOBB::Classify( const IvPlane& plane )
{
    IvVector3 xNormal = ::Transpose(mRotation)*plane.mNormal;
    float r = mExtents.x*IvAbs(xNormal.x) + mExtents.y*IvAbs(xNormal.y)
        + mExtents.z*IvAbs(xNormal.z);

    float d = plane. Test(mCenter);
    if (IvAbs(d) < r)
        return 0.0f;
    else if (d < 0.0f)
        return d + r;
    else
        return d - r;
}
```

12.3.5 Triangles

Source Code
Library
Filename
IvTriangle

All of the bounding objects we've discussed up until now have been approximations to our base object (assuming our object is more complex than, say, a box or a sphere). To test actual intersections between objects, we need to get right down to the basic building block of our geometry: the triangle. As before, we will be representing our triangle as the convex combination of three points.

12.3.5.1 Triangle–Triangle Intersection

A naive approach to determining triangle–triangle intersection uses the triangle–ray intersection test from Section 12.3.5.2. If one of the line segments composing an edge of one triangle intersects the other triangle, then the two triangles are intersecting. While this works, there are faster methods. Two commonly used approaches are by Möller [146] and Held [77]. However, if we are only concerned with determining whether intersection exists, and not the segment (or point) of intersection, then there is a faster way, concurrently discovered by two groups of researchers: Shen et al. [72] and Guigue and Devillers [68].

Figure 12.22 shows the situation. Taking the first triangle P, composed of points P_0, P_1, and P_2, we compute its plane equation. Recall that the plane equation for a normal $\mathbf{n} = (a, b, c)$ and a point on the plane $P_0 = (x_0, y_0, z_0)$ is

$$0 = ax + by + cz - (ax_0 + by_0 + cz_0)$$

or

$$0 = ax + by + cz + d$$

In this case, the plane normal is computed from $(P_1 - P_0) \times (P_2 - P_0)$ and normalized, and the plane point is P_0.

Now we take our second triangle Q, composed of points Q_0, Q_1, and Q_2. We plug each point into P's plane equation and test whether all three lie on the same side of the plane. This is true if all three results have the same sign. If they do, there is no intersection and we quit. Otherwise, we store the results d_0, d_1, and d_2 generated from the plane equation for each point and continue.

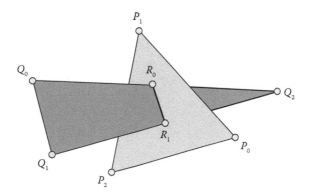

Figure 12.22. Triangle intersection.

We now need to test whether the rearranged triangles overlap by checking the intervals where their edges cross the common line between the two planes. If the interval for P is $[i, j]$ and Q is $[k, l]$, then there is intersection if the intervals overlap, producing the line segment $\overline{R_0 R_1}$. Other algorithms compute these intervals directly. However, there is a way to test this implicitly.

First, we rearrange P's vertices such that the lone vertex (the one that lies in its own half-space of Q) is first, or P_0. We also permute Q's vertices so that P_0 will "see" them in counterclockwise order. We then do the same for triangle Q, rearranging its vertices such that its lone vertex is first, and permuting P's vertices into counterclockwise order relative to the new Q_0.

Now, we make use of a signed distance test to check for interval overlap. If the signed distance between $\overline{Q_0 Q_1}$ and $\overline{P_0 P_2}$ is negative, then there is no overlap. Similarly, if the signed distance between $\overline{P_0 P_1}$ and $\overline{Q_0 Q_2}$ is negative, there is no overlap. Otherwise, the two triangles intersect.

We compute the signed distance between two edges by comparing the distance between two parallel planes, each containing one of the line segments. The normal $\mathbf{n}$ for these planes can be computed by taking the cross product between the segment vectors, say $\mathbf{n} = (Q_0 - Q_1) \times (P_0 - P_2)$. Then, we can compute the signed distance between each plane and the origin by taking the dot product of the plane normal with a point on each plane (i.e., $d_0 = \mathbf{n} \cdot Q_0$ and $d_1 = \mathbf{n} \cdot P_0$). Then, the signed distance between the planes is just $d_0 - d_1$, or $\mathbf{n} \cdot (Q_0 - P_0)$.

Note that this will not work if the two lines are parallel. Most of the cases where this might occur are culled out during the initial steps. The one case remaining is if the two triangles are coplanar. This is handled by projecting them to 2D and doing a simple test.

12.3.5.2 Triangle–Ray Intersection

There are two possible approaches to determining triangle–ray intersection. The first is to use the plane equation for the triangle (computed from the three vertices) and determine the intersection point of the ray with the plane (if any). We can then use a point-in-triangle test to determine whether the intersection lies within the triangle.

While a relatively simple approach, it has some disadvantages. First of all, we need to either store the plane equation or, if we're short on space, compute it every time we wish to do the intersection test. Second, it's a two-pass algorithm: compute the plane intersection, and then test whether it's in the triangle. Fortunately, we have an alternative. The following approach, presented by Möller and Trumbore [147], uses affine combinations to compute the ray–triangle intersection.

We define our triangle as having vertices V_0, V_1, and V_2. We can define two edge vectors $\mathbf{e}_0$ and $\mathbf{e}_1$ (Figure 12.23), where

$$\mathbf{e}_0 = V_1 - V_0$$
$$\mathbf{e}_1 = V_2 - V_0$$

Recall that the point V_0 with the vectors $\mathbf{e}_0$ and $\mathbf{e}_1$ can be used to create an affine combination that spans the plane of the triangle, with barycentric coordinates (u, v). So, the formula for

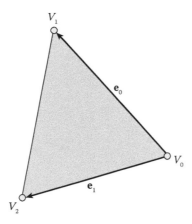

Figure 12.23. Affine space of triangle.

a point $T(u, v)$ on the plane is

$$T(u, v) = V_0 + u\mathbf{e}_0 + v\mathbf{e}_1$$
$$= V_0 + u(V_1 - V_0) + v(V_2 - V_0)$$

Rearranging terms, we get

$$T(u, v) = (1 - u - v)V_0 + uV_1 + vV_2$$

We want the contribution of each point to be nonnegative, so for a point inside the triangle,

$$u \geq 0$$
$$v \geq 0$$
$$u + v \leq 1$$

If u or $v < 0$, then the point is on the outside of one of the two axis edges. If $u + v > 1$, the point is outside the third edge. So, if we can compute the barycentric coordinates for the intersection point $T(u, v)$, we can easily determine whether the point is outside the triangle.

To compute the u, v coordinates of the intersection point, the result of the line equation $L = P + t\mathbf{d}$ will equal a solution to the affine combination $T(u, v)$ (Figure 12.24). So,

$$P + t\mathbf{d} = (1 - u - v)V_0 + uV_1 + vV_2$$

We can express this as a matrix product:

$$\begin{bmatrix} -\mathbf{d} & V_1 - V_0 & V_2 - V_0 \end{bmatrix} \begin{bmatrix} t \\ u \\ v \end{bmatrix} = P - V_0$$

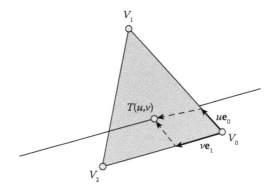

Figure 12.24. Barycentric coordinates of line intersection.

Using Cramer's rule, or row reduction, we can solve this matrix equation for (t, u, v). The final result is

$$t = \frac{\mathbf{q} \cdot \mathbf{e}_2}{\mathbf{p} \cdot \mathbf{e}_1}$$

$$u = \frac{\mathbf{p} \cdot \mathbf{s}}{\mathbf{p} \cdot \mathbf{e}_1}$$

$$v = \frac{\mathbf{q} \cdot \mathbf{d}}{\mathbf{p} \cdot \mathbf{e}_1}$$

where

$$\mathbf{e}_1 = V_1 - V_0$$
$$\mathbf{e}_2 = V_2 - V_0$$
$$\mathbf{s} = P - V_0$$
$$\mathbf{p} = \mathbf{d} \times \mathbf{e}_2$$
$$\mathbf{q} = \mathbf{s} \times \mathbf{e}_1$$

The final algorithm includes checks for division by zero and intersections that lie outside the triangle.

```
bool
TriangleIntersect( const IvVector3& v0, const IvVector3& v1,
                   const IvVector3& v2, const IvRay& ray )
{
  // test ray direction against triangle
  IvVector3 e1 = v1 - v0;
  IvVector3 e2 = v2 - v0;
  IvVector3 p =  ray.mDirection.Cross(e2);
  float a = e1.Dot(p)

  // if result zero, no intersection or infinite intersections
  // (ray parallel to triangle plane)
```

```
  if ( IsZero(a) )
    return false;

  // compute denominator
  float f = 1.0f/a;

  // compute barycentric coordinates
  IvVector3 s = ray.mOrigin - v0;
  u = f*s.Dot(p)
  if (u < 0.0f || u > 1.0f) return false;

  IvVector3 q = s.Cross(e1);
  v = f*ray.mDirection.Dot(q);
  if (v < 0.0f || u+v > 1.0f) return false;

  // compute line parameter
  t = f*e2.Dot(q);

  return (t >= 0);
}
```

Parameters *u*, *v*, and *t* can be returned if the barycentric coordinates on the triangle or the parameter for the exact point of intersection are needed.

12.3.5.3 Triangle–Plane Intersection

We covered triangle–plane intersection when we discussed triangle–triangle intersection. We take our triangle, composed of points P_0, P_1, and P_2, and plug each point into the plane equation. If all three lie on the same side of the plane, then there is no intersection. Otherwise, there is, and if we desire, we can find the particular line segment of intersection, as described earlier. If there is no intersection, the signed distance is the plane equation result of minimum magnitude.

12.4 A Simple Collision System

Now that we have some methods for testing intersection between various primitive types, we can make use of them in a practical system. The example we'll consider is collision detection. Rather than building a fully general collision system, we'll do only as much as we need to for a basic game—in our case, we'll use a submarine game as our example. This is to keep things as simple as possible and to illustrate various points to consider when building your own system. It's also good to keep in mind that a particular subsystem of a game, whether it is collision or rendering, needs only to be as accurate as the game calls for. Building a truly flexible collision system that handles all possible situations may be overkill and eat up processing time that could be used to do work elsewhere.

12.4.1 Choosing a Base Primitive

The first step in building the system is to choose the base bounding shape for our objects. We'll see in the following sections how we can use a hierarchy of bounding primitives to get a better fit to the object's surface, but for now we'll consider only one per object. Which primitive we choose depends highly on the expected topology we're trying to approximate with it. For example, if we're writing a pool game, using bounding spheres for our balls makes perfect sense. However, for a human character bounding spheres are not a good choice because one axis of the object is far longer than the other two—not a good fit. In particular, getting characters through an interior space might be a tricky proposition unless all your doorways and hallways are at least 6 ft wide.

Considering that our object is made of triangles, using them should give us the most accurate results. However, while they are cheap as a one-on-one test, it would be costly to test every possible triangle–triangle combination between two objects. This becomes more feasible when we have some sort of culling hierarchy to whittle down the possible triangle pairs to a few contenders—we'll discuss that in more detail shortly. However, if we can get a good fit with a simpler bounding volume, we can get a reasonably accurate measure of collision by doing a volume–volume test without having to do the full triangle–triangle test.

Since AABBs change size depending on the object's orientation, they are not usually a good choice for a base bounding primitive. They are more often used as a culling test, such as in the sweep-and-prune system described in Section 12.4.4.

Among the primitives we've discussed, this leaves us with capsules and OBBs. Which we choose depends on our performance requirements and how angular our objects are. If we have mostly boxy objects—like tanks—capsules or even lozenges won't provide very compelling collisions. An OBB is a better shape to choose for this situation. For our case, however, submarines and torpedoes are both generally sausage shaped. If we had to go with a single bounding object that approximates a submarine, capsules are an excellent choice.

12.4.2 Bounding Hierarchies

Source Code
Demo
Hierarchy

Unless our objects are almost exactly the shape of the bounding primitive (such as our pool ball example), then there are still going to be places where our test indicates intersection where there is visibly no collision. For example, the conning tower of our submarine makes the bounding capsule encompass a large area of empty space at the top of the hull. Suppose a torpedo is heading toward our submarine and through that area. Instead of harmlessly passing over the hull as we would expect from the visual evidence, it will explode because we have detected a collision with the inaccurately large bounding region.

The solution is to use a set of bounding primitives to get a better approximation to the surface of the object. In our submarine example, we could use one capsule for the main hull and one for the conning tower. If we are willing to allow a slightly forgiving system, we could ignore the conning tower for the purposes of collision and get a very nice fit with the hull capsule. Or we could go the more detailed route and add one for the conning tower, as well as a third for the periscope (Figure 12.25). To check for intersection, we test each bounding primitive for the first object against all the primitives in the second, much as we would have done for the triangles.

To speed this up, we can keep our original bounding capsule and use it as a rough test before checking further. Better still, we can generate bounding spheres for each object and test against those instead. It's a very cheap test and can do a great job of culling large numbers of cases. We could also generate bounding spheres for each of our smaller capsules and use these spheres in preliminary culling steps before checking individual capsule pairs.

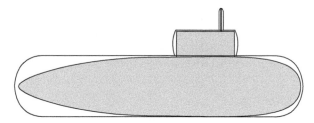

Figure 12.25. Using multiple bounding objects.

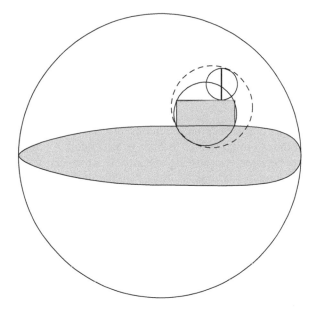

Figure 12.26. Using bounding hierarchy.

This gives us a bounding hierarchy for our object (Figure 12.26). We compare the top-level bounding spheres first. Only if they are intersecting do we then move on to the lower level of sphere check and capsule check. This can cull out a large number of cases and make it much more likely that we'll be testing only the two lower-level capsules that are actually intersecting.

We can take this technique of using bounding hierarchies further. For example, if we want to do triangle–triangle intersection testing, we can build a hierarchy to perform coarser but cheaper intersection tests. If two objects are intersecting, we can traverse the two hierarchies until we get to the two intersecting triangles (there may be more than two if the objects are concave). Obviously, we'll want to create much larger hierarchies in this case. Generating them so that they are as efficient as possible—they both cull well and have a reasonably small tree size—is not a simple task. Gottschalk et al. [60] provide some information for building OBB trees, while Ericson [41] covers the general cases.

Spheres, capsules, AABBs, and OBBs have all been used as primitives for culling bounding hierarchies. Most tests have been done for hierarchies with triangles as leaf nodes.

Gottschalk et al. [60] demonstrate that OBBs work better than both AABBs and spheres if our objects have static geometry. However, if we're constantly deforming our vertices—for example, with skinned character models—recomputing the OBBs in the hierarchy is an expensive step. Using spheres or AABBs can be a better choice in this circumstance.

12.4.3 Dynamic Objects

So far we have been using intersection tests assuming that our objects don't move between frames. This is clearly not so. In games, objects are constantly moving, and we need to be careful when we use static tests to catch collisions between moving objects.

For example, in one frame we have two objects moving toward each other, clearly heading for a collision somewhere in the center of the screen (Figure 12.27a). Ideally, in the next frame we want to catch a snapshot of them just as they collide or are slightly intersecting. However, if we take too large a simulation step, they may pass partially through each other (Figure 12.27b). Using a frame-by-frame static test, we will miss the initial collision. Worse yet, if we take a larger step, the two objects will pass right through each other, and we'll miss the collision entirely.

One way to catch this is to sweep our bounding primitives along a path and then test intersection between the swept primitives that we've generated. A simple example of this is testing intersection between two moving spheres. If we sweep a sphere along a line segment, we get—no surprise—a capsule. Based on the two objects' velocities, we can generate capsules for each object and test for intersection. If one is found, then we know the two objects may collide somewhere between frames and we can investigate further.

We generally have to worry about this problem only when the relative velocities of objects are large enough or the frame times are long enough that one object can move, relative to another, farther than half its thickness in the direction of travel. For example, a tank with

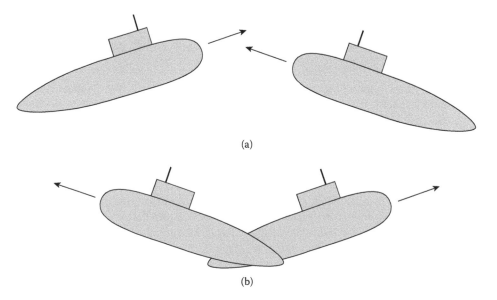

(a)

(b)

Figure 12.27. (a) Potential collision and (b) partially missed collision.

a speed of 30 km/h moves about 0.12 m/frame, assuming 60 frames/s. If the tank is 10 m long, its movement is miniscule compared to its total length and we can probably get away with static testing. Suppose, however, that we fire a 1 m long missile at that tank, traveling at 120 km/h. We also have a bug in our rendering code that causes us to drop to 10 frames/s, giving us a travel distance of 3 1/3 m. The missile's path crosses through the tank at an angle and is already through it by the next frame. This may seem like an extreme example, but in collision systems it's often best to plan for the extreme case.

Walls, since they are infinitely thin, also insist on a dynamic test of some kind. In a first-person shooter you don't want your players using a cheat to teleport through a wall by moving too fast. One way to handle this is to do a simple test of the player's path versus the nearest wall plane. Another is to create a plane for each wall with the normals pointing into the room; if a plane test shows that the object is on the negative side of the plane, then it's no longer in the room.

Submarines are large and move relatively slowly for their size, so for this collision system we don't need to worry about this issue. However, it is good to be aware of it. For more information on managing dynamic tests, see Millington [109] or Eberly [35].

12.4.4 Performance Improvements

Now that we've handled questions of which bounding shapes to use on our objects and how to achieve a tighter fit even with simple primitives, we'll consider ways of improving our performance. The main way we'll approach this is to cut down on intersection tests. We've already handled this to some extent at the object level by using a bounding hierarchy to cut down on intersection tests between primitives. Now we want to look at the world level, by cutting down on tests between objects. For example, if two objects are relatively small and at opposite ends of the map from each other, it's a pretty good bet that they're not colliding.

Source Code
Demo
SweepPrune

The most basic way to check collisions among all objects is the following loop:

```
for each object i
    for each object j, where j <> i
        test for collision between i and j
```

There are a number of problems with this. First of all, we're doing $n(n-1)$ tests, which is an $O(n^2)$ algorithm. Half of those tests are duplicates: if we test for collision between objects 1 and 5, we'll also test for collision between 5 and 1. Also, there may be a number of objects that we wish to collide with that simply aren't moving. We don't want to test collision between two such static objects. A better loop that handles these cases is as follows:

```
for each object i
    for each object j, where j > i
        if (i is moving or j is moving)
            test for collision between i and j
```

There are other possibilities. We can have two lists: one of moving objects called Colliders and one of moving or static objects called Collidables. In the first loop we iterate through the Colliders and in the second the Collidables. Each Collider should be tagged after its turn through the loop, to ensure collision pairs aren't checked twice. Still, even with this change, we're still doing $O(nm)$ tests, where n is the number of `Colliders` and m is the number of `Collidables`. We need to find a way to further cut down the number of checks.

Most approaches involve some sort of spatial subdivision to do this. The simplest is to slice the world, along the x-axis, say, by a series of evenly spaced planes (Figure 12.28). This creates a set of slabs, bounded by the planes along the x direction, and by whatever bounds we've set for our world in the y and z directions. For each slab, we store the set of objects that intersect it. To test for collisions for a particular object, we determine which slabs it intersects and then test against only the objects in those slabs. This approach can be extended to other spatial subdivisions, such as a grid or voxel-based system.

One of the disadvantages of the regular spatial subdivisions is that they don't handle clumping very well. Let's consider slabs again. If our world is fairly sparse, there may be large numbers of slabs with no objects in them, and a very few with most of the objects in them. We still may end up doing a large number of checks within each slab, which is the problem we were trying to avoid.

There is another possibility used by a number of collision detection systems, known as the *sweep-and-prune* method. It is similar to the separating axis test that we used for OBBs (it's also related to some scan line rasterization algorithms). Instead of using a regular grid

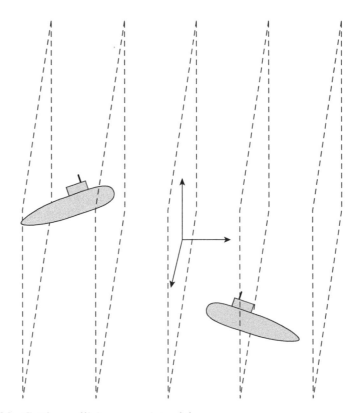

Figure 12.28. Cutting collision space into slabs.

for our world, we'll use the extents of our objects as our grid. For each object, we project its extents onto the x-axis. To keep things efficient, we can use our root-level bounding sphere to compute our extents, which for a sphere with center C and radius r gives us an interval of $[c_x - r, c_x + r]$.

Given the extent endpoint pairs for each object, we'll mark them with a pointer to the object and indicate for each value whether it is the low (*start*) or high (*finish*) endpoint. Finally, we sort all endpoints from low to high.

Once the sorted list of endpoints is created, the collision detection process runs as follows:

```
for each endpoint do
   if a start point
   if object is moving
      check collisions against all objects in list
   else
      check collisions against moving objects in list
   add corresponding object to list
else if a finish point
   remove corresponding object from list
```

Figure 12.29 shows how this works. We sweep from left to right along the x-axis and use the sorted endpoints to test intersections of intervals before the more complex intersection tests.

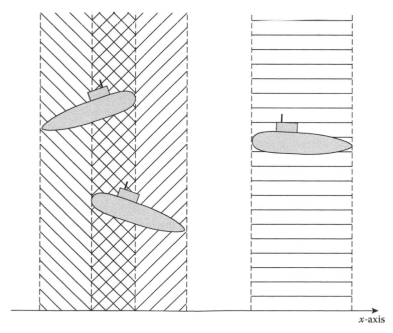

Figure 12.29. Dividing collision space by sweep-and-prune.

Normally this would be an $\Omega(n \log n)$ algorithm due to the sorting operation. However, if the time step is small enough, the relative position of the objects won't have changed that much from frame to frame—this is referred to as *temporal coherence*. Any changes that do happen will be rare but localized. Therefore, if we use a sorting algorithm that works best on mostly sorted lists, such as bubble or insertion sort, we can get linear time for our sort and hence an $O(n)$ algorithm. Another alternative to get $O(n)$ time is to use a radix sort, assuming that we have a small number of possible positions and they are easily bucketed.

This algorithm is still not as efficient as it might be. If our objects are highly localized (or clumped) in the x direction, but separated in the y direction, then we still may be doing a high number of unnecessary intersection tests. To solve this, we can extend our data structure to perform sweep-and-prune in the y and z directions as well, effectively creating a dynamic voxel space.

12.4.5 Related Systems

The other two systems we mentioned earlier were ray casting, for picking and AI tests, and frustum culling. Both systems can benefit from the techniques described in our collision system, in particular the use of bounding hierarchies and spatial partitioning.

Consider the case of ray casting. Instead of testing the ray directly against the object, we can take the ray and pass it through the hierarchy until (if we desire) we get the exact triangle of intersection. Further culling of testing can be done by using a spatial partitioning system such as voxels or k-d trees to consider only those objects that lie in the areas of the spatial partitioning that intersect the ray.

When handling frustum culling, the most basic approach involves testing an object against the six frustum planes. If, after this test, we determine that the object lies outside one of the planes, then we consider it outside the frustum and do not render it. As with ray casting, we can improve performance by using a bounding hierarchy at progressive levels to remove obvious cases. We can also use a spatial partition again, and consider only objects that lie in the areas of the partition within the view frustum.

However, there is one aspect of frustum culling of which we need to be careful. This also applies to any intersection test that requires determining whether we are inside a convex object. Consider the situation shown in Figure 12.30. The bounding sphere is near the corner of the view frustum and clearly intersecting two planes. By using the scheme described, this sphere would be considered as intersecting the frustum, but it is clearly not. An alternative is shown in Figure 12.31a. Instead of using the frustum, we trace around the frustum with the bounding sphere to get a rounded, larger frustum.[1] This represents the maximum extent that a bounding sphere can have and still be inside the frustum. Instead of testing the sphere, we can test its center against this shape. In practice, we can just push out the frustum planes by the sphere radius (Figure 12.31b), which is close enough. Similar techniques can be used for other bounding objects; see Akenine-Möller et al. [1] and Watt and Policarpo [155] for more details.

---

[1] This process is also known as convolution.

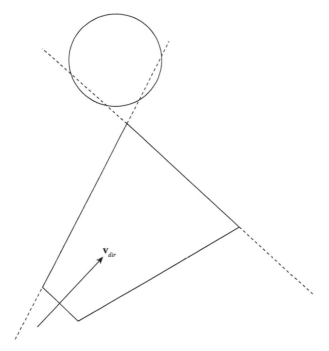

Figure 12.30. False positive for frustum intersection.

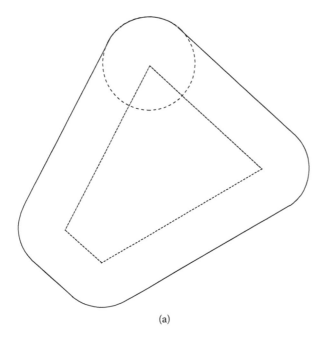

(a)

Figure 12.31. (a) Expanding view frustum for simpler inclusion test.   (*Continued*)

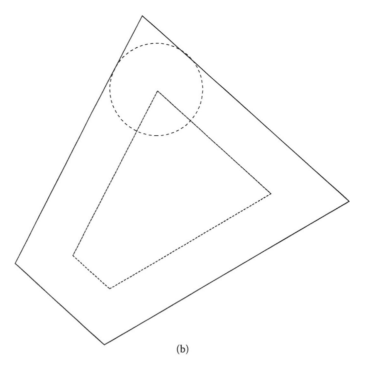

(b)

Figure 12.31. (Continued) (b) Expanding view frustum for simpler inclusion test.

12.4.6 Section Summary

The proceeding material should give some sense of the decisions that have to be made when handling collision detection or other systems that involve object intersection: pick base primitives, choose when you'll use them, consider whether to manage dynamic intersections, and cull unnecessary tests. However, this shouldn't be taken as the only approach. There are many other possible algorithms that handle much more complex cases than these. For example, there are systems, such as the University of North Carolina's I-COLLIDE, that track closest pairs of objects. This allows for considerable culling of intersection tests. There are also more sophisticated methods for managing spatial partitions, such as portals, octrees, BSP trees, and k-d trees. Whether the algorithmic complexity is necessary will depend on the application.

12.5 Chapter Summary

Testing intersection between geometric primitives is a standard part of any interactive application. This chapter has presented a few examples to provide a taste of how such algorithms are created. Most derive from a careful use of the basic properties of vectors and points as presented in Chapter 2. Using our intersection methods wisely allows us to build an efficient system for detecting collision between objects, casting rays for AI visibility checks and picking, and frustum culling.

For those who are interested in reading further, a more thorough presentation of geometric distance and intersection methods can be found in Schneider and Eberly [133]. These techniques fall under a general class of algorithms known as computational geometry; good references are Preparata and Shamos [125] and O'Rourke [115]. Two different approaches to building collision detection systems can be found in van den Bergen [149] and Ericson [41]. Finally, use of intersection techniques in rendering, plus information on more complex spatial partitioning techniques, can be found in both Akenine-Möller et al. [1] and Watt and Policarpo [155].

⓭ Rigid-Body Dynamics

13.1 Introduction

In many games, we move our objects around using a very simple movement model. In such a game, if we hold down the up arrow key, for example, we apply a constant forward translation, once a frame, to the object until the key is released, at which point the object immediately stops moving. Similarly, we can apply a constant rotation to the object if the left arrow key is held, and again, it stops upon release. This is fine for something with fast action, like a platform game or a first-person shooter, where we want quick response to our input. As soon as we hit a key, our character starts moving and stops immediately upon release. This can be thought of as an application of the theories of Aristotle, where pushing or pulling an object immediately affects its speed.

But suppose we want to do a more realistically styled game, for example, a submarine game. Submarines don't start and stop on a dime. When the propeller starts turning, it takes some time for the submarine to start forward. And they don't really have instantaneous brakes—when the engine is shut off, they will drift for quite a while before stopping. Turning is much the same—they will respond slowly to application of the rudder and then straighten out over time.

Even in a fast-action game, we may want to model how objects in the world react to our main character. When we push an object, we don't expect it to stop instantly when we stop pushing, nor do we expect it to keep moving forever. If we knock a chair over, we don't expect it to fall straight back and then stick to the floor; we expect it to turn, depending on where we hit it, and then bounce and possibly roll once. We want the game world to react to our character as the real world reacts to us, in a physically correct manner.

For both of these cases, we will want a better model of movement, known as a physically based simulation. One chapter is hardly enough space to encompass this broad topic, which covers the preceding effects as well as objects deforming due to contact, fluid simulation, and soft-body simulations such as cloth and rope. Instead, we'll concentrate on a simplified problem that is useful in many circumstances: objects that don't deform (known as *rigid bodies*) and move based on Newton's laws of motion (known as *dynamics*). We'll discuss techniques for translating rigid bodies through space in a physically based manner (linear dynamics) and then how to encompass rotational effects (rotational dynamics). Finally, we'll discuss some methods for resolving contacts and dealing with simple constrained movement within our simulation, again covering linear and rotational effects in turn.

The convention in physics is to represent some vector quantities by capital letters. To maintain compatibility with physics texts, we will use the same notation and assume that the reader can distinguish between such quantities and the occasional matrix by context.

13.2 Linear Dynamics

13.2.1 Moving with Constant Acceleration

Let's consider our object's movement through our game world as a function $X(t)$, which represents the position of the object for every time t. If we plot just the x values against t for the simple motion model described above, we would end up with a graph similar to that in Figure 13.1. Notice that we travel in a straight line for a while and then turn sharply in another direction, or we hold position. This is like our piecewise linear interpolation, except that in this case, the future x values are unknown; they are determined by the input of the player. For a given frame i, this can be represented by a line equation

$$X_i(h_i) = X_i + h_i \mathbf{v}_i$$

where X_i represents the position at the start of frame i, $\mathbf{v}_i$ is a vector generated from the player input that points along each line segment, and h_i is our frame time. We'll simplify things further by considering just the function on the first line segment, from time $t \geq 0$:

$$X(t) = X_0 + t\mathbf{v}_0$$

where $X_0 = X(0)$.

If we take the derivative of this function with respect to t, we end up with

$$\frac{d\mathbf{X}}{dt} = \mathbf{X}'(t) = \mathbf{v}_0 \tag{13.1}$$

This derivative of the position function is known as *velocity*, which is usually measured in meters per second, or m/s. For our simple motion model, we have a constant velocity across each segment. If we continue taking derivatives, we find that the second derivative of our position function is zero, which is what we'd expect when our velocity is constant. As mentioned, this motion model is known as *kinematics*.

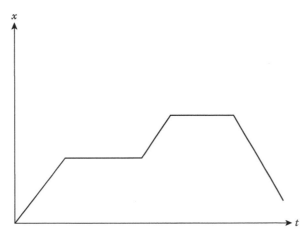

Figure 13.1. Graph of current motion model, showing x coordinate of particle as a function of time.

Now let's assume that our second derivative, instead of being zero, is a constant nonzero function. To achieve this, we'll change our velocity function to

$$\mathbf{v}(t) = \mathbf{v}_0 + t\mathbf{a} \qquad (13.2)$$

Now $\mathbf{v}(t)$ is also an affine function, this time with a constant derivative vector $\mathbf{a}$, called acceleration, or

$$\frac{d\mathbf{v}}{dt} = \mathbf{v}'(t) = \mathbf{a} \qquad (13.3)$$

The units for acceleration are usually measured in meters per second squared, or m/s^2.

Our original function $X(t)$ used a constant $\mathbf{v}_0$, so now we'll need to rewrite it in terms of $\mathbf{v}(t)$. Since $\mathbf{v}$ is changing at a constant rate across our time interval, we can instead use the average velocity across the interval, which is just one-half the starting velocity plus the ending velocity, or

$$\bar{\mathbf{v}} = \frac{1}{2}(\mathbf{v}_0 + \mathbf{v}(t))$$

Substituting this into our original $X(t)$ gives us

$$X(t) = X_0 + t\left[\frac{1}{2}(\mathbf{v}_0 + \mathbf{v}(t))\right]$$

Substituting in for $\mathbf{v}(t)$ gives the final result of

$$X(t) = X_0 + t\mathbf{v}_0 + \frac{1}{2}t^2\mathbf{a} \qquad (13.4)$$

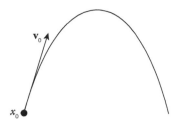

Figure 13.2. Parabolic path of object with initial velocity and affected only by gravity.

Our equation for position becomes a quadratic equation, and our velocity is represented as a linear equation:

$$P_i(t) = P_i + t\mathbf{v}_i + \frac{1}{2}t^2\mathbf{a}_i$$

$$\mathbf{v}_i(t) = \mathbf{v}_i + t\mathbf{a}_i$$

So, given a starting position and velocity and an acceleration that is constant over the entire interval $[0, t]$, we can compute any position within the interval. As an example, let's suppose we have a projectile, with an initial velocity $\mathbf{v}_0$ and initial position P_0. We represent acceleration due to gravity by the constant g, which is $9.8\,\mathrm{m/s^2}$. This acceleration is applied only downward, or in the $-z$ direction, so $\mathbf{a}$ is the vector $(0, 0, -g)$. If we plot the z component as a function of t, then we get a parabolic arc, as seen in Figure 13.2. This function will work for any projectile (assuming we ignore air friction), from a thrown rock (low initial velocity) to a cannonball (medium initial velocity) to a bullet (high initial velocity).[1]

Within our game, we can use these equations on a frame-by-frame basis to compute the position and velocity at each frame, where the time between frames is h_i. So, for a given frame $i + 1$,

$$X_{i+1} = X_i + h_i\mathbf{v}_i + \frac{1}{2}h_i^2\mathbf{a}_i$$

$$\mathbf{v}_{i+1} = \mathbf{v}_i + h_i\mathbf{a}_i$$

This process of motion with nonzero acceleration is known as *dynamics*.

13.2.2 Forces

One question that has been left open is how to compute our acceleration value. We do so based on a vector quantity known as a *force*. Forces cause change in an object's motion, pushing or pulling it around, either to speed it up or slow it down. So, for example, to throw a ball, your hand and arm exert a certain force on it, to begin its motion through the air. That force, when applied, produces an acceleration directly proportional to the object's mass, measured in kilograms. The proportional relationship is shown in Newton's second law of motion:

$$\mathbf{F} = m\mathbf{a}$$

The unit for force ends up being $\mathrm{kg\text{-}m/s^2}$ or newton (N), in homage to its creator.

---

[1] In most cases, this last is approximated by a line equation for efficiency reasons.

In the previous section we represented gravity as an acceleration, but in truth, it is a force whose value is always proportional to the mass of the object. For an object with mass m on the earth, its magnitude is mg and its direction points to the center of the earth. In games and other small-scale simulations, we usually assume the world is locally flat, and so the gravity vector points in the $-z$ direction. Other possible forces include the friction caused by air or water molecules pushing against an object to slow it down, or the thrust generated by a rocket engine or propeller, or simply the *normal force* of the ground pushing up to counteract gravity (there has to be such a force, otherwise we'd sink into the earth). In general, if something is pushing or pulling on an object, there is a force there.

Usually we have more than one force applied to an object at a time. Taking our ball example, we have the initial force when the ball is thrown, force due to gravity, and forces due to air resistance and wind. After the ball leaves your hand, that pushing force will be removed, leaving only gravity and air effects. Forces are vectors, so in both cases we can add all forces on an object together to create a single force that encapsulates their total effect on the object. We then scale the total force by $1/m$ to get the acceleration for Equation 13.4.

For simplicity's sake, we will assume for now that our forces are applied in such a way that we have no rotational effects. In Section 13.4 we'll discuss how to handle such cases.

13.2.3 Linear Momentum

As we've seen, the relationship between acceleration and velocity is

$$\mathbf{a} = \frac{d\mathbf{v}}{dt}$$

There is a corresponding related entity $\mathbf{P}$ for a force $\mathbf{F}$, which is

$$\mathbf{F} = m\mathbf{a} = m\frac{d\mathbf{v}}{dt} = \frac{d\mathbf{P}}{dt}$$

The quantity $\mathbf{P} = m\mathbf{v}$ is known as the *linear momentum* of the object, and it represents the tendency for an object to remain in its current linear motion. The heavier the object or faster it is moving, the greater the force needed to change its velocity. So, while a pebble at rest is easier to kick aside than a boulder, this is not necessarily true if the pebble is shot out of a gun.

An important property of Newtonian physics is the conservation of momentum. Suppose we take a collection of objects and treat them as a single system of objects. Now consider only the forces within the system, that is, only those forces acting between objects. Newton's third law of motion states that for every action, there is an equal and opposite reaction. So, for example, if you push on the ground due to gravity, the ground pushes back just as much, and the forces cancel. Due to this, within the system, pairwise forces between objects will cancel and the total force is zero. If the external force is 0 as well, then

$$\mathbf{F} = \frac{d\mathbf{P}}{dt} = 0$$

so $\mathbf{P}$ is constant. No matter how objects may move within the system, the total momentum must be conserved. This property will be useful to us when we consider collisions.

13.2.4 Moving with Variable Acceleration

There is a problem with the approach that we've been taking so far: we are assuming that total force, and hence acceleration, is constant across the entire interval. For more complex simulations this is not the case. For example, it is common to compute a drag force proportional to but opposite in direction to velocity:

$$\mathbf{F}_{drag} = -m\rho\mathbf{v} \tag{13.5}$$

This can provide a simple approximation to air friction; the faster we go, the greater the friction force. The quantity ρ in this case controls the magnitude of drag. An alternative example is if we wish to model a spring in our system. The force applied depends on the current length of the spring, so the force is dependent on position:

$$\mathbf{F}_{spring} = -kX$$

The spring constant k fulfills a similar role to ρ: it controls the proportion of force dependent on the position. In both of these cases, since acceleration is directly dependent on the force, it will vary over the time interval as velocity or position vary. It is no longer constant. So for these cases, Equations 13.2 and 13.4 are incorrect.

In order to handle this, we'll have to use an alternative approach. We begin by deriving a function for velocity in terms of any acceleration. Rewriting Equation 13.3 gives us

$$d\mathbf{v} = \mathbf{a}\,dt$$

To find $\mathbf{v}$ we take the indefinite integral or antiderivative of both sides:

$$\int d\mathbf{v} = \int \mathbf{a}\,dt$$

For example, if we assume as before that $\mathbf{a}$ is constant, we can move it outside the integral sign:

$$\int d\mathbf{v} = \mathbf{a}\int dt$$

And integrating gives us

$$\mathbf{v} = t\mathbf{a} + \mathbf{c}$$

We can solve for $\mathbf{c}$ by using our velocity $\mathbf{v}_0$ at time $t = 0$:

$$\mathbf{c} = \mathbf{v}_0 - 0 \cdot \mathbf{a}$$
$$= \mathbf{v}_0$$

So, our final equation is as before:

$$\mathbf{v}(t) = \mathbf{v}_0 + t\mathbf{a}$$

We can perform a similar integration for position. Rewriting Equation 13.1 gives

$$dX = \mathbf{v}(t)dt$$

We can substitute Equation 13.2 into this to get

$$dX = \mathbf{v}_0 + t\mathbf{a}\ dt$$

Integrating this, as we did with velocity, produces Equation 13.4 again.

For general equations we perform the same process, reintegrating $d\mathbf{v}$ to solve for $\mathbf{v}(t)$ in terms of $\mathbf{a}(t)$. So, using our drag example, we can divide Equation 13.5 by the mass m to give acceleration:

$$\mathbf{a} = \frac{d\mathbf{v}}{dt} = -\rho\mathbf{v}(t)$$

Rearranging this and integrating gives

$$\int d\mathbf{v} = \int -\rho\mathbf{v}(t)dt$$

We can consult a standard table of integrals to find that the answer in this case is

$$\mathbf{v}(t) = \mathbf{v}_0 e^{-\rho t}$$

where, as before, $\mathbf{v}_0 = \mathbf{v}(0)$.

While this particular equation was relatively straightforward, in general calculating an exact solution is not as simple as the case of constant acceleration. First of all, differential equations in which the quantity we're solving for is part of the equation are not always easily—if at all—solvable by analytic means. In many cases, we will not necessarily be able to find an exact equation for $\mathbf{v}(t)$, and thus not for $X(t)$. And even if we can find a solution, every time we change our simulation equations, we'll have to integrate them again, and modify our simulation code accordingly. Since we'll most likely have many different possible situations with many different applications of force, this could grow to be quite a nuisance. Because of both these reasons, we'll have to use a numerical method that can approximate the result of the integration.

13.3   Numerical Integration

13.3.1   Definition

The solutions for $\mathbf{v}$ and X that we're trying to integrate fall under a class of differential equation problems called *initial value problems*. In an initial value problem, we know the following about a function $\mathbf{y}(t)$:

1. An initial value of the function $\mathbf{y}_0 = \mathbf{y}(t_0)$.

2. A derivative function $\mathbf{f}(t, \mathbf{y}) = \mathbf{y}'(t)$.

3. A time interval h.

The problem we're trying to solve is, given these parameters, what is the value at $\mathbf{y}(t_0 + h)$? For our purposes, this actually becomes a series of initial value problems: At each frame our previous solution becomes our new initial value $\mathbf{y}_i$, and our interval h_i will be based on the current frame time. Once computed, our new solution will become the next initial value $\mathbf{y}_{i+1}$. More specifically, the initial value $\mathbf{y}_i$ is our current position X_i and current velocity $\mathbf{v}_i$, stored in a single 6-vector as

$$\mathbf{y}_i = \left[\begin{array}{c} X_i \\ \mathbf{v}_i \end{array} \right]$$

So, how do we evaluate the derivative function $\mathbf{f}(t, \mathbf{y})$? This will be another vector quantity:

$$\mathbf{f}(t, \mathbf{y}) = \left[\begin{array}{c} \mathbf{X}'_i \\ \mathbf{v}'_i \end{array} \right]$$

The value of our derivative for X_i is our current velocity $\mathbf{v}_i$. Our derivative for $\mathbf{v}_i$ is the acceleration, which is based on the current total force. To compute this total force, it is convenient to create a function called `CurrentForce()`, which takes X and $\mathbf{v}$ as arguments and combines any forces derived from position and velocity with any constant forces, such as those created from player input. We'll represent this as $\mathbf{F}_{tot}(t, X, \mathbf{v})$ in our equations. So, given our current state, the result of our function $\mathbf{f}(t, \mathbf{y})$ will be

$$\mathbf{y}' = \mathbf{f}(t, \mathbf{y}) = \left[\begin{array}{c} \mathbf{v}_i \\ \mathbf{F}_{tot}(t_i, X_i, \mathbf{v}_i)/m \end{array} \right]$$

The function $\mathbf{f}(t, \mathbf{y})$ is important in understanding how we can solve this problem. For every point $\mathbf{y}$ it returns a derivative $\mathbf{y}'$. This represents a vector field, where every point has a corresponding associated vector. To get a sense of what this looks like, let's take as an example a planet revolving in a perfectly circular orbit. Figure 13.3 shows a two-dimensional (2D) plot of the vector field of position and velocity, accentuating certain lines of flow. If we start at a particular point and follow the vector flow, this will trace out one possible solution (or level curve) to the differential equation, starting at that initial value.

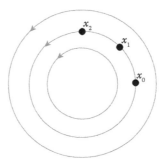

Figure 13.3. Orbit example, showing some level curves and idealized integration path.

This gives us a sense of what our general approach will be. We'll start at $\mathbf{y}_i$ and then, using our derivative function, take steps in time to generate new samples that approximate the function, until we generate an approximation for $\mathbf{y}_{i+1}$. In a way, we are doing the opposite of what we were doing when we were interpolating. Instead of generating an approximation to an unknown function based on known sample points, we're generating approximate sample points based on the derivative of an unknown function. Different integration techniques are different forms of this approach, some more accurate than others.

13.3.2 Euler's Method

Assuming our current time is t and we want to move ahead h in time, we could use Taylor's series to compute $\mathbf{y}(t+h)$:

$$\mathbf{y}(t+h) = \mathbf{y}(t) + h\mathbf{y}'(t) + \frac{h^2}{2}\mathbf{y}''(t) + \cdots + \frac{h^n}{n!}\mathbf{y}^{(n)}(t) + \cdots$$

We can rewrite this to compute the value for time step $i+1$, where the time from t_i to t_{i+1} is h_i:

$$\mathbf{y}_{i+1} = \mathbf{y}_i + h_i\mathbf{y}_i' + \frac{h_i^2}{2}\mathbf{y}_i'' + \cdots + \frac{h_i^n}{n!}\mathbf{y}_i^{(n)} + \cdots$$

This assumes, of course, that we know all the values for the entire infinite series at time step i, which we don't—we have only $\mathbf{y}_i$ and $\mathbf{y}_i'$. However, if h_i is small enough and all values of $\mathbf{y}_i''$ are bounded, we can use an approximation instead:

$$\begin{aligned} \mathbf{y}_{i+1} &\approx \mathbf{y}_i + h_i\mathbf{y}_i' \\ &\approx \mathbf{y}_i + h_i\mathbf{f}(t_i, \mathbf{y}_i) \end{aligned}$$

Another way to think of this is that we have a function $\mathbf{f}(t_i, \mathbf{y}_i)$ that, given a time t_i and initial value $\mathbf{y}_i$, can compute tangents to the unknown function's curve. We can start at our known initial value, and step h_i distance along the tangent vector to get to the next approximation point in the vector field (Figure 13.4).

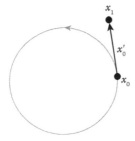

Figure 13.4. Orbit example, showing Euler step.

Separating out position and velocity gives us

$$X_{i+1} \approx X_i + h_i X_i'$$

$$\approx X_i + h_i \mathbf{v}_i$$

$$\mathbf{v}_{i+1} \approx \mathbf{v}_i + h_i \mathbf{v}_i'$$

$$\approx \mathbf{v}_i + h_i \mathbf{F}_{tot}(t_i, X_i, \mathbf{v}_i)/m$$

This is known as *Euler's method*.

To use this in our game, we start with our initial position and velocity. At each new frame, we grab the difference in time between the previous frame and current frame and use that as h_i. To compute $\mathbf{f}(t_i, \mathbf{y}_i)$ for the velocity, we use our `CurrentForce()` method to add up all of the forces on our object and divide the result by the mass to get our acceleration. Plugging in our current values, we use the preceding formulas to generate our new position and velocity. In code, this looks like the following:

```
void
SimObject::Integrate( float h )
{
    IvVector3 accel;

    // compute acceleration
    accel = CurrentForce( mTime, mPosition, mVelocity ) / mMass;
    // clear small values
    accel.Clean();

    // compute new position, velocity
    mPosition += h*mVelocity;
    mVelocity += h*accel;
    // clear small values
    mVelocity.Clean();
}
```

It's important to compute the new velocity after the new position in this case, so that we don't overwrite the velocity prematurely.

Note that we clear near-zero values in the new velocity. This prevents little shifts in position due to tiny changes in velocity, such as those generated after an object has slowed down due to drag. While technically accurate, they can be visually distracting, so after a certain point we clamp our velocity to zero. The same is done with acceleration.

For many cases, this works quite well. If our time steps are small enough, then the resulting approximation points will lie close to the actual function and we will get good results. However, the ultimate success of this method is based on the assumption that the slope at the current point is a good estimate of the slope over the entire time interval h. If not, then the approximation can drift off the function, and the farther it drifts, the worse the tangent approximation can get. We can see this with our orbit example in Figure 13.5. The first step in our approximation takes us to an orbit with a larger radius, and the next step to a larger radius still. Once the error grows, in many cases further steps don't get us back, and we continue to drift off of the actual solution.

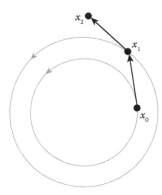

Figure 13.5. Orbit example, showing continuation of Euler's method.

For Euler's method, we say that the error is directly dependent on the time step, or $O(h)$. So, one potential solution to this problem is to decrease the time step, for example, take a step of $h/2$, followed by another step of $h/2$. While this may solve some cases, we may need to take a smaller time step, say $h/4$. And this may still lead to significant error. In the meantime, we are grinding our simulation to a halt while we recalculate quantities four or eight, or however many times for a single frame.

So, what's happening here? First, some situations that can lead to problems with Euler's method are characterized by large forces. If we examine the remaining terms of the Taylor expansion,

$$\frac{h_i^2}{2}\mathbf{y}_i'' + \cdots + \frac{h_i^n}{n!}\mathbf{y}_i^{(n)} + \cdots$$

we can see why this could cause a problem. When we set up our approximation, we assumed that h_i was small and $\mathbf{y}_i''$ bounded. A large force leads to a large acceleration, which leads to a larger difference between our approximation and the actual value. Larger values of h_i will magnify this error. Also, if the force changes quickly, this means that the magnitude of the velocity's second derivative is high, and so we can run into similar problems with velocity. This is known as truncation error, and as we can see, due to the $h_i^2/2$ factor in the the second derivative term, the truncation error for Euler's method is $O(h^2)$. Accumulating this across all iterations we end up with the global error $O(h)$.

However, our particular example falls into a class of differential equations known as *stiff* systems. Situations that can lead to stiffness problems are often characterized by large spring and damping forces, such as in a stiff spring (hence the name). Such systems tend to have terms with rapidly decaying values, such as $e^{-\rho t}$—exactly the situation with our orbit example. These terms tend to 0 as t approaches infinity but, as we've seen, won't always converge with a numerical method. The larger ρ is, the smaller h must be. This can also affect systems where we wouldn't expect the term to contribute that much. For example, suppose the solution to our system is $y(t) = 1 + e^{-200t}$. As t increases from 0, $y(t)$ quickly approaches 1. However, approximating this with a numerical method without taking care to control the error can lead the e^{-200t} term to dominate the calculations, which leads to invalid results.

Due to these issues, Euler's method is not a very robust integrator. It is, however, quite cheap and easy to implement, which is why a lot of simple physics engines use it. Fortunately, there are other methods that we can try.

Source Code
Demo

Force

13.3.3 Runge–Kutta Methods

So far we've been using the derivative at the beginning of the interval as our estimate of the average tangent. A better possibility may be to take the derivative in the middle of the interval. To do this, we first use Euler's method to take a step halfway into the interval; that is, we integrate using a step size of $h/2$. Given our estimated position and velocity at the halfway point, we calculate $\mathbf{f}(t, \mathbf{y})$ at this location. We then go back to our original starting location, and use the derivatives we calculated at the midpoint to move across the entire interval. This method is known as the *midpoint method*.

Figure 13.6 shows how this works with our original function. In Figure 13.6a, the arrow shows our initial half-step, and the line our estimated tangent. Figure 13.6b uses the tangent we've calculated with our full time step, and our final location. As we can see, with this method we are following much closer to the actual solution and so our error is much less than before. The order of the error for the midpoint method is dependent on the square of the time step, or $O(h^2)$, which for values of h less than 1 is better than Euler's method. Instead of approximating the function with a line, we are approximating it with a quadratic.

While the midpoint method does have better error tolerance than Euler's method, as we can see from our example, it still drifts off of the desired solution. To handle this, we'll have to consider some methods with better error tolerances still.

Both the midpoint method and Euler's method fall under a larger class of algorithms known as *Runge–Kutta methods*. Whereas both of our previous techniques used a single estimate to compute a tangent for the entire interval, others within the Runge–Kutta family compute multiple tangents at fixed time steps across the interval and take their weighted average.

One possibility is to take the derivative at the end of the interval, and average with the derivative at the beginning. Like the midpoint method, we can't actually compute the derivative at the end of the interval, so we'll approximate it by performing normal Euler

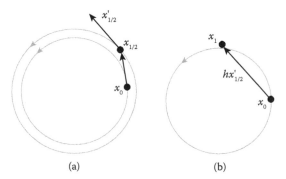

(a)   (b)

Figure 13.6. (a) Orbit example, showing first step of midpoint method: getting the midpoint derivative. (b) Orbit example, stepping with midpoint derivative to next estimate.

integration and computing the derivative at that point. This is known as the *modified Euler's method*. Interestingly, the error for this approach is still $O(h^2)$, due to the fact that we're taking an inaccurate measure of the final derivative. Another approach is *Heun's method*, which takes 1/4 of the starting derivative, and 3/4 of an approximated derivative 2/3 along the step size. Again, its error is $O(h^2)$, or no better than the midpoint method.

The standard $O(h^4)$ method is known as *Runge–Kutta order four*, or simply RK4. RK4 can be thought of as a combination of the midpoint method and modified Euler, where we weight the midpoint tangent estimates higher than the endpoint estimates. Representing this with our function notation, we get

$$\mathbf{u}_1 = h_i \mathbf{f}(t_i, \mathbf{y}_i)$$

$$\mathbf{u}_2 = h_i \mathbf{f}\left(t_i + \frac{h_i}{2}, \mathbf{y}_i + \frac{1}{2}\mathbf{u}_1\right)$$

$$\mathbf{u}_3 = h_i \mathbf{f}\left(t_i + \frac{h_i}{2}, \mathbf{y}_i + \frac{1}{2}\mathbf{u}_2\right)$$

$$\mathbf{u}_4 = h_i \mathbf{f}(t_i + h_i, \mathbf{y}_i + \mathbf{u}_3)$$

$$\mathbf{y}_{i+1} = \mathbf{y}_i + \frac{1}{6}[\mathbf{u}_1 + 2\mathbf{u}_2 + 2\mathbf{u}_3 + \mathbf{u}_4]$$

Clearly, improved accuracy doesn't come without cost. To perform standard Euler requires calculating a result for $f(t, \mathbf{y})$ only once. Midpoint, modified Euler, and Heun's need two calculations, and RK4 takes four. While achieving the level of error tolerance of RK4 would require many more evaluations of Euler's method, using RK4 still adds both complexity and increased simulation time that may not be necessary. It does depend on your application, but for simple rigid-body simulations with fast frame rates and low accelerations, Euler's method or one of the other two Runge–Kutta methods will probably be suitable.

13.3.4 Verlet Integration

There is another class of integration methods, known as *Verlet methods*, that is commonly used in molecular dynamics. Verlet methods have come to the attention of the games community because they can be useful in simulating collections of small, unoriented masses known as particles—in particular, when constrained distances between particles are required [85]. Such systems of constrained particles can simulate soft objects such as cloth, rope, and dead bodies (this last one is also known as rag-doll physics).

Source Code
Demo
Force

The most basic Verlet method can be derived by adding the Taylor expansion for the current time step to the expansion for the previous time step:

$$\mathbf{y}(t+h) + \mathbf{y}(t-h) = \mathbf{y}(t) + h\mathbf{y}'(t) + \frac{h^2}{2}\mathbf{y}''(t) + \cdots$$

$$+ \mathbf{y}(t) - h\mathbf{y}'(t) + \frac{h^2}{2}\mathbf{y}''(t) - \cdots$$

Solving for $\mathbf{y}(t+h)$ gives us

$$\mathbf{y}(t+h) = 2\mathbf{y}(t) - \mathbf{y}(t-h) + h^2\mathbf{y}''(t) + O(h^4)$$

Rewriting in our stepwise format, we get

$$\mathbf{y}_{i+1} = 2\mathbf{y}_i - \mathbf{y}_{i-1} + h_i^2\mathbf{y}_i''$$

This gives us an $O(h^2)$ solution for integrating position from acceleration, without involving velocity at all. This can be a problem if we want to use velocity elsewhere in our calculations, but we can estimate it as

$$\mathbf{v}_i = \frac{(X_{i+1} - X_i)}{2h_i}$$

One question may be, how do we find the first $\mathbf{y}_{i-1}$? The standard method is to start the process off with one pass of standard Euler or other Runge–Kutta method and store the initial position and integrated position. From there we'll have two positions to apply to our Verlet integration.

Standard Verlet has a few advantages: It is time invariant, which means that we can run it forwards and then backwards and end up in the same place. Also, the lack of velocity means that we have one less quantity to calculate. Because of this, it is often used for particle systems, which generally are not dependent on velocity. However, if we want to apply friction based on velocity or when we want to handle spinning rigid objects, the lack of velocity and angular velocity makes it more difficult. There are ways around this, as described in Jakobson [85], but in most cases it will be easier to use a method that allows us to track both velocity terms. One other disadvantage is that our velocity estimation is (1) not very accurate and (2) one time step behind our position.

If you wish to use Verlet methods and require velocity, you have two choices. *Leapfrog Verlet* tracks velocity, but at half a time step off from the position calculation:

$$\mathbf{v}\left(t+\frac{h}{2}\right) = \mathbf{v}\left(t-\frac{h}{2}\right) + h\mathbf{a}(t)$$

$$X(t+h) = X(t) + h\mathbf{v}\left(t+\frac{h}{2}\right)$$

Like with standard Verlet, we can start this off with a Runge–Kutta method by computing velocity at a half-step and proceed from there. If velocity on a whole step is required, it can be computed from the velocities, but as with standard Verlet, one time step behind position:

$$\mathbf{v}_i = \frac{(\mathbf{v}_{i+1/2} - \mathbf{v}_{i-1/2})}{2}$$

As with standard Verlet, leapfrog Verlet is an $O(h^2)$ method.

The third, and most accurate, Verlet method is *velocity Verlet*:

$$X(t+h) = X(t) + h\mathbf{v}(t) + \frac{h^2}{2}\mathbf{a}(t)$$

$$\mathbf{v}(t+h) = \mathbf{v}(t) + \frac{h}{2}[\mathbf{a}(t) + \mathbf{a}(t+h)]$$

Unlike with the previous Verlet methods, we now have to compute the acceleration twice: once at the start of the interval and once at the end. This can be done in a stepwise manner by

$$\mathbf{v}_{i+1/2} = \mathbf{v}_i + \frac{h_i}{2\mathbf{a}_i}$$

$$X_{i+1} = X_i + h_i \mathbf{v}_{i+1/2}$$

$$\mathbf{v}_{i+1} = \mathbf{v}_{i+1/2} + \frac{h_i}{2\mathbf{a}_{i+1}}$$

In between the position calculation and the velocity calculation, we recompute our forces and then the acceleration $\mathbf{a}_{i+1}$. Note that in this case the forces can be dependent only on position, since we have added only half of the acceleration contribution to velocity. In the case of molecular dynamics or particles, this isn't a problem since most of the forces between them will be positional, but again, for rigid-body problems this is not the case.

While Verlet integration has good stability characteristics, its main problem for our purposes is the estimated velocity, as mentioned above. While it works well for particle systems, it isn't as good for rigid bodies. As such, we'll look elsewhere for our solution.

13.3.5 Implicit Methods

All the methods we've described so far integrate based on the current position and velocity. They are called *explicit methods* and make use of known quantities at each time step, for example, Euler's method:

$$\mathbf{y}_{i+1} = \mathbf{y}_i + h\mathbf{y}_i'$$

But as we've seen, even higher-order explicit methods don't handle extreme cases of stiff equations very well.

Implicit methods make use of quantities from the next time step:

$$\mathbf{y}_{i+1} = \mathbf{y}_i + h_i\mathbf{y}_{i+1}'$$

This particular implicit method is known as *backward Euler*. The idea is that we are going to grab the derivative at our destination rather than at our current position. That is, we are going to find a $\mathbf{y}_{i+1}$ with the derivative that, if we were to run the simulation backwards, would end up at $\mathbf{y}_i$.

Implicit methods don't add energy to the system, but instead lose it. This doesn't guarantee us more accuracy, but it does avoid simulations that spin out of control—instead, they'll dampen down to an equilibrium state. Since, in most cases, we're going to add a damping factor anyway, this is a small price to pay for a more stable simulation. An example of using this is our old orbit example (Figure 13.7). Here we see the effect of losing energy—instead of spiraling outward, we spiral inward toward the center of the orbit. Better than Euler's method, but still not ideal.

This sounds good in theory, but in practice, how do we calculate $\mathbf{y}_{i+1}'$? One way is to solve for it directly. For example, let's consider air friction. In this example, our force is directly dependent on velocity, but in the opposing direction. Considering only velocity,

$$\mathbf{v}_{i+1} = \mathbf{v}_i - h\rho\mathbf{v}_{i+1}$$

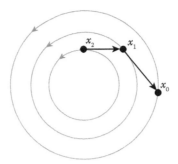

Figure 13.7. Implicit Euler. The arrows point backwards to indicate that we are getting the derivative from the next time step.

Solving for $\mathbf{v}_{i+1}$ gives us

$$\mathbf{v}_{i+1} = \frac{\mathbf{v}_i}{1+h\rho}$$

We can't always use this approach. Either we will have a function too complex to solve in this manner, or we'll be experimenting with a number of functions and won't want to take the time to solve each one individually. Another way is to use a *predictor–corrector* method. We move ahead one step using an explicit method to get an approximation. Then we use that approximation to calculate our $\mathbf{y}'_{i+1}$. This will be more accurate than the explicit method alone, but it does involve twice the number of calculations, and we're depending on the accuracy of the first approximation to make our final calculation.

Another, more accurate approach is to rewrite the equation so that it can be solved as a linear system. If we represent $\mathbf{y}_{i+1}$ as $\mathbf{y}_i + \Delta\mathbf{y}_i$, and ignore the factor t, we can rewrite backwards Euler as

$$\mathbf{y}_i + \Delta\mathbf{y}_i = \mathbf{y}_i + h_i\mathbf{f}(\mathbf{y}_i + \Delta\mathbf{y}_i)$$

or

$$\Delta\mathbf{y}_i = h_i\mathbf{f}(\mathbf{y}_i + \Delta\mathbf{y}_i)$$

We can approximate $\mathbf{f}(\mathbf{y}_i + \Delta\mathbf{y}_i)$ as $\mathbf{f}(\mathbf{y}_i) + \mathbf{f}'(\mathbf{y}_i)\Delta\mathbf{y}_i$. Note that $\mathbf{f}'(\mathbf{y}_i)$ is a matrix since $\mathbf{f}(\mathbf{y}_i)$ is a vector. Substituting this approximation, we get

$$\Delta\mathbf{y}_i \approx h_i(\mathbf{f}(\mathbf{y}_i) + \mathbf{f}'(\mathbf{y}_i)\Delta\mathbf{y}_i)$$

Solving for $\Delta\mathbf{y}_i$ gives

$$\Delta\mathbf{y}_i \approx \left(\frac{1}{h_i}\mathbf{I} - \mathbf{f}'(\mathbf{y}_i)\right)^{-1}\mathbf{f}(\mathbf{y}_i)$$

In most cases, this linear system will be sparse, so it can be solved in near-linear time. More information can be found in Witkin and Baraff [158].

While implicit methods do have some characteristics that we like—they're good for forces that depend on stiff equations—they do tend to lose energy and may dampen more than we might want. Again, this is better than explicit Euler, but it's not ideal. They're also more complex and more expensive than explicit Euler. Fortunately, there is a solution that provides the simplicity of explicit Euler with the stability of implicit Euler.

13.3.6 Semi-Implicit Methods

Up to this point, we have been treating position and velocity as independent variables while integrating; that is, we act as if they are one six-element vector that gets integrated at once. However, the fact is that position is dependent on how velocity changes. We can make use of this relationship and create a very stable integrator for dynamics. The trick is to run an explicit Euler step for velocity, and then an implicit Euler step for position:

$$\mathbf{v}_{i+1} \approx \mathbf{v}_i + h_i \mathbf{v}_i'$$

$$\approx \mathbf{v}_i + h_i \mathbf{F}_{tot}(t_i, X_i, \mathbf{v}_i)/m$$

$$X_{i+1} \approx X_i + h_i X_i'$$

$$\approx X_i + h_i \mathbf{v}_{i+1}$$

Note that the position update is using the *new* velocity, not the old one. This is called *semi-implicit* or *symplectic* Euler. Note that position is integrated using implicit Euler, which makes this particularly good for position-dependent forces. Thus, this method gives us the advantages of both explicit and implicit methods, plus it also has an additional advantage: it conserves energy over time, which keeps things very stable.

Let's look at our orbit example again, this time using semi-implicit Euler (Figure 13.8). We note that it follows the path exactly, rather than converging or diverging. Admittedly, this example is a bit contrived, but it shows the power of using a semi-implicit method.

Because it is a first-order Euler method it's still not as accurate in some cases as RK4, but it is cheap and stable. And in games, it's far more important to have a stable solution than a 100 percent correct one. This integration technique is also very easy to adapt to rotational dynamics. This makes it suitable for most of our needs beyond the most egregious cases, and thus will be the method we use for our examples.

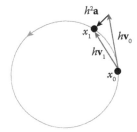

Figure 13.8. Semi-implicit Euler. The gray arrows indicate the original velocity and its modification by acceleration.

13.4 Rotational Dynamics

13.4.1 Definition

The equations and methods that we've discussed so far allow us to create physical simulations that modify an object's position. However, one aspect of dynamics we've passed over is simulating changes in an object's orientation due to the application of forces, or *rotational dynamics*. When discussing rotational dynamics, we use quantities that are very similar to those used in linear dynamics. Comparing the two,

Linear	Rotational
Position X	Orientation Ω or $\mathbf{q}$
Velocity $\mathbf{v}$	Angular velocity ω
Force $\mathbf{F}$	Torque τ
Linear momentum $\mathbf{P}$	Angular momentum $\mathbf{L}$
Mass m	Inertia tensor $\mathbf{J}$

We'll discuss each of these quantities in turn.

13.4.2 Orientation and Angular Velocity

Orientation we have seen before; we'll represent it by a matrix Ω or a quaternion $\mathbf{q}$. The angular velocity ω represents the change in orientation. It is a vector quantity, where the vector direction is the axis we rotate around to effect the change in orientation, and the length of the vector represents the rate of rotation around that axis, in radians per second.

The orientation and angular velocity are applied to an object around a point known as the *center of mass*. The center of mass can be defined as the point associated with an object where, if you apply a force at that point, it will move without rotating. One can think of it as the point where the object would perfectly balance. Figure 13.9 shows the center of mass for some common objects. The center of mass for a seesaw is directly in the center, as we'd expect. The center of mass for a hammer, however, is closer to one end than the other, since the head of the hammer is more massive than the handle.

For our objects, we'll assume that we have some sense of where the center of mass is—it's set by either the artist or some other means. One possibility discussed shortly is to compute the center of mass directly from our model data. Other choices are to use the

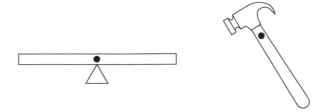

Figure 13.9. Comparing centers of mass. The seesaw balances close to the center, while the hammer has a center of mass closer to the end.

local model origin or the bounding box center (or centroid) as an approximation. Once the center of mass is determined, it is usually convenient to translate our object so that we can treat the local model origin as the center of mass, and therefore use the same orientation and position representation for both simulation and rendering.

It is possible to convert from angular velocity to linear velocity. Given an angular velocity ω, and a point at displacement $\mathbf{r}$ from the center of mass, we can compute the linear velocity at the point by using the equation

$$\mathbf{v} = \omega \times \mathbf{r} \tag{13.6}$$

This makes sense if we look at a rotating sphere. If we look at various points on the sphere (Figure 13.10a), their linear velocity is orthogonal to both the axis of rotation and their displacement vector, and this corresponds to the direction of the cross product. The length of $\mathbf{v}$ will be

$$\|\mathbf{v}\| = \|\omega\|\|\mathbf{r}\| \sin \theta$$

where θ is the angle between ω and $\mathbf{r}$. This also makes sense. As the rate of rotation $\|\omega\|$ increases, we'd expect the linear velocity of each point on the object to increase. As we move out from the equator, a rotating point has to move a longer linear distance in order to maintain the same angular velocity relative to the center (Figure 13.10b), so as $\|\mathbf{r}\|$ increases, $\|\mathbf{v}\|$ will increase. Finally, the linear velocity of a point as we move from the equator to the poles will decrease to zero (Figure 13.10c), and the quantity $\sin \theta$ provides this.

13.4.3 Torque

Up until now we've been simplifying our equations by applying forces only at the center of mass, and therefore generating only linear motion. On the other hand, if we apply an off-center force to an object, we expect it to spin. The rotational force created, known as *torque*, is directly dependent on the location where the force is applied. The farther away from the center of mass we apply a given force, the larger the torque. To compute torque, we take the cross product of the vector from the center of mass to the force application point, and with the corresponding force (Figure 13.11), or

$$\tau = \mathbf{r} \times \mathbf{F} \tag{13.7}$$

The direction of τ combined with the right-hand rule tells us the direction of rotation the torque will attempt to induce. If you align your right thumb along the direction of torque, your curled fingers will indicate the direction of rotation—if the vector is pointing toward you, this is counterclockwise around the axis of torque. The magnitude of τ provides the magnitude of the corresponding torque.

To compute the total torque, we need to compute the corresponding torque for each application of force, and then add them up. Adding the offsets and taking the cross product of the resulting vector with the total force will not compute the correct result, as shown by Figure 13.12. The sum of the offsets is $\mathbf{0}$, producing a torque of $\mathbf{0}$, which is clearly not the case—the true total torque as shown will start the circle rotating counterclockwise.

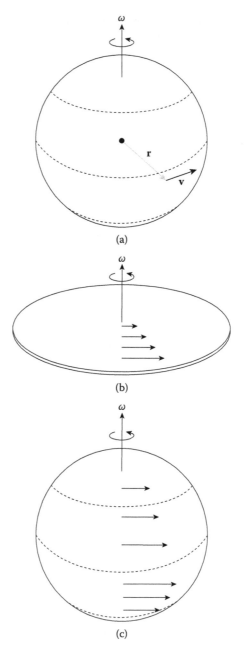

Figure 13.10. (a) Linear velocity of points on the surface of a rotating sphere. Velocity is orthogonal to both angular velocity vector and displacement vector from the center of rotation. (b) Comparison of speed of points on surface of rotating disk. Points farther from the center of rotation have larger linear velocity. (c) Comparison of speed of points on surface of rotating sphere. Points closer to the equator of the sphere have larger linear velocity.

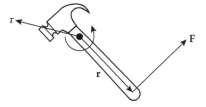

Figure 13.11. Computing torque. Torque is the cross product of displacement vector and force vector.

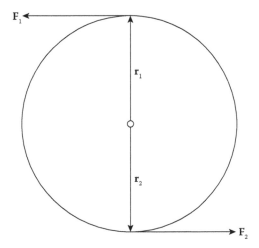

Figure 13.12. Adding two torques. If forces and displacements are added separately and then the cross product is taken, total torque will be 0. Each torque must be computed and then added together.

13.4.4 Angular Momentum and Inertia Tensor

Recall that a force $\mathbf{F}$ is the derivative of the linear momentum $\mathbf{P}$. There is a related quantity $\mathbf{L}$ for torque, such that

$$\tau = \frac{d\mathbf{L}}{dt}$$

Like linear momentum, the angular momentum $\mathbf{L}$ describes how much an object tends to stay in motion, but in rotational motion rather than linear motion. The higher the angular momentum, the larger the torque needed to change the object's angular velocity. Recall that linear momentum is equal to the mass of the object times its velocity. Angular momentum is similar, except that we use angular velocity, and the rotational equivalent of mass, the inertia tensor matrix:

$$\mathbf{I} = \mathbf{J}\omega \tag{13.8}$$

Why use a matrix $\mathbf{I}$ instead of a scalar, as we did with mass? The problem is that while shape has no effect (other than, say, for friction) on the general equations for linear dynamics, it does have an effect on how objects rotate. Take the classic example of a figure skater in a spin. As she starts the spin, her arms are out from her sides, and she has a low angular velocity. As she brings her arms in, her angular velocity increases until she opens her arms again to gracefully pull out of the spin. Torque is near zero in this case (ignoring some minimal friction from the ice and air), so we can consider angular momentum to be constant. Since angular velocity is clearly changing and mass is constant, the shape of the skater is the only factor that has a direct effect to cause this change.

So, to represent this effect of shape on rotation, we use a 3×3 symmetric matrix, where

$$
\mathbf{I} =
\begin{bmatrix}
I_{xx} & -I_{xy} & -I_{xz} \\
-I_{xy} & I_{yy} & -I_{yz} \\
-I_{xz} & -I_{yz} & I_{zz}
\end{bmatrix}
$$

We need these many factors because, as we've said, rotation depends heavily on shape and each factor describes how the rotation changes around a particular axis. The diagonal elements are called the *moments of inertia*. If we're in the correct coordinate frame, then the nondiagonal elements, or *products of inertia*, are 0. For such a frame, the axes are called the *principal axes*. For example, if the object is symmetric, the principal axes lie along the axes of symmetry and through the center of mass. We'll see next how to handle the case if our object is *not* in the principal axes frame.

The following are some examples of simple inertia tensors for objects with constant density and mass m:

- Sphere (radius of r):

$$
\begin{bmatrix}
\frac{2}{5}mr^2 & 0 & 0 \\
0 & \frac{2}{5}mr^2 & 0 \\
0 & 0 & \frac{2}{5}mr^2
\end{bmatrix}
$$

- Solid cylinder (main axis aligned along x, radius r, length d):

$$
\begin{bmatrix}
\frac{1}{2}mr^2 & 0 & 0 \\
0 & \frac{1}{4}mr^2 + \frac{1}{12}md^2 & 0 \\
0 & 0 & \frac{1}{4}mr^2 + \frac{1}{12}md^2
\end{bmatrix}
$$

- Box ($x_{dim} \times y_{dim} \times z_{dim}$):

$$
\begin{bmatrix}
\frac{1}{12}m(y_{dim}^2 + z_{dim}^2) & 0 & 0 \\
0 & \frac{1}{12}m(x_{dim}^2 + z_{dim}^2) & 0 \\
0 & 0 & \frac{1}{12}m(x_{dim}^2 + y_{dim}^2)
\end{bmatrix}
$$

For many purposes, these can be reasonable approximations. If necessary, it is possible to compute an inertia tensor and center of mass for a generalized model, assuming a constant density. A number of methods have been presented to do this, in increasing refinement [16, 37, 88, 110]. The general concept is that in order to compute these quantities, we need to do a solid integral across our shape, which is a triple integral across three dimensions. If we assume constant density, then for a polytope this is equivalent to adding up tetrahedra, where each tetrahedron consists of one of the polygonal faces and a shared central point. Code to perform this operation is available at www.geometrictools.com, for those who desire it.

13.4.5 Integrating Rotational Quantities

Source Code
Demo
Torque

As with linear dynamics, we use our angular velocity to update to our new orientation. Ideally, we could use Euler's method directly and compute our new orientation as

$$\mathbf{\Omega}_{i+1} = \mathbf{\Omega}_i + h\omega_i$$

However, this won't work, mainly because we are trying to combine vector and matrix quantities. What we need to do is compute a matrix that represents the derivative and use that with Euler's method.

Recall that the column vectors of a rotation matrix are three orthonormal vectors. We need to know how each vector will change with time; that is, we need the linear velocity at each vector tip. What we want to do is convert the angular velocity into linear velocities that affect each of our basis vectors. We can apply Equation 13.6 to each of our basis vectors to compute this, and then use the matrix generated to integrate orientation. One way would be to take the cross product of ω with each column vector, but instead we can take our three angular velocity values, and create a skew symmetric matrix $\tilde{\omega}$, where

$$\tilde{\omega} = \begin{bmatrix} 0 & -\omega_3 & \omega_2 \\ \omega_3 & 0 & -\omega_1 \\ -\omega_2 & \omega_1 & 0 \end{bmatrix} \tag{13.9}$$

If we multiply this by our current orientation matrix, this will take the cross product of ω with each column vector, and we end up with the derivative of orientation in matrix form. Using this with Euler's method, we end up with

$$\mathbf{\Omega}_{n+1} = \mathbf{\Omega}_n + h(\tilde{\omega}_n \mathbf{\Omega}_n) \tag{13.10}$$

If we're using a quaternion representation for orientation, we use a similar approach. We take our angular velocity vector and convert it to a quaternion $\mathbf{w}$, where

$$\mathbf{w} = (0, \omega)$$

We can multiply this by one-half of our original quaternion to get the derivative in quaternion form, giving us, again with Euler's method,

$$\mathbf{q}_{n+1} = \mathbf{q}_n + h\left(\frac{1}{2}\mathbf{w}_n\mathbf{q}_n\right) \tag{13.11}$$

A derivation of this equation is provided by Witkin and Baraff [158] or Hanson [71], for those who are interested.

Using either of these methods allows us to integrate orientation. As far as updating angular velocity, computing acceleration for rotational dynamics is rather complicated, so we won't be using angular acceleration at all. Instead, since torque is the derivative of angular momentum, we'll integrate the torque to update angular momentum, and then compute the angular velocity from that. As when we integrated force, we'll need a function to compute total torque across the entire interval, called `CurrentTorque()`. For both methods, we'll have to modify our input variables to take into account orientation and angular velocity, as well as position and velocity.

To find the angular velocity, we rewrite Equation 13.8 to solve for ω:

$$\omega = \mathbf{I}^{-1}\mathbf{L} \tag{13.12}$$

When computing the angular velocity in this way, there is one detail that needs to be managed carefully. The inertia tensor is in the model space of the object. However, angular momentum is integrated from torque, which is computed in world space, and we want our resulting angular velocity to also be in world space. To keep things consistent, we need a way to convert our model space $\mathbf{I}^{-1}$ to world space. If we're using a rotation matrix to represent orientation, we can use it to transform $\mathbf{L}$ from world to model space, apply the inverse inertia tensor, and then transform back into world space. So, for a given time step,

$$\omega_{i+1} = \mathbf{\Omega}_{i+1}\mathbf{I}^{-1}\mathbf{\Omega}_{i+1}^{T}\mathbf{L}_{i+1} \tag{13.13}$$

If we're using quaternions, the most efficient way to handle this is to convert our quaternion to a matrix, and then compute Equation 13.13.

Using semi-implicit Euler and quaternions, the full code for handling rotational quantities looks like the following:

```
// compute new angular momentum, orientation
mAngMomentum += h*CurrentTorque( mTranslate, mVelocity,
                                     mRotate, mAngVelocity);
mAngMomentum.Clean();

// update angular velocity
IvMatrix33 rotateMat(mRotate);
IvMatrix33 worldMomentsInverse =
        rotateMat*mMomentsInverse*::Transpose(rotateMat);
mAngVelocity = worldMomentsInverse*mAngMomentum;
mAngVelocity.Clean();
IvQuat w = IvQuat( 0.0f, mAngVelocity.x,
                   mAngVelocity.y, mAngVelocity.z );
mRotate += h*0.5f*w*mRotate;
mRotate.Normalize();
mRotate.Clean();
```

13.5   Collision Response

Up to this point, we haven't considered collisions. Our objects are moving gracefully through the world, speeding up or slowing down as we adjust our forces—all of which is accurately modeled, except that the objects go right through each other. Not a very realistic or fun game. Instead, we'll need a way to simulate the two objects bouncing away from each other due to the collision. We can do so by using the methods we've discussed in Chapter 12 in combination with some new techniques.

13.5.1 Contact Generation

For the purposes of this discussion, we'll assume a simple collision model, where the objects are convex and there is a single collision point. To perform our collision response properly, we have to know two things about the collision. The first is the point of contact between the two objects A and B—in other words, the point on the objects where they just touch (Figure 13.13). Since the two objects are just touching, there is a tangent plane that passes between the two, which also intersects both at that point. This is represented in the figure as a line. The second thing we need to know is the normal $\hat{n}$ to that plane. We'll choose our normal to point from A, the first object, to B, the second.

Our main problem in figuring out collision location is that we're trying to detect collisions within an interval of time. In one time step, two objects may be completely separate; in the next, they are colliding. In fact, in most cases when collision is detected, we have missed the initial point of collision and the objects are already interpenetrating (Figure 13.14). Because of this, there is no single point of collision.

One possibility for finding the exact point when initial collision occurs is to do a binary search within the time interval. We begin by running our simulation and then testing for collisions. If we find one, and the two objects involved are interpenetrating, we step the entire simulation back half a time step and check again. If there is still penetration, we go back a quarter of the original time step; otherwise, we go forward a quarter of the original time step. We keep doing this, ratcheting time forward or back by smaller and smaller intervals until we get an exact point of collision (unlikely) or we reach a certain level of iteration. At the end of the search, we'll either have found the exact collision point or be reasonably close.

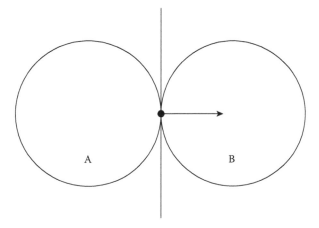

Figure 13.13. Point of collision. At the moment of impact between two convex objects, there is a single point of collision. Also shown is the collision plane and its normal.

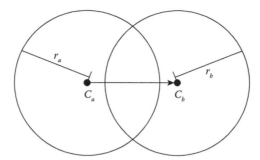

Figure 13.14. Penetrating objects. Determining penetration distance and collision normal.

This technique has a few flaws. First of all, it's slow. Chances are that every time you get a collision, you'll need to run the simulation at least two or three additional times to get a point where the objects are just touching. In addition, in order for detection to be perfectly accurate, you need to rerun the simulation for all the objects, because their position at the time of the collision will be slightly different than their position at the end of the time interval. This may affect which objects are colliding. So, you need to run the simulation back, determine the collision point, apply the collision response, and then run the simulation forward until you hit another collision, do another binary search, and so on. In the worst case, with many colliding objects, your simulation will get bogged down, and you'll end up with long frame times. The accuracy of this method may be suitable for offline simulation, but it's not good for interactivity.

Another possibility is to ignore it, approximate the contact point and normal, and let the collision response push the two objects apart. This can work, but if the response is too slow, the two objects may remain interpenetrated for a while. This can look quite odd and may ruin the illusion of reality.

The third alternative begins by looking at the overlap between the two objects. The longest distance along that overlap is known as the penetration distance. We can push the two objects apart by the penetration distance until they just touch, and then use the point and normal from that intersection for collision calculations.

For example, take two spheres (Figure 13.14), with centers C_a and C_b and radii r_a and r_b. If we subtract one center C_a from the other center C_b, we get the direction for our collision normal. The penetration distance p is then the sum of the two radii minus the length of this vector, or

$$p = (r_a + r_b) - \|C_b - C_a\| \tag{13.14}$$

We can move each sphere in opposite directions along this normal by the distance $p/2$, which will move them to a position where they just touch. This assumes that both objects can move—if one is not expected to move, like a boulder or a church, we translate the other object by the entire normal length. So, for two moving objects A and B, the formula is

```
mTranslate -= 0.5f*penetration*centerDiff;
other->mTranslate += 0.5f*penetration*centerDiff;
```

Once we've pushed them apart, the collision point is where our center difference vector crosses the boundary of the two spheres. We can compute this point by halving the difference vector and adding it to the old C_a. We finish up by normalizing the difference vector to get our collision normal.

Handling penetration distance for capsules is just as simple. Instead of using the center points to compute the collision normal, we use the closest points on the line segments that define each capsule. The penetration distance becomes the sum of the radii minus the distance between these points. For bounding boxes, Eberly [35] provides a method that computes the penetration distance between two oriented boxes.

This technique does have some flaws. First, pushing the two objects apart by the entire penetration distance may look too abrupt. Instead, we can push them apart by a fraction of the penetration distance and assume that the collision response will separate them the rest of the way. The slight interpenetration will only be noticeable for one or two frames. Second, if objects are moving fast enough and the collision is detected too late, the two objects may pass through each other. If this case is not handled in the collision detection, we will get some very odd results when the objects are pushed apart. Finally, because we're pushing objects away from each other instantaneously, we may end up with situations where two objects collide, and one of them is moved into a third, causing a new interpenetration. Because we may have already tested for collision between the second pair of objects, we'll miss this collision. If we're expecting a large number of collisions between close objects, this simple system may not be practical.

As a final note on contact generation, usually the collision detection system will generate a pair of contact features, one for each object, per collision. There may be multiple contacts per object (think of a book resting on its edge, or even its face), and there may be dependencies between many objects that control how contacts are resolved (think of a stack of boxes). We'll briefly discuss how to manage such problems later, but for our main thread of discussion we'll concentrate on single points of contact.

13.5.2 Linear Collision Response

Source Code
Demo
LinCollision

Whatever method we use, we now have two of the properties of the collision we need to compute the linear part of our collision response: a collision normal $\hat{\mathbf{n}}$ and a collision point P. The other two elements are the incoming velocities of the two objects, $\mathbf{v}_a$ and $\mathbf{v}_b$. Using this information, we are finally ready to compute our collision response.

The technique we'll use is known as an *impulse-based* system. The idea is that near the time of collision, the forces and position remain nearly constant, but there is a discontinuity in the velocity. At one point in time, the velocities of the objects are heading toward one another; in the next infinitesimal moment later, they are heading away. How much and in what relation the velocities change depends on the magnitude and direction of the incoming velocities, the direction of the collision normal, and the masses of the two objects.

Let's look again at the simple case of our two spheres A and B (Figure 13.15a). For now, let's assume their masses are equal. We again see our two incoming velocities $\mathbf{v}_a$ and $\mathbf{v}_b$ and our collision normal $\hat{\mathbf{n}}$. The idea is that we want to modify our velocity by an *impulse* that is normal to the point of collision. The impulse will act to push the two objects apart—if the masses are equal, it will be equal in magnitude, but opposite in direction for each object. So, we need to generate a scale factor j for our collision normal, and then add the scaled

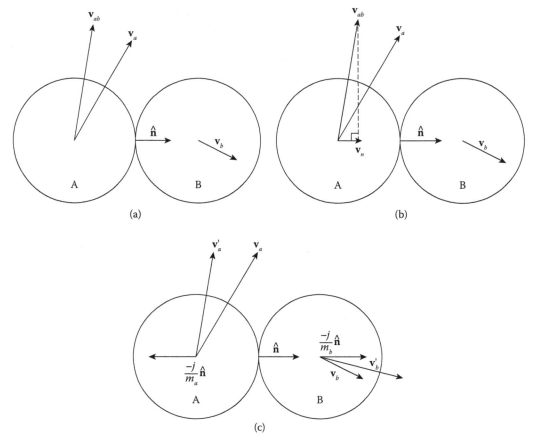

Figure 13.15. (a) Computing collision response. Calculating relative velocity. (b) Collision response. Computing relative velocity along normal. (c) Collision response. Adding impulses to create outgoing velocities.

collision normal $j\hat{\mathbf{n}}$ and $-j\hat{\mathbf{n}}$ to $\mathbf{v}_a$ and $\mathbf{v}_b$ to get our outgoing velocities. So, in order to compute the impulse vector, we need to compute this factor j.

To begin our computation, we need the relative velocity $\mathbf{v}_{ab}$, which is just $\mathbf{v}_a - \mathbf{v}_b$ (Figure 13.15a). From that, we'll compute the amount of relative velocity that is applied along the collision normal (Figure 13.15b). Recall that the dot product of any vector with a normalized vector gives the projection along the normal vector, which is just what we want. So,

$$\mathbf{v}_n = (\mathbf{v}_{ab} \bullet \hat{\mathbf{n}})\hat{\mathbf{n}}$$

At this point, we do one more test to see if we actually need to calculate an impulse vector. If the relative velocity along the collision normal is negative, then the two objects are heading away from each other and we don't need to compute an impulse. We can break out of the collision response code and proceed to the next collision. Otherwise, we continue with computing j.

In order to compute a proper impulse, two conditions need to be met. First of all, we need to set the ratio of the outgoing velocity along the collision normal to the incoming velocity. We do this by setting a *coefficient of restitution* ϵ:

$$\mathbf{v}'_n = -\epsilon \mathbf{v}_n$$

or

$$(\mathbf{v}'_a - \mathbf{v}'_b) \cdot \hat{\mathbf{n}} = -\epsilon(\mathbf{v}_a - \mathbf{v}_b) \cdot \hat{\mathbf{n}} \tag{13.15}$$

Each object will have its own value of ϵ. This simulates two different physical properties. First of all, when one object collides with another, some energy is lost, usually in the form of heat. Second, if the object is somewhat soft or sticky, or *inelastic*, the bonding forces between it and its target will decrease the outgoing velocities. Elastic in this case doesn't refer to the stretchiness of the object, but how resilient it is. A superball is not very malleable, but has very elastic collisions. So, the quantity ϵ represents how much energy is lost and how elastic the collision between the two objects is. If both objects have an ϵ of 1, then they will bounce away from each other with the same relative velocity they had coming in. If both objects have an ϵ of 0, they will stick together like two clay balls and move as one. Values in between will give a linear range of elastic responsiveness. Values greater than 1 or less than 0 are not permitted. An ϵ greater than 1 would add energy into the system, so a ball bouncing on a flat surface would bounce progressively higher and higher. An ϵ less than 0 means that the objects would be highly attracted to each other upon collision and would lead to undesirable interpenetrations.

Even if energy is not quite conserved (technically it is, but we're not tracking the heat loss), then momentum is. Because of this, the total momentum of the system of objects before and after the collision needs to be equal. So,

$$m_a \mathbf{v}_a + j\hat{\mathbf{n}} = m_a \mathbf{v}'_a$$

or

$$\mathbf{v}'_a = \mathbf{v}_a + \frac{j}{m_a}\hat{\mathbf{n}} \tag{13.16}$$

Similarly,

$$m_b \mathbf{v}_b - j\hat{\mathbf{n}} = m_b \mathbf{v}'_b$$

or

$$\mathbf{v}'_b = \mathbf{v}_b - \frac{j}{m_b}\hat{\mathbf{n}} \tag{13.17}$$

With this, we finally have all the pieces that we need. If we substitute Equations 13.16 and 13.17 into Equation 13.15 and solve for j, we get the final impulse factor equation:

$$j_a = \frac{-(1 + \epsilon_a)\mathbf{v}_{ab} \cdot \hat{\mathbf{n}}}{\left(\frac{1}{m_a} + \frac{1}{m_b}\right)} \tag{13.18}$$

The equation for j_b is similar, except that we substitute ϵ_b for ϵ_a.

Now that we have our impulse values, we substitute them back into Equations 13.16 and 13.17, respectively, to get our outgoing velocities (Figure 13.15c). Note the effect of mass on the outgoing velocities. As we expect, as the mass of an object grows larger, it grows more resistant to changing its velocity due to an incoming object. This is counteracted by j, which grows as relative velocity increases, or as the combined masses increase.

Our final algorithm for collision response between two spheres is as follows:

```
float radiusSum = mRadius + other->mRadius;
collisionNormal = other->mTranslate - mTranslate;
float distancesq = collisionNormal.LengthSquared();
// if distance squared < sum of radii squared, collision!
if ( distancesq <= radiusSum*radiusSum )
{
    // handle collision
    // penetration is distance - radii
    float distance = ::IvSqrt(distancesq);
    penetration = radiusSum - distance;
    collisionNormal.Normalize();

    // collision point is average of penetration
    collisionPoint = 0.5f*(mTranslate + mRadius*collisionNormal)
            + 0.5f*(other->mTranslate - other->mRadius*collisionNormal);

    // push out by penetration
    mTranslate -= 0.5f*penetration*collisionNormal;
    other->mTranslate += 0.5f*penetration*collisionNormal;

    // compute relative velocity
    IvVector3 relativeVelocity = mVelocity - other->mVelocity;

    float vDotN = relativeVelocity*collisionNormal;
    if (vDotN < 0)
        return;

    // compute impulse factor
    float modifiedVel = vDotN/(1.0f/mMass + 1.0f/other->mMass);
    float j1 = -(1.0f+mElasticity)*modifiedVel;

    float j2 = -(1.0f+other->mElasticity)*modifiedVel;

    // update velocities
    mVelocity += j1/mMass*collisionNormal;
    other->mVelocity -= j2/other->mMass*collisionNormal;
}
```

In this simple example, we have interleaved the sphere collision detection with the computation of the collision point and normal. This is for efficiency's sake, since both use the sum of the two radii and the difference vector between the two centers for their computations. As mentioned above, a more complex collision system will generate contact pairs to be fed to the collision response system.

13.5.3 Rotational Collision Response

Source Code
Demo
RotCollision

This is all well and good, but most objects are not spheres, which means that they have a visible orientation. When one collides with another at an offset to the center of mass, we would expect some change in angular velocity as well as linear velocity. In addition, any incoming angular velocity should affect the collision as well. A cue ball with spin (or English) applied causes a much different effect on a target pool ball than a cue ball with no spin—and the cue ball's response is different as well.

As with linear and rotational dynamics, the way we handle rotational collision response is very similar to how we handle linear collision response. We need to modify only a few equations and recalculate our impulse factor j.

One modification we have to make is the effect of angular velocity on the incoming velocity. Up to this point, we've assumed that when the two objects strike each other, their surfaces are not moving, so the velocity at the collision point is simply the linear velocity. However, if one or both of the objects are rotating, then there is an additional velocity factor applied at the point of collision, as one surface passes by the other. Recall that Equation 13.6 allows us to take an angular velocity ω and a displacement from the center of mass $\mathbf{r}$ and compute the linear velocity contributed by the angular velocity at the point of displacement. Adding this to the original incoming velocities, we get

$$\bar{\mathbf{v}}_a = \mathbf{v}_a + \omega_a \times \mathbf{r}_a$$

$$\bar{\mathbf{v}}_b = \mathbf{v}_b + \omega_b \times \mathbf{r}_b$$

Now the relative velocity $\mathbf{v}_{ab}$ at the collision point becomes

$$\mathbf{v}_{ab} = \bar{\mathbf{v}}_a - \bar{\mathbf{v}}_b$$

and Equation 13.15 becomes

$$(\bar{\mathbf{v}}_a' - \bar{\mathbf{v}}_b') = -\epsilon(\bar{\mathbf{v}}_a - \bar{\mathbf{v}}_b) \tag{13.19}$$

The other change needed is that in addition to handling linear momentum, we also need to conserve angular momentum. This is a bit more complex than the equations for linear motion, but the general concept is the same. The outgoing angular momentum should equal the sum of the incoming angular momentum and any momentum imparted by the collision. For object A, this is represented by

$$\mathbf{I}_a \omega_a + \mathbf{r}_a \times j\hat{\mathbf{n}} = \mathbf{I}_a \omega_a' \tag{13.20}$$

or

$$\omega_a' = \omega_a + \mathbf{I}_a^{-1}(\mathbf{r}_a \times j\hat{\mathbf{n}}) \tag{13.21}$$

For object B, this is

$$\mathbf{I}_b \omega_b - \mathbf{r}_b \times j\hat{\mathbf{n}} = \mathbf{I}_b \omega_b' \tag{13.22}$$

or

$$\omega_b' = \omega_b - \mathbf{I}_a^{-1}(\mathbf{r}_b \times j\hat{\mathbf{n}}) \tag{13.23}$$

Just as with linear collision response, we can substitute Equations 13.21 and 13.23 into 13.19, and together with Equations 13.16 and 13.17, solve for j to get

$$j = \frac{-(1+\epsilon)\mathbf{v}_{ab} \cdot \hat{\mathbf{n}}}{\left(\frac{1}{m_a} + \frac{1}{m_b}\right) + \left[(\mathbf{I}_a^{-1}(\mathbf{r}_a \times \hat{\mathbf{n}})) \times \mathbf{r}_a + (\mathbf{I}_b^{-1}(\mathbf{r}_b \times \hat{\mathbf{n}})) \times \mathbf{r}_b\right] \cdot \hat{\mathbf{n}}} \tag{13.24}$$

Using this modified j value we calculate new angular momenta using Equations 13.20 and 13.22, and from that calculate angular velocity as we did with angular dynamics, using Equation 13.8. We use this same j for our linear collision response as well. And of course, as before, we'll use different ϵs for the two objects.

We change our linear collision–handling code in three places to achieve this. First of all, the relative velocity collision incorporates incoming angular velocity, as follows:

```
// compute relative velocity
IvVector3 r1 = collisionPoint - mTranslate;
IvVector3 r2 = collisionPoint - other->mTranslate;
IvVector3 vel1 = mVelocity + Cross( mAngularVelocity, r1 );
IvVector3 vel2 = other->mVelocity + Cross( other->mAngularVelocity, r2 );
IvVector3 relativeVelocity = vel1 - vel2;
```

Then, we add angular factors to our calculation for j, as follows:

```
// compute impulse factor
float denominator = (1.0f/mMass
    + 1.0f/other->mMass)*(collisionNormal.Dot(collisionNormal));

// compute angular factors
IvVector3 cross1 = Cross(r1, collisionNormal);
IvVector3 cross2 = Cross(r2, collisionNormal);
cross1 = mWorldMomentsInverse*cross1;
cross2 = other->mWorldMomentsInverse*cross2;
IvVector3 sum = Cross(cross1, r1) + Cross(cross2, r2);
denominator += (sum.Dot(collisionNormal));
float modifiedVel = vDotN/denominator;
```

Finally, in addition to linear velocity, we recalculate angular velocity, as follows:

```
// update angular velocities
mAngularMomentum += Cross(r1, j1*collisionNormal);
mAngularVelocity = mWorldMomentsInverse*mAngularMomentum;
other->mAngularMomentum += Cross(r2, j2*collisionNormal);
other->mAngularVelocity = mWorldMomentsInverse*other->mAngularMomentum;
```

13.5.4 Extending the System

Everything up to this point will provide a reasonable rigid-body simulation, with moving and colliding bodies. However, there may be some additional features we may want to add.

The following present some possible solutions for expanding and extending our simple system.

13.5.4.1 Friction

Source Code
Demo
Friction

Another factor in changing an object's motion during a collision is the frictional force between the two objects. For example, the transfer of English between billiard balls is due to the friction between the balls as they strike. A simple way to simulate this is to use the Coloumb friction model. In this model, there are two cases to consider—if the objects are moving relative to each other (*dynamic* or *kinetic* friction), or if they're not (*static friction*). Static friction opposes a tangential force up to a certain threshold, after which the object starts moving and dynamic friction applies. For simplicity's sake we'll consider only the dynamic case, particularly since we're only assuming objects with nonzero relative velocity.

Figure 13.16 shows the updated situation. We've added a new unit vector $\hat{\mathbf{t}}$, which is orthogonal to $\hat{\mathbf{n}}$. Its direction is the projection of $\mathbf{v}_{ab}$ onto the tangent line between the two objects, or

$$\hat{\mathbf{t}} = \frac{\mathbf{v}_{ab} - (\mathbf{v}_{ab} \cdot \hat{\mathbf{n}})\hat{\mathbf{n}}}{\|\mathbf{v}_{ab} - (\mathbf{v}_{ab} \cdot \hat{\mathbf{n}})\hat{\mathbf{n}}\|}$$

Our frictional force $\mathbf{F}_f$ will oppose the relative velocity along the tangent line, so

$$\mathbf{F}_f = -f_k \hat{\mathbf{t}}$$

The magnitude of $\mathbf{F}_f$ is proportional to any external forces along the collision normal. This is intuitive if we think of the friction increasing as we push two objects together—the harder we push, the harder it is to slide them against each other. The proportion is controlled by the dynamic friction constant μ_k, so

$$f_k = \mu_k \|\mathbf{F}_n\|$$

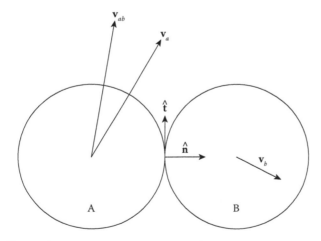

Figure 13.16. Computing collision response with friction.

In our impulse collision system, $\|\mathbf{F}_n\|$ gets replaced by the impulse generated by the collision, or

$$f_k = \mu_k \left\| \frac{j}{m} \hat{\mathbf{n}} \right\|$$

So the impulse produced by dynamic friction is

$$\mathbf{j}_f = -\mu_k \left\| \frac{j}{m} \right\| \hat{\mathbf{t}}$$

Two cases to be careful of here are when $\mathbf{v}_{ab}$ is 0, or $\mathbf{v}_{ab}$ and $\hat{\mathbf{n}}$ are parallel. In the former case, the objects are not moving relative to each other. In the latter, they are hitting dead on. In both cases, there won't be any dynamic friction due to the relative velocity, though in the first case there may be some static friction due to external forces, which may become dynamic friction if the static friction threshold is exceeded.

13.5.4.2 Resting Contact

The methods we described above handle the case when two objects are heading toward each other along the collision normal. Obviously, if they're heading apart, we don't need to consider these methods—they are separating. However, if their relative velocity along the normal is 0, then we have what is called a *resting contact*. A simple example of a resting contact is a box sitting on the floor; it has no downward velocity, and yet it is in contact with the floor.

While in general we wouldn't expect that we would have to handle a resting contact, consider the case when the box is being affected by gravity. After one time step it will have a downward velocity into the floor, and then we'll have to handle it as a colliding contact. However, doing so will lead to the box leaping up into the air as we subtract out the initial velocity and then add the response due to the impulse. The box will fall again due to gravity, and then bounce up, and we'll get a very jittery result. Obviously, we'd like to deal with the resting contact before this occurs.

One solution is to compute a force that counteracts the force of gravity. This is known as a *constraint force*, as we're constraining the box from passing through the floor. This is certainly a reasonable solution in the absence of other forces, but suppose we now have two boxes stacked on top of one another. We'll need some way to transfer that constraint force up to the next box to make sure they both don't move, in addition to preventing interpenetration between the boxes. When using constraint forces, things can get very complicated very fast.

A less accurate but more tractable alternative is to use a modification of our impulse method. This is known as a *microimpulse engine*, as our impulses due to resting contact will be very small. The key to a microimpulse engine is to add the right amount of correction to ensure that objects don't pass through each other and don't bounce. Millington [109] detects the case that we described above by comparing the velocity generated from the current frame to the object's current velocity. If it's less, then we continue with normal collision resolution; otherwise, we know it's the resting case. Catto [22] does something similar, but uses an iterative process to lower the impulse value (see below). In either case, it requires only minor tweaks to our basic algorithm to get some very nice results. Note that while we have ignored static friction in our simple model, in this case we will have to deal with it—to keep boxes from sliding down shallow inclines, for example.

13.5.4.3 Constraints

As mentioned, resting contact can be thought of as a constraint on our system, as it is preventing us from pushing an object through a surface. There are other constraints we can set up similarly. For example, suppose we have a collection of particles, and we want to keep each of them a fixed distance away from their neighbors, say in a grid. This is particularly useful when trying to simulate cloth. We can also have joint contraints, which keep two points coincident while allowing the remainder of the objects to swing free. And the list goes on. Any case that describes a fixed relationship between two objects can be modeled as a constraint.

Constraints are particularly useful in modeling a class of objects known as *soft bodies*. We've already mentioned cloth, above. Similar principles can be applied to simulate rope. When we build a simple hierarchical system, we get a skeleton that can be used to simulate a dead or unconscious figure, known as rag-doll physics. Therefore, contraints are extremely powerful in creating a new sort of interaction in our world.

We could implement these constraints as springs, but as we've seen, stiff springs cause us a lot of problems when integrating. An alternative is to compute the exact force to keep the two objects constrained, as was suggested with resting contact. However, as before, with multiple objects this can get quite complex and requires yet another system to be added to our simulation engine.

Fortunately, impulses can work in this case, too. As mentioned, collision and resting contact are just two kinds of constraint. To model others, we just need to compute the necessary impulse to keep the two objects from breaking the constraint condition and no more. This has the noted advantage that it works well with our existing impulse system for collisions and resting contacts. It's also usually simpler to compute an impulse that keeps two objects constrained than a force, as we're removing one level of indirection from position and orientation.

For those interested, details for building various types of constraint systems can be found in Catto [21], Jakobson [85], Millington [109], and Witkin and Baraff [158].

13.5.4.4 Multiple Points

The final issue we'll discuss is how to manage multiple constraints and contacts, both on one object and across multiple objects. In reality, our constraint forces and contact impulses are occurring simultaneously, so the most accurate way to handle this is to build a large system of equations and solve for them all at once. This is usually a quite complex process, in both constructing the equations and solving them. While it often ends up as a linear system, using Gaussian elimination is too expensive due to the large numbers of equations involved. Instead, an iterative process such as the Gauss–Seidel or Jacobi method is used. In principle, this is similar to Newton's method in that it involves computing an initial approximation and then refining that approximation to converge on the final answer.

An alternative, suggested in different ways by Catto [22] and Millington [109], is to continue to update impulses sequentially. However, instead of updating once per contact pair, we take a page from the iterative methods and update each pair as necessary, until a certain level of convergence is reached. Millington's method is to iterate through the contact pairs, finding the ones with the deepest penetration and resolving them first. One set of pairs may be revisited because it is affected by one or more other sets of pairs. In this way the impulses are iteratively adjusted until hopefully they converge on a reasonable solution.

Catto's method, on the other hand, involves updating the impulse values at each contact pair for several iterations, then applying the impulses when done. This has the advantage that it can cut down on jitter. Normally, impulses are required to be positive, so what happens is that any correction in the negative direction will be clamped to zero. This means that we can get overcorrection where objects bounce into the air briefly and then settle back down, much as we saw with resting contact. Instead, Catto recommends accumulating the impulse value, including the incorporation of negative values. He has also found that doing this while clamping the accumulated impulse is equivalent to an iterative matrix method known as *projected Gauss–Seidel*, which is a common variant used for solving constraint systems. This provides an excellent mathematical justification for this approach.

As before, details on solving these issues can be found in Catto [22], Jakobson [85], Millington [109], and Witkin and Baraff [158]. Golub and Van Loan [57] have information on Gauss–Seidel and Jacobi methods.

13.6 Efficiency

Now that we have a simple simulation system, some notes on using it efficiently may be appropriate. The first rule is that this is a game. Don't waste time with any more processing power than you need to get the effect you want. While a fully realistic simulation may be desirable, it can't take too much processing power away from the other subsystems, for instance, graphics or artificial intelligence. How resources are allocated among subsystems in a game depends on the game's focus. If a simpler solution will come close enough to the appearance of realism, then it is sometimes better to use that instead.

One way to reduce the amount of resources used is to simplify the problem. So far we've been assuming that we're building a truly 3D game, where the objects need to move in three degrees of freedom. If, however, you were building a tank game, it's highly unlikely that the tank would leave the ground. In most cases, land warfare games take place on a 2D map, with some height variation, so with the exception of projectiles, the entire situation is really a 2D problem. You don't have to consider gravity, as angular dynamics is constrained to just rotation around z, and thus you really need only one factor for your moments of inertia. This considerably simplifies the angular dynamics equations. The same is true for a first-person shooter; in general, characters will interact as cylinders sliding on a flat floor, with vertical walls as boundaries. In this case, we can simplify the collision problem to circles on a 2D plane.

Another way to improve efficiency is to run simulation code only on some of the objects in the world. For example, we could restrict full simulation to those objects that are visible or near the player. We could use a simplified simulation model for the other objects or not move them at all. We could also not simulate objects that aren't currently moving, and begin simulation only when forces are applied or another object collides with them. When using this technique, we need to be careful about discontinuities in the simulation. We don't want a falling object that passes out of view to stop in midair, only to start falling again when it's visible again. Nor do we want objects to jerk, move strangely, or jump position as one simulation model ceases and another takes over. While managing these discontinuities can be tricky, using such restrictions can also gain quite a performance boost.

Simplifying the forces computed during simulation is another place to find speed improvements. We've alluded to this before. In a truly complete simulation we would compute a gravitational force, a normal force to keep the object from sinking through the ground, and a static frictional force to keep the object from sliding down any inclines. In most cases, we can assume that the sum of all these forces is zero and ignore them completely. We really haven't covered friction in any detail, but it's a similar case. We could compute a complex equation for an object that handles all contact points, current surface area, and whether we are moving or at rest, or we could just use a drag coefficient multiplied by velocity. If your game calls for the full friction model, then by all means do it, but in many cases, it can be overkill.

13.7 Chapter Summary

The use of physical simulation is becoming an important part of providing realistic motion in games and other interactive applications. In this chapter we have described a simple physical simulation system, using basic Newtonian physics. We covered some techniques of numeric integration, starting with Euler's method, and discussed their pros and cons. Using these integration techniques, we have created a simple system for linear and rotational rigid-body dynamics. Finally, we have shown how we can use the results of our collision system to generate impulses for collision response.

The system we've presented is a very simple one—we've barely scratched the surface of what is possible in terms of physical simulation. For those who are interested in proceeding further, Millington [109] presents the gradual development of a simple physics engine that is suitable for game engines. Eberly [37] presents a more complete look at the mathematics in game physics, including the use of physics in graphics shaders. Burden and Faires [19] and Golub and Ortega [58] have more descriptions of numerical integration techniques and managing error bounds. Finally, Witkin and Baraff [158], Jakobson [85], and Catto [21] describe different methods for building constraint systems.

References

[1] Tomas Akenine-Möller, Eric Haines, and Naty Hoffman. *Real-Time Rendering*. CRC Press, Boca Raton, FL, 3rd edition, 2008.

[2] Tony Albrecht. Pitfalls of object oriented programming. http://research.scee.net/files/presentations/gcapaustralia09/Pitfalls_of_Object_Oriented_Programming_GCAP_09.pdf.

[3] AMD. AMD developer support web site. http://www.amd.com.

[4] American National Standards Institute and Institute of Electrical and Electronic Engineers. IEEE standard for floating-point arithmetic. IEEE Standard 754-2008, New York, 2008.

[5] Amy Williams, Steve Barnes, H. Keith Morley, and Peter Shirley. An efficient and robust ray-box intersection algorithm. *Journal of Graphics Tools*, 10(1):49–54, 2005.

[6] Howard Anton and Chris Rorres. *Elementary Linear Algebra: Applications Version*. John Wiley & Sons, New York, 11th edition, 2014.

[7] ARM. ARM developer support web site. http://www.arm.com.

[8] Sheldon Axler. *Linear Algebra Done Right*. Springer-Verlag, New York, 2nd edition, 1997.

[9] Martin Baker and Michael Norel. EuclideanSpace web site. http://www.euclideanspace.com.

[10] Richard H. Bartels, John C. Beatty, and Brian A. Barsky. *An Introduction to Splines for Use in Computer Graphics and Geometric Modeling*. Morgan Kaufman Publishers, San Francisco, 1987.

[11] J. F. Blinn and M. E. Newell. Clipping using homogeneous coordinates. In *Computer Graphics (SIGGRAPH '78 Proceedings)*, pages 245–251. ACM, New York, 1978.

[12] Jim Blinn. *A Trip Down the Graphics Pipeline*. Morgan Kaufmann Publishers, San Francisco, 1996.

[13] Jim Blinn. *Notation, Notation, Notation*. Morgan Kaufmann Publishers, San Francisco, 2002.

[14] Jonathan Blow. Hacking quaternions. *Game Developer*, March 2002.

[15] Jonathan Blow. Understanding slerp, then not using it. *Game Developer*, February 2004.

[16] Jonathan Blow and Atman J. Binstock. How to find the inertia tensor (or other mass properties) of a 3D solid body represented by a triangle mesh. Technical report, http://number-none.com, 2004.

[17] W. Boehm. Inserting new knots into b-spline curves. *Computer Aided Design*, 12(4):199–201, 1980.

[18] W. Boehm. On cubics: A survey. *Computer Graphics and Image Processing*, 19:201–226, 1982.

[19] Richard L. Burden and J. Douglas Faires. *Numerical Analysis*. PWS Publishing Company, Boston, 5th edition, 1993.

[20] Thomas Busser. Polyslerp: A fast and accurate polynomial approximation of spherical linear interpolation (slerp). *Game Developer*, February 2004.

[21] Erin Catto. Iterative dynamics with temporal coherence. Technical report, Crystal Dynamics, 2005.

[22] Erin Catto. Fast and simple physics using sequential impulses. Game Developers Conference 2006 Tutorial: Physics for Game Programmers, 2006.

[23] Arthur Cayley. *The Collected Mathematical Papers of Arthur Cayley*. Cambridge University Press, Cambridge, 1889–1897.

[24] Michael F. Cohen and John R. Wallace. *Radiosity and Realistic Image Synthesis*. Morgan Kaufman Publishers, San Francisco, 1993.

[25] R. L. Cook and K. E. Torrance. A reflectance model for computer graphics. *ACM Transactions on Graphics*, 1(1):7–24, 1982.

[26] T. N. Cornsweet. *Visual Perception*. Academic Press, New York, 1970.

[27] R. Courant and D. Hilbert. *Methods of Mathematical Physics*, volume 1. Wiley-VCH, Germany, 1989 (reprint).

[28] M. Cyrus and J. Beck. Generalized two- and three-dimensional clipping. *Computers and Graphics*, 3:23–28, 1978.

[29] Bruce Dawson. Comparing floating point numbers, 2012 edition. http://randomascii. wordpress.com/2012/02/25/comparing-floating-point-numbers-2012-edition/.

[30] Bruce Dawson. That's not normal: The performance of odd floats. http://randomascii.wordpress.com/2012/05/20/thats-not-normalthe-performance-of-odd-floats/.

[31] Eugene d'Eon and David Luebke. Advanced techniques for realistic real-time skin rendering. In Hubert Nguyen, editor, *GPU Gems 3*, pages 293–345. Addison-Wesley, Reading, MA, 2007.

[32] Tony deRose. Three-dimensional computer graphics: A coordinate-free approach. Technical report, University of Washington, 1993.

[33] Rene Descartes. *La geometrie (The Geometry of Rene Descartes)*. Dover Publications, New York, 1954.

[34] Sim Dietrich. Attenuation maps. In Mark DeLoura, editor, *Game Programming Gems*, pages 543–548. Charles River Media, Hingham, MA, 2000.

[35] David H. Eberly. *3D Game Engine Design*. Morgan Kaufmann Publishers, San Francisco, 2001.

[36] David H. Eberly. Rotation representations and performance issues. Technical report, Geometric Tools, 2002.

[37] David H. Eberly. *Game Physics*. Morgan Kaufmann Publishers, San Francisco, 2003.

[38] David H. Eberly. Eigensystems for 3×3 symmetric matrices (revisited). Technical report, Geometric Tools, 2006.

[39] David S. Ebert, F. Kenton Musgrave, Darwyn Peachey, Ken Perlin, and Steven Worley. *Texture and Modelling: A Procedural Approach*. Morgan Kaufmann, San Francisco, 3rd edition, 2003.

[40] Wolfgang Engel, editor. *GPU Pro: Advanced Rendering Techniques*. CRC Press, Boca Raton, FL, 2010.

[41] Christer Ericson. *Real-Time Collision Detection*. Morgan Kaufmann, San Francisco, 2004.

[42] Gerald Estrin. Organization of computer systems—The fixed plus variable structure computer. In *Proceeding of the Western Joint Computer Conference*, pages 33–40, 1960.

[43] Euclid. *The Elements*. Dover Publications, New York, 1956.

[44] Cass Everitt. Interactive order-independent transparency. Technical report, NVIDIA, 2001.

[45] Randima Fernando, editor. *GPU Gems: Programming Techniques, Tips, and Tricks for Real-Time Graphics*. Addison-Wesley, Reading, MA, 2004.

[46] Ronald A. Fisher and Frank Yates. *Statistical Tables for Biological, Agricultural and Medical Research*. Oliver and Boyd, London, 1938.

[47] Agner Fog. C++ vector class library. http://www.agner.org/optimize/#vectorclass.

[48] Agner Fog. Optimizing software in c++. Technical report, Technical University of Denmark, 2014.

[49] Tom Forsyth. Premultiplied alpha. http://home.comcast.net/~tom_forsyth/blog.wiki.html.

[50] H. Freeman and R. Shapira. Determining the minimum-area encasing rectangle for an arbitrary closed curve. *Communications of the ACM*, 8(7):409–413, 1975.

[51] Stephen H. Friedberg, Arnold J. Insel, and Lawrence E. Spence. *Linear Algebra*. Prentice-Hall, Englewood Cliff, NJ, 1979.

[52] Fabian Giesen. Phong normalization factor derivation. http://www.farbrausch.de/~fg/stuff/phong.pdf.

[53] Andrew S. Glassner, editor. *An Introduction to Ray Tracing*. Academic Press, Boston, 1989.

[54] Andrew S. Glassner. *Principles of Digital Image Synthesis*. Morgan Kaufmann Publishers, San Francisco, 1994.

[55] Ron Goldman. *Rethinking Quaternions: Theory and Computation*. Morgan and Claypool Publishers, San Rafael, CA, 2010.

[56] Ronald N. Goldman. Decomposing linear and affine transformations. In David Kirk, editor, *Graphics Gems III*, pages 108–116. Academic Press, San Diego, 1992.

[57] Gene H. Golub and Charles F. Van Loan. *Matrix Computations*. Johns Hopkins University Press, Baltimore, MD, 1993.

[58] Gene H. Golub and James M. Ortega. *Scientific Computing and Differential Equations: An Introduction to Numerical Methods*. Academic Press, Boston, 1992.

[59] Larry Gonick and Woollcott Smith. *The Cartoon Guide to Statistics*. Harper Collins, New York, 1993.

[60] S. Gottschalk, M. C. Lin, and D. Manocha. Obbtree: A hierarchical structure for rapid interference detection. In *Computer Graphics (SIGGRAPH '96 Proceedings)*, pages 171–180, 1996.

[61] Jens Gravesen. The length of bezier curves. In *Graphics Gems V*, pages 199–205. Academic Press, San Diego, CA, 1998.

[62] Kris Gray. *The Microsoft DirectX 9 Programmable Graphics Pipeline*. Microsoft Press, Redmond, WA, 2003.

[63] Jason Gregory. *Game Engine Architecture*. AK Peters/CRC Press, Boca Raton, FL, 2nd edition, 2014.

[64] Gil Gribb and Klaus Hartmann. Fast extraction of viewing frustum planes from the worldview-projection matrix, 2001. http://www8.cs.umu.se/kurser/5DV051/HT12/lab/plane_extraction.pdf.

[65] Charles M. Grinstead and J. Laurie Snell. *Introduction to Probability*. American Mathematical Society, Providence, RI, 2003.

[66] Khronos Group. OpenGL 4.5 reference card. https://www.khronos.org/files/opengl45-quick-reference-card.pdf.

[67] Brian Guenter and Richard Parent. Computing the arc length of parametric curves. *IEEE Computer Graphics and Applications*, 10(3):72–78, 1990.

[68] Philippe Guigue and Olivier Devillers. Fast and robust triangle-triangle overlap using orientation predicates. *Journal of Graphics Tools*, 8(1):25–32, 2003.

[69] William Hamilton. On quaternions, or on a new system of imaginaries in algebra. *Philosophical Magazine*, 1844–1850 (available online).

[70] A. Hanson and H. Ma. Parallel transport approach to curve framing. Technical Report 425, Indiana University Computer Science Department, 1995.

[71] Andrew Hanson. *Visualizing Quaternions*. Morgan Kaufmann, San Francisco, 2006.

[72] Hao Shen, Phen Ann Heng, and Zesheng Tang. A fast triangle-triangle overlap test using signed distances. *Journal of Graphics Tools*, 8(1):17–24, 2003.

[73] Donald Hearn and M. Pauline Baker. *Computer Graphics*. Prentice-Hall, Upper Saddle River, NJ, 2nd edition, 1996.

[74] Paul Heckbert. Texture mapping polygons in perspective. Technical report, New Institute of Technology, 1983.

[75] Paul Heckbert and Henry Moreton. Interpolation for polygon texture mapping and shading. In David Rogers and Rae Earnshaw, editors, *State of the Art in Computer Graphics: Visualization and Modeling*, pages 101–111. Springer-Verlag, Berlin, 1991.

[76] Chris Hecker. Under the hood/behind the screen: Perspective texture mapping (series). *Game Developer Magazine*, 1995–1996.

[77] Martin Held. Erit—A collection of efficient and reliable intersection tests. *Journal of Graphics Tools*, 2(4):25–44, 1997.

[78] John L. Hennessy and David A. Patterson. *Computer Architecture: A Quantitative Approach*. Morgan Kaufmann Publishers, San Francisco, 5th edition, 2011.

[79] Naty Hoffman. Background: Physics and math of shading. In *SIGGRAPH 2013 Course: Physically Based Shading in Theory and Practice*, 2013.

[80] Wiliam George Horner. A new method of solving numerical equations of all orders, by continuous approximation. *Philosophical Transactions*, 308–335, 1819.

[81] John F. Hughes. Personal communication, 1993.

[82] John F. Hughes, Andries van Dam, Morgan McGuire, David F. Sklar, James D. Foley, Steven K. Feiner, and Kurt Akeley. *Computer Graphics: Principles and Practice*. Addison-Wesley, Reading, MA, 3rd edition, 2013.

[83] Institute of Electrical and Electronics Engineers. IEEE standard for binary floating-point arithmetic. ANSI/IEEE Standard 754-1985, New York, 1985.

[84] Intel. Intel developer support web site. http://developer.intel.com.

[85] Thomas Jakobson. Advanced character physics. In *Proceedings of Game Developers Conference*, 2001.

[86] David Jones. Good practice in (pseudo) random number generation for bioinformatics applications. Technical report, UCL Bioinformatics Group, 2010.

[87] William Kahan. Lecture notes on the status of IEEE-754, 1996. Postscript file accessible electronically at http://http.cs.berkeley.edu/~wkahan/ieee754status/ieee754.ps.

[88] Michael Kallay. Computing the moment of inertia of a solid defined by a triangle mesh. *Journal of Graphics Tools*, 11(2):51–57, 2006.

[89] Brano Kamen. Maximizing depth buffer range and precision. http://outerra.blogspot.com/2012/11/maximizing-depth-buffer-range-and.html.

[90] Donald E. Knuth. *The Art of Computer Programming: Seminumerical Algorithms*. Addison-Wesley, Reading, MA, 3rd edition, 1993.

[91] Doris H. U. Kochanek and Richard H. Bartels. Interpolating splines with local tension, continuity, and bias control. In *Computer Graphics (SIGGRAPH '84 Proceedings)*, pages 33–41, 1984.

[92] Sébastien Lagarde and Charles de Rousiers. Moving frostbite to physically based rendering, 2014. http://www.frostbite.com/2014/11/moving-frostbite-to-pbr/.

[93] Pierre L'Ecuyer. Tables of linear congruential generators of different sizes and good lattice structure. *Mathematics of Computation*, 68(225):249–260, 1999.

[94] Pierre L'Ecuyer and Richard Simard. Testu01: A C library for empirical testing of random number generators. *ACM Transactions on Mathematical Software*, 33(4): 2007.

[95] Eric Lengyel. Fundamentals of grassman algebra. Game Developers Conference 2012, 2012. http://www.terathon.com/gdc12_lengyel.pdf.

[96] Yu-Dong Liang and Brian Barsky. A new concept and method for line clipping. *ACM Transactions on Graphics*, 3(1):1–22, 1984.

[97] D. Malacara. *Color Vision and Colorimetry: Theory and Applications*. SPIE Press, Bellingham, WA, 2nd edition, 2011.

[98] George Marsaglia. Random numbers fall mainly in the planes. *Proceedings of the National Academy of Sciences USA*, 61:25–28, 1968.

[99] George Marsaglia. Remarks on choosing and implementing random number generators. *Communications of the ACM*, 36(7):105–108, 1993.

[100] George Marsaglia. Yet another RNG. *Sci. Stat. Math.*, August 1, 1994.

[101] George Marsaglia. Xorshift RNGs. *Journal of Statistical Software*, 8(14):1–9, 2003.

[102] George Marsaglia and Wai Wan Tsang. Some difficult-to-pass tests of randomness. *Journal of Statistical Software*, 7(3):1–9, 2002.

[103] George Marsaglia and Arif Zaman. A new class of random number generators. *Annals of Applied Probability*, 1(3):462–480, 1991.

[104] George Marsaglia and Arif Zaman. The kiss generator. Technical report, Department of Statistics, Florida State University, 1993.

[105] Marta Löfstedt and Tomas Akenine-Möller. An evaluation framework for ray-triangle intersection algorithms. *Journal of Graphics Tools*, 10(2):13–26, 2005.

[106] Makoto Matsumoto and Takuji Nishimura. Mersenne Twister: A 623-dimensionally equidistributed uniform pseudorandom number generator. *ACM Transactions on Modelling and Computer Simulation*, 8:3–30, 1998.

[107] Morgan McGuire and Louis Bavoil. Weighted blended order-independent transparency. *Journal of Computer Graphics Techniques (JCGT)*, 2(2):122–141, 2013.

[108] Gary McTaggert. Half-Life 2/Valve source shading. Game Developers Conference 2004, 2004.

[109] Ian Millington. *Game Physics Engine Development*. CRC Press, Boca Raton, FL, 2nd edition, 2010.

[110] Brian Mirtich. Fast and accurate computation of polyhedral mass properties. *Journal of Graphics Tools*, 1(2):31–50, 1996.

[111] Hubert Nguyen. Casting shadows. *Game Developer Magazine*, March 1999.

[112] nVidia. nVidia developer support web site. http://developer.nvidia.com.

[113] Michael Oren and Shree K. Nayar. Generalization of Lambert's reflectance model. In *Proceedings of the 21st Annual Conference on Computer Graphics and Interactive Techniques (SIGGRAPH '94)*, pages 239–246, 1994.

[114] Joseph O'Rourke. Finding minimal enclosing boxes. *International Journal of Computer and Information Sciences*, 14(3):183–199, 1985.

[115] Joseph O'Rourke. *Computational Geometry in C*. Cambridge University Press, Cambridge, United Kingdom, 2000.

[116] Lewis Padgett. Mimsy were the borogroves. In *Science Fiction Hall of Fame*, volume 1, Doubleday & Company Inc., Garden City, NY, 1943.

[117] Rick Parent. *Computer Animation: Algorithms and Techniques*. Morgan Kaufmann Publishers, San Francisco, 3rd edition, 2012.

[118] Stephen K. Park and Keith W. Miller. Random number generators: Good ones are hard to find. *Communications of the ACM*, 31(10):1192–1201, 1988.

[119] Ken Perlin. An image synthesizer. In *Computer Graphics (SIGGRAPH '85 Proceedings)*, pages 287–296, 1985.

[120] Matt Pharr, editor. *GPU Gems 2: Mapping Computational Concepts to GPUs*. Addison-Wesley, Reading, MA, 2005.

[121] Matt Pharr and Greg Humphreys. *Physically Based Rendering: From Theory to Implementation*. Morgan Kaufmann, San Francisco, 2004.

[122] Bui Tuong Phong. Illumination for computer generated pictures. *Communications of the ACM*, 18(6):311–317, 1975.

[123] Thomas Porter and Tom Duff. Compositing digital images. In *Proceedings of the 11th Annual Conference on Computer Graphics and Interactive Techniques (SIGGRAPH '84)*, pages 253–259, 1984.

[124] Charles Poynton. Charles Poynton's color FAQ. http://www.poynton.com/.

[125] Franco P. Preparata and Michael Ian Shamos. *Computational Geometry: An Introduction*. Springer-Verlag, Berlin, 1991.

[126] William H. Press, Saul A. Teukolsky, William T. Vetterling, and Brian P. Flannery. *Numerical Recipes in C : The Art of Scientific Computing*. Cambridge University Press, New York, 3rd edition, 2007.

[127] Ravi Ramamoorthi and Pat Hanrahan. An efficient representation for irradiance environment maps. In *Proceedings of the 28th Annual Conference on Computer Graphics and Interactive Techniques (SIGGRAPH '01)*, pages 497–500, 2001.

[128] Nathan Reed. On vector math libraries. http://www.reedbeta.com/blog/2013/12/28/on-vector-math-libraries.

[129] Robert J. Simpson, editor. The OpenGL ES® shading language. Technical report, Khronos Group, Inc., 2014.

[130] David F. Rogers. *An Introduction to NURBS: With Historical Perspective*. Morgan Kaufmann Publishers, San Francisco, 2000.

[131] David F. Rogers and J. Alan Adams. *Mathematical Elements for Computer Graphics*. McGraw-Hill, New York, 1990.

[132] Randi Rost. *OpenGL® Shading Language*. Addison-Wesley Professional, Reading, MA, 2004.

[133] Philip J. Schneider and David H. Eberly. *Geometric Tools for Computer Graphics*. Morgan Kaufmann Publishers, San Francisco, 2002.

[134] I. Schrage. A more portable Fortran random number generator. *ACM Transactions on Mathematical Software*, 5(2):132–138, 1979.

[135] Mark Segal and Kurt Akeley. The OpenGL® graphics system: A specification (version 4.5 (core profile)—October 30, 2014). Technical report, Khronos Group, Inc., 2014.

[136] Ken Shoemake. Animating rotation with quaternion curves. In *Computer Graphics (SIGGRAPH '85 Proceedings)*, volume 19, pages 245–254, 1985.

[137] Ken Shoemake. Quaternion calculus for animation. In *Math for SIGGRAPH (ACM SIGGRAPH '89 Course Notes 23)*, pages 187–205, 1989.

[138] Ken Shoemake and Tom Duff. Matrix animation and polar decomposition. In *Proceedings of Graphics Interface '92*, pages 258–264, 1992.

[139] William Stallings. *Computer Organization and Architecture*. Prentice Hall, Upper Saddle River, NJ, 9th edition, 2012.

[140] Stephen Vincent and David Forsey. Fast and accurate parametric curve length computation. *Journal of Graphics Tools*, 6(4):29–40, 2001.

[141] Dan Sunday. Distance between lines and segments with their closest point of approach. Technical report, http://geometryalgorithms.com, 2001.

[142] I. E. Sutherland. Sketchpad: A man–machine graphical communications system. In *IFIPS Proceedings of the Spring Joint Computer Conference*, 1963.

[143] I.E. Sutherland and G.W. Hodgeman. Reentrant polygon clipping. *Communications of the ACM*, 17(1):32–42, 1974.

[144] Steve Theodore. Why be normal? *Game Developer Magazine*, October 2004.

[145] Andy Thomason. Faster quaternion interpolation using approximations. In Kim Pallister, editor, *Game Programming Gems 5*. Charles River Media, Hingham, MA, 2005.

[146] Tomas Möller. A fast triangle–triangle intersection test. *Journal of Graphics Tools*, 2(2):25–30, 1997.

[147] Tomas Möller and Ben Trumbore. Fast, minimum storage ray/triangle intersection. *Journal of Graphics Tools*, 2(1):21–28, 1997.

[148] Ken Turkowski. Filters for common resampling tasks. In Andrew S. Glassner, editor, *Graphics Gems*, pages 147–165. Academic Press Professional, San Diego, 1990.

[149] Gino van den Bergen. *Collision Detection in Interactive 3D Environments*. Morgan Kaufmann Publishers, San Francisco, 2003.

[150] James M. Van Verth. Using the covariance matrix for better fitting bounding objects. In Andrew Kirmse, editor, *Game Programming Gems 4*. Charles River Media, Hingham, MA, 2004.

[151] James M. Van Verth. Spline-based time control for animation. In Kim Pallister, editor, *Game Programming Gems 5*. Charles River Media, Hingham, MA, 2005.

[152] Sebastiano Vigna. An experimental exploration of Marsaglia's xorshift generators, scrambled. *Computing Research Repository*, abs/1402.6246, 2014.

[153] Sebastiano Vigna. Further scramblings of Marsaglia's xorshift generators. *Computing Research Repository*, abs/1404.0390, 2014.

[154] David R. Warn. Lighting controls for synthetic images. In *Computer Graphics (SIGGRAPH '83 Proceedings)*, 1983.

[155] Alan Watt and Fabio Policarpo. *3D Games: Real-Time Rendering and Software Technology*, volume 1. Addison-Wesley, Harlow, UK, 2001.

[156] E. Welzl. Smallest enclosing disks (balls and ellipsoids). In H. Maurer, editor, *Lecture Notes in Computer Science, New Results and New Trends in Computer Science*, volume 555, pages 359–370. Springer-Verlag, New York, 1991.

[157] Lance Williams. Pyramidal parametrics. In *Computer Graphics (SIGGRAPH '83 Proceedings)*, 1983.

[158] Andrew Witkin and David Baraff. Physically based modelling: Principles and practice, SIGGRAPH 2001 course notes, 2001.

[159] George Wohlberg. *Digital Image Warping*. IEEE Computer Society Press, Los Alamitos, CA, 1990.

Index

P